Semiconductor-Laser Physics

Springer

Berlin
Heidelberg
New York
Barcelona
Budapest
Hong Kong
Lor.don
Milan
Paris
Santa Clara
Singapore
Tokyo

Weng W. Chow   Stephan W. Koch
Murray Sargent III

Semiconductor-Laser Physics

With 180 Figures

 Springer

Dr. Weng W. Chow
Sandia National Laboratories, Albuquerque, NM 87185, USA

Professor Dr. Stephan W. Koch
Fachbereich Physik und Wissenschaftliches Zentrum für Materialwissenschaften
Philipps-Universität Marburg, Mainzergasse 33, D-35032 Marburg, Germany

Professor Dr. Murray Sargent III
Optical Sciences Center, University of Arizona, Tucson, AZ 85721, USA

The Library of Congress has cataloged the first printing as follows:

Chow, W. W. (Weng W.), 1948- Semiconductor-laser physics / W.W. Chow, S. W. Koch, M. Sargent III. p. cm.
Includes bibliographical references and index. ISBN 3-540-57614-2 (Berlin: acid-free paper). – ISBN 0-387-57614-2
(New York: acid-free paper)
1. Semiconductor lasers. I. Koch, S. W. (Stephan W.) II. Sargent, Murray. III. Title.
QC689.55.S45C48   1994   621.36'61–dc20   93-46745

First Edition 1994
Corrected Printing 1997
ISBN-13: 978-3-642-64752-9   e-ISBN-13: 978-3-642-61225-1
DOI: 10.1007/978-3-642-61225-1

© Springer-Verlag Berlin Heidelberg 1994
Softcover reprint of the hardcover 1st edition 1994

Typesetting: Camera ready by the authors
Cover design: *design & production* GmbH, Heidelberg

SPIN 10643054   54/3144 – 5 4 3 2 1 0 –

To

Ruth, Rita, and Helga

PREFACE

One might wonder if a book on the physics of semiconductor lasers could offer new insights into these devices, in view of the fact that they have been developed for more than thirty years and are very successfully used in countless CD players, laser printers, and many other applications. One would think that such incredibly important devices must be thoroughly understood and that their performance must be completely predictable. However with closer examination, one easily finds numerous shortcomings in the usual simple approaches used for modeling semiconductor lasers. Since one can often attribute these shortcomings to insufficient consideration of the physical mechanisms involved in laser operation, we focus the presentation in this book on the underlying physics and operational principles of semiconductor lasers. The optical and electronic properties of the semiconductor medium are analyzed in detail, including quantum confinement and strain-related band-structure modifications. We carefully distinguish between global laser properties and local characteristics of the gain medium. Under many conditions we can describe the local gain medium in terms of its optical susceptibility (gain/absorption and refractive index) and the susceptibility derivatives, which appear, for example, in the usual expression for the linewidth/antiguiding factor and in multiwave mixing. The global laser properties are described by a self-consistent coupling of the material polarization to the light field in the cavity.

The book starts with a discussion of semiconductor laser diodes, including basic laser devices such as edge emitters and vertical-cavity surface-emitting lasers (VCSELs). Gain and index guiding and the relevant concepts of semiconductor physics are summarized. This includes elementary aspects of band structures, the quantum mechanics of the semiconductor medium, carrier distributions, and quantum confinement effects. A detailed discussion of quantum mechanical models for the local semiconductor gain and refractive index is presented, starting with the elementary free carrier theory, and then incorporating Coulomb effects such as bandgap renormalization, interband Coulomb attraction, as well as collision and dephasing processes. Many-body gain models are developed both for bulk semiconductors and for semiconductor quantum wells. The effects of band mixing and gain in realistic quantum wells are discussed and a large number of results for different unstrained and strained semiconductor heterostructures are shown.

Different approaches to a semiconductor laser theory are presented, starting with a semiclassical description based on the multimode Maxwell equations. A single-mode laser theory and stability analysis are developed and the concept of injection locking is discussed. Multimode operation is analyzed including multiwave mixing, sidemode interactions, and mode locking. A quantum mechanical analysis of the laser action includes single-mode field quantization, quantum and semiconductor Langevin equations, and a calculation of power spectra and the laser linewidth. The chapters on transverse effects and arrays cover effects like wave guiding, self-focusing, as well as unstable resonators, array configurations, phase locking, and instabilities. The book closes with a chapter on the recently developed nonequilibrium theory that treats the light-field dynamics and the material kinetics in a fully quantum-mechanical way.

In general, this book discusses semiconductor lasers from the viewpoint of a physicist, combining input from condensed matter theory, laser physics, and quantum optics. A basic theme is to give a systematic and consistent quantum mechanical analysis of the semiconductor laser medium. With increasingly stringent approximations, we recover a hierarchy of simpler theories, such as the free-carrier theory and the simple linear-density gain rate-equation models. The microscopic approach provides a more general treatment and allows us to determine the ranges of validity of the simpler approaches. In addition to presenting the concepts and theory, we give numerous numerical illustrations for bulk semiconductors, quantum wells, and strained-layer systems, and we apply these theories to a variety of laser problems.

In writing this book we have benefited from numerous interactions with many colleagues and students. Special thanks go to H. Haug, who pioneered the early microscopic analysis of semiconductor lasers and has been a continuing source for stimulating collaborations and discussions. It is a pleasure to thank K. Henneberger and F. Jahnke for collaborations on the nonequilibrium laser theory, R. Binder and A. Knorr for the pulse propagation work, M. Pereira for work on strained-layer gain models, J. Moloney and P. Ru for collaboration on laser instabilities and unstable resonators, and M. Lindberg for numerous discussions and collaboration on coupled-cavity configurations. Finally we thank the Aspen Center for Theoretical Physics, where this project was started, as well as AFOSR, Sandia National Laboratories, and OCC for financial support.

Albuquerque, NM   *W.W. Chow*
Tucson, AZ   *S.W. Koch*
Tucson, AZ   *M. Sargent III*
October 1993

CONTENTS

Chapter 1
SEMICONDUCTOR LASER DIODES

In 1961, the concept of a semiconductor laser was introduced by Basov *et al.* who suggested that stimulated emission of radiation could occur in semiconductors by the recombination of carriers injected across a *p-n* junction. The first semiconductor lasers appeared in 1962, when three laboratories independently achieved lasing. After that, progress was slow for several reasons. One reason was the need to develop a new semiconductor technology. Semiconductor lasers could not be made from silicon where a mature fabrication technology existed. Rather, they require direct bandgap materials which were found in compound semiconductors which at the time were less well understood. There were also problems involving high threshold currents for lasing, which limited laser operation to short pulses at cryogenic temperatures, and low efficiency, which led to a high heat dissipation. A big stride toward solving the above problems was made in 1969, with the introduction of heterostructures. In a heterostructure laser, one replaces the simple *p-n* junction with multiple semiconductor layers of different compositions. The immediate impact to laser performance due to heterostructures was reinforced over the years with better laser designs and better control of the growth processes.

Two factors are largely responsible for the transformation of semiconductor lasers from laboratory devices operating only at cryogenic temperatures into practical opto-electronic components capable of running continuously at room temperature. One is the exceptional and fortuitous close lattice match between AlAs and GaAs, which allows heterostructures consisting of layers of different compositions of $Al_x Ga_{1-x} As$ to be grown. The second is the presence of several important opto-electronic applications where semiconductor lasers are uniquely well suited because they have the smallest size (several cubic millimeters), highest efficiency (often as much as 50%), and the longest life of all existing lasers. This enables the field of semiconductor lasers to draw the attention and resources that are necessary for its development.

One such application is optical-fiber communication, where device design is simplified by the fact that the laser output can be modulated simply by modulating the injection current. Gigahertz information transmission rates are now possible. Optical fiber communication also motivated the development of semiconductor lasers at $1.3\mu m$, where optical fiber loss is minimum, and at $1.5\mu m$, where dispersion in the fiber is minimum. The need for repeaters led to the development of laser amplifiers, the introduction of underwater optical communication lines necessitate improvements in device reliability, and frequency multiplexing of transmissions led to distributive feedback (DFB) and Bragg reflector (BR) lasers for frequency stability.

There are other applications of semiconductor lasers as well. The optical memory (audio and video discs) industry has generated a large enough demand of semiconductor lasers to help in reducing laser cost. High power semiconductor lasers are being used in printers and copiers. When even higher power is reliably available, the list of applications will expand to include, for e.g., line-of-sight communications, laser radar and fuzing. Schemes for increasing laser power are plentiful. They involve widening the active region (broad-area lasers), phase locking many narrow active regions (arrays and external cavity lasers) and optically pumping a solid state laser with a stack of semiconductor lasers.

Advances are continually being made. For example, linear laser arrays have evolved into surface emitting two-dimensional arrays, which are of interest in optical data processing and computing. In addition to high output power, external cavity lasers can also be designed to have very narrow linewidths. While a semiconductor laser by itself does not make a good high energy pulse laser because its gain medium performs badly in terms of energy storage due to a short carrier recombination time, it can be used to efficiently pump a Q-switched solid state laser. Diode pumped YAG lasers have produced high energy (over 1 joule) short (tens of nanoseconds) pulses. They also have the advantages over conventional flashlamp pumped ones by being more efficient and compact. Tiny "microlasers" only half to several wavelengths long can be fabricated with a variety of special properties. So, we see that semiconductor lasers have many applications. With important applications come better lasers, which generate more applications, which in turn support the development of even better and more versatile lasers.

Section 1-1 describes the basic p-n junction diode commonly used in microelectronics. Section 1-2 shows how the junction can be configured to obtain an inversion and optical gain. Section 1-3 describes the heterostructures used to confine the carriers and light in the transverse direction (that along the current flow), and the various fabrication techniques. Section 1-4 discusses gain and index guiding methods, section 1-5 introduces the concept of the recently developed semiconductor microlasers, and section 1-6 shows observed laser output power as a function of current, revealing a rolloff presumably due to increased temperature at high current levels. Section 1-7 shows how buried-heterostructure laser frequency spectra vary with input current, ranging from multimode operation near threshold to

single-mode operation and then back to multimode operation for sufficiently high currents. Section 1-8 discusses transverse mode structure and
filamentation.

The chapter concludes with a simple phenomenological laser theory
based on a linear gain approximation that describes some of the experimentally observed behavior, most notably the markedly homogeneously
broadened gain saturation. This characteristic is surprising at first, since
the semiconductor band structure (see Figs. 2-1 and 2-2) automatically
produces a wide range of transition frequencies, that is, inhomogeneously
broadening, and we see in later chapters that the homogeneous character
results from strong carrier-carrier scattering. The simple theory is very
widely used in the literature and may be readily used in situations that are
insensitive to the more precise nature of the gain medium. It is the first of
a hierarchy of semiconductor gain models presented in this book; others
include the free-carrier model of Chap. 3 and the many-body models of
Chaps. 4 through 6.

1-1. The Diode

Since semiconductor lasers consist of various kinds of diode structures,
we review how a basic diode works. The simplest kind of diode consists of
a p-n junction (Fig. 1-1). This is a p-doped semiconductor layer, which is
a semiconductor layer doped with acceptor impurities, in contact with an
n-doped layer, which is one doped with donor impurities. The donors and
acceptors introduce impurity levels as depicted in Fig. 1-2. In an n-doped
medium, the donor levels lie well within a thermal energy $k_B T$ of the conduction band, so that the levels are effectively ionized, yielding conduction
electrons, i.e., negatively charged current carriers, which gives the name n-
doped. Here k_B is Boltzmann's constant and T is the absolute temperature.

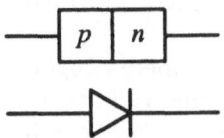

Fig. 1-1. (top) Simple diode consisting of a p-n junction, that is,
a layer of p-doped semiconductor in contact with one of n-
doped. (bottom) electrical symbol for a diode indicating the
direction (here left to right) in which current flows readily.

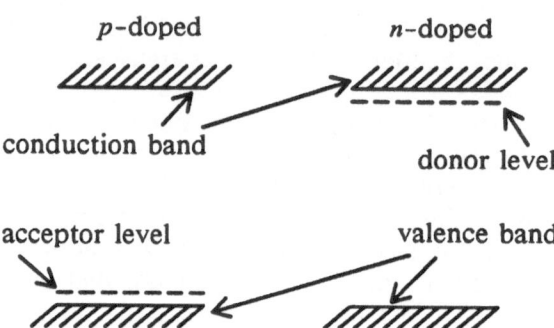

Fig. 1-2. Energy level diagram for p-doped and n-doped semi-conductor media. In p-doped media, acceptor impurities intro-duce a level very near the valence band and contribute positively charged current carriers (holes). In n-doped media, donor impurities introduce a level very near the conduction band and contribute negatively charged current carriers (electrons).

n-doped media can conduct well compared to intrinsic semiconductors, which do not have impurity levels and derive conductivity only by carriers that manage to bridge the whole band gap, an energy large compared to the thermal energy. In the GaAs semiconductor, a popular donor impurity is Se from column VI in the periodic table, which has one more valence electron (6) than As, which is in column V.

Similarly in a p-doped medium, the acceptor impurity level lies well within a thermal energy of the valence band, allowing an electron from the valence band to fall into the level, thereby creating a hole in the valence band. This creates positively charged current carriers, which gives the name p-doped. In the GaAs semiconductor, a popular acceptor impurity is Zn from column II, which has one less valence electron (2) than Ga from column III. Note that the doped media by themselves are electrically neu-tral, that is, they are not charged positively or negatively, even though they have current carriers.

Putting the two kinds of media together changes both the carrier and the charge distributions in the vicinity of the junction as shown in Figs. 1-3a and 1-3b, respectively. The conduction electrons in the n-doped medium spy the holes over in the p-doped medium and diffuse across the junction to fill up the holes. Similarly the valence holes diffuse across the junction in the reverse direction, eliminating the conduction electrons. This diffusion process causes the n-doped material to become positively charged near the junction, and the p-doped material to become negatively

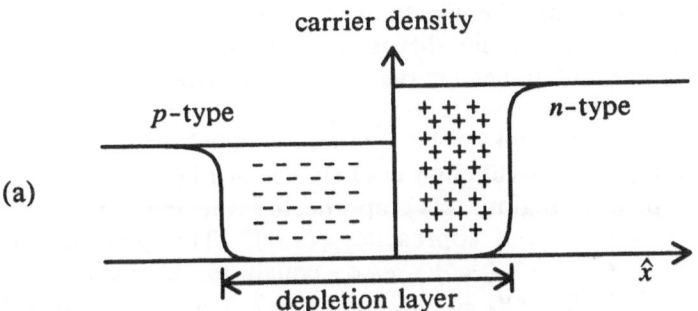

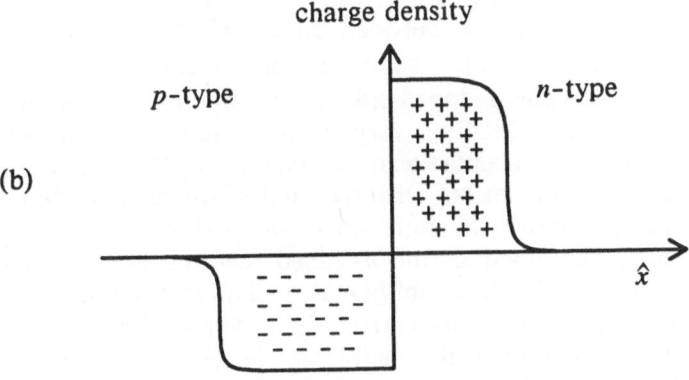

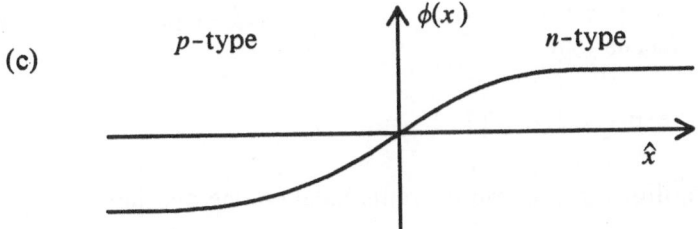

Fig. 1-3. (*a*) carrier density as a function of distance along the direction *x* of current flow. (*b*) corresponding charge density. (*c*) corresponding electric potential function $\phi(x)$.

charged. In equilibrium the Coulomb attraction between these charged regions prevents further carrier diffusion, resulting in a *depletion layer* in which there is substantial net charge but no positive or negative current carriers. This layer can range from 10^2 to 10^4 Å in width.

This state of affairs looks a bit like an extended electric dipole and is described by the potential function $\phi(x)$ shown in Fig. 1-3c. On the $x > 0$ side of the depletion region, $\phi(x)$ approaches the constant value $\phi(\infty)$, while on the $x < 0$ side, it approaches $\phi(-\infty)$. This gives the potential difference $\Delta\phi_0 = \phi(\infty) - \phi(-\infty)$. $e\Delta\phi_0$ equals to the band-gap energy, which in Si is about 0.7 eV, so that $\Delta\phi_0 = 0.7$ Volts. In pure GaAs the band-gap energy is 1.424 eV, and in $Al_x Ga_{1-x} As$ it is larger, depending on the Al concentration x.

To get a simple feel for the current-voltage characteristics of the diode, we consider the balance between diffusion and Coulomb attraction in a bit more detail. Let J_h^{gen} be the small current consisting of holes generated thermally on the n-doped side of the junction and moving across the junction. This is admittedly a very small current, since the probability of excitation is roughly proportional to $\exp[-\varepsilon_g/k_B T]$, where ε_g is the bandgap energy. Nevertheless, whatever holes are generated on the n-doped side are swept across the junction to the p side due to the potential difference. Holes generated in the n-doped medium are called *minority carriers*, since they are far less common than the electrons in the conduction band, which are the *majority carriers* in n-type media.

Similarly holes generated thermally on the p-doped side lead to a current J_h^{rec} moving in the opposite direction from J_h^{gen}. In contrast to holes generated on the p-doped side, those generated on the n-type side have to fight an uphill battle to get across the junction, that is, they have to overcome the potential barrier of height $\Delta\phi_0$ equal to the band-gap energy. This means that

$$J_h^{rec} \propto \exp(-e\Delta\phi_0/k_B T) \ . \tag{1}$$

In thermal equilibrium the two currents balance one another, that is

$$J_h^{rec} = J_h^{gen} \ . \tag{2}$$

Now let us impress a voltage V across the junction. This leads to the modified potential drop $\Delta\phi$ given by

$$\Delta\phi = \Delta\phi_0 + V. \tag{3}$$

While a nonzero V does not affect J_h^{gen} appreciably, since that current already flows rapidly "downhill", it does change J_h^{rec} to be

$$J_h^{rec} \propto \exp(-e\Delta\phi/k_B T) \; . \tag{4}$$

Combining Eqs. (1), (2), and (4), we find

$$J_h^{rec} \propto J_h^{gen} \exp(-eV/k_B T) \; ,$$

which gives the hole current

$$J_h = J_h^{rec} - J_h^{gen} = J_h^{gen}[\exp(eV/k_B T) - 1] \; .$$

Conduction electrons generated on either side of the junction produce similar currents, giving the total current

$$J = [J_e^{gen} + J_h^{gen}][e^{eV/k_B T} - 1] \; . \tag{5}$$

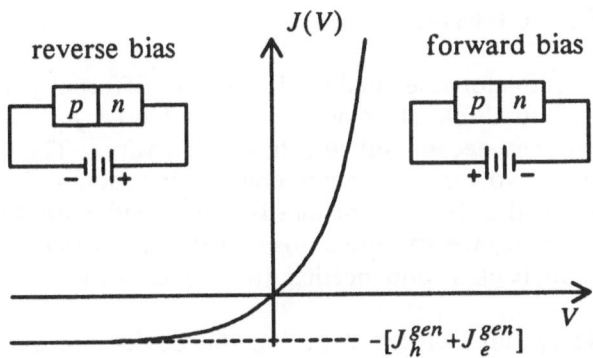

Fig. 1-4. Current flow across a p-n junction as a function of the voltage V across the junction.

This equation is dramatically different from Ohm's law, as is illustrated by the current/voltage curve in Fig. 1-4. If V is negative (called reverse biased), only the small minimum current $-(J_e^{gen} + J_h^{gen})$ flows. On the other hand, if V is positive, the current flows readily. So readily, in fact, that one needs to include a current-limiting resistor (see Fig. 1-5) to keep from destroying the diode.

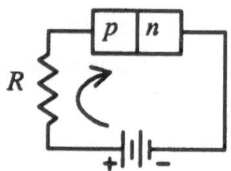

Fig. 1-5. Diagram showing how a resistor can be used to limit the current through a forward-biased diode. If the resistance R is sufficiently large, the diode plays a negligible role in determining the magnitude and statistics of the flowing current. The voltage drop across the diode is somewhat greater than the bandgap energy/e.

As we discuss next, in laser diodes, a p-i-n structure is typically used, in which the i stands for *intrinsic*, which means that no doping is used. The intrinsic layer is fabricated to be very thin, typically less than 1000Å, and is crucial to the success of the laser diode.

1-2. Basic Laser Device

The semiconductor laser looks different to different people. An electrical engineer may think of it as a forward biased p-n junction, while a crystal grower may see mainly the heterostructure. The laser fabricator's view probably involves too much engineering detail for the solid-state physicist. Since this book is primarily concerned with the theory of the semiconductor laser, we structure this and the following sections to provide the construction background needed for useful theoretical models of laser operation.

The basic features of the experimental device are shown in Fig. 1-6, which is a vastly simplified diagram of a semiconductor laser. A semiconductor laser is usually fabricated by growing a p-doped layer on top of a n-doped semiconductor substrate. Current is injected via two electrodes, one of which is electrically connected to a heat sink. Lasing occurs in the active (gain) region between the electrodes as indicated by the shaded area in Fig. 1-6. This shaded area represents the depletion region in a simple p-n junction or the specially fabricated intrinsic layer in a heterostructure laser (see Sec. 1-3). The optical resonator is formed by two parallel facets that are made by cleaving the substrate along crystal planes. Owing to the high gain in the active region, resonator facets are often left uncoated,

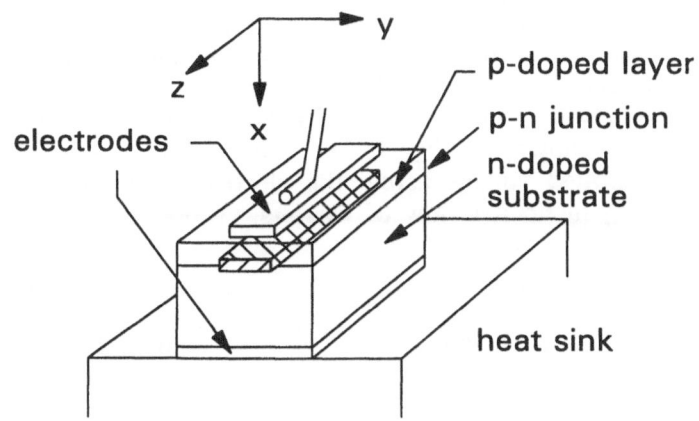

Fig. 1-6. Diagram of a semiconductor laser. The active region is indicated by the shaded area.

which gives a Fresnel reflectivity of about 30%, since the semiconductor index of refraction is about 3.5 compared to unity outside. Our laser is mounted p-side up. For more efficient heat removal, semiconductor lasers are sometimes mounted p-side down, so that the active region is closer to the heat sink. Typical sizes of the active region are 1000 Å thick by 10μm wide by 250μm long. Much smaller lasers can be fabricated easily with interesting special properties as we discuss in later chapters.

Several refinements may be incorporated into the laser of Fig. 1-6 to attain certain desirable characteristics: for e.g., low threshold, cw operation, operation at high temperature, narrow linewidth (even single mode) spectra, or high output power. All semiconductor lasers are now fabricated with heterostructures in order to have the low thresholds necessary for cw or room temperature operation. The heterostructure widths may be chosen to produce either a bulk or a quantum-well gain medium. In most semiconductor lasers on the market, the lateral variations ($\hat{y}$ direction in Fig. 1-6) of the light are *gain guided*, that is limited in $\hat{y}$ extent to the region having appreciable current flow. Some lasers are fabricated more intricately to achieve index guiding, which allows them to operate in a single mode. Combinations of high and low reflection coatings are often used to optimize the optical resonator quality factor, Q. Non-absorbing facet technology, which increases the facet damage threshold, is instrumental in the development of high-power single-mode lasers. Some semiconductor lasers are fabricated with narrow stripes for single transverse-mode operation, while others are fabricated with broad active areas for high output power.

    To see how an inversion is created at a *p-n* junction, we plot the energy bands and electron occupation as a function of position in the transverse $\hat{x}$ direction, i.e., perpendicular to the junction plane. These are given approximately by adding $-e\phi(x)$ of Fig. 1-3c to the valence and conduction bands as shown in Fig. 1-7a. This figure shows that in the absence of an applied voltage across the electrodes, the Fermi energy is constant, resulting in no net flow of carriers. More importantly, there is

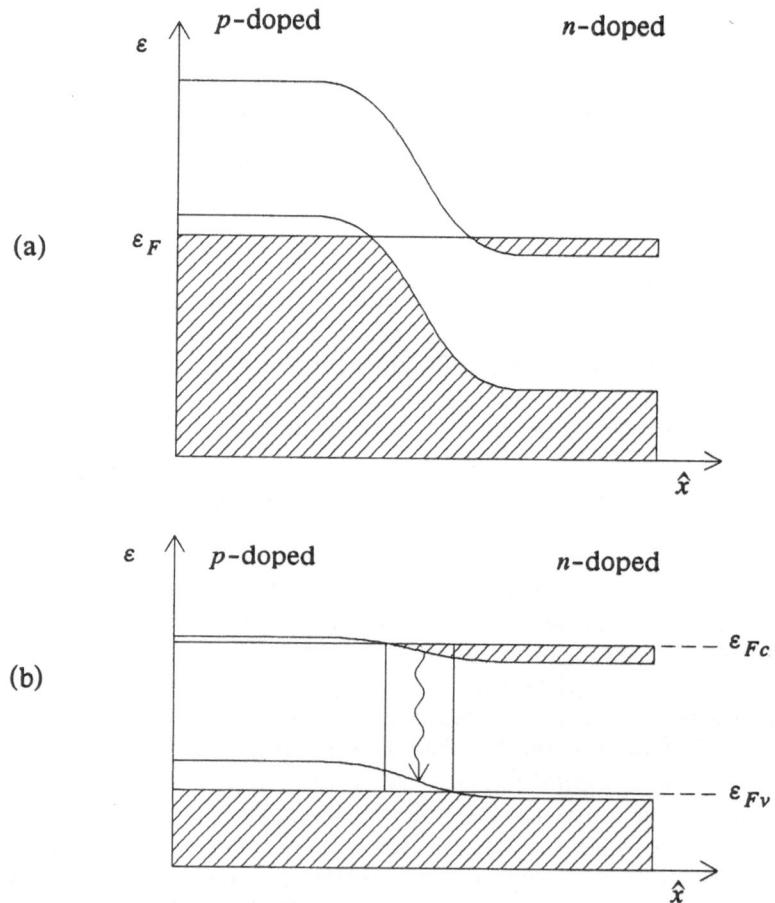

Fig. 1-7. Electron energy and occupation (*a*) without an applied voltage and (*b*) with a forward biased applied voltage.

no region containing both electrons in the conduction band and holes in the valence band, which is necessary to obtain an inverted population. When a voltage is applied so that the p-doped region is positive relative to the n-doped region, the electron energies are altered as shown in Fig. 1-7b. The voltage drop across the junction is reduced (forward biased). When the forward bias approximately equals the band-gap potential, an inverted or active region is created within the junction. Inside this region, stimulated emission occurs due to electron-hole recombination. At steady state, the inversion is maintained by the injection of carriers, via the electrodes, by an external power supply. Lasing occurs when the rate of stimulated emission due to electron-hole recombination approximately equals the total rate of optical losses.

1-3. Heterostructures

The first semiconductor lasers were homostructure devices in that each laser was fabricated with only one semiconductor material. These lasers had high threshold current densities, even when operated at low temperature, where the gain is higher and the carrier density necessary for reaching transparency is lower than at room temperature. In addition to not working at room temperature, these lasers also could not operate cw.

The methods which drastically improve semiconductor laser performance may be understood with the following discussion. A laser's threshold gain is determined by the unity round-trip condition

$$R_1 R_2 \exp[2(\Gamma G_{th} - \alpha_{abs})L] = 1 , \tag{6}$$

where R_1 and R_2 are the facet reflectivities, Γ is the confinement factor, which is a measure of the overlap between the lasing mode and the active region cross section, L is the laser length, and α_{abs} accounts for the optical losses. Solving Eq. (6) for G_{th}, we have

$$G_{th} = \frac{1}{\Gamma}\left[\alpha_{abs} - \frac{ln(R_1 R_2)}{2L}\right] . \tag{7}$$

To a good approximation, and this is verified by experiments and theory, the peak gain in a bulk semiconductor varies more or less linearly with carrier density. Such a linear relation may be written as

$$G = A_g(N - N_g) , \tag{8}$$

where A_g is the gain coefficient and N_g is the carrier density needed to reach transparency in the gain medium. A_g and N_g depend on both the gain material (via relaxation rates, band structure, etc.) and the laser configuration (via the lasing frequency, temperature, etc). In the more phenomenological approaches, one takes their values from experiment. In later chapters we derive more precise gain formulas, which when reduced to the form of Eq. (8) give the functional form of A_g and N_g. Note that although Eq. (8) displays no tuning dependence, it clearly shows that with too few carriers, the medium absorbs radiation instead of amplifying it.

Also to a good approximation, the injected carrier density may be related to the injection current density by

$$N = \frac{J\eta}{e\gamma_{nr}d} , \tag{9}$$

where e is the electron charge, γ_{nr} is the recombination rate, d is the active region thickness (or depth), and η is the quantum efficiency with which the injected carriers arrive in the active region and contribute to the inversion. Since N is inversely proportional to d, the gain (8) increases as d decreases for a given J. More specifically, solving Eq. (8) for N using the threshold gain of Eq. (7), we find the threshold injection current density

$$J_{th} = \frac{e\gamma_{nr}d}{\eta} \left[N_g + \frac{\alpha_{abs} - ln(R_1 R_2)/2L}{A_g \Gamma} \right] . \tag{10}$$

We note that the threshold current density is a strong function of the active region thickness. In a homostructure laser, d is the distance traveled by a conduction electron going from the n-doped region to the p-doped region before it recombines with a hole. In a homostructure GaAs laser, $d \simeq 1\mu m$. Reduction of this thickness reduces the threshold current density proportionately, unless Γ is changed.

The method that is now generally adopted for decreasing the active layer thickness involves blocking the carrier flow with a layer of material that has a higher band-gap energy than the active region. The resulting structure is called single heterostructure if only one blocking layer is used, and double heterostructure if a blocking layer is used on either side of the active region (Fig. 1-8). With a heterostructure laser, the thickness of the active region is determined during growth, and active region thicknesses of 0.1 μm or less can readily be achieved.

Heterostructures may only be grown with crystals with sufficiently similar lattices. For example, one may use GaAs and AlAs because both are face-centered cubic crystals, with almost equal lattice constants of 5.652 Å and 5.662 Å at room temperature, respectively. One can then grow

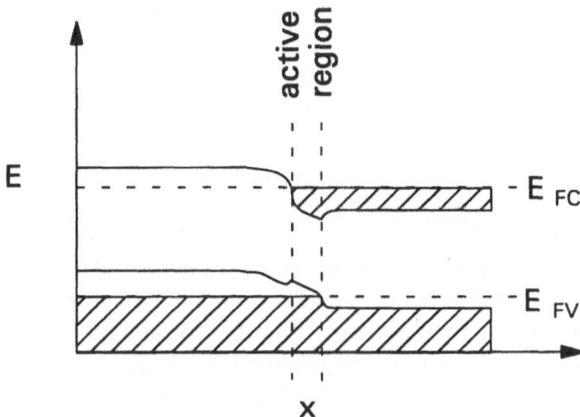

Fig. 1-8. Real space energy structure of a double heterostructure semiconductor diode.

layers of $(GaAs)_{1-x}(AlAs)_x$ which is also written as $Al_x Ga_{1-x} As$. For $x > .45$, this compound has an indirect band gap, which is not useful as a gain medium. For smaller values of x, the bandgap energy of such a layer is given empirically by

$$\varepsilon_g = 1.424 + 1.266x + 0.266x^2 . \tag{11}$$

For two materials that can form a stable heterostructure, the larger band-gap material usually has a lower refractive index. According to experimental data, the refractive index of $Al_x Ga_{1-x} As$ may be approximated by,

$$n = 3.590 - 0.710x + 0.091x^2 . \tag{12}$$

Therefore, the double heterostructure also provides an optical waveguide for the laser field, resulting in a higher confinement factor. Equally important, because of their wider band gap, the blocking layers are transparent to the laser field, thus reducing optical losses. The improvement in laser performance due to the introduction of heterostructures is largely responsible for making semiconductor lasers into practical devices.

As an example, one might sandwich a 1000 Å layer of $Al_{.05}Ga_{.95}As$ in between a layer of p-$Al_{.3}Ga_{.7}As$ and a layer of n-$Al_{.3}Ga_{.7}As$. The intrinsic active layer lases at a wavelength $\lambda = 0.83$ μm and has a gain of about 200 cm^{-1}. For a stripe width of 5 μm and a length of 250 μm and uncoated facet reflectivities of 32%, this gives $J_{th} = 50$ mA.

Once lasing is achieved, the next goal is usually to increase the output power. For the double heterostructure laser just described, scaling to higher power is a problem. The reason is that both the carriers and the laser mode are confined to within the same thin region. While we would like the carriers to be in a thin layer to maximize its density, we would also like the radiation field to be in a thick layer to ensure that its intensity is below the material damage threshold. It turns out we can have both with more complicated heterostructure configurations. Examples are the large optical cavity (LOC) structure shown in Fig. 1-9a, or the separate confinement heterostructure (SCH) shown in Fig. 1-9b. These heterostructures involve either one or two barrier layers for carrier confinement, and two cladding layers for optical confinement. The flexibility of the LOC and SCH designs makes them widely used in semiconductor lasers.

Present state of the art fabrication techniques allow one to reduce the active layer thickness even to the dimension of the order of or less than an electron deBroglie wavelength, which is about 120 Å in GaAs. We then have a quantum-well laser where the carriers are confined to a square well in the transverse dimension and free in the other two. As later chapters show, the change from a three-dimensional to a two-dimensional free-particle density of states causes a quantum-well gain medium to behave very differently from a bulk gain medium. Another useful property of a quantum-well layer is that it is thin enough to form stable heterostructures with semiconductors of noticeably different lattice constants. The necessary deformation (strain) in the quantum-well lattice structure produces stress in the neighborhood of the interface which significantly alters the band structure. The change in band structure can occur in the direction that reduces laser threshold current density. This and other features of strained layer quantum-well lasers make it an interesting and attractive device for further development.

Additional improvements of the semiconductor laser performance appear possible if one reduces the dimensionality of the gain medium even further than in the quasi-two-dimensional quantum wells. Instead of having carrier confinement only in one space dimension, one may produce structures where the quantum confinement occurs in two or even all three space dimensions. These quasi-one-dimensional or quasi-zero-dimensional nanostructures are referred to as quantum wires or quantum dots, respectively. Simple density-of-state arguments (Sec. 2-4) indicate that the reduced dimensionality leads to a more efficient inversion and, hence, to the possibility of ultra low threshold laser operation. However, more recent studies show that the Coulomb interaction effects among the charge carriers become increasingly more important for a decreased dimensionality of the semiconductor structure. These Coulomb effects seem to, at least partially, remove the advantages gained by the modified density of states. Furthermore, the manufacturing of quantum-wire or quantum-dot laser structures is still in its infancy. Therefore, besides the brief analysis of the basic quantum confinement effects in Sec. 2-4, we do not discuss the potentially very interesting quantum-wire or quantum-dot laser devices in

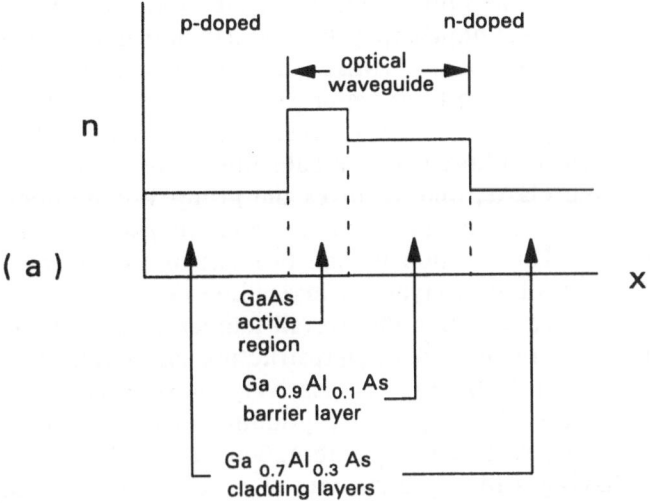

(a)

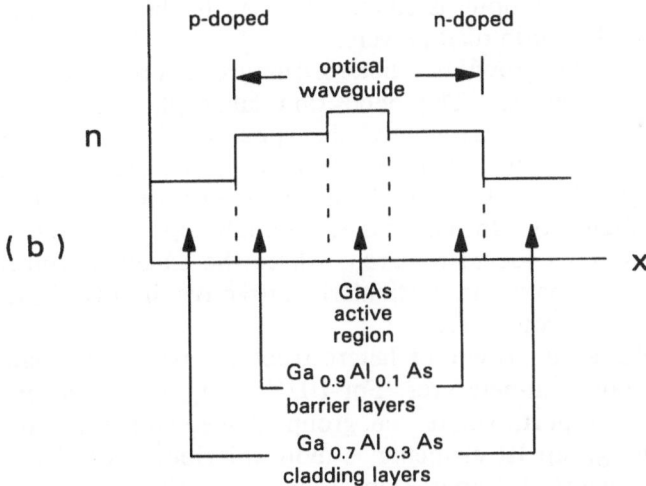

(b)

Fig. 1-9. (*a*) large optical cavity (LOC) structure. (*b*) separate confinement heterostructure (SCH).

this book. For more information we refer the interested reader to the literature at the end of this chapter.

We now summarize the epitaxy technology that allows one to fabricate heterostructures. The techniques may be divided into liquid phase epitaxy (LPE), molecular beam epitaxy (MBE) and chemical vapor deposition (CVD). CVD is also referred to as vapor phase epitaxy (VPE).

LPE involves the growth of the heterostructure layers by the precipitation of material from a solution onto a substrate. For example, to grow a III/V compound like GaAs, one dissolves the group V compound, As, and the desired dopant (e.g. Zn for p-doping and Se for n-doping) in the molten solvent, Ga. The solution is then brought into contact with a substrate wafer. Growth of thin layers of crystalline GaAs on the substrate is brought about by either reducing the overall temperature, or by cooling the substrate. Almost all conventional heterostructure lasers may be fabricated by LPE. The necessary apparatus is relatively inexpensive and simple to operate. LPE produces high quality crystalline material that has a high radiative to non-radiative carrier recombination ratio. However, it has the disadvantage of having a high growth rate of around $100\text{Å}/s$, which causes difficulty in controlling thickness uniformity and reproducibility. As a result the thickness of layers that can accurately be grown is 1000Å or more which prevents the fabrication of quantum-well structures. In comparison with the other epitaxy methods LPE produces the worst surface and interface qualities. The technique is commonly used in the mass fabrication of lasers for use in CD-audio disk players.

MBE involves the growth of heterostructure layers by evaporation of material onto a substrate. The deposition takes place under ultra high vacuum (10^{-11} Torr). Consequently, the apparatus is very expensive and difficult to operate. An advantage of MBE is the slow deposition rate ($1\text{Å}/s$), which gives good thickness control, quality of interfaces, uniformity and reproducibility. In most cases, the high apparatus and operational costs of MBE limit its use to research, where the excellent control during the growth process makes it particularly attractive in the fabrication of novel structures and prototypes.

CVD or VPE is the growth of heterostructure layers with materials that are gaseous at room temperature. For III/V compounds, one method involves hydride transport, where the group V elements are introduced as hydrides and the group III elements as monochlorides. Another method is Metal-Organic Chemical Vapor Deposition (MOCVD), where the metals are introduced in the form of metal-organic alkyl (di-ethyl or dimethyl) compounds, which are liquids at normal temperature. The bubbling of hydrogen through the liquids produces a vapor that is saturated with the compounds, which is then passed through the growth apparatus. Compared to hydride transport, MOCVD provides a more controllable supply of gas so that composition changes are more easily made. Both methods do not require high vacuums, so that apparatus and operational costs are only moderately expensive. However, the gases used are highly toxic, and the need for safety equipment adds to the cost of a MOCVD facility.   The growth procedure is not as difficult as in MBE.   Excellent thickness control, interface quality, uniformity and reproducibility are still achievable because of the slow 30 Å/s growth rate.   MOCVD is the industrial expitaxy technique for the fabrication of quantum-well lasers.

1-4. Gain and Index Guiding

In the lateral dimension ($\hat{y}$ of Fig. 1-6), i.e., in the junction plane and perpendicular to the laser axis, the laser field may be confined by either gain guiding or index guiding.   In a gain-guided laser, the lateral extent of the laser field is determined by the lateral spatial carrier distribution at the p-n junction, while for an index-guided laser, the lateral field profile is determined by a fabricated waveguide.   In this section, we describe some different methods for gain and index guiding, and discuss the differences in gain- and index-guided modes.

In general there are three common methods for making gain-guided stripes.   One method uses an insulating material for current confinement, the second method uses a reversed-bias p-n junction, and the third method uses a semi-insulating layer produced by proton-bombardment or oxygen-implantation.   Regardless of the method of stripe definition, a gain-guided mode has a diverging phase front.   However, the lateral width of a gain-guided mode remains constant throughout the laser because of the absorption that occurs where the carrier density falls below the value needed for transparency in the semiconductor.   Early semiconductor lasers were all gain guided, and because they are easier to fabricate than index-guided lasers, gain-guided lasers still dominate today's market.

Figure 1-10a shows an oxide-insulated stripe contact laser.   The main features are the n-doped substrate, the heterostructure, and an insulating layer of SiO_2 with a channel etched in it.   In practice, several details are necessary for the laser to work properly.   A GaAs capping layer is needed to provide good electrical contact with the metallic electrode, which does not bond well to AlGaAs.   Also, we need the lateral current confinement to continue into the p-doped layers between the insulating and the gain regions.   We accomplish this by first minimizing the dopant densities in the p-doped layers to reduce electrical conductivity and then diffusing Zn into these layers within the volume that is directly below the channel.   Zn diffusion has the net effect of increasing the p-dopant density and thus increasing electrical conductivity directly below the channel.   Figure 1-10b shows an alternate design that still uses an oxide insulating layer but does not require Zn diffusion.   The shallow mesa stripe laser requires etching away as much as possible the regions of the p-doped layers that is outside the channel.   In this way the p-doped layers may be highly doped to

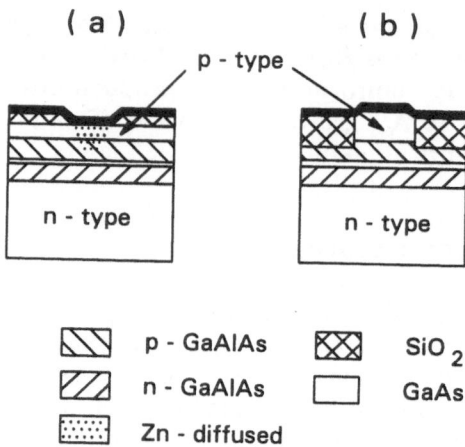

Fig. 1-10. *a*) Oxide-insulated stripe laser, *b*) Shallow mesa stripe laser.

optimize electrical conductivity without affecting current confinement. In any case the use of oxide-insulating material for stripe definition requires some careful fabrication steps, especially because of the need to minimize strain at the interfaces. Oxide-insulated stripe contact lasers tend to have higher thresholds.

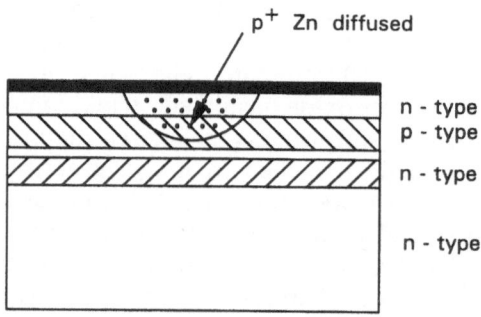

Fig. 1-11. Zn-diffused planar stripe laser.

Figure 1-11 depicts a Zn-diffused planar stripe laser that does not use an insulating layer for lateral carrier confinement. Instead, it has an *n*-doped capping layer with a *p*-doped channel created by Zn diffusion in the middle. Outside of the Zn-diffused channel, we have *p-n* junctions between the *n*-doped GaAs capping layer and the *p*-doped AlGaAs cladding layer. When a forward bias is applied to the active region, these p-n junctions become reverse biased which then prevents the flow of current. Stripe definition by Zn diffusion is the most difficult of the three methods to implement.

Regions of relatively high electrical resistance may also be made without using an oxide-insulating material. Semi-insulating regions may be created by either proton-bombardment or oxygen-implantation. In both cases, the incident ions enter the unmasked regions of the semiconductor and create crystalline defects that act as localized recombination centers, trapping the injected carriers. Figure 1-12a shows a laser made with a

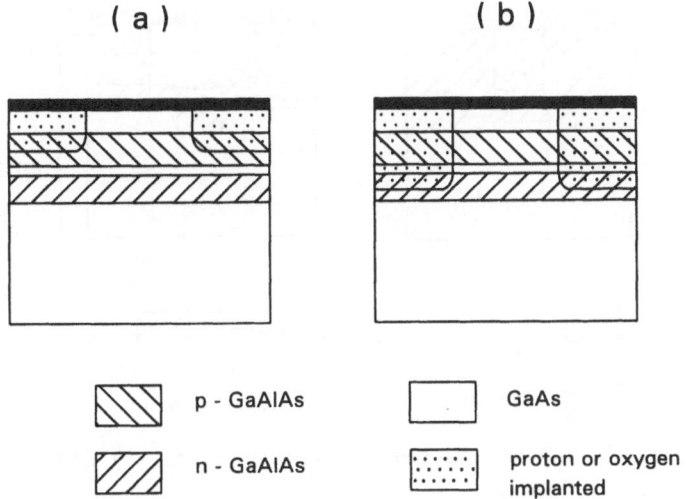

Fig. 1-12. a) Shallow bombarded or implanted stripe. b) Deep bombarded or implanted stripe.

shallow bombarded or implanted stripe. This method of stripe definition has the advantage of being the simplest to implement. However, the method is based on damaging a certain volume of semiconductor material to reduce its electrical conductivity. The damaged material gives rise to strains that may cause refractive index inhomogeneities within the laser mode volume. Output beam quality may suffer because of this, especially in a deep bombarded or implanted stripe laser (Fig. 1-12b). Another problem is that self annealing may occur in the damage regions, thus causing a reduction in lateral current confinement with time.

When a narrow laser linewidth is desired, one usually uses index guiding. Here, an optical waveguide is fabricated into the laser chip. An index-guided mode is characterized by a flat phase front. Two factors are involved in producing better spectral properties. First, the optical waveguide may be made sufficiently narrow to support only the lowest-order lateral mode. Second, a flat phase front reduces the fraction of spontaneous emission (which is emitted over 4π radians) contributing to the laser field.

(a)   (b)

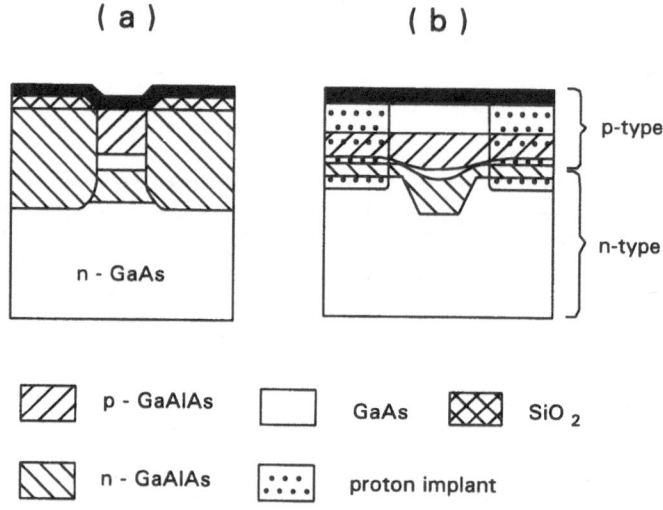

Fig. 1-13. (a) Buried-heterostructure laser. (b) Channeled-substrate planar laser.

Index-guided lasers vary considerably in complexity. A conceptually simple design is the Buried-Heterostructure (BH) laser shown in Fig. 1-13a. The active region is surrounded by p- and n-doped AlGaAs. The lateral boundaries of the waveguide are defined by the regrowth n-doped AlGaAs regions which, because of their larger bandgap and lower refractive index, contribute to carrier and light confinement. A problem with a BH laser is the narrow ($\simeq 1\mu m$) waveguide width necessary for single lateral mode operation. This limits the output power because of facet damage due to high intensities. In principle, one can increase the waveguide width and yet maintain single mode operation by reducing the refractive index difference between the waveguide and the regrowth regions. In practice, the composition control needed to do so is sufficiently stringent to make reproducible results questionable. The sensitivity to the composition of the AlGaAs regrowth regions is because of the GaAs substrate which can effect the effective index distribution substantially.

Fortunately, solutions for obtaining high power single lateral mode lasers exist. A relatively direct method involves the use of non-absorbing mirrors. We note that the region immediately behind a laser facet has low carrier density because it does not lie directly under the electrode. The low carrier density leads to optical absorption. Etching away this region and regrowing with higher bandgap AlGaAs reduces the optical absorption. BH lasers with non-absorbing mirrors (sometimes called window striped lasers), capable of producing more than 100 mW single mode output, have recently become commercially available.

The fabrication of a BH laser is very complicated and time consuming because it involves two growth processes. A simpler index guided laser to fabricate is the Channeled-Substrate (CS) laser shown in Fig. 1-13b. It uses the fact that LPE enables the growth of layers that will eventually fill in grooves present in a substrate. One starts with a substrate with a channel etched in the middle. The deposition of material for the heterostructure is such that the channel in the substrate is filled in as shown in the figure. Single mode operation with a $3\mu m$ active region has been achieved.

1-5. Semiconductor Microlasers

For many applications and improved performance characteristics it is desirable to reduce the volume of the gain medium and to miniaturize the entire semiconductor laser structure. A significant step in this direction is achieved with semiconductor microlaser structures. Two examples of such structures are shown in Fig. 1-14. Conceptually the simplest structure, and possibly the simplest semiconductor laser structure overall, consists of a thin layer of active material between two highly reflecting mirrors (Fig. 1-14a). Such a microlaser is often called *Vertical Cavity Surface Emitting Laser*, or VCSEL, because the laser cavity is vertical to the chip and the light emission comes out of the surface, in contrast to conventional semiconductor lasers, where the cavity is horizontal and the light comes out of the edge of the structure. Therefore, these conventional lasers are sometimes called edge emitters.

The entire VCSEL structure is grown vertically by using MBE or MOCVD techniques to sequentially deposit the different layers onto the substrate. One starts with the bottom mirror, followed by the gain region, and then comes the top mirror. The gain region itself may be either a relatively thick layer, so that it can be modelled as bulk material, or it may be a (small) number of quantum wells. Patterning the entire wafer surface into small areas yields dense two-dimensional arrays of VCSELs. Typically, a single VCSEL has a 5 - 10 μm diameter and a total height of a few microns.

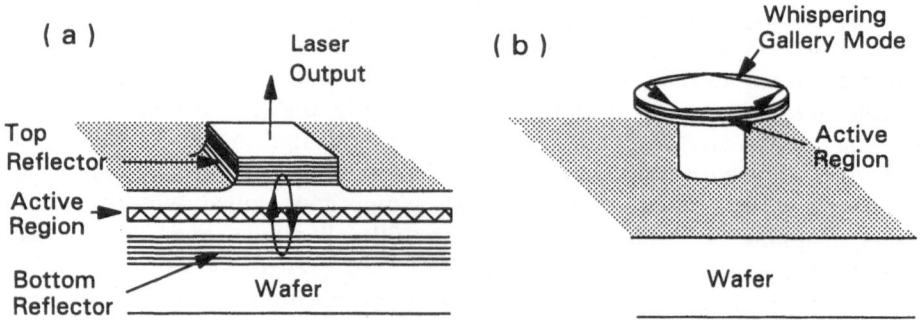

Fig. 1-14.   Schematic drawing of two semiconductor microlaser structures. (*a*) The VCSEL (vertical cavity surface emitting laser) consists of an active medium between two highly reflecting mirrors.   These mirrors are Bragg reflectors made of alternating layers of semiconductor material with slightly different refractive indices. (*b*) The microdisk whispering-gallery-mode laser consists of a thin layer of gain material between passive semiconductor cladding material.   The laser mode is a whispering-gallery mode which propagates around the edge of the disk.   It experiences almost total internal reflection at the boundary between the semiconductor gain material (refractive index $n_b \simeq 3.5$) and its surrounding medium ($n_b \simeq 1$).

By choosing the appropriate material composition, layer thickness, and number of layers, the Bragg mirror of the VCSELs can be designed to have high reflectivity (around 99%) only in a relatively narrow spectral region. This high reflectivity is needed for lasing because the thin layer of active medium yields relatively little gain per cavity round trip. Since the energetic spacing between the longitudinal modes in the microcavities is quite large, this mirror design flexibility makes it possible to realize the situation where only one longitudinal cavity mode exists in the spectral region of high reflectivity. Hence, these VCSELs closely approach the ideal case of a single mode semiconductor laser.

The mircodisk laser structure in Fig. 1-14*b* is also grown by MBE or MOCVD techniques, followed by appropriate patterning, and sequential etching.   The microdisk typically consists of InGaAs active material between InGaAsP barriers. The disk is supported by a post of InP material.

1-6. Output Power-Current Characteristics

We now move on to the discussion of some general semiconductor laser behavior. A motivation for this discussion is to find experimental observations that help us understand the physics of a semiconductor gain medium. Our goal as theoreticians is to arrive at a self-consistent theory that is qualitatively and quantitatively consistent with experiments. Our hope is that such a theory will give us the predictive power to design better semiconductor lasers.

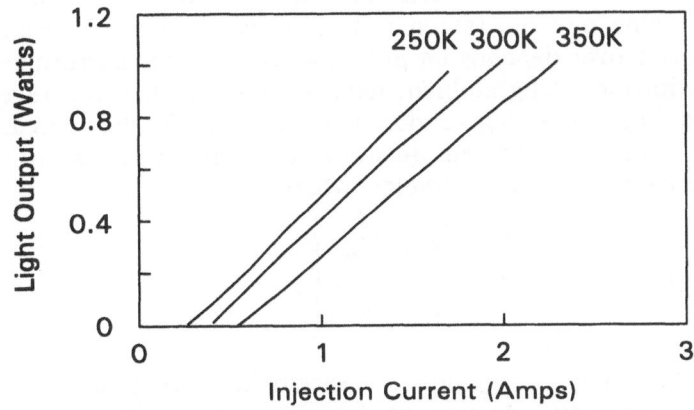

Fig. 1-15. Experimental L-I curves for GaAs quantum-well laser. The output power was measured from a $100\mu m$ core optical fiber that was butt coupled to the laser so that the actual optical power at the laser facet was approximately twice as high. (Courtesy of W. Ng, Hughes Research Lab.)

We begin with Fig. 1-15, which shows typical output power versus injection current, or L-I curves for a semiconductor quantum-well laser. Each curve can be characterized by two parameters: the threshold current density and the slope efficiency. A low threshold is important because it reduces the input electrical power that is not converted into laser radiation. According to Eq. (10), the threshold current density depends on the optical resonator (via the facet reflectivities, cavity length, confinement factor and active layer thickness), and the gain material (via gain coefficient, carrier density at transparency, carrier decay rates). In terms of the latter, a quantum-well gain medium is found to be better than a bulk gain medium, and there is reason to believe that a strain-layer quantum-well gain medium will perform better than an unstrained one.

The slope efficiency, which is the laser efficiency excluding the injection power needed to achieve threshold, is high in semiconductor lasers when compared to other types of lasers. For example, quantum-well laser slope efficiencies typically equal 50%. This translates to 1 photon produced for every 2 injection electrons, after threshold is reached. This is an impressive number that makes semiconductor lasers competitive in many opto-electrical applications. It is also an interesting number because it indicates an efficient extraction of electrical power that is only possible when the laser field is able to interact with essentially the entire carrier distribution. In other words, a semiconductor gain medium basically saturates more or less homogeneously, even though the band structure contributes large inhomogeneous broadening (see Figs. 2-1 and 2-2).

Note that at high injection current, $L-I$ curves can show noticeable roll-over. One possible reason is heating of the laser in which case the degree of roll-over depends on pulse duration. Laser performance can also degrade with increasing ambient temperature, which is not revealed in the simple Eq. (8). The degradation is specifically in the increase in lasing threshold. Therefore, we can quantify the temperature sensitivity of semiconductor lasers by a T_0 parameter, where

$$I_{th}(T_2) = I_{th}(T_1) \exp\left[\frac{T_2 - T_1}{T_0}\right],$$ (13)

where $I_{th}(T_1)$ and $I_{th}(T_2)$ are the threshold currents at temperatures T_1 and T_2, respectively. At present, high T_0 lasers tend to be graded-index (GRIN) single quantum-well lasers. The graded-index structure helps in the capture of carriers in the active region, and this is especially important at high temperatures where the injection electrons are, on the average, more energetic. The quantum-well structure also helps in increasing T_0. The reason has to do with the two-dimensional band structure of quantum-well structures which makes laser performance less sensitivity to the changes in the carrier energy distributions with temperature than the three-dimensional band structure of bulk gain media. We expand on this last point in latter chapters. The values of T_0 varies from around 70° C in bulk semiconductor lasers to as high as over 250° C in quantum-well lasers.

1-7. Frequency Spectrum

The implication from the high slope efficiency that semiconductor lasers saturate homogeneously would not be surprising except that spectral data indicate differently. Figure 1-16 shows the spectra of a laser for increas-

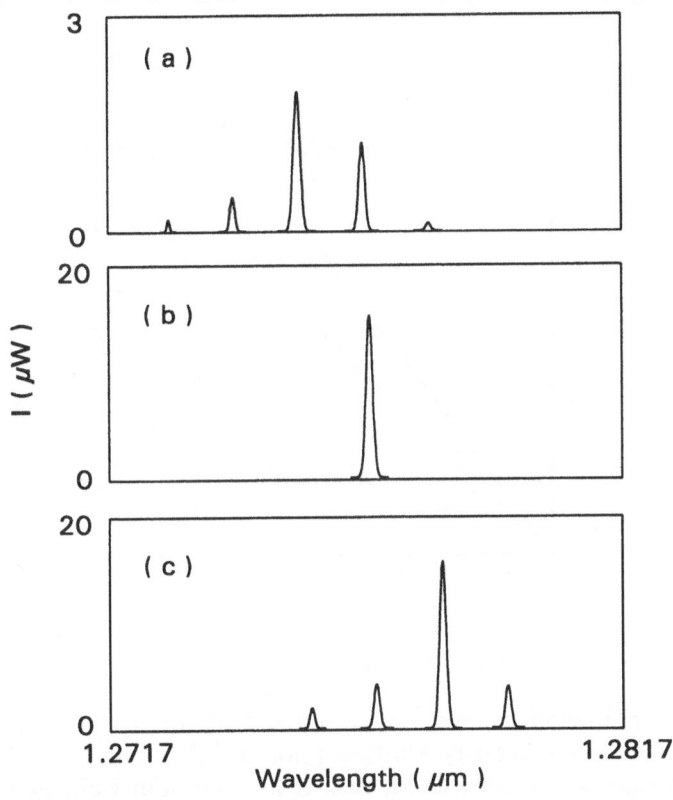

Fig. 1-16. Spectra of a InGaAsP buried heterostructure laser. The excitation is I/I_{th} = 1.1 (*a*), 1.6 (*b*) and 2. (*c*). (Courtesy of W. Ng, Hughes Research Lab.)

ing injection current. Note that the spectrum is multimode close to threshold because the high semiconductor spontaneous emission rate leads to relatively high intensities in the amplified spontaneous emission (ASE) modes, which are actually not above the lasing threshold. The spectrum becomes single mode at higher current because of mode competition. It is interesting to note that the spectrum reverts back to multimode output at even higher current levels. This multi-longitudinal mode output is only possible in an inhomogeneously broadened gain medium. Since most types of lasers may be unambiguously classified as being either homogeneously or inhomogeneously broadened, the dual character exhibited by semiconductor lasers makes their physics particularly interesting.

As we can make increasingly stable semiconductor lasers, it becomes interesting to have a feel of the ultimate linewidth of a single-mode semiconductor laser. This linewidth is due to fluctuations in the laser field

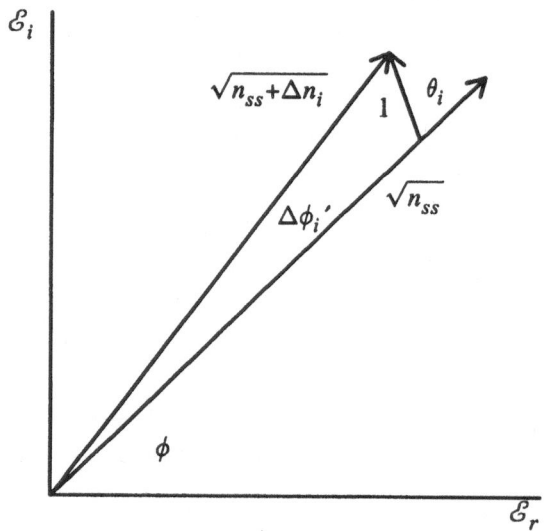

Fig. 1-17. Changes in the phase of a laser field vector due to spontaneously emitted photons.

caused by spontaneous emission. Figure 1-17 depicts the situation. The addition of a spontaneously emitted photon, which has an arbitrary phase relative to that of the laser field, results in a random walk of the tip of the laser field vector. The field amplitude remains at essentially the square of the photon number, while the phase fluctuates freely, eventually assuming all values between 0 and 2π. The diffusion of phase leads to a vanishing laser field vector sum, and the rate of decay of the ensemble average of the field vector is a measure of the spontaneous emission linewidth of the laser. According to this picture, the laser linewidth is given by the Schawlow-Townes formula,

$$\Delta\nu_{S-T} = \frac{A}{n_{ss}}, \qquad (14)$$

where A is the spontaneous emission coefficient into the lasing mode and n_{ss} is the steady-state photon number of the lasing mode.

The linewidth of the semiconductor laser has an additional contribution that comes from fluctuations in the refractive index caused by fluctuations in the carrier density. Because of gain clamping, intensity fluctuations have negligible direct effect on the linewidth, but they do cause fluctuations in the carrier density. Since the refractive index change due to carrier density at the gain peak is large in a semiconductor gain medium, the density fluctuations cause substantial index fluctuations, which, in turn, lead to phase fluctuations. This results in the following increase in the fundamental laser linewidth

$$\Delta \nu = (1 + \alpha^2)\, \Delta \nu_{S-T} \, , \tag{15}$$

where α is the linewidth-enhancement factor which is a measure of the change in the medium refractive index for a corresponding change in the gain. Hence intensity fluctuations contribute *indirectly* to the linewidth, even though their direct contribution is negligible. In two-level media, the indirect contribution is also negligible, since the change in index of refraction goes through zero at the gain maximum. The change in saturation caused by intensity fluctuations is thus multiplied by zero, unless the laser is forced to operate away from the gain maximum.

1-8. Transverse Mode Structure

Let us move onto another aspect of semiconductor laser behavior by discussing the subject of laser field profile. In the transverse ($\hat{x}$) direction, the laser field is index guided by the heterostructure. The optical guide is usually made sufficiently narrow to support only one transverse mode. Since the required guide thickness is approximately 1 micron, the transverse beam divergence may be as large as 30 degrees. However, one should remember that if the field is diffraction limited in this direction, in principle it may be collimated, expanded, focused, etc. to any desired shape with conventional optics. Of course, doing so may be impractical because the needed optical elements are likely to be considerably larger than the laser.

In the lateral ($\hat{y}$) direction, the optical-field confinement is often weaker, leading to substantial astigmatism in the output. Single-mode operation is still possible with a narrow gain or index stripe width. However for high-power operation, the lateral dimension has to be wide in order to prevent material damage due to high optical intensities. The lateral mode profile then depends more on the gain medium than is the case with the transverse mode profile, and a wide stripe laser usually operates multimode. The onset of multimode operation is hastened by self-focusing, which is caused by the saturation of the carrier-induced refractive index change. To understand this effect, we begin by looking at the case of low laser intensity. Here, because of the lack of gain saturation, the carrier distribution follows the injection current distribution.

As is shown in later chapters, the carrier induced refractive index decreases with increasing carrier density so that the resulting refractive index distribution tends to defocus the laser beam, i.e. we have an antiguide. The effect of the carrier induced refractive index is reversed at high laser intensity. Owing to gain saturation a spatial hole is burned into the center of the distribution by the laser field. This leads to variations in the index distribution, which results in the focusing of the laser field. The focused laser field burns a deeper hole in the carrier distribution which in turn leads to stronger focusing. This is called self-focusing. Eventually, the self-focusing is balanced by diffraction and gain. The final intensity profile then consists of several narrow *'bumps'* which are commonly referred to as filaments.

One method to prevent filamentation and yet have a broad emitting region is to fabricate a semiconductor laser with a gain region that is made up of several narrow gain strips, i.e. an array. Semiconductor laser arrays provide high power with relatively good to very good beam quality. In a good array, each laser in the array is made sufficiently narrow to operate single mode. The frequency locking of the lasers leads to the coherent combination of their output. If the lasers are locked in-phase with one another, then the far-field intensity pattern is a single lobe, with an on-axis intensity that scales as the square of the number of lasers in the array.

Of course, the perfect array described above does not exist for all operating conditions. Each laser does not always operate single mode, especially when we try to maximize optical power. The frequency locking is often incomplete because of differences between the lasers that cannot be overcome by mutual coupling. Even when the lasers are locked, they tend to operate with phase differences that are different from zero. The most likely phase difference is π between adjacent lasers, which results in a double-lobe far-field distribution.

An understanding of the frequency locking mechanisms is important for the development of better arrays. Theories for doing so includes coupled (super) mode theory, which treats the array as a superposition of the individual laser modes, to array mode theory, which treats the array as a single complex resonator. Both approaches have their advantages and weaknesses, and they are discussed in detail in a later chapter of this book.

1-9. Phenomenological Gain Model

Semiconductor laser models have been developed to study the various semiconductor laser phenomena discussed in the previous sections. At present, most of these models use a rate equation to describe the time evolution of the total carrier density

$$\frac{d}{dt}N = \frac{J\eta}{ed} - BN^2 - \gamma_{nr}N - \frac{GI}{\hbar\nu},\qquad(16)$$

where B is the radiative bimolecular carrier recombination (spontaneous emission) coefficient, γ_{nr} is the non-radiative carrier decay rate, $\hbar\nu$ is the laser photon energy, G is the gain, and I is the laser intensity. The primary assumption is that intraband carrier-carrier scattering is sufficiently fast, i.e., faster than the stimulated emission and interband carrier decay rates, so that the laser medium saturates homogeneously. Otherwise one has to track the population of each electron and hole state, which is a non-trivial task. There are instances where doing so is necessary, for example, when modeling femtosecond experiments in semiconductors, or when investigating spectral hole-burning effects in a laser.

In later chapters we derive the equation of motion (16) from basic principles, thereby establishing its range of validity. Here it's instructive to solve for the steady-state value of N, which is the one given by setting $\dot{N} = 0$. Although more accurate gain formulas can be used, we substitute Eq. (8) into Eq. (16) with $\dot{N} = 0$, and solve for $N - N_g$. For simplicity assuming that the BN^2 term is included approximately in the $\gamma_{nr}N$ term, we find

$$N - N_g = \frac{J\eta/ed\,\gamma_{nr} - N_g}{1 + IA_g/\hbar\nu\gamma_{nr}} = \frac{J\eta/ed\,\gamma_{nr} - N_g}{1 + I/I_{sat}}.\qquad(17)$$

Equation (17) suggests that the gain (8) saturates in a way proportional to $1/(1 + I/I_{sat})$, where the saturation intensity $I_{sat} = \hbar\nu\gamma_{nr}/A_g$. Thus although Eq. (8) is a linear function of the carrier density N, it is a nonlinear function of the laser intensity I.

Equation (6) gives the unity round-trip condition for threshold operation and, in fact, for steady-state operation in general. Stated differently, it requires that the *saturated-gain equals the losses*. It ensures that as the laser intensity I builds up due to increasing injection current, the saturation compensates for the increase in J, so that both G and hence N remain at their threshold values. This situation is often called *gain clamping*. Using this fact, we can solve Eq. (16) in steady-state for the laser intensity to find

$$I = \frac{\hbar\nu\eta}{G_{th}\,ed}\,[J - J_{th}]\,, \tag{18}$$

where we used Eq. (9) at the threshold. G_{th} is given by Eq. (7) and J_{th} by Eq. (10), respectively. Eq. (18) predicts the linear power output versus current seen in the major portions of the graphs in Fig. 1-15.

This simple approach assumes that the intensity I is constant inside the laser cavity. Since the facet reflectivities are low, e.g., around 30%, this uniform intensity assumption may be inadequate. Accordingly, we can solve Eq. (16) together with the propagation equation

$$\frac{dI}{dz} = [\Gamma\,G\,(N) - \alpha_{abs}]I\,, \tag{19}$$

where α_{abs} accounts for absorption losses in the medium and Γ is the confinement (or fill) factor which measures the overlap between the laser mode and the gain region. The coupling between Eqs. (16) and (19) is via the saturated gain $G\,(N)$. If the laser phase is of interest, we need to replace Eq. (19) with the amplitude propagation equation

$$\frac{\partial \mathcal{E}}{\partial z} = \frac{i}{2nK_0}\frac{\partial^2 \mathcal{E}}{\partial y^2} + \Gamma\left[\frac{G}{2} + iK_0\delta n\right]\mathcal{E}\,, \tag{20}$$

where $\mathcal{E}$ is the complex field amplitude and $G\,(N)$ and $\delta n(N)$ couple the medium and field equations. This equation allows for diffraction in the lateral direction. For both Eqs. (19) and (20), we need to know $G\,(N)$.

The simplest formula to use for $G\,(N)$ is Eq. (8). We might further suppose that the carrier-induced change in index is given by

$$\delta n = -\,R\,A_g\,N\,\frac{n}{2K_0}\,, \tag{21}$$

where R is an *antiguiding* factor (note that the host index contribution is given by n_b). R is closely related to the linewidth enhancement factor α in Eq. (15), which is not surprising since they both represent effects due to changes in the index of refraction. The precise relationship is given in Sec. 3-6. The formulas (8) and (21) for G and δn form the phenomenological gain model. They are admittedly very crude: if we change any of the various operating parameters such as laser tuning or temperature, the A_g, N_g, and R factors have to be changed appropriately. A lot depends on what we expect to do with such formulas. If we want to explain some linear variations about some point in the laser parameter space for which the phenomenological constants A_g, N_g, and R are known either from measurement or from a more fundamental theory, Eqs. (8) and (21) can be useful. However if we need a more general formulation, such as that for deriving the index variation supposedly modeled by Eq. (21) from the gain of Eq. (8) using the Kramers-Kronig relations, we will be sorely disappointed: for such an application, we have to know the explicit frequency dependence of A_g and N_g. As a practical matter, to carry out such a derivation, we find it is better to replace these equations with general ones that arise naturally from a more complete theory.

Although Eq. (8) is a reasonable approximation for some purposes, one must be careful in using it because depending on the problem, the peak gain may not be the desired quantity. For example in studies involving spatial field profiles or laser amplifiers, we need to know the gain for a fixed frequency. In this case, the semiconductor laser gain varies sublinearly with carrier density. A more important failure of the phenomenological approach is that Eq. (21) is very likely to be incorrect. The little experimental data available contain enough discrepancies to cast doubt on the claim that the carrier-induced refractive index varies linearly with carrier density. More careful calculations that include many-body Coulomb effects clearly show that R depends on the carrier density. In addition the phenomenological coefficients have to be modified somehow even for something as simple as a change in temperature. We would like to have simple theories that automatically include temperature variations as well as tuning and carrier-dependent effects. Finally, neither Eq. (8) nor (21) is accurate for quantum-well gain.

We do not wish to give the impression that the simple phenomenological approach is totally incorrect or not useful. In many problems the difference between a linear or a sublinear carrier density dependence of bulk semiconductor gain is negligible. We feel that Eq. (8) together with the carrier density rate equation and the appropriate field equation have contributed significantly to the early understanding of semiconductor laser behavior, and their simplicity continues to make the phenomenological approach a viable method for studying some aspects of semiconductor lasers.

Maybe we should liken the phenomenological approach to walking a tightrope: so long as you follow the narrow path for which the phenomenology was strung up, you will be OK, even if a bit shaky. But if you stray off that narrow path without the appropriate flying apparatus, you will fall into the abyss. The purpose of this book is to provide the theory needed to traverse general terrain without making such precipitous errors.

References

Adams, M. J. (1981), *An Introduction to Optical Waveguides*, John Wiley, New York.

Agrawal, G. A., and N. K. Dutta (1986), *Long-Wavelength Semiconductor Lasers*, Van Nostrand Reinhold Co., New York.

Pulfrey, P. I., N. G. Tarr (1989), *Introduction to Microelectronic Devices*, Prentice Hall, Englewood Cliffs, New Jersey.

Streifer, W., R. D. Burnham, T. L. Paoli, and D. R. Scifres (1984), Laser Focus/Electro-optics, June 1984.

Thompson, G. H. B. (1980), *Physics of Semiconductor Lasers*, Wiley & Sons, New York.

Wilson, J. and J. F. B. Hawkes (1983), *Optoelectronics: An Introduction*, Prentice Hall, Englewood Cliffs, New Jersey.

Yariv, A. (1975), *Quantum Electronics*, 2nd Edition, Wiley & Sons, New York.

For more information about the physics of the electronic and optical properties of quantum wires and quantum dots, see

Banyai, L., and S. W. Koch (1993), *Semiconductor Quantum Dots*, World Scientific Series in Atomic, Molecular and Optical Physics - Vol. 2, World Scientific Publ., Singapore.

Haug, H., and S. W. Koch (1993), *Quantum Theory of the Optical and Electronic Properties of Semiconductors*, 2nd Edition, World Scientific Publ. , Singapore.

Quantum dot and quantum wire laser concepts are discussed, e.g., in

Arakawa, Y., K. Vahala and A. Yariv (1986), Surf. Sci. **174**, 155;

Asada, M., Y. Miyamoto, and Y. Suematsu (1986), IEEE J. Quantum Electron. **22**, 1915;

Vahala, K. (1988), IEEE J. Quantum Electron. **24**,523;

Kapon, E., J. P. Harbison, R. Bhat, and D. M. Hwang (1989), *p.* 49 in *Optical Switching in Low-Dimensional Systems*, H. Hazug and L. Banyai, eds., NATO ASi Series *B*, Vol. 194, Plenum, New York.

Microcavity lasers are discussed, e.g., in

Slusher, R. E. (1993), Optics and Photonics News, *p.* 8, February 1993.

Slusher, R. E., and Y. Yamamoto (1993), Physics Today, June 1993.

These reviews give also references to the original papers.

Chapter 2
BASIC CONCEPTS

Chapter 1 summarizes various semiconductor laser phenomena and discusses a simple phenomenologically based approach for modeling a few of the experimental observations. The advantage of simple models is that they provide a feel for some of the most pervasive semiconductor laser characteristics, such as the decrease in lasing threshold with decreasing thickness of the active region. Using such simple models is analogous to viewing a forest from a great distance with low quality binoculars. We are able to recognize the outlines of the forest and distinguish it from the surrounding land, but we can not see clearly enough to tell what kind of trees are in the forest or even if it is tropical or temperate. With the phenomenological model, important parameters such as carrier effective masses, temperature, and band structure must all be implicitly accounted for with phenomenological "constants" that change markedly as soon as the parameters themselves change. Important physics needed for a more general description is that the electrons obey Fermi-Dirac statistics and that as charged particles they interact strongly with themselves and with the lattice. The task of incorporating the relevant physics into the theory is the subject of this and the next four chapters.

In this chapter, we lay the theoretical foundation for the development of a theory for the semiconductor laser. We begin by discussing some basic aspects of the semiconductor band structure, i.e., the basic energy bands of electrons for the unexcited semiconductor (zero carrier density). A more detailed band-structure analysis including the modifications caused by quantum confinement and strain effects is presented in Chap. 6.

Section 2-2 discusses CGS and MKS units, both of which are used extensively in the semiconductor laser literature. The problem is that MKS has been used traditionally for lasers, while CGS is often used in semiconductor theory. Hence the marriage of the fields requires a familiarity with both systems of units.

Excitation of an electron from one band to a higher one, leaves behind a hole that interacts with the electron via the Coulomb potential. This modifies the electron energy in fundamental ways, which are discussed in detail in Chaps. 4 to 6. In particular for the high densities of excited electrons (and holes) which are needed to have gain, the scattering of these charge carriers among themselves leads to Fermi-Dirac distributions of the carrier probabilities as functions of temperature and density, as discussed in Sec. 2-3. A major improvement in the modeling of the semiconductor laser replaces Eq. (1.8) with a simple gain formula based on these distributions, while keeping Eq. (1.16) for N, and noting that N itself is given by summing the Fermi-Dirac distributions over all momentum and spin states. This model is called the free-carrier model, since other than existing inside a crystal and undergoing many collisions, the carriers are free to move about, unhindered by interactions that form excitons at low densities and reduce the band gap at higher densities.

Section 2-4 introduces the concept of quantum confinement and Sec. 2-5 makes contact with the laser electric field by outlining a derivation of the slowly varying electromagnetic-field equations. This shows how the field amplitude and phase are influenced by an induced polarization of the medium. Section 2-6 begins our discussion of the reverse, namely how the field induces this polarization according to the laws of the quantum mechanics of a semiconductor medium. This lays the foundations for Chaps. 3 through 6, which derive the polarization of the semiconductor medium with increasing levels of accuracy and complexity.

2-1. Elementary Aspects of Band Structures

In a simple picture of a semiconductor, an electronic state is identified by its momentum, $\mathbf{k}$, and z-component of spin, s_z. The allowed electronic energies are a result of the interaction of the electron with the regular lattice of ions. The resulting band structure describes how the energy of an electron in the ionic lattice is related to the carrier momentum in the absence of other mobile carriers. Figure 2-1 shows as an example the electronic band structure for GaAs. We see that it is quite complicated. There are regions with continuous distributions (or *bands*) of energies and regions where electronic states are forbidden (*bandgaps*). The figure identifies two types of bands: *conduction* bands, which consist of unoccupied states; and *valence* bands, which consist of occupied states. GaAs is an example of a *direct band-gap* semiconductor, for which the conduction-band energy minimum and the valence-band energy maximum have the *same* momentum. If the band extrema occur at different momentum values, the semiconductor has an *indirect* band gap. Most III/V and II/VI compounds (the numerals refer to columns in the Periodic Table) are direct band-gap materials. Examples of indirect band-gap materials are Si and Ge (both column IV), and AlAs (III/V). Direct band-gap materials tend to have high radiative transition rates, whereas indirect band-gap materials do not. Since the abscissa in Fig. 2-1 covers the full Brillouin zone, it gives the complete description of the electronic band structure. In general, a

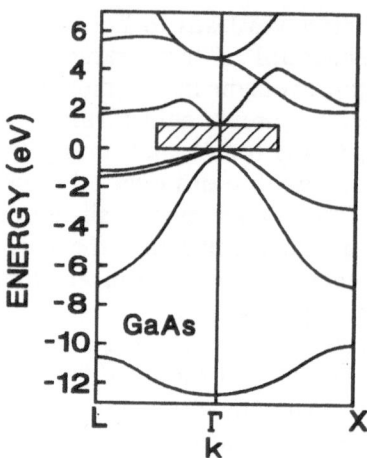

Fig. 2-1. Band structure of GaAs. The hatched region is the region of interest for optical transitions. Γ, X and L are high symmetry points in the first Brillouin zone, the Γ point is the zone center.

semiconductor electronic band structure has numerous bands with asymmetric shapes and sometimes several energy maxima and minima.

Fortunately for optical transitions with frequencies in the visible or near infrared, it is often sufficient to consider only a small portion of the band structure shown in Fig 2-1. This simplification is due to two factors. One is that optical transitions are direct transitions, i.e., the momenta of the initial and final electronic states are essentially equal. This is because from the conservation of momentum, the difference in electron momenta must equal the momentum of the photon involved in the transition. The photon momentum is

$$\hbar K = \frac{\hbar \omega n}{c} , \qquad (1)$$

where $\hbar \omega$ is the photon energy, K is its wavevector, n is the refractive index of the semiconductor, and c is the speed of light. For GaAs, $\hbar \omega \simeq 1.4 eV$ and $n \simeq 3.6$, so that the photon wavevector, $K \simeq 2.54 \times 10^7/m$, which is negligible on the scale ($\simeq 10^{10}/m$) of the electronic band structure. Therefore, we only need to consider a narrow region of the band structure around the band-gap minimum, where optical transitions are most likely to occur. If the region of interest is sufficiently narrow, it is often reasonable to approximate the energy bands in that region as symmetric and parabolic in shape. We discuss deviations from this simple parabolic band approximation later in this book.

Another important simplifying factor is that all low lying, completely filled bands may be ignored since they do not contribute directly to the optical transitions in the frequency range of interest. Hence, the electronic band structure that we have to consider usually involves only a very small portion of the entire band structure indicated by the hatched area in Fig. 2-1.

Particularly when we get into the theory of strained quantum-well devices (Chap. 6), it is useful to know roughly why there are three valence bands and one conduction band in Fig. 2-1. The single conduction band in Fig. 2-1 results from a $4s$-state of the GaAs "molecule", while the three valence bands come from a $4p$-state. More precisely, there are two spin states for each $\mathbf{k}$ in the conduction band. The spin-orbit interaction in the valence band leads to the total angular momentum J with values $J = 3/2$ and $J=1/2$. The lowest valence band in Fig. 2-1 corresponds to the two spin states of $J=1/2$, the highest valence band corresponds to the states $m_J = \pm 3/2$, and the middle valence band (degenerate with the highest at $k = 0$) corresponds to $m_J = \pm 1/2$ belonging to $J = 3/2$. The lowest band is variously known as the split-off band, the spin-orbit band, or simply as SOB (no prejudice intended!) The upper valence bands are called the heavy-hole and light-hole bands corresponding to the reciprocals of their curvatures, as described below.

Most semiconductor lasers may be described by a band structure consisting of one conduction band and several valence bands. Sometimes, even a simple two-band model (one conduction and one valence band) is sufficient to illustrate the physics of semiconductor laser behavior. In this chapter we limit our discussions to such a two-band model. Generalization to the case of multiple valence bands involves introducing a valence band index, something we do in the later chapters.

In the absence of dopant atoms, thermal energy, and pumping processes such as interaction with an optical field, the valence bands of a semiconductor are completely full and the conduction band is empty. As such no states are free for electrons to move to within their respective band, and hence no current can flow. The band structure is calculated for this case of unexcited electrons. As discussed above, if the portion of the bandstructure of interest is sufficiently small, we may use the *effective-mass approximation*, where the detailed conduction and valence band structures are approximated by the simple parabolae

$$\varepsilon_{c\mathbf{k}} = \frac{\hbar^2 k^2}{2m_c} + \varepsilon_g \, , \tag{2}$$

$$\varepsilon_{v\mathbf{k}} = \frac{\hbar^2 k^2}{2m_v} \, . \tag{3}$$

Here m_c and m_v are the effective masses of the electrons in the conduction and valence bands, respectively, and ε_g is the band-gap energy in the absence of excited electrons. The band structure diagram is shown in Fig.

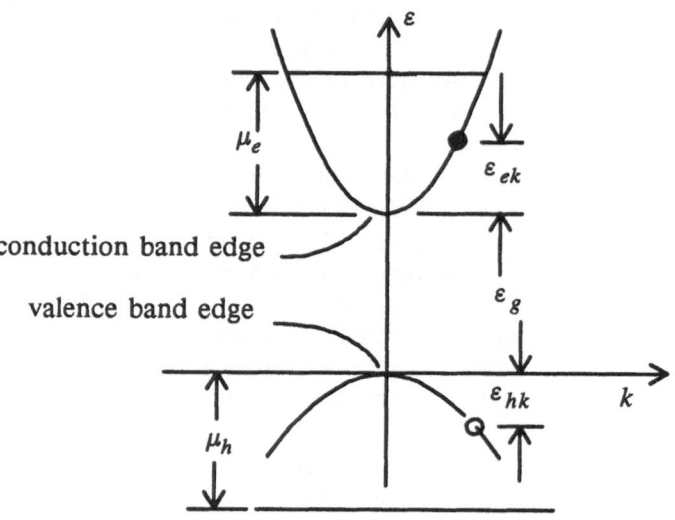

Fig. 2-2. Two-band model of a direct band-gap semiconductor. The zero in energy is choosen to be at the maximum of the valence band (the valence band edge). μ_e and μ_h are the quasiequilibrium chemical potentials described in Sec. 2-3.

2-2. The effective masses are defined by the reciprocals of the band curvatures, that is, by the second derivatives

$$\frac{1}{m_i} = \frac{1}{\hbar^2} \frac{\partial^2}{\partial k^2} \varepsilon_{i\mathbf{k}} \bigg|_{k=0} \qquad i = c, v \tag{4}$$

of the conduction- or valence-band energies. As a consequence of the negative valence-band curvature, m_v is negative, while m_c is positive.

If an electron in the full valence band absorbs light, it is excited into the empty conduction band leaving behind a missing electron in the valence band. For simplicity, we refer to the conduction electrons simply as electrons and the missing valence-band electrons as holes. Since all states are vacant in the conduction band except the one occupied by the excited electron, a tiny one-electron current can flow in the conduction band. As such the excited electron is called a charge *carrier*. Similarly, an electron in the valence band can move into the hole, which moves the hole, so that the hole too is a charge carrier with a charge $+e$ opposite that of the electron.

For an optical (vertical) transition, the transition energy at the carrier momentum **k** is given by

$$\hbar\omega_{\mathbf{k}} = \varepsilon_{e\mathbf{k}} + \varepsilon_{h\mathbf{k}} + \varepsilon_g \, , \tag{5}$$

where the electron and hole energies are given by

$$\varepsilon_{e\mathbf{k}} = \frac{\hbar^2 k^2}{2m_e}$$

$$\varepsilon_{h\mathbf{k}} = \frac{\hbar^2 k^2}{2m_h} \, , \tag{6}$$

and m_e and m_h are the effective masses of the electron and hole, respectively. The electron mass m_e equals m_c. In this electron-hole description of a semiconductor, the energy of the hole may be thought of as the energy of the completely filled valence band minus the energy of the valence band with a vacant electronic state. In this case, an increase in the hole momentum leads to an increase in the hole energy. Therefore whereas the effective electron mass in the valence band is negative, the effective hole mass is positive. The relationship between m_h and m_v requires taking into consideration the Coulomb interaction among carriers, which is of course different for a completely filled valence band and for a valence band with a vacancy.

The resonance energies for the optical transitions can be changed by the Coulomb interaction, which for low densities leads to the creation of *excitons*. Here the Coulomb attraction can bind an excited electron and hole pair into an exciton, which is a hydrogen-like "atom" with a finite lifetime. The exciton lives are terminated through electron-hole recombination, which transfers the exciton energy to light (*radiative recombination*), or to the lattice, impurities, etc. (*nonradiative recombination*). By replacing the proton mass by the reduced electron-hole mass, we can use the Bohr hydrogen model to describe an exciton. The radius of the lowest exciton state is given by the exciton Bohr radius (in CGS units)

$$a_0 = \frac{\hbar^2 \epsilon_b}{e^2 m_r},$$

(7)

and the energy of the lowest state is given by exciton Rydberg energy

$$\varepsilon_R = \frac{\hbar^2}{2m_r a_0^2} = \frac{e^2}{2\epsilon_b a_0} = \frac{e^4 m_r}{2\epsilon_b^2 \hbar^2},$$

(8)

where ϵ_b is the background dielectric constant and m_r is the reduced electron-hole mass defined by

$$\frac{1}{m_r} = \frac{1}{m_e} + \frac{1}{m_h}.$$

(9)

In GaAs, $a_0 \simeq 124$ Å compared to 0.5 Å in the H atom, and $\varepsilon_R = 4.2\ meV$, which is tiny compared with 13.6 eV for the H atom and small compared to room-temperature thermal energy $k_B T = 25\ meV$. Whether excitons are important in the description of semiconductor behavior depends on a_0 compared to the screening length and ε_R compared to $k_B T$. The screening length is a measure of the effectiveness of the screening of the Coulomb interaction between two carriers by other carriers. As the carrier density increases (due to an injection current or optical absorption), the Coulomb potential becomes increasingly screened, and for sufficiently high densities the excitons are completely ionized. Chapter 4 discusses screening further.

There is some discussion about the values of m_e and m_h to use for GaAs. For example, one may see $m_e = 1.127 m_r$, $m_h = 8.82 m_r$, and $m_r = 0.05896 m_0$, where m_0 is the mass of the electron in free space. Alternatively, in this book we usually use $m_e = 1.176 m_r$ (= 0.0665 m_0), $m_h = 6.669 m_r$ (= 0.377 m_0), and $m_r = 0.05653 m_0$, which agrees with the Luttinger Hamiltonian discussion in Chap. 6 provided one uses the heavy-hole mass m_{hh} for m_h. The corresponding light-hole mass is $m_{lh} = 0.09 m_0$. Part of the problem arises in attempting to use a two-band theory when two valence bands participate. For example, the reduced mass m_r given by Eq. (8) based on measurements of ε_R and ϵ_b probably do not agree with that based on using the heavy-hole effective mass as gleaned from the Luttinger theory. We see in the density of states discussion [see Eq. (22)] how to determine what fraction of the total carriers reside in the heavy and light-hole bands when the bands are degenerate at $k = 0$. Crystal strain introduced by MBE growth with lattice mismatch can mix all three (six) valence bands in ways that destroy the accuracy of parabolic band effective mass approximations altogether. Nevertheless, the simplicity of the two-band model merits our careful consideration.

At this point, we need to remind ourselves that the carrier-density-independent band structure discussed so far assumes that only one conduction-band electron and one valence-band hole are present in the semiconductor. In the presence of more electrons and holes, many-body interactions cause the band structure to change. This band structure change is a result of changes in the Coulomb repulsion among carriers within the same band, due to screening and exchange interactions. Exchange interactions are a consequence of the quantum statistics. One effect of the many-body carrier-carrier interactions is a reduction in the optical transition energy with increasing carrier density. This change is called *band-gap renormalization*, and modifies Eq. (5) to be

$$\hbar\omega_{\mathbf{k}} = \varepsilon_{e\mathbf{k}} + \varepsilon_{h\mathbf{k}} + \varepsilon_{g0} + \delta\varepsilon_g = \varepsilon + \varepsilon_{g0} + \delta\varepsilon_g , \qquad (10)$$

where the renormalization energy $\delta\varepsilon_g$ is often assumed to be independent of electron momentum. Band-gap renormalization explains why the laser diode typically oscillates at frequencies just below the zero-density band-gap energy, ε_g, and has consequences in predictions concerning the laser linewidth and antiguiding. Coulomb attraction between electrons and holes affects semiconductor behavior by reshaping the semiconductor gain spectrum in a way called *Coulomb enhancement*. These many-body effects are discussed in Chap. 5.

2-2. Units

As we can see from Eqs. (7) and (8), the question of the choice of units arises early in any discussion on semiconductor lasers. If we deal with laser physics alone, we would encounter no arguments with using MKS units. However semiconductor physicists often use CGS units. Hence we have the problem of using CGS versus MKS units.

For this book, we propose the following compromise. The exciton Bohr radius often plays the role of a characteristic length scale in semiconductors and the Rydberg energy is the natural energy unit. Expressing results in terms of a_0 and ε_R helps to sidestep the units problem. This is important because semiconductor lasers are of interest in both solid-state physics and engineering communities. Whenever important results are given that have to reveal the underlying units, we give them in both forms. In particular, in MKS units, the e^2 in Eqs. (7) and (8) should be replaced by

$$e^2 \rightarrow e^2/4\pi\epsilon_0 \,, \tag{11}$$

where ϵ_0 is the MKS permittivity of free space.

2-3. Fermi-Dirac Distributions

A very important effect of the Coulomb interaction is *carrier-carrier* scattering. This has a counterpart in gas lasers known as velocity-changing collisions, but it is a much stronger effect in semiconductors. For densities high enough to get gain (about 2×10^{18} cm^{-3}), the excitons are ionized and the carrier-carrier scattering drives the electron and hole distributions each into Fermi-Dirac distributions provided external forces like light fields vary little in the carrier-carrier scattering time of 0.1 picoseconds or less. These distributions are called *quasiequilibrium* distributions because they result from the equilibration of the carriers within their bands, but not among bands. In true thermodynamic equilibrium, the electrons are described by a single Fermi-Dirac distribution, which for typical temperatures gives filled valence bands and empty conduction bands, that is, a semiconductor in its ground state. Quasiequilibrium occurs on a time scale long compared to the carrier-carrier scattering time but short compared to interband relaxation times, which are on the order of nanoseconds. Quasi-equilibrium can be maintained in a steady state by pumping the electrons from the valence band to the conduction band by injecting a current or by applying a sufficiently strong optical field.

The rapid carrier equilibration into Fermi-Dirac distributions greatly simplifies the analysis of the semiconductor medium. Instead of having to follow the carrier densities on an individual **k** basis, we may only need to determine the total carrier density N. The individual **k**-dependent carrier population probabilities are then given by

$$\boxed{\; f_{\alpha\mathbf{k}} = \frac{1}{e^{\beta[\varepsilon_{\alpha\mathbf{k}} - \mu_\alpha]} + 1} \;} \tag{12}$$

where $\alpha = e$ for electrons, $\alpha = h$ for holes, $\beta = 1/k_B T$, k_B is Boltzmann's constant, T is the absolute temperature, and μ_α is the carrier quasichemical potential, which is chosen to yield the total carrier density N. We measure the quasichemical potentials from their respective band edges. From Eq. (12), we see that independent of temperature, when $\varepsilon_{\alpha\mathbf{k}_c} = \mu_\alpha$, the probability to have a carrier in the band α with the momentum $\mathbf{k}_c$ is $\frac{1}{2}$. For $k < k_c$, the occupation probability is therefore $> \frac{1}{2}$. A negative chemical potential indicates that band α does not contain enough carriers to fill any

state with $\frac{1}{2}$ probability.   At $T = 0\ K$, the chemical potential equals the
Fermi energy, which is the upper most level filled by carriers.

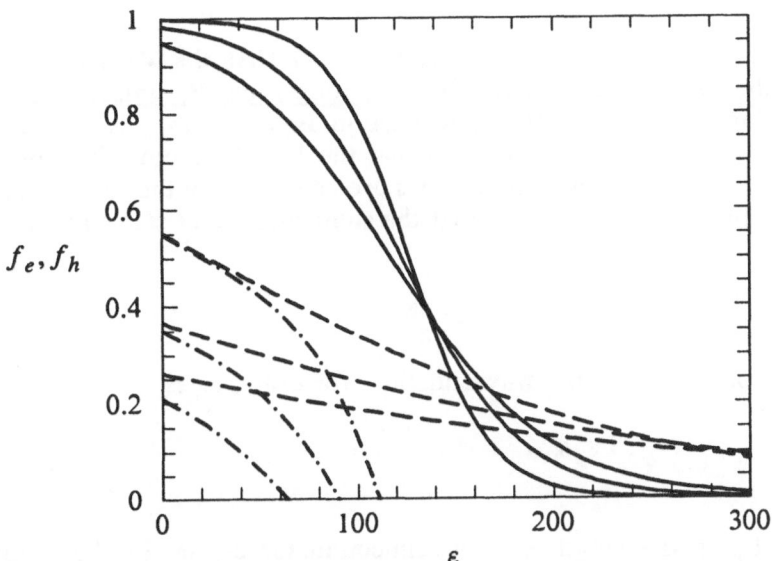

Fig. 2-3.   Fermi-Dirac distributions (12) versus reduced mass
energy $\varepsilon = \hbar^2 k^2 / 2m_r$ for an effective mass $m_e = 1.176m_r$ (solid
lines), an effective mass $m_h = 6.669m_r$ (dashed lines), $m_r = 0.05653m_0$, along with the sum $f_e + f_h - 1$ (dot-dashed lines),
which enters the free-carrier gain formula (inversion corresponds
to positive values of $f_e + f_h - 1$).   For each group, the top curve
is for $T = 200\ K$, the middle for $T = 300\ K$, and the bottom for $T = 400\ K$.   In all cases, the chemical potentials are chosen so that
Eq. (13) gives the total carrier density $N = 3 \times 10^{18}\ cm^{-3}$.

Figure 2-3 is a plot of the Fermi-Dirac distributions for different temperatures.   At $T = 0K$, the Fermi-Dirac distribution is a step function,
such that all states with energy below the chemical potential are completely
filled and those above are completely empty.   As the temperature is increased carriers begin to occupy energetically higher states.   Except for
high temperatures, the changes in the occupation of the states occur primarily in the energy range $\pm k_B T$ around μ_α.

As mentioned earlier, the chemical potential is determined by the temperature and by the total number of carriers.   In the absence of doping, the
total carrier density is the same for electrons and holes.   Denoting this by
N, μ_α is determined by the condition

$$N = \frac{1}{V} \sum_{states} f_{\alpha\mathbf{k}} , \tag{13}$$

where V is the volume of the sample. At first the V seems suspicious, since the result should not depend strongly on V, unless one or more dimensions are smaller than the exciton Bohr radius. However, we see presently that the $\Sigma_{states} \propto V$, so that the V cancels out. In a bulk semiconductor the electron or hole states are specified by the momenta, k_x, k_y, k_z and spin component s_z, so that the summation over states gives

$$N_\alpha = \frac{1}{V} \Sigma_{k_x} \Sigma_{k_y} \Sigma_{k_z} \Sigma_{s_z} f_{\alpha\mathbf{k}} . \tag{14}$$

The x-component of the wavefunction, for example, is

$$\psi_{k_x}(x) = \frac{1}{L_x} e^{ik_x x} , \tag{15}$$

where L_x is the length of the semiconductor crystal in the x direction. Assuming periodic boundary conditions within the semiconductor, k_x is quantized according to

$$k_x = \frac{2\pi n}{L_x} , \tag{16}$$

where n is an integer ranging from $-\infty$ to ∞. For L_x sufficiently large, we can assume an essentially continuous range of values for k_x. Then we can replace the summation over k_x by an integral

$$\sum_x^k \rightarrow \int_{-\infty}^{\infty} dk_x \, \frac{dn}{dk_x} , \tag{17}$$

where

$$\frac{dn}{dk_x} = \frac{L_x}{2\pi}$$

is the number of states within the interval k_x and $k_x + dk_x$. Using a similar argument for the other two components of $\mathbf{k}$ gives

$$N_\alpha = \frac{1}{V} \sum_{\mathbf{k}} f_{\alpha\mathbf{k}} = 2 \cdot (2\pi)^{-3} \int_{-\infty}^{\infty} dk_x \int_{-\infty}^{\infty} dk_y \int_{-\infty}^{\infty} dk_z f_\alpha(\mathbf{k}) , \qquad (18)$$

where the factor of 2 comes from the s_z summation and the $1/V$ has been cancelled out by the product $L_x L_y L_z$. If the integrand is spherically symmetric, then

$$\int_{-\infty}^{\infty} dk_x \int_{-\infty}^{\infty} dk_y \int_{-\infty}^{\infty} dk_z \to \int_0^{\infty} dk\, 4\pi k^2 , \qquad (19)$$

where $k^2 = k_x^2 + k_y^2 + k_z^2$, so that

$$N_\alpha = \int_0^{\infty} dk\, \mathscr{D}_\alpha(k) f_\alpha(k) , \qquad (20)$$

where $\mathscr{D}_\alpha(k) = (k/\pi)^2$ is the electron or hole momentum density of states of a bulk semiconductor and it gives the number of states between k and $k + dk$.

We can convert the integration variable from k in Eq. (20) to the energy $\varepsilon_\alpha = \hbar^2 k^2 / 2m_\alpha$, so that

$$N_\alpha = \int_0^{\infty} d\varepsilon_\alpha\, \mathscr{D}_\alpha(\varepsilon_\alpha) f_\alpha(\varepsilon_\alpha) , \qquad (21)$$

where

$$\mathscr{D}_\alpha(\varepsilon_\alpha) = \frac{(2m_\alpha/\hbar^2)^{3/2}}{2\pi^2} \sqrt{\varepsilon_\alpha} \qquad (22)$$

is the electron or hole energy density of states in a bulk semiconductor and it gives the number of states between ε and $\varepsilon + d\varepsilon$. When it is obvious as to whether we are referring to the energy or momentum density of states, the functional dependence is usually omitted. The energy density of states is often more useful than the momentum density of states because transitions involve states that are within a range of energy instead of momentum. Specifically, the initial and final states that contribute strongest to a transition are those separated by $\hbar\omega \pm \hbar\gamma$, where ω and γ are the transition frequency and linewidth, respectively. Since the optical transition frequency (5) involves the reduced mass m_r of Eq. (9) instead of the individual carrier masses m_α, it is most convenient to transform to the energy coordinate $\varepsilon = \hbar^2 k^2/2m$.

We can use Eq. (21) to check the validity of the two-band model, i.e., the use of the heavy hole band alone. In quasiequilibrium, both valence bands share the same chemical potential, μ_h. The total carrier density for the holes is the sum of the light and heavy hole band densities

$$N_h = N_{lh} + N_{hh} = \frac{(2m_{lh}/\hbar^2)^{3/2} + (2m_{hh}/\hbar^2)^{3/2}}{2\pi^2} \int_0^\infty \frac{d\varepsilon\,\sqrt{\varepsilon}}{e^{\beta(\varepsilon - \mu_h)} + 1}.$$

Hence

$$\frac{N_{lh}}{N_{hh}} = \left[\frac{m_{lh}}{m_{hh}}\right]^{3/2}. \tag{23}$$

For the values $m_{hh} = 0.377 m_0$ and $m_{lh} = 0.09 m_0$, this gives 12% of the holes in the light-hole band, which is small enough to neglect for our simpler modeling.

In general, the determination of the chemical potential for a given carrier density is not straightforward because the integrals in Eqs. (20) and (21) cannot be evaluated analytically. The chemical potential can always be found by iteration, with the necessary integrations done numerically. However there are approximate analytic expressions relating the chemical potential to carrier density for some cases. One such case occurs when only the high energy tail of the Fermi-Dirac distribution is within the band. In other words, the chemical potential lies sufficiently far inside the band gap (is sufficiently negative) that

$$\varepsilon_{\alpha k} - \mu_\alpha \gg k_B T. \tag{24}$$

Then the exponential term may dominant the 1 in the denominator of Eq. (12) and

$$f_{\alpha k} \simeq e^{\beta\mu_\alpha}\, e^{-\beta\varepsilon_{\alpha k}}, \tag{25}$$

which is a Maxwell-Boltzmann distribution. Equation (21) then can be readily evaluated as (setting $q = \beta\hbar^2/2m_\alpha$ for typographical simplicity)

$$N_\alpha \simeq \frac{1}{\pi^2}e^{\beta\mu_\alpha}\int_0^\infty dk\, k^2 e^{-qk^2} = -\frac{1}{\pi^2}e^{\beta\mu_\alpha}\frac{\partial}{\partial q}\int_0^\infty dk\, e^{-qk^2}$$

$$= -\frac{1}{\pi^2}\frac{\sqrt{\pi}}{2}e^{\beta\mu_\alpha}\frac{\partial}{\partial q}\frac{1}{\sqrt{q}} = \frac{e^{\beta\mu_\alpha}}{4(\pi\beta\hbar^2/2m_\alpha)^{3/2}}. \tag{26}$$

Hence

$$e^{\beta\mu_\alpha} = 4N_\alpha\,[\pi\beta\hbar^2/2m_\alpha]^{3/2} = \bar{N}_\alpha\,, \tag{27}$$

or

$$\beta\mu_\alpha = ln\,[\,\bar{N}_\alpha\,]\,, \tag{28}$$

where we call $\bar{N}_\alpha$ the normalized total carrier density. Noting from Eq. (8) that $\hbar^2/2m_r = \varepsilon_R\, a_0^2$, we can write $\bar{N}_\alpha$ in terms of the exciton Bohr radius and Rydberg energy as $4(N_\alpha a_0^3)(\pi\beta\varepsilon_R\, m_r/m_\alpha)^{3/2}$. The Maxwell-Boltzmann distribution is sometimes a good approximation for the hole distribution because of the high density of states in the valence band. Figure 2-4 compares the Fermi-Dirac and Maxwell-Boltzmann distributions for a couple of cases.

For $T \to 0$, $\beta = 1/k_B T \to \infty$, for which the Fermi-Dirac distribution becomes a step function, truncating the integral (21) to

$$N_\alpha = \frac{(2m_\alpha/\hbar^2)^{3/2}}{2\pi^2}\int_0^{\mu_\alpha} d\varepsilon_\alpha\,\sqrt{\varepsilon_\alpha} = \frac{(2m_\alpha\mu_\alpha/\hbar^2)^{3/2}}{3\pi^2}.$$

Inverting this, we have the chemical potential

$$\mu_\alpha = (3\pi^2 N_\alpha)^{2/3}\hbar^2/2m_\alpha = (3\pi^2 N_\alpha a_0^3)^{2/3}\varepsilon_R\, m_r/m_\alpha. \tag{29}$$

Hence at $T = 0$, the bulk-medium chemical potential in exciton Rydbergs equals the $3\pi^2$ times the number of carriers in a cubic exciton Bohr radius raised to the 2/3 power.

Using a series representation for the Fermi-Dirac function, a resummation, and a Padé approximation, one can derive [see, for example, Haug and Koch (1993)] the following analytic approximation for $\mu_\alpha(N_\alpha,T)$

$$\beta\mu_\alpha \simeq ln\bar{N}_\alpha + K_1 ln(K_2\bar{N}_\alpha + 1) + K_3\bar{N}_\alpha\,, \tag{30}$$

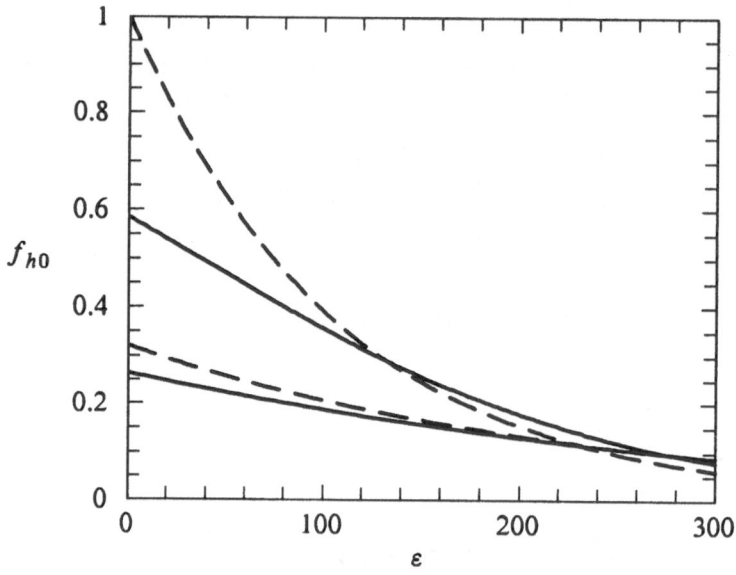

Fig. 2-4. Fermi-Dirac (solid lines) and Maxwell-Boltzmann (dashed lines) distributions as functions of carrier energy $\varepsilon = \hbar^2 k^2/2m_r$, in meV, where m_r is the reduced mass, for $N = 3 \times 10^{18}$ cm^{-3} and $T = 300\ K$, which give gain, since μ_e is sufficiently positive. The upper curves are for the carrier mass $m_h = 4.1335m_r$ (which gives a Boltzmann $\mu_h = 0$; a Fermi-Dirac $\mu_h = 8.91\ meV$) and the lower are for $m_h = 8.82m_r$ (Boltzmann $\mu_h = -29.38\ meV$; Fermi-Dirac $\mu_h = -26.53\ meV$). The curves are all normalized for a bulk semiconductor, i.e., the areas under $\sqrt{\varepsilon}f_h(\varepsilon)$ are all equal. As one would guess from Eq. (12), the Maxwell-Boltzmann distribution is a poor approximation unless μ_h is sufficiently negative.

where the constants $K_1 = 4.8966851$, $K_2 = 0.04496457$, and $K_3 = 0.1333760$. This is good for all except very strongly degenerate ($T \simeq 0$) situations. Comparing Eqs. (30) and (28), we see that the lead term $ln\,N_\alpha$ in Eq. (30) is just the classical (Maxwell-Boltzmann) result. The chemical potential given by Eq. (30) is plotted in Fig. 2-5. Within the drawing accuracy, the result is indistinguishable from the exact chemical potential obtained as a numerical solution of Eq. (20), showing that Eq. (30) yields an excellent approximation for the range $-\infty < \beta\mu_\alpha \leq 30$. Numerous examples of electron and hole Fermi-Dirac distributions are given in Secs. 3-4 and 3-5 for typical laser operating conditions.

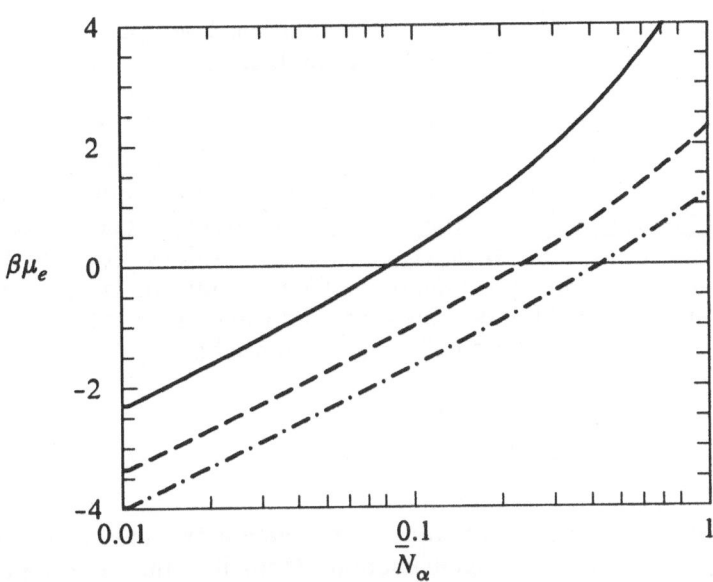

Fig. 2-5. Chemical potential μ_α for a three-dimensional Fermi gas as a function of the normalized total carrier density $N_\alpha = 4(Na_0^3)(\pi\beta\varepsilon_R m_r/m_\alpha)^{3/2}$ for $m_\alpha = 2m_r$, $3m_r$, and $4m_r$, $\varepsilon_R = 4.2$ meV, and $a_0 = 124.3\,\text{Å}$ (appropriate for GaAs).

2-4. Quantum Confinement

GaAs does not occur in nature and as such can be considered a "designer material". Thanks to the crystal growth techniques discussed in Sec. 1-3, one can not only determine the composition of semiconductors with remarkable precision, but can also determine their shape virtually on an atomic scale. In particular, it is possible to fabricate microstructures so small that their electronic and optical properties deviate substantially from those of bulk materials. The onset of pronounced quantum confinement effects occur when one or more dimensions of a structure becomes comparable to the characteristic length scale of the elementary excitations. Quantum confinement may be in one spatial dimension, as in *quantum wells*, in two spatial dimensions as in *quantum wires*, or in all three spatial dimensions as in *quantum dots*. The confinement modifies the allowed energy states of the crystal electrons and changes the density of states. In this section, we introduce the basic properties of quantum confined structures that we use in the next three chapters on laser gain. In Chap. 6, we discuss the finer but still important modifications to the quantum-well bandstructure.

We begin our discussion with quantum wells, which are the most developed of the quantum-confined structures. Quantum-well lasers are commercially available, while quantum-wire and quantum-dot lasers are still in the research stages. An understanding of the basic effects is best obtained by considering *ideal quantum confinement* conditions, for which the elementary excitations are completely confined inside the microstructure and the electronic wavefunctions vanish beyond the surfaces. For this idealized situation, we can write the confinement potential as

$$V_{con}(z) = \begin{cases} 0 & |z| < L_c/2 \\ \infty & |z| > L_c/2 \end{cases}.$$ (31)

In the xy plane there is no quantum confinement and the carriers can move freely. The electron eigenfunction (actually the envelope of the eigenfunction as discussed in Chap. 6) can be separated as

$$\psi_{n,\mathbf{k}_\perp}(\mathbf{r}) = \phi_{\mathbf{k}_\perp}(\mathbf{r}_\perp)\,\zeta_n(z)\,,$$ (32)

where the z and transverse components obey the Schrödinger equations

$$\left[-\frac{\hbar^2}{2m_z}\frac{d^2}{dz^2} + V_{con}(z) \right] \zeta_n(z) = E_n\,\zeta_n(z)\,,$$ (33)

and

$$-\frac{\hbar^2}{2m_\perp}\nabla_\perp^2 \phi(\mathbf{r}_\perp) = E_{\mathbf{k}_\perp}\phi(\mathbf{r}_\perp)\,,$$ (34)

respectively. For simplicity, we assume that the bulk-material bandstructure can be described by parabolic energy bands that are characterized by the effective masses m_z and $m_\perp$ As shown in Chap. 6, m_z and $m_\perp$ are equal for the conduction bands. They differ for the valence bands, which leads to the interesting property of mass reversal. Equation (34) describes a two-dimensional free particle with eigenfunctions

$$\phi_{\mathbf{k}_\perp}(\mathbf{r}_\perp) = \frac{1}{\sqrt{L_c}}e^{\pm i\mathbf{k}_\perp\cdot\mathbf{r}_\perp}$$ (35)

and eigenvalue

$$E_{\mathbf{k}_\perp} = \frac{\hbar^2 \mathbf{k}_\perp^2}{2m_\perp} . \tag{36}$$

Because of the infinite confinement potential, we have the boundary conditions

$$\zeta_n(L_c/2) = \zeta_n(-L_c/2) = 0 , \tag{37}$$

which lead to the even and odd solutions of Eq. (33)

$$\zeta_n(z) = \sqrt{\frac{2}{L_c}} \cos(k_n z) , \; n \text{ even} , \tag{38}$$

$$\zeta_n(z) = \sqrt{\frac{2}{L_c}} \sin(k_n z) , \; n \text{ odd} , \tag{39}$$

where the wave numbers k_n are given by

$$k_n = n\pi/L_c , \tag{40}$$

and the bound state energies E_n are given by

$$E_n = \frac{\hbar^2 k_n^2}{2m_z} = \frac{\pi^2 \hbar^2 n^2}{2m_z L_c^2} . \tag{41}$$

Adding the energies of the motion in the xy plane and in the z-direction, we find the total energy of the electron subjected to one-dimensional quantum confinement to be

$$E = \frac{\hbar^2 n^2 \pi^2}{2m_z L_c^2} + \frac{\hbar^2 k_\perp^2}{2m_\perp} , \tag{42}$$

where $n = 1, 2, 3, \dots$, indicating a succession of energy subbands, i.e., energy parabola $\hbar^2 k_\perp^2/2m_\perp$ separated by $\hbar^2\pi^2/2m_z L_c^2$. The different subbands are labeled by the quantum numbers n. Figure 2-6 depicts the energy eigenstates.

Realistically, we can only fabricate finite confinement potentials, so that

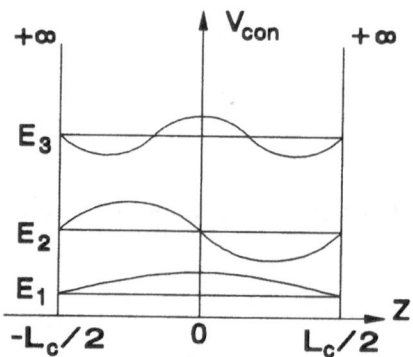

Fig. 2-6. Energy eigenfunctions and eigenvalues of an infinite one-dimensional square potential well.

$$V_{con}(z) = \begin{cases} 0, & |z| < L_c/2 \\ V_c, & |z| > L_c/2 \end{cases}.$$ (43)

The analysis follows closely the treatment of the infinite potential, with the Schrödinger equation for the x-y motion being unchanged. However, solutions in the z-direction can no longer be determined analytically. Equation (33) now has to be solved separately in the regions, i) $|z| < L_c/2$, ii) $z > L_c/2$, and iii) $z < -L_c/2$. In region i), the solutions are given by Eqs. (38) and (39), while in regions ii) and iii) they are

$$\varsigma(z) = C_{\pm}\, e^{K_z z}\,,$$ (44)

where

$$K_z^2 = \frac{2m_z}{\hbar^2}(V_c - E_z)\,.$$ (45)

Since the wavefunction has to be normalizable we have to pick the decaying solutions in Eq. (44). Also, we have to match the solutions and their derivatives at the interfaces $\pm L_c/2$. This yields the even states

$$\varsigma_{2n}(z) = \begin{cases} B \cos k_z z & |z| \le L_c/2 \\ C\, e^{-K_z |z|} & |z| > L_c/2 \end{cases}$$ (46)

with the condition

$$\sqrt{E_z} \tan \left[\sqrt{\frac{m_z E_z}{2\hbar^2}} L_c \right] = \sqrt{V_c - E_z} \, , \qquad (47)$$

whose solution gives the energy of the even states.
   Similarly the odd-states wave functions are given by

$$\zeta_{2n}(z) = \begin{cases} A \sin k_z z & |z| \le L_c/2 \\ C e^{-K_z |z|} & |z| > L_c/2 \end{cases} \qquad (48)$$

with the condition

$$-\sqrt{E_z} \cot \left[\frac{\sqrt{m_z E_z}}{2\hbar^2} L_c \right] = \sqrt{V_c - E_z} \, , \qquad (49)$$

whose solution gives the energy of the odd states. Figure 2-7 depicts the solutions of a finite one-dimensional square well.

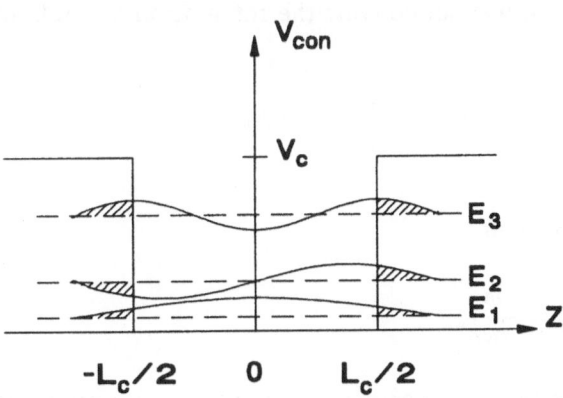

Fig. 2-7. Energy eigenfunctions and eigenvalues of a finite one-dimensional square well.

   An important difference between bulk and quantum-confined structures is in the density of states. Following the steps taken for the bulk material in the previous section, the sum over states in the quantum well may be approximated by the integral

$$\sum_{states} \rightarrow \sum_n \frac{2A}{(2\pi)^2} \int_{-\infty}^{\infty} dk_x \int_{-\infty}^{\infty} dk_y \ , \tag{50}$$

where the factor of 2 comes from the spin summation and the volume of the material is $V = wA$. If the band structure is symmetric in the xy plane, then

$$\int_{-\infty}^{\infty} dk_x \int_{-\infty}^{\infty} dk_y = \int_0^{\infty} dk_{\perp} 2\pi k_{\perp}, \tag{51}$$

where $k_{\perp}^2 = k_x^2 + k_y^2$, so that

$$\frac{1}{V} \sum_{states} \rightarrow \sum_n \int_0^{\infty} dk_{\perp} \mathscr{D}(k_{\perp}) \ , \tag{52}$$

where $\mathscr{D}(k_{\perp}) = k/(\pi w)$ is the 2-dimensional electron or hole momentum density of states giving the number of states between $k_{\perp}$ and $k_{\perp} + dk_{\perp}$. Using Eq. (36), we can convert the integration variable into energy, so that

$$\frac{1}{V} \sum_{states} \rightarrow \sum_n \int_0^{\infty} d\epsilon \ \mathscr{D}(\epsilon) \ , \tag{53}$$

where the constant

$$\mathscr{D}(\varepsilon) = \frac{m_z}{\pi w \hbar^2} \tag{54}$$

gives the number of states per unit volume between ε and $\varepsilon + d\varepsilon$.

The two-dimensional carrier density integral can be evaluated analytically [see Haug and Koch (1993)], yielding the chemical potential

$$\beta\mu_{\alpha} = ln[e^{\pi\beta\hbar^2 N_w / m_{\alpha}} - 1] \tag{55}$$

which gives

$$\beta\mu_\alpha = ln[e^{2\pi\beta N_\alpha a_0{}^2 \varepsilon_R\, m/m_\alpha} - 1]. \tag{56}$$

For $T \rightarrow 0$, this reduces to

$$\mu_\alpha = \pi N_\alpha \hbar^2/m_\alpha \tag{57}$$

or

$$\mu_\alpha = 2\pi N_\alpha a_0^2 \varepsilon_R\, m/m_\alpha . \tag{58}$$

Armed with these ideas, lets turn to a more complete discussion of the quantum well.

2-5. Slowly-Varying Maxwell Equations

We are now in a position to consider how a semiconductor gain medium interacts with a laser field. Most theoretical problems involving lasers may be treated using the semiclassical approximation according to which one describes the laser field classically and the gain medium quantum mechanically. Figure 2-8 summarizes the steps involved in the appli-

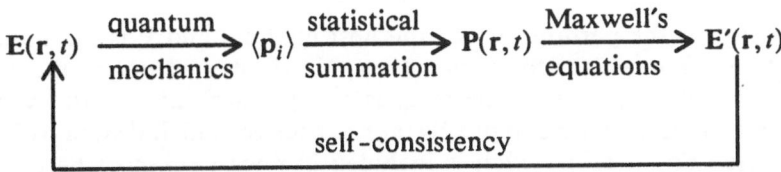

Fig. 2-8. Self-consistent semiclassical theory of a laser field $E(r,t)$ interacting with a quantum-mechanical medium. The assumed electric field $E(r,t)$ induces a polarization of the medium that, in turn, drives the field self consistently.

cation of semiclassical laser theory. One begins by calculating the microscopic electric-dipole moments induced in the gain medium by a given laser field, $E(r,t)$. These dipoles are summed to yield a macroscopic polarization $P(r,t)$ of the gain medium. This polarization then drives the laser field, $E'(r,t)$ according to Maxwell's equations. Self-consistency is imposed by the condition, $E(r,t) = E'(r,t)$. In this section we complete the Maxwell equation part of this scheme by finding out how a slowly-varying electric field amplitude is affected by the polarization of the medium. Section 2-6 outlines the quantum mechanical method for deriving the other part of the scheme, namely how the field induces the polarization of the medium according to the quantum mechanics of the semiconductor.

We describe the laser field by Maxwell's equations (in MKS units)

$$\nabla \cdot \mathbf{B} = 0 \tag{59}$$

$$\nabla \cdot \mathbf{D} = 0 \tag{60}$$

$$\nabla \times \mathbf{E} = -\frac{\partial \mathbf{B}}{\partial t} \tag{61}$$

$$\nabla \times \mathbf{H} = \frac{\partial \mathbf{D}}{\partial t} . \tag{62}$$

In these equations, the magnetic flux, $\mathbf{B}$, and the magnetic field, $\mathbf{H}$ are related by the constitutive relation

$$\mathbf{B} = \mu \mathbf{H} , \tag{63}$$

where μ is the permeability of the host medium. The displacement electric field, $\mathbf{D}$, is given by

$$\mathbf{D} = \epsilon \mathbf{E} = \epsilon_0 \mathbf{E} + \mathbf{P} , \tag{64}$$

where ϵ is the permittivity of the host medium and $\mathbf{P}$ is the induced polarization. For a semiconductor laser, the host medium is the lattice. The permeability, $\mu = \mu_0$, where μ_0 is the permeability of the vacuum. The semiconductor lattice typically has a background index of refraction, $n \simeq 3.5$ [see Eq. (1.12)], which is included in ϵ via $\epsilon = n^2 \epsilon_0$, where ϵ_0 is the permittivity in vaccuum. The polarization, $\mathbf{P}$, gives the gain and carrier-induced refractive index, and is induced by the laser field interacting with the electrons in the conduction and valence bands.

Combining the curl of Eq. (61) with Eq. (62), gives

$$\nabla \times \nabla \times \mathbf{E} = \nabla(\nabla \cdot \mathbf{E}) - \nabla^2 \mathbf{E} = -\frac{\mu_0 \partial^2 \mathbf{D}}{\partial t^2} . \tag{65}$$

Since most light field vectors vary little along the directions in which they point, $\nabla \cdot \mathbf{E} \simeq 0$. For example, a plane-wave field is constant along the direction it points, causing its $\nabla \cdot \mathbf{E}$ to vanish identically. Using Eq. (64) for $\mathbf{D}$, we get the wave equation

$$-\nabla^2 \mathbf{E} + \left(\frac{n}{c}\right)^2 \frac{\partial^2 \mathbf{E}}{\partial t^2} = -\frac{\mu_0 \partial^2 \mathbf{P}}{\partial t^2}$$

(66)

where c is the speed of light in vacuum and we use $\mu_0 \epsilon = (n/c)^2$.

For the purposes of calculating the gain and index of the medium, we consider a laser field of the simple plane-wave form

$$\mathbf{E}(z,t) = \tfrac{1}{2}\hat{x} E(z)e^{i[Kz - \nu t - \phi(z)]} + \text{c.c.},$$

(67)

where $E(z)$ and $\phi(z)$ are the real field amplitude and phase shift that vary little in an optical wavelength, ν is the field frequency in radians/second and $\exp(iKz)$ accounts for most of the spatial variation in the laser field. We choose a monochromatic plane travelling wave because it allows us to illustrate the necessary physics of the semiconductor gain medium with the minimum of algebra. We use a plane wave in calculating the local properties of a gain medium, where the volume element considered can always be made sufficiently small compared to the transverse variations in the laser field. The local gain and refractive index are needed for beam propagation and wave optical studies. A limitation to using a monochromatic field is that we cannot deal with the coherent response of a gain medium to multimode fields. Accordingly, we use other forms of the electric field in later chapters.

The laser field induces a polarization in the medium,

$$\mathbf{P}(z,t) = \tfrac{1}{2}\hat{x} \mathscr{P}(z)e^{i[Kz - \nu t - \phi(z)]} + \text{c.c.},$$

(68)

where $\mathscr{P}(z)$ is a complex polarization amplitude that varies little in a wavelength. It is related to the complex susceptibility of the medium by

$$\mathscr{P}(z) = \epsilon \chi(z) E(z) .$$

(69)

Substituting Eqs. (67) and (68) into the wave equation (66), we find

$$-\frac{d^2 E}{dz^2} - 2i\left(K - \frac{d\phi}{dz}\right)\frac{dE}{dz} + \left[\left(K - \frac{d\phi}{dz}\right)^2 + i\frac{d^2\phi}{dz^2} - \left(\frac{n\nu}{c}\right)^2\right]E = \mu_0 \nu^2 \mathscr{P} .$$

(70)

This equation simplifies considerably because E and $d\phi/dz$ vary little in a wavelength, so that terms containing d^2E/dz^2, $d^2\phi/dz^2$ and $(dE/dz)(d\phi/dz)$

may be neglected. This gives

$$-2iK\frac{dE}{dz} + K^2E - 2KE\frac{d\phi}{dz} - \left(\frac{n\nu}{c}\right)^2 E = \mu_0\nu^2\mathscr{P} \,,$$

i.e.,

$$\frac{dE}{dz} - iE\frac{d\phi}{dz} = \frac{i\mu_0\nu^2}{2K}\mathscr{P} = \frac{i\nu}{2n\epsilon_0 c}\mathscr{P} = \frac{iK}{2}\chi E \,, \tag{71}$$

where we use $K = \nu n/c$. Equating the real and imaginary parts, we find the *self-consistency equations*

$$\frac{dE}{dz} = -\frac{\nu}{2\epsilon_0 nc}\text{Im}[\mathscr{P}(z)] = -\frac{K}{2}\chi''(z)E(z) \tag{72}$$

$$\frac{d\phi(z)}{dz} = -\frac{\nu}{2\epsilon_0 ncE}\text{Re}[\mathscr{P}(z)] = -\frac{K}{2}\chi'(z) \tag{73}$$

where $\chi = \chi' + i\chi''$. Self-consistency refers to the requirement that the field parameters ultimately appearing in the formulas for $\mathscr{P}(z)$ are taken to be the very same as the parameters in Eq. (69).

Two useful parameters for characterizing a gain medium are the gain and the carrier-induced refractive index change. The amplitude gain is defined as

$$\frac{dE}{dz} = gE \,, \tag{74}$$

where in general, g is a function of E and equals one half the intensity gain G of Eq. (1.8). The gain has units of inverse length. Comparing Eq. (72) to (74), we find the local gain to be

$$g = -\frac{K}{2}\chi'' \,. \tag{75}$$

To find the carrier-induced refractive index δn, note that the wave number of the laser field given by Eq. (67) is

$$K - \frac{d\phi}{dz} = (n + \delta n)K_0 , \tag{76}$$

where n is the refractive index of the lattice, δn is the change due to the carriers, and K_0 is the wave number in vacuum. Since $K = nK_0$, we have

$$\frac{d\phi}{dz} = -K_0\delta n . \tag{77}$$

Combining this with Eq. (73), we have the relative index change

$$\frac{\delta n}{n} = \frac{\chi'}{2} . \tag{78}$$

2-6. Quantum Mechanics of the Semiconductor Medium

In this section, we introduce the second quantized (or Fock) representation to treat the semiconductor gain medium. There are two reasons for going beyond elementary quantum mechanics, which treats the wave function as a simple complex function. First, we have to account for the fact

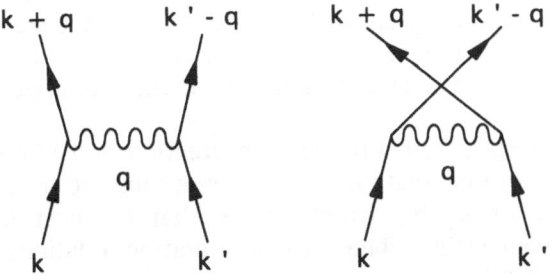

Fig. 2-9. Direct and exchange electron-electron scattering. The facts that these two events are experimentally indistinguishable and that their scattering amplitudes have opposite signs are automatically taken into account in the second quantized representation.

that electrons are indistinguishable as depicted in Fig. 2-9. Similarly the holes are indistinguishable particles. The calculation of the scattering cross section of two electrons has to involve the correct average of the two experimentally indistinguishable events, direct and exchange scattering. We can do that with the properly antisymmetrized wavefunctions, but the second quantized representation takes care of the book keeping more conveniently.

The second reason for going beyond elementary quantum mechanics is because in the electron-hole picture, particle number is not conserved: we can simultaneously create or annihilate electron-hole pairs. This is very inconvenient to treat with elementary quantum mechanical methods. On the other hand, it is straightforward with the use of second quantization.

In the second quantized representation, the wave function of an electron is replaced by a *field operator*

$$\hat{\psi}(\mathbf{r},t) = \sum_{\lambda} \sum_{\mathbf{k}} \sum_{s_z} a_{\lambda \mathbf{k} s_z}(t) \, \phi_{\lambda \mathbf{k} s_z}(\mathbf{r}) \quad , \tag{79}$$

where $\phi_{\lambda \mathbf{k} s_z}(\mathbf{r})$ is the single-particle eigenfunction for an electron in the semiconductor and $a_{\lambda \mathbf{k} s_z}$ is the *annihilation operator* for the electron in that state, which we specify by the band index, $\lambda = c$ or v, momentum, $\mathbf{k}$ and z-component of spin, s_z. Its Hermitean adjoint, $a^{\dagger}_{\lambda \mathbf{k} s_z}$, *creates* an electron in the same state.

For the sake of clarity, we write the operators with all their indices appearing explicitly. Starting with Chap. 3, we usually incorporate the spin variable into $\mathbf{k}$ for typographical simplicity. In that case, the subscript $\mathbf{k}$ represents the three-dimensional momentum vector $\mathbf{k}$ and two possible spin directions, $s_z = \pm\frac{1}{2}$. The summation over $\mathbf{k}$ then involves summations over k_x, k_y, k_z, and s_z.

As Fermion operators, the electron creation and annihilation operators obey anticommutation relations. These relations are a consequence of the Pauli exclusion principle, which states that at most one Fermion can occupy any given state. The anticommutation relations for the electron creation and annihilation operators are

$$[a_{\lambda \mathbf{k} s_z}, a_{\lambda' \mathbf{k}' s_z'}]_+ = [a^{\dagger}_{\lambda \mathbf{k} s_z}, a^{\dagger}_{\lambda' \mathbf{k}' s_z'}]_+ = 0 \tag{80}$$

$$[a_{\lambda \mathbf{k} s_z}, a^{\dagger}_{\lambda' \mathbf{k}' s_z'}]_+ = \delta_{\lambda \lambda'} \delta_{\mathbf{k} \mathbf{k}'} \delta_{s_z s_z'} \quad , \tag{81}$$

where for two operators, A and B, the anticommutator is defined by

$$[A, B]_+ = AB + BA .$$

The combination $a^\dagger_{\lambda k s_z} a_{\lambda k s_z}$ is the number operator for an electron in band λ with momentum k and s_z. The eigenstates for $a^\dagger_{\lambda k s_z} a_{\lambda k s_z}$ are $|0_{\lambda k s_z}\rangle$ and $|1_{\lambda k s_z}\rangle$, which are the states containing no electron and one electron, respectively. These eigenstates when operated on by the creation, annihilation and number operators give

$$a_{\lambda k s_z} |0_{\lambda k s_z}\rangle = a^\dagger_{\lambda k s_z} |1_{\lambda k s_z}\rangle = 0$$

$$a_{\lambda k s_z} |1_{\lambda k s_z}\rangle = |0_{\lambda k s_z}\rangle$$

$$a^\dagger_{\lambda k s_z} |0_{\lambda k s_z}\rangle = |1_{\lambda k s_z}\rangle$$

$$a^\dagger_{\lambda k s_z} a_{\lambda k s_z} |0_{\lambda k s_z}\rangle = 0$$

$$a^\dagger_{\lambda k s_z} a_{\lambda k s_z} |1_{\lambda k s_z}\rangle = |1_{\lambda k s_z}\rangle , \qquad (82)$$

where the first equation expresses the fact that it is impossible to create an electron in an already filled state. This is a consequence of the Pauli exclusion principle.

In the electron-hole representation for a two-band model, we define the hole creation operator

$$b^\dagger_{-k,-s_z} \equiv a_{v k s_z} . \qquad (83)$$

This equation indicates that the annihilation of a valence-band electron with a given momentum and z-component of spin corresponds to the creation of a hole with the opposite momentum and z-component of spin. Note that for clarity in Eq. (83) we use a comma between the $-k$ and $-s_z$ subscripts, although it is probably clear without the comma since it does not make sense to subtract a spin quantum number from a k. Similarly the hole annihilation operator is given by

$$b_{-k-s_z} \equiv a^\dagger_{v k s_z} . \qquad (84)$$

The hole operators also obey anticommutation relationships, so that the probability of finding a particular valence-band electron becomes

$$\langle a^\dagger_{v\mathbf{k}s_z} a_{v\mathbf{k}s_z} \rangle = 1 - \langle b^\dagger_{-\mathbf{k},-s_z} b_{-\mathbf{k},-s_z} \rangle \ , \tag{85}$$

where the brackets $\langle ... \rangle$ are used to indicate an expectation (or quantum mechanically averaged) value. As expected, the probability of finding a valence electron is one *minus* the probability of finding a hole. In the electron-hole representation, we use the term electrons to refer to conduction-band electrons and holes to refer to valence-band holes. The electron annihilation operator is

$$a_{\mathbf{k}s_z} = a_{c\mathbf{k}s_z} \tag{86}$$

and the electron creation operator is

$$a^\dagger_{\mathbf{k}s_z} = a^\dagger_{c\mathbf{k}s_z} \ . \tag{87}$$

In the second quantized representation, the operators representing physical quantities are given by

$$\mathcal{O}_2 = \int d^3r_1 \int d^3r_2 \cdots \int d^3r_N \, \hat{\psi}^\dagger(\mathbf{r}_N)...\hat{\psi}^\dagger(\mathbf{r}_2)\, \hat{\psi}^\dagger(\mathbf{r}_1)\mathcal{O}_1\hat{\psi}(\mathbf{r}_1)\, \hat{\psi}(\mathbf{r}_2)...\hat{\psi}(\mathbf{r}_N) \ , \tag{88}$$

where $\mathcal{O}_1$ and $\mathcal{O}_2$ are the operators in the first and second quantized representations, and we have assumed a N-particle system and the $\psi(r_k)$ are given by Eq. (79). In particular, for N non-interacting electrons, the Hamiltonian is

$$\mathcal{H}_{kin} = \int d^3r_1 \int d^3r_2 \cdots \int d^3r_N \, \hat{\psi}^\dagger(\mathbf{r}_N)\cdots\hat{\psi}^\dagger(\mathbf{r}_2)\, \hat{\psi}^\dagger(\mathbf{r}_1)$$

$$\times \sum_{n=1}^{N} \frac{p_n^2}{2m_e} \hat{\psi}(\mathbf{r}_1)\hat{\psi}(\mathbf{r}_2)\cdots\hat{\psi}(\mathbf{r}_N) \ , \tag{89}$$

where the subscript, kin, indicates that this is the kinetic energy part of the interacting case. Using Eq. (79) in Eq. (89), we find that

$$\mathcal{H}_{kin} = \sum_{\lambda} \sum_{k} \sum_{s_z} \varepsilon_{\lambda k} a^{\dagger}_{\lambda k s_z} a_{\lambda k s_z} \; . \tag{90}$$

Evaluating the band summation and restricting ourselves again to the two-band approximation gives

$$\mathcal{H}_{kin} = \sum_{k} \sum_{s_z} \left[\varepsilon_{ck} a^{\dagger}_{ck s_z} a_{ck s_z} + \varepsilon_{vk} a^{\dagger}_{vk s_z} a_{vk s_z} \right] . \tag{91}$$

In the electron-hole representation, Eq. (91) becomes

$$\mathcal{H}_{kin} = \sum_{k} \sum_{s_z} \left[\varepsilon_{ck} a^{\dagger}_{k s_z} a_{k s_z} + \varepsilon_{vk} \left(1 - b^{\dagger}_{-k,-s_z} b_{-k,-s_z} \right) \right] . \tag{92}$$

Since the origin of energy is arbitrary, the constant term, $\Sigma_{k s_z} \varepsilon_{vk}$ is usually left out. Then

$$\mathcal{H}_{kin} = \sum_{k} \sum_{s_z} [\varepsilon_g + \varepsilon_{ek}] a^{\dagger}_{k s_z} a_{k s_z} + \sum_{k} \sum_{s_z} \varepsilon_{hk} b^{\dagger}_{k s_z} b_{k s_z} \; , \tag{93}$$

where we used the quadratic relationship between $\varepsilon_{\lambda k}$ and k discussed in Sec. 2-1 and ε_{ek} and ε_{hk} are given by Eq. (6). In going from Eq. (91) to Eq. (93) we set $m_h = -m_v$, where m_v is the valence electron effective mass and m_h is the hole effective mass. This is only true for noninteracting electrons. When the Coulomb interactions among electrons are taken into account, the relationship between m_v and m_h is more complicated (see Sec. 4-1). However the kinetic energy part of the total Hamiltonian can still be put in the form of Eq. (93).

The translation to the second quantized representation is not always so simple. For example, as described in more detail in Chap. 4, the algebra is somewhat more involved in going from

$$\mathcal{H}_C = \int d^3r_1 \int d^3r_2 \cdots \int d^3r_N \, \hat{\psi}^\dagger(\mathbf{r}_N) \cdots \hat{\psi}^\dagger(\mathbf{r}_2) \, \hat{\psi}^\dagger(\mathbf{r}_1)$$

$$\times \sum_{\substack{i,j \\ i \neq j}} \frac{e^2}{\epsilon_b |\mathbf{r}_i - \mathbf{r}_j|} \, \hat{\psi}(\mathbf{r}_1) \hat{\psi}(\mathbf{r}_2) \cdots \hat{\psi}(\mathbf{r}_N) , \tag{94}$$

to

$$\mathcal{H}_C = \frac{1}{2} \sum_{\substack{kk',q \\ q \neq 0}} V_q \, (a^\dagger_{k+q} a^\dagger_{k'-q} a_{k'} a_k + b^\dagger_{k+q} b^\dagger_{k'-q} b_{k'} b_k$$

$$- 2a^\dagger_{k+q} b^\dagger_{k'-q} b_{k'} a_k) , \tag{95}$$

where $\mathcal{H}_C$ is the Coulomb interaction part of the many–body Hamiltonian and

$$V_q = \frac{1}{V} \int d^3r \, e^{-i\mathbf{q}\cdot\mathbf{r}} \, V(r) = \frac{4\pi e^2}{\epsilon_b V q^2} \tag{96}$$

is the Fourier transform of the Coulomb potential energy. However, once we are in the second quantized representation, the physical interpretation of the individual terms becomes very apparent. In the Coulomb interaction Hamiltonian (95), the first term accounts for the repulsion between electrons, the second term for the repulsion between holes, and the third term for the attraction between an electron and a hole.

An important physical quantity that is represented by an operator is the charge distribution

$$n(\mathbf{r}) = \hat{\psi}^\dagger(\mathbf{r}) \hat{\psi}(\mathbf{r}) = \frac{1}{V} \sum_{kk'} e^{i(\mathbf{k}-\mathbf{k}')\cdot\mathbf{r}} a^\dagger_{k'} a_k = \sum_q n_q \, e^{i\mathbf{q}\cdot\mathbf{r}} , \tag{97}$$

where

$$n_q = \frac{1}{V} \sum_k a^\dagger_{k-q} a_k \tag{98}$$

is the Fourier amplitude of the density distribution operator. As discussed in Chap. 4, the determination of the density distribution involves the solution of the equation of motion for n_q.

To compute expectation values for an N-particle system, we need to choose a basis set. For our problem a convenient basis is the one made up of the eigenstates of $a^\dagger_{ks_z} a_{ks_z}$ and $b^\dagger_{ks_z} b_{ks_z}$. These are the products

$$|\{n_i\}\rangle = |n_{ek_1 s_{z1}}\rangle |n_{ek_2 s_{z2}}\rangle \cdots |n_{hk_1 s_{z1}}\rangle |n_{hk_2 s_{z2}}\rangle \cdots$$

$$= |n_{ek_1 s_{z1}} n_{ek_2 s_{z2}} \cdots n_{hk_1 s_{z1}} n_{hk_2 s_{z2}} \cdots \rangle . \tag{99}$$

These products give the occupancy of every state in the portion of the band structure of interest. For example, Eq. (99) represents the eigenstate with $n_{ek_1 s_{z1}}$ electrons in conduction band state 1, $n_{ek_2 s_{z2}}$ electrons in conduction band state 2, and so on. Because of the Pauli exclusion principle, the n's are either 0 or 1.

Any state of the system can then be written as

$$|\psi\rangle = \sum_{\{n_i\}} c_{\{n_i\}} |\{n_i\}\rangle , \tag{100}$$

where the summation is over all permutations of n_i's and $c_{i\{n_i\}} = \langle \{n_i\}|\psi\rangle$ is the probability amplitude of finding the semiconductor in the eigenstate $|\{n_i\}\rangle$. If we know the state vector for the semiconductor, then the expectation value of an operator $\mathcal{O}$ is

$$\langle \mathcal{O} \rangle = \langle \psi| \mathcal{O} |\psi\rangle . \tag{101}$$

On the other hand if the system is not in a single state, then the expectation value is

$$\langle \mathcal{O} \rangle = \sum_j P_j \langle \psi_j | \mathcal{O} | \psi_j \rangle \, , \tag{102}$$

where P_j is the probability that the semiconductor is described by the state vector $|\psi_j\rangle$. One should not make the mistake of associating P_j with the quantum mechanical uncertainty given by $c_{\{n_i\}}$. The lack of knowledge that led to P_j is usually classical in origin and in most cases is due to the lack of information on the initial state vector. Inserting the identity operator, $\Sigma_{\{n_i\}} |\{n_i\}\rangle \langle \{n_i\}|$, between $\mathcal{O}$ and $|\psi_j\rangle$, we find

$$\langle \mathcal{O} \rangle = \sum_{\{n_i\}} \sum_j P_j \langle \psi_j | \mathcal{O} | \{n_i\}\rangle \langle \{n_i\} | \psi_j \rangle$$

$$= \sum_{\{n_i\}} \sum_j \langle \{n_i\} | \psi_j \rangle P_j \langle \psi_j | \mathcal{O} | \{n_i\}\rangle$$

$$= \mathrm{Tr}(\rho \mathcal{O}) = \mathrm{Tr}(\mathcal{O} \rho) \, , \tag{103}$$

where ρ is the density operator

$$\rho = \sum_j P_j |\psi_j\rangle \langle \psi_j| \, . \tag{104}$$

This density operator is useful for describing lasers because collisions, recombination, and randomness (incoherence) in the excitation processes creating the inversion prohibit a precise knowledge of the state vector of the system. The diagonal elements of the density operator

$$\langle \{n_i\} | \rho | \{n_i\}\rangle = \sum_j P_j \langle \{n_i\} | \psi_j \rangle \langle \psi_j | \{n_i\}\rangle = \sum_j P_j |c_{\{n_i\}}|^2 \tag{105}$$

give the probability of finding the system in the eigenstate $|\{n_j\}\rangle$. The off-diagonal elements of the density operator

$$\langle\{n_i\}|\rho|\{n_j\}\rangle = \sum_l P_l \langle\{n_i\}|\psi_l\rangle\langle\psi_l|\{n_j\}\rangle = \sum_l P_l c^*_{\{n_j\}} c_{\{n_i\}} . \quad (106)$$

contain information concerning the relative phases (coherence) between probability amplitudes.

The statistical average over possible state vectors tends to destroy the coherence in the system, so that whenever collisions or pump effects dominate, we are likely to have a diagonal density operator. In a semiconductor laser, the rapid intraband collisions usually dominate the dynamics within each band. These collisions tend to drive the carrier distributions into quasi-equilibrium distributions. As a result, the electron and hole density operators are often to a very good approximation

$$\rho_e = \frac{1}{Z_e}\exp\left[-\beta\sum_k\sum_{s_z}(\varepsilon_{ek}-\mu_e)a^\dagger_{ks_z}a_{ks_z}\right] \quad (107)$$

$$\rho_h = \frac{1}{Z_h}\exp\left[-\beta\sum_k\sum_{s_z}(\varepsilon_{hk}-\mu_h)b^\dagger_{ks_z}b_{ks_z}\right], \quad (108)$$

respectively. Here the partition function for the conduction electrons is

$$Z_e = \mathrm{tr}\left\{\exp\left[-\beta\sum_k\sum_{s_z}(\varepsilon_{ek}-\mu_e)a^\dagger_{ks_z}a_{ks_z}\right]\right\}$$

$$= \prod_k\prod_{s_z}\left[\langle0_{ks_z}|e^{-\beta(\varepsilon_{ek}-\mu_e)a^\dagger_{ks_z}a_{ks_z}}|0_{ks_z}\rangle\right.$$

$$\left. + \langle1_{ks_z}|e^{-\beta(\varepsilon_{ek}-\mu_e)a^\dagger_{ks_z}a_{ks_z}}|1_{ks_z}\rangle\right]$$

which yields

$$Z_e = \prod_{k} \prod_{s_z} [1 + e^{-\beta(\varepsilon_{ek} - \mu_e)}] . \tag{109}$$

Similarly,

$$Z_h = tr\left\{ exp\left[-\beta \sum_{k} \sum_{s_z} (\varepsilon_{hk} - \mu_h) b_{ks_z}^\dagger b_{ks_z} \right] \right\}$$

$$= \prod_{k} \prod_{s_z} [1 + e^{-\beta(\varepsilon_{hk} - \mu_h)}] \tag{110}$$

is the partition function for the holes. These density operators give the Fermi-Dirac distributions for the carrier distributions. For example, the probability of finding an electron with momenta k and s_z is

$$\sum_{\{n_{k's'_z}\}} \langle \{n_{k's'_z}\} 1_{ks_z} | \rho_e | \{n_{k's'_z}\} 1_{ks_z} \rangle$$

$$= \frac{1}{Z_c} e^{-\beta(\varepsilon_{ek} - \mu_e)} \prod_{k'} \prod_{s'_z} [1 + e^{-\beta(\varepsilon_{ek'} - \mu_e)}] = \frac{1}{e^{\beta(\varepsilon_{ek} - \mu_e)} - 1} ,$$

$$\tag{111}$$

where $\{n_{k's'_z}\}$ represents all possible permutations of the eigenstates except the one with momenta k and s_z.

The dynamics for the expectation value given by Eq. (103) may reside in the operator $\mathcal{O}$ or in the density operator ρ. The former corresponds to the Heisenberg picture of quantum mechanics, and the latter corresponds to the Schrödinger picture of quantum mechanics. They are equivalent in the sense that both pictures give the same expectation values. In this book we choose to work mostly with the Heisenberg Picture because it turns out to be a more convenient approach in the many-body treatment presented in the later chapters. In the Heisenberg Picture, the operator $\mathcal{O}$ obeys the equation of motion

$$i\hbar \, \frac{d\mathcal{O}}{dt} = [\mathcal{O}, \mathcal{H}] \, ,$$

(112)

where the commutator

$$[A, B] \equiv [A, B]_- \equiv AB - BA \, ,$$

and $\mathcal{H}$ is the total Hamiltonian. In the Schrödinger picture, the density operator obeys the Schrödinger equation

$$i\hbar\dot{\rho} = [\mathcal{H}, \rho] \, .$$

(113)

For the semiconductor laser, the total Hamiltonian consists of terms describing the kinetic energy, the interaction between that laser field and the carriers and the Coulomb interaction among carriers. In the electron-hole picture, we have

$$\mathcal{H} = \mathcal{H}_{kin} + \mathcal{H}_C$$

$$- \sum_{\mathbf{k}} \sum_{s_z} \left[\mu_{\mathbf{k}} a^{\dagger}_{\mathbf{k}s_z} b^{\dagger}_{-\mathbf{k},-s_z} + \mu^*_{\mathbf{k}} b_{-\mathbf{k},-s_z} a_{\mathbf{k}s_z} \right] E(z,t) \, ,$$

(114)

where the last term accounts for the dipole interaction between the laser field and the carriers, $-P \cdot E$, with the active medium polarization given by the operator

$$P = \sum_{\mathbf{k}} \sum_{s_z} \left[\mu_{\mathbf{k}} a^{\dagger}_{\mathbf{k}s_z} b^{\dagger}_{-\mathbf{k},-s_z} + \mu^*_{\mathbf{k}} b_{-\mathbf{k},-s_z} a_{\mathbf{k}s_z} \right] \, .$$

(115)

In Eqs. (114) and (115) $\mu_{\mathbf{k}}$ is the dipole matrix element between the valence and conduction band. This matrix element, and details of the dipole coupling between realistic bands are discussed in Sec. 6-3.

The link between the classical laser field and the quantum mechanical semiconductor gain medium is via the polarization

$$P(z,t) = \langle P \rangle = \text{tr}\{P\rho\} \, .$$

(116)

Using Eqs. (68) and (116), the polarization amplitude appearing in the slowly varying amplitude and phase equations (72) and (73) is

$$\mathcal{P}(z) = 2e^{-i[Kz \,-\, \nu t \,-\, \phi(z)]} \sum_{\mathbf{k}} \sum_{s_z} \mu_{\mathbf{k}}^* p_{\mathbf{k}s_z} \,, \tag{117}$$

where

$$p_{\mathbf{k}s_z} = \langle\, b_{-\mathbf{k},-s_z}\, a_{\mathbf{k}s_z} \,\rangle \,. \tag{118}$$

A calculation in the framework of semiclassical laser theory then involves solving the medium equations of motion for $p_{\mathbf{k}s_z}$. These equations are derived using the Heisenberg equation of motion and they are likely to consist of coupled equations for $p_{\mathbf{k}s_z}$, and the carrier populations $\langle\, a_{\mathbf{k}'s_z'}^{\dagger}\, a_{\mathbf{k}'s_z'} \,\rangle$ and $\langle\, b_{\mathbf{k}'s_z'}^{\dagger}\, b_{\mathbf{k}'s_z'}^{\dagger} \,\rangle$. The derivation and solution of the medium equations is often complicated. Fortunately, approximations may be made. The next three chapters discuss these approximations. In these chapters we show that different levels of sophistication exist on how we treat the semiconductor gain medium. There is still uncertainty as to how rigorous one needs to be to correctly analyze a particular phenomenon, which is one of the reasons why the study of semiconductor lasers is still interesting.

References

Haug, H., and S. W. Koch (1993), *Quantum Theory of the Optical and Electronic Properties of Semiconductors*, 2nd Edition, World Scientific Publ., Singapore.

Meystre, P., and M. Sargent III (1991), *Elements of Quantum Optics*, 2nd Edition, Springer Verlag, Heidelberg.

Chapter 3
FREE-CARRIER THEORY

A discussed in the previous two chapters, the laser field and the semi-conductor gain medium are coupled by the gain and the carrier-induced refractive index, or equivalently by the induced complex susceptibility. To determine these quantities, we need to solve the quantum mechanical gain medium equations of motion for the polarization of the medium. The evolution of these equations is driven by the system Hamiltonian. Equation (2.114) gives the total quantum mechanical Hamiltonian for the semiconductor medium, aside from interactions between the carriers and phonons, and injection current pumping. It contains contributions from the kinetic energies, the many-body Coulomb interactions and the electric-dipole interaction between the carriers and the laser field. Since a complete theory using the full Hamiltonian is relatively complicated, one often makes approximations that allow one to begin with a tractable treatment that is reasonably accurate and hopefully contains the most important effects. By gradually eliminating the approximations, one can work toward increasing rigorous treatments. In this book, we take such an approach.

We begin by assuming that the charged particle interactions are sufficiently fast compared to the field transients to be treated as reservoir interactions that establish intraband thermodynamic quasiequilibrium Fermi-Dirac distributions. For simplicity, we neglect many-body effects due to Coulomb interactions between the carriers that renormalize the bandgap energy and the electric-dipole interaction energy. As such, we treat the carriers as ideal Fermi gases, which labels our present theory as a "free-carrier" theory. Many-body and band-structure effects are added in Chaps. 4, 5 and 6. On the crudest level, it is the strong reservoir interactions (carrier-carrier scattering) that gives what appears to be a markedly inhomogeneously broadened transition its homogeneously-broadened saturation behavior. On the other hand, the wide tuning characteristics reveal aspects of the underlying inhomogeneously broadened transition.

In this chapter, we derive a polarization of a semiconductor medium that can explain this dual nature and at the same time track the medium's response to temperature and carrier density variations. The free-carrier model is fairly accurate for describing the bandfilling aspects of the semiconductor laser under most normal operating conditions. We also use the two-band approximation discussed around Eq. (2.60). For the bulk GaAs semiconductor, the contributions of the heavy-hole band dominates that of the light-hole band (see Fig. 2-1) since it has a much larger density of states (2.22). One can get a rough approximation of the ratio of the total number of carriers in the light-hole band to that in the heavy-hole band by using the ratio of the respective densities of states $(m_{lh}/m_{hh})^{3/2}$ [see Eq. (2.23)]. For $m_{lh} = 0.09m_0$ and $m_{hh} = 0.377m_0$, this gives about 12%, which seems reasonable to neglect since at $k = 0$ the bands are degenerate. This situation changes in strained quantum-well systems for which the light-hole band can be lifted above the heavy-hole band.

Section 3-1 describes the free-carrier Hamiltonian and derives the free-carrier equations of motion both in the Heisenberg and Schrödinger pictures. Section 3-2 introduces the quasi-equilibrium approximation, which enables us to solve for the gain and carrier-induced refractive index in terms of Fermi-Dirac distributions. In this approximation the k-dependent carrier populations and polarizations are assumed to adiabatically follow temporal variations of the total carrier density and the electric field envelope. Section 3-3 uses the free-carrier gain equation to predict gain spectra in bulk and in quantum wells. The width of the gain spectrum is shown to depend strongly on the total chemical potential. Section 3-4 gives predictions and explanations of the dependence of gain on the total carrier density. Dependence on temperature are discussed in Sec. 3-5 and dependence on applied light intensity (saturation) in Sec. 3-6. The chapter closes with Sec. 3-7 on free-carrier predictions of the carrier-induced refractive index. This subject is only treated briefly here because an accurate description of the index really requires the more complete analysis of the carrier-carrier Coulomb interactions given in Chap. 4.

3-1. Free-Carrier Equations of Motion

The free-carrier theory assumes that the primary effect of the carrier-carrier Coulomb interaction $\mathcal{H}_C$ given by Eq. (2.95) is to maintain equilibrium carrier distributions within the conduction and valence bands. We can account for this effect by treating $\mathcal{H}_C$ as a reservoir interaction instead of a dynamical interaction and simply replace the carrier distributions appearing in our gain and index equations with appropriate Fermi-Dirac distributions. This leaves us with the simpler Hamiltonian

$$\mathcal{H} = \sum_{\mathbf{k}} \mathcal{H}_{\mathbf{k}}, \qquad (1)$$

where the individual k-dependent Hamiltonians are given by

$$\mathscr{H}_{\mathbf{k}} = \left[\varepsilon_{g0} + \frac{\hbar^2 k^2}{2m_e}\right] a_{\mathbf{k}}^\dagger a_{\mathbf{k}} + \frac{\hbar^2 k^2}{2m_h} b_{-\mathbf{k}}^\dagger b_{-\mathbf{k}} - [\mu_{\mathbf{k}} a_{\mathbf{k}}^\dagger b_{-\mathbf{k}}^\dagger + \mu_{\mathbf{k}}^* b_{-\mathbf{k}} a_{\mathbf{k}}]E(z,t) .$$

$$(2)$$

Here we have used the two-band approximation and absorbed the spin index into $\mathbf{k}$, so that $\Sigma_{\mathbf{k}}$ is actually $\Sigma_{\mathbf{k}s}$, and $a_{\mathbf{k}}$, $a_{\mathbf{k}}^\dagger$, $b_{\mathbf{k}}$ and $b_{\mathbf{k}}^\dagger$ are actually $a_{\mathbf{k}s}$, $a_{\mathbf{k}s}^\dagger$, $b_{\mathbf{k}s}$ and $b_{\mathbf{k}s}^\dagger$, respectively. The unrenormalized band gap energy is ε_{g0}; we use ε_g for the renormalized value obtained from many-body theory (see Chap. 4).

We need to calculate the polarization of the medium given by Eq. (2.116). This involves the "dipole" expectation value (here with s lumped into $\mathbf{k}$)

$$p_{\mathbf{k}} = \langle b_{-\mathbf{k}} a_{\mathbf{k}} \rangle .$$

$$(3)$$

As we find out shortly, the equation of motion for $p_{\mathbf{k}}$ using either the Heisenberg or Schrödinger pictures is coupled to the electron and hole number operator expectation values

$$n_{e\mathbf{k}} = \langle a_{\mathbf{k}}^\dagger a_{\mathbf{k}} \rangle ,$$

$$(4)$$

$$n_{h\mathbf{k}} = \langle b_{-\mathbf{k}}^\dagger b_{-\mathbf{k}} \rangle .$$

$$(5)$$

In the Heisenberg picture, the derivation of the equations of motion for the bilinear operators in Eqs. (3) through (5) involves the evaluation of commutators as dictated by the Heisenberg equation of motion (2.112). First note that any bilinear product of Fermion operators for $\mathbf{k}$ commutes with any bilinear product of Fermion operators for $\mathbf{k}'$, where $\mathbf{k}' \neq \mathbf{k}$. This follows immediately because four anticommuting exchanges are involved and $(-1)^4 = 1$. Hence the Heisenberg equation of motion for $\mathcal{O}_{\mathbf{k}}$ equal to any of the bilinear operators appearing in Eqs. (3) through (5) simplifies as

$$\dot{\mathcal{O}}_{\mathbf{k}} = \frac{i}{\hbar}[\mathscr{H}, \mathcal{O}_{\mathbf{k}}] = \frac{i}{\hbar}[\mathscr{H}_{\mathbf{k}}, \mathcal{O}_{\mathbf{k}}] .$$

$$(6)$$

This simplification does not occur for the $V(\mathbf{q})$ terms in the carrier-carrier Hamiltonian $\mathscr{H}_C$ of Eq. (2.95).

For example, the derivation of the equation of motion for $b_{-\mathbf{k}} a_{\mathbf{k}}$ involves evaluating the commutator $[b_{-\mathbf{k}} a_{\mathbf{k}}, a_{\mathbf{k}}^\dagger b_{-\mathbf{k}}^\dagger]$. Noting that

$$a_{\mathbf{k}}^{\dagger} b_{-\mathbf{k}}^{\dagger} b_{-\mathbf{k}} a_{\mathbf{k}} = a_{\mathbf{k}}^{\dagger} a_{\mathbf{k}} b_{-\mathbf{k}}^{\dagger} b_{-\mathbf{k}}$$

$$b_{-\mathbf{k}} a_{\mathbf{k}} a_{\mathbf{k}}^{\dagger} b_{-\mathbf{k}}^{\dagger} = b_{-\mathbf{k}} b_{-\mathbf{k}}^{\dagger} a_{\mathbf{k}} a_{\mathbf{k}}^{\dagger} = (1 - b_{-\mathbf{k}}^{\dagger} b_{-\mathbf{k}})(1 - a_{\mathbf{k}}^{\dagger} a_{\mathbf{k}}) \ , \tag{7}$$

we have

$$[b_{-\mathbf{k}} a_{\mathbf{k}}, a_{\mathbf{k}}^{\dagger} b_{-\mathbf{k}}^{\dagger}] = (1 - b_{-\mathbf{k}}^{\dagger} b_{-\mathbf{k}})(1 - a_{\mathbf{k}}^{\dagger} a_{\mathbf{k}}) - a_{\mathbf{k}}^{\dagger} a_{\mathbf{k}} b_{-\mathbf{k}}^{\dagger} b_{-\mathbf{k}}$$

$$= 1 - b_{-\mathbf{k}}^{\dagger} b_{-\mathbf{k}} - a_{\mathbf{k}}^{\dagger} a_{\mathbf{k}} \ . \tag{8}$$

The other commutators may be similarly evaluated to give the Heisenberg equations of motion

$$\frac{d}{dt} b_{-\mathbf{k}} a_{\mathbf{k}} = -i\omega_{\mathbf{k}} b_{-\mathbf{k}} a_{\mathbf{k}} - \frac{i}{\hbar} \mu_{\mathbf{k}} (a_{\mathbf{k}}^{\dagger} a_{\mathbf{k}} + b_{-\mathbf{k}}^{\dagger} b_{-\mathbf{k}} - 1) E(z,t) \tag{9}$$

$$\frac{d}{dt} a_{\mathbf{k}}^{\dagger} a_{\mathbf{k}} = \frac{i}{\hbar} \mu_{\mathbf{k}} a_{\mathbf{k}}^{\dagger} b_{-\mathbf{k}}^{\dagger} E(z,t) + \text{h.a.} = \frac{d}{dt} b_{-\mathbf{k}}^{\dagger} b_{-\mathbf{k}} \ , \tag{10}$$

where h.a. stands for Hermitian adjoint, $\hbar\omega_{\mathbf{k}}$ is the transition energy

$$\hbar\omega_{\mathbf{k}} = \varepsilon_{g0} + \varepsilon_{e\mathbf{k}} + \varepsilon_{h\mathbf{k}} = \varepsilon_{g0} + \frac{\hbar^2 k^2}{2m_e} + \frac{\hbar^2 k^2}{2m_h} = \varepsilon_{g0} + \frac{\hbar^2 k^2}{2m_r} \ , \tag{11}$$

and m_r is the reduced mass given by Eq. (2.9). Note that the adjoint has to appear in Eq. (10), since the number operator is Hermitian but the dipole operator $a_{\mathbf{k}}^{\dagger} b_{-\mathbf{k}}^{\dagger}$ is not. Furthermore, we expect the electron and hole number operators to be affected by radiative transitions identically, since these transitions either create both an electron and a hole, or they annihilate one of each. Taking the expectation values given in Eqs. (3) through (5), we have

$$\dot{p}_{\mathbf{k}} = -i\omega_{\mathbf{k}} p_{\mathbf{k}} - \frac{i}{\hbar} \mu_{\mathbf{k}} (n_{e\mathbf{k}} + n_{h\mathbf{k}} - 1) E(z,t) \ , \tag{12}$$

$$\dot{n}_{e\mathbf{k}} = \dot{n}_{h\mathbf{k}} = \frac{i}{\hbar} (\mu_{\mathbf{k}} p_{\mathbf{k}}^* - \mu_{\mathbf{k}}^* p_{\mathbf{k}}) E(z,t) \ . \tag{13}$$

These equations of motion are missing terms describing pumping and relaxation processes. While the effects of current injection may be readily incorporated, we have to think harder about collisions. The important collisions involve carrier-carrier, carrier-phonon, and carrier-impurity (lattice imperfection) scattering, which are not explicitly present in the free-carrier Hamiltonian (1); the carrier-carrier scattering terms *are* included in the many-body Hamiltonian (2.95). Carrier recombination via spontaneous emission is also not taken into account since we did not quantize the radiation field. For now, we include these effects phenomenologically so that Eqs. (12) and (13) become

$$\dot{p}_{\mathbf{k}} = - i\omega_{\mathbf{k}} \, p_{\mathbf{k}} - \frac{i}{\hbar} \mu_{\mathbf{k}} (n_{e\mathbf{k}} + n_{h\mathbf{k}} - 1) E(z,t) + \dot{p}_{\mathbf{k}}\big|_{col} \qquad (14)$$

$$\dot{n}_{\alpha\mathbf{k}} = \Lambda_{\alpha\mathbf{k}} - B_{\mathbf{k}} n_{e\mathbf{k}} n_{h\mathbf{k}} - \gamma_{nr} n_{\alpha\mathbf{k}} + \dot{n}_{\alpha\mathbf{k}}\big|_{col} + \frac{i}{\hbar} (\mu_{\mathbf{k}} p_{\mathbf{k}}^* - \mu_{\mathbf{k}}^* p_{\mathbf{k}}) E(z,t) \qquad (15)$$

Here $\Lambda_{\alpha\mathbf{k}}$, $\alpha = e$ or h, is the pump rate due to an injection current, γ_{nr} is the nonradiative decay constant due to capture by vacancies due to defects in the semiconductor, $B_{\mathbf{k}}$ is the radiative recombination (spontaneous emission) rate constant, and $\dot{p}_{\mathbf{k}}\big|_{col}$ and $\dot{n}_{\alpha}\big|_{col}$ are the collision contributions. In particular carrier-carrier and carrier-phonon scattering drive the distribution $n_{\alpha\mathbf{k}}$ toward the Fermi-Dirac distribution of Eq. (2.12). In the simplest approximation the collision contribution $\dot{p}_{\mathbf{k}}\big|_{col}$ in the polarization equation describes polarization decay (dephasing) according to

$$\dot{p}_{\mathbf{k}}\big|_{col} \simeq - \gamma \, p_{\mathbf{k}} \, . \qquad (16)$$

While rapidly suppressing deviations from the Fermi-Dirac distribution, the scattering does not change the total carrier density N of Eq. (2.13). Hence we can write Eq. (2.13) more generally as

$$N = \frac{1}{V} \sum_{\mathbf{k}} f_{e\mathbf{k}} = \frac{1}{V} \sum_{\mathbf{k}} f_{h\mathbf{k}} = \frac{1}{V} \sum_{\mathbf{k}} n_{e\mathbf{k}} = \frac{1}{V} \sum_{\mathbf{k}} n_{h\mathbf{k}} \qquad (17)$$

where we should remember that the spin summation is included in the k-summation. Accordingly summing Eq. (15), we find the equation of motion

$$\dot{N} = \Lambda - \gamma_{nr} N - \frac{1}{V} \sum_{\mathbf{k}} B_{\mathbf{k}} n_{e\mathbf{k}} n_{h\mathbf{k}} - \frac{i}{\hbar V} \sum_{\mathbf{k}} (\mu_{\mathbf{k}} p_{\mathbf{k}}^* - \mu_{\mathbf{k}}^* p_{\mathbf{k}}) E(z,t) \, , \qquad (18)$$

where the injection current pump Λ is given by

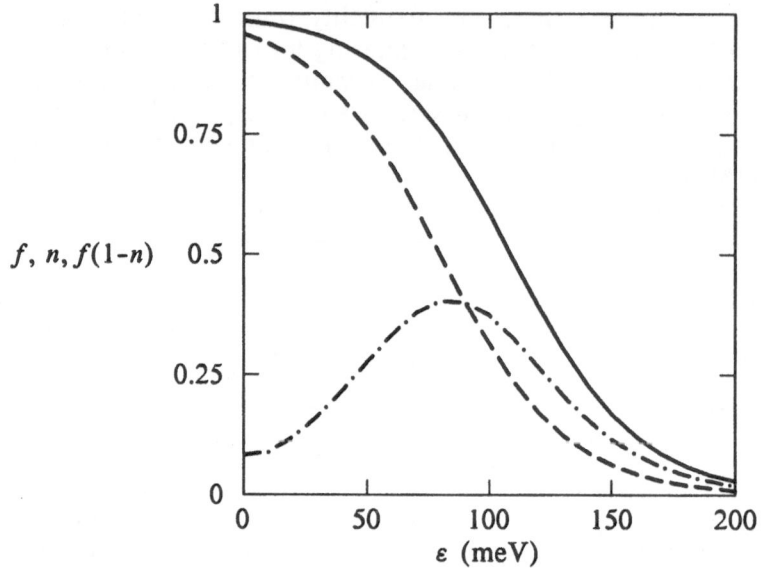

Fig. 3-1. Example of pump blocking. The pump generated distribution $f_{\mathbf{k}0}$ is shown as solid line, the actual carrier distribution $n_{\mathbf{k}}$ inside the active region is shown as dashed line, and the effective pump rate $f_{\mathbf{k}0} (1 - n_{\mathbf{k}})$ is plotted as line with short and long dashes, respectively. The x-axis is given in units of the energy $\varepsilon = \hbar^2 k^2 / 2m_r$.

$$\Lambda = \eta J / ed , \qquad (19)$$

η is the total quantum efficiency that the injected carriers contribute to the inversion, J is the current density, e is the charge of an electron, and d is the thickness of the active region. One can safely assume that by the time the injected carriers reach the active region, they collide often enough to be in equilibrium with one another. Therefore we have

$$\Lambda_{\alpha \mathbf{k}} = \frac{\eta_{tr} J}{ed N_0} f_{\alpha \mathbf{k}0} (1 - n_{\alpha \mathbf{k}}), \qquad (20)$$

where η_{tr} is the "transport" part of the quantum efficiency, giving the efficiency that the injected carriers reach the active region. Furthermore, N_0 and $f_{\alpha \mathbf{k}0}$ are the values of N and $f_{\alpha \mathbf{k}}$, respectively, in the absence of an electromagnetic field $[E(z,t) = 0]$. The presence of carriers inside the active region, with the distribution $n_{\alpha\mathbf{k}}$, reduces the efficiency of the pumping since each quantum state can be occupied only by one carrier. An example of this pump blocking is shown in Fig. 3-1. We see that only the high energy part of the carriers generated by the pump source can actually enter the active region.

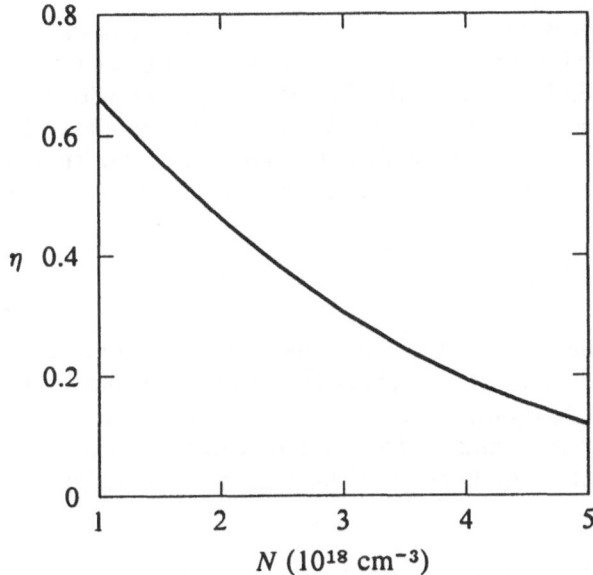

Fig. 3-2. Effective quantum efficiency η (in units of η_{tr}/N_0) for fixed pump rate and increasing carrier density N inside the active region.

The pump blocking effects can be included in a definition of the total quantum efficiency,

$$\eta = \frac{\eta_{tr}}{N_0} \sum_{\mathbf{k}} f_{\alpha\mathbf{k}0} \, (1 - n_{\alpha\mathbf{k}}) \, . \tag{21}$$

For fixed pump rate, i.e. fixed $f_{\alpha\mathbf{k}0}$, the quantum efficiency decreases with increasing carrier density in the active region, as shown in Fig. 3-2.

As is shown in Chap. 4, a more accurate account of intraband scattering results in Boltzmann equations for the carrier populations, where the scattering terms couple the different $\mathbf{k}$ states. One then has an infinite set of coupled nonlinear differential equations for the gain medium. The solution of these equations is nontrivial. Fortunately the problem may often be simplified by noting that in terms of the carrier populations, the net effect of intraband scattering is to return the electron and hole distributions to equilibrium. Then, one way to approximate the intraband scattering terms by writing

$$\dot{n}_{\alpha,\mathbf{k}}\big|_{col} = -\gamma_\alpha(n_{\alpha,\mathbf{k}} - f_{\alpha,\mathbf{k}}) , \qquad (22)$$

where $\alpha = e$ for electrons, $\alpha = h$ for holes, and $f_{\alpha,\mathbf{k}}$ is the Fermi-Dirac distribution satisfying Eq. (17) for a saturated total carrier density, i.e., one in the presence of the laser field. The rates γ_α with which a perturbed carrier distribution returns to equilibrium are sometimes referred to as the cross-relaxation rates. The approximation (22) is consistent with the polarization decay approximation of Eq. (16), provided we use the dipole decay constant

$$\gamma = \tfrac{1}{2}(\gamma_e + \gamma_h) . \qquad (23)$$

In semiconductor gain media, the scattering rates are typically of the order of $10^{13}/s$. Except for strong laser fields that induce stationary nonequilibrium distributions, the scattering rates are sufficiently large to dominate any other mechanism that tries to cause the carrier distributions to deviate from quasiequilibrium. The free-carrier model takes the limit of $\gamma_\alpha \to \infty$, so that

$$n_{e\mathbf{k}} \simeq f_{e\mathbf{k}} \quad \text{and} \quad n_{h\mathbf{k}} \simeq f_{h\mathbf{k}} . \qquad (24)$$

Consequently the polarization dynamics is usually eliminated adiabatically as discussed in Sec. 3-2. Of course, both the γ_α and free-carrier models are approximations.

The total radiative recombination rate is

$$\Gamma_{rr} = \frac{1}{V} \sum_{\mathbf{k}} B_{\mathbf{k}} f_{e\mathbf{k}} f_{h\mathbf{k}} . \qquad (25)$$

For small N the Fermi-Dirac distributions have negative chemical potentials and can be approximated by Maxwell-Boltzmann distributions. Equations (2.25) and (2.27) show that these distributions are proportional to N, so that for small N Eq. (25) can be approximated by BN^2. In fact, using the same approach as in Eq. (2.26), we set $q = \beta\hbar^2/2m$ for typographical simplicity and find

$$\Gamma_{rr} \simeq \frac{B}{V} \sum_{\mathbf{k}} f_{e\mathbf{k}} f_{h\mathbf{k}} = \frac{B}{\pi^2} e^{\beta\mu_e} e^{\beta\mu_h} \int_0^{\infty} dk\ k^2 e^{-qk^2}$$

$$= -\frac{B\bar{N}_e\bar{N}_h}{\pi^2} \frac{\partial}{\partial q} \int_0^{\infty} dk\ e^{-qk^2} = -\frac{B\bar{N}_e\bar{N}_h}{\pi^2} \frac{\sqrt{\pi}}{2} \frac{\partial}{\partial q} \frac{1}{\sqrt{q}} = \frac{B\bar{N}_e\bar{N}_h}{4(\pi\beta\hbar^2/2m_r)^{3/2}}$$

$$= B\ N^2(\pi\beta\hbar^2 m_r/2m_e m_h)^{3/2}\ . \tag{26}$$

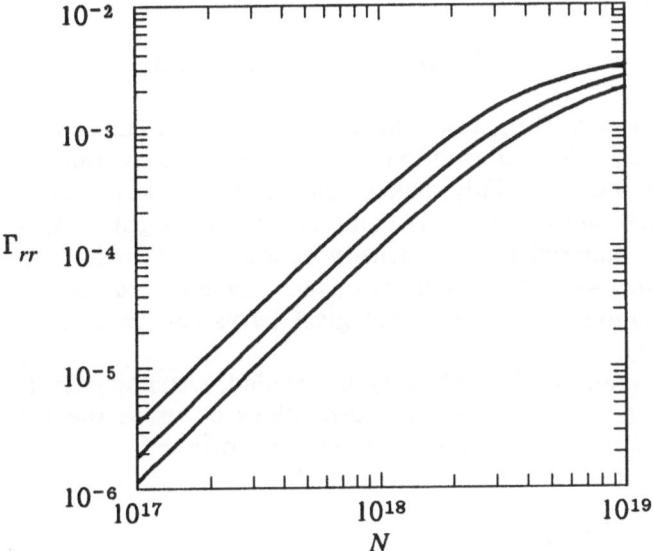

Fig. 3-3. Radiative-recombination (spontaneous emission) coefficient Γ_{rr} versus total carrier density. A slope of two (looks like unity) represents a quantity proportional to N^2, so we see that for small N, $\Gamma_{rr} \propto N^2$.

This is illustrated in Fig. 3-3, where we see that for $N < 10^{18}$ cm^{-3} that Γ_{rr} is proportional to N^2 and that the coefficient decreases as $T^{-3/2}$. However for larger N, the rate of increase is less than N^2, due to the fact that the Fermi-Dirac limit $f_{e\mathbf{k}}, f_{h\mathbf{k}} \leq 1$ comes into play.

At present, most semiconductor lasers operate with either undoped or lightly doped gain regions. The results so far are for an undoped gain region. To generalize to the lightly doped case, we need to take into account the difference between the total electron density N_e and the total hole density N_h. In a p-doped gain medium, $N_h = N_e + N_A$, where N_A is the acceptor density, and in a n-doped gain medium, $N_e = N_h + N_D$, where N_D is the donor density. Also, in the absence of an injection current, there is a residual carrier density due to the dopants. For example, the spontaneous emission term is $B_{\mathbf{k}}(n_{e\mathbf{k}} - f_{D\mathbf{k}})n_{h\mathbf{k}}$ for a n-doped gain medium, and $B_{\mathbf{k}}n_{e\mathbf{k}}(n_{h\mathbf{k}} - f_{A\mathbf{k}})$ for a p-doped gain medium, where $f_{D\mathbf{k}}$ and $f_{A\mathbf{k}}$ are the Fermi-Dirac distributions for the carriers due to the donor and acceptor populations. The extension to a lightly doped gain medium is straightforward and the details are left as an exercise for the reader. For a heavily doped gain region, the band structure is modified by the doping, and laser transitions are possible between free carrier and $\mathbf{k}$-independent bound impurity states. Since heavily doped gain regions are seldom encountered, we do not discuss this situation further.

Density Matrix Derivation

Before solving Eqs. (14) and (15) for various cases, let us see how to derive the equations of motion (12) and (13) using the density operator equation of motion. This makes contact with a large literature that uses this approach and offers some complementary insights. Appendix A gives background material for the density operator in terms of the simple two-level system. We deal initially with the contributions given by the Hamiltonian (1) alone; the phenomenological terms can be added in as for Eqs. (14) and (15).

No anticommutation relations are needed. Analogously to the simplification in Eq. (7), since no $\mathbf{kk'}$ correlations occur in the Hamiltonian (1), the density operator ρ factors into the outer product

$$\rho = \prod_{\mathbf{k}} \rho^{\mathbf{k}} , \tag{27}$$

where we use a superscript to get the $\mathbf{k}$ up out of the way of upcoming matrix subscripts. The Schrödinger equation of motion (2.113) reduces to the solution of the individual $\mathbf{k}$-dependent equations of motion

$$\dot{\rho}^{\mathbf{k}} = -\frac{i}{\hbar}[\mathscr{H}_{\mathbf{k}}, \rho^{\mathbf{k}}] . \tag{28}$$

The $\mathbf{k}$-dependent subspace is spanned by the four eigenvectors of the form $|n_e n_h \mathbf{k}\rangle$, namely

$$|3\rangle \equiv |11\mathbf{k}\rangle\ ,\quad |2\rangle \equiv |10\mathbf{k}\rangle\ ,$$

$$|1\rangle \equiv |01\mathbf{k}\rangle\ ,\quad |0\rangle \equiv |00\mathbf{k}\rangle\ ,$$

(29)

where the number j in $|j\rangle$ is the decimal version of the binary number $n_e n_h$. For the energy-zero choice in Eq. (2), these states are ordered according to increasing energy eigenvalue, with $|0\rangle$ having the lowest

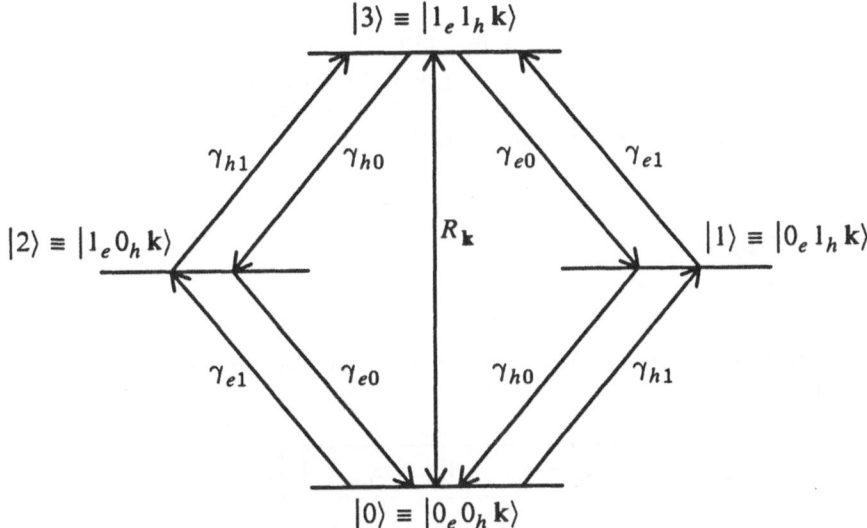

Fig. 3-4. Two-band semiconductor level diagram showing rate constants and directions of probability flow due to carrier-carrier scattering and to stimulated emission and absorption. The carrier-carrier scattering rate constant corresponding to Eq. (22) for losing an αth carrier is given by $\gamma_{\alpha 0} = \gamma_\alpha(1 - f_{\alpha\mathbf{k}})$, and that for gaining an αth carrier is given by $\gamma_{\alpha 1} = \gamma_\alpha f_{\alpha\mathbf{k}}$. $R_\mathbf{k}$ is the rate of stimulated emission and absorption between the highest and lowest states.

(zero) energy, and $|3\rangle$ having the highest (see Fig. 3-4).

The four states imply a 4×4 density matrix, which appears to be a lot more complicated than Eqs. (9) and (10). However note that the Hamiltonian (1) is only able to connect the lowest state $|0\rangle \equiv |00\mathbf{k}\rangle$ to the highest state $|3\rangle \equiv |11\mathbf{k}\rangle$, which limits the number of equations we have to consider. Carrier-carrier scattering couples all four states as well as coupling states of differing k. The states $|1\rangle$ and $|2\rangle$ do not occur in a two-level system, since there is no way to take away the built-in hole of an excited two-level system.

We start by calculating $\dot{\rho}_{30} \equiv \langle 11\mathbf{k}|\dot{\rho}^{\mathbf{k}}|00\mathbf{k}\rangle$ using Eq. (28). We find

$$\dot{\rho}_{30} = -\frac{i}{\hbar}\langle 11\mathbf{k}|\mathcal{H}_{\mathbf{k}}\rho^{\mathbf{k}} - \rho^{\mathbf{k}}\mathcal{H}_{\mathbf{k}}|00\mathbf{k}\rangle$$

$$= -i\omega_{\mathbf{k}}\rho_{30} + \frac{i\mu_{\mathbf{k}}E(z,t)}{\hbar}\langle 11\mathbf{k}|a_{\mathbf{k}}^{\dagger}b_{-\mathbf{k}}^{\dagger}\rho^{\mathbf{k}} - \rho^{\mathbf{k}}a_{\mathbf{k}}^{\dagger}b_{-\mathbf{k}}^{\dagger}|00\mathbf{k}\rangle$$

$$= -i\omega_{\mathbf{k}}\rho_{30} + \frac{i\mu_{\mathbf{k}}E(z,t)}{\hbar}[\rho_{00} - \rho_{33}] . \tag{30}$$

To make Eq. (30) resemble Eq. (12), we need to figure out how the probability difference $\rho_{00} - \rho_{33}$ is given in terms of $n_{e\mathbf{k}}$ and $n_{h\mathbf{k}}$. Since there is either 0 or 1 electron in the $\mathbf{k}$ state and either 0 or 1 hole in the $\mathbf{k}$ state, we have the normalization condition

$$\sum_{j=0}^{3} \rho_{jj} = 1 . \tag{31}$$

Solving this for the unexcited-state probability ρ_{00}, we have the probability difference

$$\boxed{\rho_{33} - \rho_{00} = n_{e\mathbf{k}} + n_{h\mathbf{k}} - 1} , \tag{32}$$

where

$$n_{e\mathbf{k}} = \rho_{22} + \rho_{33} = \mathrm{tr}_{hole}\{\rho^{\mathbf{k}}\} , \tag{33}$$

is the probability of having an electron in the state $\mathbf{k}$ *independent* of whether there's a hole. Similarly $n_{h\mathbf{k}} = \mathrm{tr}_{electron}\{\rho^{\mathbf{k}}\}$, which is the probability of having a hole in the state $\mathbf{k}$ independent of whether there is an electron.

Equation (32) reduces Eq. (30) to

$$\dot{\rho}_{30} = -i\omega_{\mathbf{k}}\rho_{30} + \frac{i\mu_{\mathbf{k}}E(z,t)}{\hbar}[(\rho_{33} + \rho_{22}) + (\rho_{33} + \rho_{11}) - 1]$$

$$= -i\omega_{\mathbf{k}}\rho_{30} + \frac{i\mu_{\mathbf{k}}E(z,t)}{\hbar}[n_{e\mathbf{k}} + n_{h\mathbf{k}} - 1] . \tag{34}$$

Finally noting that

$$p_{\mathbf{k}} = \text{tr}\{\rho^{\mathbf{k}} b_{-\mathbf{k}} a_{\mathbf{k}}\} = \rho_{30} \ , \tag{35}$$

we see that Eq. (34) is just Eq. (12) in a slightly different notation.

Spontaneous emission, often called radiative recombination in semiconductors, is the decay of an upper state to a lower state. For the semiconductor, the excited state $|3\rangle \equiv |11\mathbf{k}\rangle$ decays radiatively to the unexcited state $|0\rangle \equiv |00\mathbf{k}\rangle$, with the spontaneous emission rate $-B_{\mathbf{k}}\rho_{33}$. To use the $n_{e\mathbf{k}}$ and $n_{h\mathbf{k}}$ values, we write ρ_{33} as

$$\rho_{33} = \text{tr}\{\rho^{\mathbf{k}} a_{\mathbf{k}}^{\dagger} b_{-\mathbf{k}}^{\dagger} b_{-\mathbf{k}} a_{\mathbf{k}}\} = \text{tr}\{\rho^{\mathbf{k}} a_{\mathbf{k}}^{\dagger} a_{\mathbf{k}} b_{-\mathbf{k}}^{\dagger} b_{-\mathbf{k}}\}$$

$$\simeq n_{e\mathbf{k}} n_{h\mathbf{k}} \ , \tag{36}$$

where the factored approximation [used in Eq. (18)] is reasonable since for radiative recombination to occur it's necessary to have both a nonzero conduction-electron probability as well as a nonzero hole probability to give the electron a place to fall into. In this approximation, the lower-state probability ρ_{00} is given by

$$\rho_{00} = \text{tr}\{\rho^{\mathbf{k}} b_{-\mathbf{k}} a_{\mathbf{k}} a_{\mathbf{k}}^{\dagger} b_{-\mathbf{k}}^{\dagger}\} \simeq (1 - n_{e\mathbf{k}})(1 - n_{h\mathbf{k}}) \ , \tag{37}$$

which with Eq. (36) yields the probability difference (32).

In addition, we calculate the Hamiltonian contribution to the equation of motion for ρ_{33} to be

$$\dot{\rho}_{33} = -i\hbar^{-1}\langle 11\mathbf{k}|\mathcal{H}_{\mathbf{k}}\rho^{\mathbf{k}}|11\mathbf{k}\rangle + \text{c.c.}$$

$$= -i\mu_{\mathbf{k}} E(z,t)\hbar^{-1}\langle 11\mathbf{k}|a_{\mathbf{k}}^{\dagger} b_{-\mathbf{k}}^{\dagger} \rho^{\mathbf{k}}|11\mathbf{k}\rangle + \text{c.c.}$$

$$= -i\mu_{\mathbf{k}} E(z,t)\hbar^{-1}\rho_{03} + \text{c.c.} \tag{38}$$

Adding the spontaneous-emission term $-B_{\mathbf{k}}\rho_{33} = -B_{\mathbf{k}} n_{e\mathbf{k}} n_{h\mathbf{k}}$ and substituting the sum into the time rate of change of Eq. (33), we find

$$\dot{n}_{e\mathbf{k}} = \dot{\rho}_{22} + \dot{\rho}_{33}$$

$$= \dot{\rho}_{22} - B_{\mathbf{k}} n_{e\mathbf{k}} n_{h\mathbf{k}} - [i\mu_{\mathbf{k}} E(z,t)\hbar^{-1}\rho_{03} + \text{c.c.}]$$

$$+ \text{ pumping and decay terms} \ . \tag{39}$$

Note that $n_{h\mathbf{k}}$ also gets the $\dot{\rho}_{33}$ combinations. The $\dot{\rho}_{22}$ contribution is dominated by carrier-carrier scattering, i.e., leads to terms like $\dot{n}_{e\mathbf{k}}|_{col}$.

These observations show how we find Eqs. (14) and (15) using the density matrix. The density-matrix calculation is longer than the Heisenberg-picture calculation, because the latter only has to deal with three operators, the electric-dipole operator $b_{-\mathbf{k}}a_{\mathbf{k}}$ and the carrier number operators, while the density-matrix approach needs to consider five matrix elements to derive the same expectation values. Furthermore we know how to generalize the Heisenberg-picture analysis to deal with the many-body Hamiltonian, while we have not yet figured out how to do it with the density matrix. Nevertheless, the density-matrix approach offers a valuable alternative perspective, particularly because it deals explicitly with the energy eigenstates and therefore allows a more precise comparison with two-level systems (see the end of Sec. 3-2). We also see more naturally why the radiative recombination is proportional to $n_{e\mathbf{k}}n_{h\mathbf{k}}$.

3-2. Quasiequilibrium Approximation

Our goal is to solve the active medium equations of motion for the gain and carrier-induced refractive index in the limit that the electric field envelope and the total carrier density vary little in the dipole lifetime $T_2 \equiv 1/\gamma$. To do that, we multiply Eq. (14) by the integrating factor $\exp[(i\omega_{\mathbf{k}} + \gamma)t]$ to find

$$\frac{d}{dt}[p_{\mathbf{k}}e^{(i\omega_{\mathbf{k}} + \gamma)t}] = -\frac{i}{\hbar}\mu_{\mathbf{k}}[n_{e\mathbf{k}} + n_{h\mathbf{k}} - 1]E(z,t)e^{(i\omega_{\mathbf{k}} + \gamma)t} , \qquad (40)$$

which can be formally integrated to give

$$p_{\mathbf{k}}(t) = -\frac{i}{\hbar}\mu_{\mathbf{k}}\int_{-\infty}^{t} dt' \, E(z,t') \, e^{(i\omega_{\mathbf{k}} + \gamma)(t'-t)}[n_{e\mathbf{k}}(t') + n_{h\mathbf{k}}(t') - 1] . \quad (41)$$

For a simple steady-state or nearly steady-state theory, we assume that the k-dependent carrier densities $n_\alpha(t)$ vary little in the time $T_2 \equiv 1/\gamma$. When in addition the field envelope varies little in T_2, we can make the *rate-equation approximation*, which consists of evaluating the carrier densities $n_{e\mathbf{k}}(t')$ and $n_{h\mathbf{k}}(t')$ and the field envelope all at the time t and removing them from the integral. This approximation is called the rate-equation approximation, because as we see shortly it leads to a rate equation for the total carrier density. For our present purposes, we consider the constant field envelope used in Eq. (2.67), namely that described by the plane wave field

$$E(z,t) = \frac{1}{2}\varepsilon(z)e^{i[Kz \, - \, \nu t \, - \, \phi(z)]} + \text{c.c.,} \tag{42}$$

although we consider more complicated fields in Chap. 7 and later chapters.

As we show in Chap. 4, fluctuations due to the carrier-carrier scattering term $\dot{n}_{\alpha\mathbf{k}}|_{col}$ have a time scale on the order of T_2. However since we consider a field envelope that varies little in the time T_2 (in fact here one that is constant), the $n_{\alpha\mathbf{k}}(t)$ are driven by the scattering into an equilibrium in which they can adiabatically track the slow time variations of the total carrier density N via Eq. (17). N varies significantly only in relatively long times like the interband relaxation time $1/\gamma_{nr}$. With this approximation, the integral in Eq. (41) can be readily performed, giving

$$p_{\mathbf{k}}(t) = -\frac{i\mu_{\mathbf{k}}}{\hbar}\varepsilon(z)\left[n_{e\mathbf{k}}(t) + n_{h\mathbf{k}}(t) - 1\right]\left\{\frac{e^{i[Kz-\nu t-\phi(z)]}}{i(\omega_{\mathbf{k}} - \nu) + \gamma} + \frac{e^{-i[Kz-\nu t-\phi(z)]}}{i(\omega_{\mathbf{k}} + \nu) + \gamma}\right\}. \tag{43}$$

Since at optical frequencies the second term in the {} of Eq. (43) has a very large denominator relative to that for the first term, we neglect it. This is called the *rotating-wave approximation* and reduces Eq. (43) to

$$\boxed{p_{\mathbf{k}}(t) = -\frac{i\mu_{\mathbf{k}}}{2\hbar}\varepsilon(z)\frac{e^{i[Kz-\nu t-\phi(z)]}}{i(\omega_{\mathbf{k}} - \nu) + \gamma}\left[n_{e\mathbf{k}}(t) + n_{h\mathbf{k}}(t) - 1\right]} \, . \tag{44}$$

By substituting Eq. (44) into Eq. (2.117), we find the complex polarization

$$\mathcal{P}(z) = -\frac{i\varepsilon(z)}{\hbar\gamma V}\sum_{\mathbf{k}}|\mu_{\mathbf{k}}|^2[n_{e\mathbf{k}} + n_{h\mathbf{k}} - 1)]\mathcal{L}(\omega_{\mathbf{k}} - \nu)\left[1 - i\frac{\omega_{\mathbf{k}}-\nu}{\gamma}\right], \tag{45}$$

where we have again used the rotating-wave approximation, and

$$\mathcal{L}(\omega_{\mathbf{k}} - \nu) = \frac{\gamma^2}{\gamma^2 + (\omega_{\mathbf{k}} - \nu)^2} \tag{46}$$

is the Lorentzian lineshape function. Substituting Eq. (44) into Eq. (18), we have the equation of motion for the total carrier density

$$\dot{N} = \Lambda - \gamma_{nr} N - \frac{1}{V} \sum_{\mathbf{k}} B_{\mathbf{k}} n_{e\mathbf{k}} n_{h\mathbf{k}}$$

$$- \frac{\epsilon(z)^2}{2\hbar^2 \gamma V} \sum_{\mathbf{k}} |\mu_{\mathbf{k}}|^2 [n_{e\mathbf{k}} + n_{h\mathbf{k}} - 1] \mathscr{L}(\omega_{\mathbf{k}} - \nu) . \qquad (47)$$

Our present calculation is also valid if the field envelope $\epsilon(z)e^{-i\phi(z)}$ is replaced with a time-varying envelope that varies little in the time T_2. This point is important for the study of relaxation oscillations and instabilities in external cavity lasers - see Chaps. 7 and 8.

To use Eq. (45) in the gain and index formulas, we need to know the carrier distributions $n_{\alpha\mathbf{k}}(t)$. The easiest thing to do is to note from the Boltzmann-equation analysis of Sec. 4-6 that the carrier-carrier scattering term $\dot{n}_{\alpha\mathbf{k}}|_{col}$ vanishes if the distributions are given by Fermi-Dirac distributions that satisfy Eq. (17), i.e., we use Eq. (24), where the chemical potential μ_α is determined from N as discussed in Sec. 2-3. This is a simple, powerful approach and we use it for most of the remainder of this chapter. The saturation of the response is then totally due to changes induced by the field in the slowly varying total carrier density $N(t)$. We call this approximation the *quasiequilibrium approximation*, since the carrier probabilities are described by quasiequilibrium Fermi-Dirac distributions. This limit does not imply that the $f_{\alpha\mathbf{k}}$ are constant in time; they only have to vary little in the time T_2.

Note that in addition to the rate-equation requirement that the field envelope vary little in the time T_2, the field intensity must not be so strong that it can burn holes in the Fermi-Dirac distributions. In contrast, the rate equation approximation could still be valid for fields that cause the carrier distributions to deviate from Fermi-Dirac distributions.

To get an inkling of what might happen for such intense fields, suppose that the carrier-carrier scattering is given approximately by Eq. (22). The γ_α terms thoroughly dominate the other pump and decay terms in Eq. (15), so we neglect them (one cannot do this when summing Eq. (15) over $\mathbf{k}$, since the carrier-carrier scattering terms sum to zero!) Substituting Eqs. (22) and (44) into Eq. (15) and solving in steady state [$\dot{n}_\alpha(t) = 0$; Eq. (43) is only valid in this limit], we have

$$n_{\alpha\mathbf{k}} \simeq f_{\alpha\mathbf{k}} - \frac{|\mu_{\mathbf{k}}|^2 \epsilon^2}{2\hbar^2 \gamma \gamma_\alpha} [n_{e\mathbf{k}} + n_{h\mathbf{k}} - 1] \mathscr{L}(\omega_{\mathbf{k}} - \nu)$$

which gives

$$n_{e\mathbf{k}} + n_{h\mathbf{k}} - 1 \simeq \frac{f_{e\mathbf{k}} + f_{h\mathbf{k}} - 1}{1 + I\mathscr{L}(\omega_{\mathbf{k}} - \nu)/I_{sc}} , \tag{48}$$

where $I = \frac{1}{2}\epsilon_0 cn|\varepsilon(z)|^2$ and $I_{sc} = \frac{1}{2}\epsilon_0 nc\hbar^2\gamma/T_{1f}|\mu_{\mathbf{k}}|^2$, and the "fast" T_1 is given by

$$T_{1f} = \frac{1}{2}\left[\frac{1}{\gamma_e} + \frac{1}{\gamma_h}\right]. \tag{49}$$

An approximation to the Boltzmann-equation theory also gives $\gamma \simeq \frac{1}{2}(\gamma_e + \gamma_h)$, Eq. (23), so that T_2 and T_1 are similar in size. In Eq. (48), the depletion of the inversion and hence the saturation of gain are described by the denominator $1 + \mathscr{L}(\omega_{\mathbf{k}} - \nu)I/I_{sc}$ and by the saturated carrier distributions $f_{e\mathbf{k}}$ and $f_{h\mathbf{k}}$. The former is due to spectral hole burning, which is the frequency-dependent saturation of the inversion by stimulated emission or absorption. The latter describes the decrease in the overall inversion due to the filling of the spectral holes by intraband collisions. Spectral hole burning along with other nonlinear effects of the gain medium play an important role in both multimode operation and very high-speed modulation of semiconductor lasers. On the other hand, using $|\mu_{\mathbf{k}}|/e$ of $\simeq 3\text{Å}$ for the dipole matrix element of GaAs, we find $I_{sc} \simeq 60MWcm^{-2}$, which is much bigger than the facet damage threshold of standard edge emitting semiconductor lasers. Therefore, $I/I_{sc} \ll 1$ and for most problems involving single-mode laser fields, the carrier distributions reduce to the Fermi-Dirac distributions in agreement with Eq. (24).

Accordingly setting $n_{\alpha\mathbf{k}} \simeq f_{\alpha\mathbf{k}}$ in the complex polarization (45) and using Eq. (2.69), we obtain the complex susceptibility

$$\chi(z) = -\frac{i}{\hbar\gamma\epsilon_b V}\sum_{\mathbf{k}} |\mu_{\mathbf{k}}|^2[f_{e\mathbf{k}} + f_{h\mathbf{k}} - 1]\mathscr{L}(\omega_{\mathbf{k}} - \nu)\left[1 - i\frac{\omega_{\mathbf{k}} - \nu}{\gamma}\right] . \tag{50}$$

Similarly the equation of motion (18) for the total carrier density becomes

$$\dot{N} = \Lambda - \gamma_{nr} N - \frac{1}{V} \sum_{\mathbf{k}} B_{\mathbf{k}} f_{e\mathbf{k}} f_{h\mathbf{k}}$$

$$- \frac{E^2}{2\hbar^2 \gamma V} \sum_{\mathbf{k}} |\mu_{\mathbf{k}}|^2 [f_{e\mathbf{k}} + f_{h\mathbf{k}} - 1] \mathscr{L}(\omega_{\mathbf{k}} - \nu) , \qquad (51)$$

where Eq. (17) relates the Fermi-Dirac distributions to the total carrier density N. Substituting $\chi(z)$ into Eqs. (2.75) and (2.78), we find the amplitude gain and carrier-induced refractive index

$$g = \frac{\nu n}{2\epsilon_b c \hbar \gamma V} \sum_{\mathbf{k}} |\mu_{\mathbf{k}}|^2 [f_{e\mathbf{k}} + f_{h\mathbf{k}} - 1] \, \mathscr{L}(\omega_{\mathbf{k}} - \nu) , \qquad (52)$$

$$\delta n = - \frac{n}{2\epsilon_b \hbar \gamma V} \sum_{\mathbf{k}} |\mu_{\mathbf{k}}|^2 [f_{e\mathbf{k}} + f_{h\mathbf{k}} - 1] \, \mathscr{L}(\omega_{\mathbf{k}} - \nu) \frac{\omega_{\mathbf{k}} - \nu}{\gamma} . \qquad (53)$$

The equation of motion (51) for N can be simplified by substituting the gain g of Eq. (52). This gives

$$\dot{N} = \Lambda - \gamma_{nr} N - \frac{1}{V} \sum_{\mathbf{k}} B_{\mathbf{k}} f_{e\mathbf{k}} f_{h\mathbf{k}} - \frac{2gI}{\hbar\nu} . \qquad (54)$$

As we see in Chap. 4, this formula is also valid for the quasiequilibrium many-body case, provided g is the many-body gain. Equation (54) is sometimes further simplified by writing the spontaneous emission term simply as $- BN^2$ as in Eqs. (1.16) and (26). This is a reasonable thing to do when using the linear-density gain model of Eq. (1.8), since then we do not have to calculate Fermi-Dirac distributions. However if we use a scheme where we deal with these distributions, the more accurate spontaneous emission value in Eq. (54) does not increase the calculation time significantly and allows for the fact that in gain media the rate is somewhere in between an N^2 and an N dependence as illustrated in Fig. 3-3.

Hence given that the carrier density is related to the carrier distributions by the Fermi-Dirac distributions, the simultaneous solution of (52) and (54) gives us the saturated gain as a function of laser intensity and injection current. The quasi-chemical potential obtained in process can then be used in Eq. (53) to calculate the carrier-induced refractive index.

Generalization to lightly doped gain regions is simple in the free-carrier model. The minority carrier density, i.e., that for the carrier with the lower density, is given by the solution of Eq. (54). The majority carrier density is just the minority carrier density plus the acceptor or donor density.

Continuous Quasiequilibrium Fermi-Dirac Distributions

We can simplify the summations over **k** by assuming a symmetric band structure with an essentially continuous distribution of states. Then similar to the density of states discussion of Sec. 2-3, we find for a bulk gain medium

$$\frac{1}{V} \sum_{\mathbf{k}} f(\mathbf{k}) \rightarrow \frac{2}{(2\pi)^3} \int_0^\infty dk \int_0^{2\pi} k\,d\phi \int_0^\pi k\sin\theta\,d\theta\, f(k,\theta,\phi)$$

$$\rightarrow \frac{2}{(2\pi)^3} \int_0^\infty dk\; 4\pi k^2\, f(k) \;, \tag{55}$$

where the factor of 2 comes from the spin summation and the last integral is valid when the function $f(\mathbf{k})$ is spherically symmetric in k, as are the quasiequilibrium Fermi-Dirac distributions. It is often more convenient to integrate over the reduced-mass energy

$$\varepsilon = \hbar^2 k^2 / 2m \;, \tag{56}$$

which enters the transition energy according to Eq. (2.10). Similar to the derivation for the carrier density of states, we find a joint density of states, $\mathcal{D}(\varepsilon)$, defined by the equations (for spherically symmetric functions)

$$\frac{1}{V} \sum_{\mathbf{k}} \rightarrow \int_0^\infty d\varepsilon\; \mathcal{D}(\varepsilon) \tag{57}$$

$$\mathcal{D}(\varepsilon) = \frac{(2m/\hbar^2)^{3/2}\sqrt{\varepsilon}}{2\pi^2} = \frac{\sqrt{\varepsilon}}{2\pi^2 a_0^3 \varepsilon_R^{\,3/2}} \;, \tag{58}$$

where the exciton Bohr radius and Rydberg energy are given by Eqs. (2.7) and (2.8), respectively. In these units, the total carrier density of Eq. (17) is given by

$$N \simeq \frac{1}{2\pi^2 a_0^3 \varepsilon_R^{3/2}} \int_0^\infty \frac{\sqrt{\varepsilon}\, d\varepsilon}{e^{\beta(\varepsilon\, m/m_\alpha - \mu_\alpha)} + 1} , \tag{59}$$

In terms of ε, we have the energy detuning

$$\hbar(\omega_\mathbf{k} - \nu) = (\hbar\omega - \varepsilon_{g0}) - (\hbar\nu - \varepsilon_{g0}) = \varepsilon - \hbar\delta , \tag{60}$$

where the field detuning relative to the bandgap is given by $\hbar\delta = \hbar\nu - \varepsilon_{g0}$.

To carry out integrations over carrier energy, we express all frequencies in meV. To express $\hbar\gamma$ in meV, we take advantage of the fact that γ is usually given in terms of its inverse, the carrier-carrier scattering time τ_s. Planck's constant $\hbar = 6.5817 \times 10^{-13}\ meV\ s$, which gives

$$\hbar\gamma = \frac{\hbar}{\tau_s} = 6.5817 \times \frac{10^{-13}}{\tau_s} . \tag{61}$$

For example, a scattering time of $\tau_s = 10^{-13}\ s$ gives $\hbar\gamma = 6.5817\ meV$. Similarly we express the frequency difference $\hbar\delta$ in meV. For room temperature, $1/\beta \simeq 25\ meV$. The optical gain coefficient given by Eq. (52) becomes

$$g = \frac{K}{4\epsilon_b \hbar\gamma \pi^2 a_0^3 \epsilon_R^{3/2}} \int_0^\infty d\varepsilon\ \frac{|\mu(\varepsilon)|^2 \sqrt{\varepsilon}}{1 + (\varepsilon - \hbar\delta)^2/(\hbar\gamma)^2} [f_e(\varepsilon) + f_h(\varepsilon) - 1] , \tag{62}$$

where the energy-dependent Fermi-Dirac distribution is given by

$$f_\alpha(\varepsilon) = \frac{1}{e^{\beta(\varepsilon\, m_r/m_\alpha - \mu_\alpha)} + 1} . \tag{63}$$

In an ideal 2D quantum-well laser, the electrons are free to move only in the plane of the active layer, so that the summation over $\mathbf{k}$ is restricted to two dimensions according to

$$\frac{1}{L^2} \sum_{\mathbf{k}} \rightarrow \frac{2}{(2\pi)^2} \int_0^\infty dk \, 2\pi k \;, \tag{64}$$

provided the function summed over is cylindrically symmetric. When written in the form given by Eq. (57), we have a two-dimensional relative density of states

$$\mathscr{D} = \frac{m_r}{\pi \hbar^2} = \frac{1}{2\pi a_0^2 \varepsilon_R} \;, \tag{65}$$

which unlike the 3-dimensional value (58) does not depend on the reduced-mass energy ε.

Comparison with the Inhomogeneously Broadened Two-Level Model

The density-matrix derivation allows us to make a fairly precise comparison between the free-carrier semiconductor medium and the inhomogeneously broadened two-level medium (see App. A), which has sometimes been used to model semiconductor gain media. One is tempted to conclude that the two models are quite similar. For example, Eq. (30) has the same form as the standard two-level dipole equation of motion

$$\dot{\rho}_{ab}(\mathbf{k}) = -i\omega(\mathbf{k})\rho_{ab}(\mathbf{k}) + \frac{i\mu_{\mathbf{k}} E(z,t)}{\hbar} [\rho_{bb}(\mathbf{k}) - \rho_{aa}(\mathbf{k})] \;, \tag{A.40}$$

where we identify the upper state $|a\rangle$ with the fully excited electron-hole state $|3\rangle \equiv |11\mathbf{k}\rangle$ and the lower state $|b\rangle$ with unexcited state $|0\rangle \equiv |00\mathbf{k}\rangle$.

However, the two-band semiconductor gain medium is an inhomogeneously broadened *four-level* medium with rapid cross relaxation due to Coulomb coupling such as carrier-carrier scattering and all four levels have appreciable probability in gain media. The two-level medium has no states corresponding to the partially excited semiconductor states $|01\mathbf{k}\rangle$ and $|10\mathbf{k}\rangle$, which involve two electrons. Alternatively stated, the two-level system is a one-electron system that guarantees that an excited electron has a built-in hole into which to fall; the Pauli exclusion principle never gets a chance to enter the picture. No such guarantee exists for a conduction electron in a semiconductor; with different effective electron and hole masses it is entirely possible to have considerable electron probability with little corresponding hole probability so that Eq. (32) is negative, which causes the medium to absorb light, rather than amplify it. This fact leads to the strong dependence of the gain on the band curvatures and their associated densities of states.

The rapid carrier-carrier scattering drives the k-dependent carrier densities toward quasiequilibrium Fermi-Dirac distributions, each characterized by the temperature, the total carrier density, and the appropriate effective mass. This strong cross relaxation plays a crucial role in giving the semiconductor gain medium a homogeneously broadened saturation character, that is, in this limit the gain tends to come from all the carriers rather than just those with k values corresponding to the optical transition.

Rapid cross relaxation is well known in two-level media, as in the case of velocity-changing collisions in gases, and also tends to give a homogeneously broadened saturation character. One asks if such an approach is sufficient to describe the semiconductor medium provided appropriate Fermi-Dirac inhomogeneous distributions for the upper and lower levels are used. The answer is no. While the two-level analysis can have different distributions for the upper and lower levels, what is needed for the semiconductor with unequal electron and hole masses is different distributions for the electron and hole probabilities. According to Eq. (36), the excited-level probability ρ_{33} is the *product* of the two distributions, which cannot be proportional to a single distribution, unless the effective masses are equal. Furthermore the unexcited-level probability ρ_{00} of Eq. (37) is the sum of four terms, one with no distribution at all, two each proportional to a distribution, and one proportional to the product of the distributions. So even for equal effective masses, the two-level model cannot simulate the free-carrier semiconductor model.

Another flawed two-level approach is based on the conduction-electron, valence-electron picture, in which case the probability inversion factor (32) becomes $f_{ek} - f_{vk}$, where $f_{vk} = 1 - f_{hk}$. It is easy to show in the quasiequilibrium limit that f_{vk} is also described by a quasiequilibrium Fermi-Dirac distribution [see Eq. (71)]. But just because the difference between two quantities equals the difference between two other quantities does *not* mean the respective quantities are individually equal to one another. So if one is tempted to identify upper-level probability $\rho_{aa}(\mathbf{k})$ with f_{ek} and the lower-level probability $\rho_{bb}(\mathbf{k})$ with f_{vk}, one is confronted with paradoxes such as the *sum* $f_{ek} + f_{vk}$ can equal 2 or 0 or any value in between! Meanwhile the two-level model for this case has to satisfy the normalization condition $\rho_{aa}(\mathbf{k}) + \rho_{bb}(\mathbf{k}) = 1$. Hence any inhomogeneously broadened two-level theory that attempts to consider $\rho_{aa}(\mathbf{k})$ and $\rho_{bb}(\mathbf{k})$ individually, instead of taking their difference, is prone to predicting bizarre behavior about quasiequilibrium semiconductor media.

For pulsed interactions that are fast compared to carrier-carrier scattering, the states $|01\mathbf{k}\rangle$ and $|10\mathbf{k}\rangle$ may not acquire appreciable probability, so that the two states $|11\mathbf{k}\rangle$ and $|00\mathbf{k}\rangle$ may suffice to describe the problem. For these extremely short coherent-transient interactions, one might conclude that the inhomogeneously broadened two-level model could provide a realistic model of a semiconductor medium. However Coulomb interactions enter to modify the picture substantially, causing coherent transients, such as photon echo, to be more intricate than they are in two-level media.

But life is not so bad. The free-carrier model is actually easier to use than an inhomogeneously broadened two-level medium with rapid cross relaxation. Furthermore the ideas we have about two-level media can be used with appropriate modifications for the semiconductor medium. In particular, the concepts of spatial and spectral hole burning are very useful. Of course, the free-carrier model, too, is based on approximations. Comparison with the full semiconductor Bloch equations of Sec. 4-3 reveals that both $E(z,t)$ and ω_k in Eqs. (14) and (15) are "renormalized" to take into account Coulomb interactions between the carriers, and Sec. 4-6 shows that the carrier-carrier scattering is described by a multidimensional Boltzmann-equation.

3-3. Semiconductor Gain

Armed with the free-carrier gain equation (62), we can investigate the dependence of semiconductor gain on the total carrier density and temperature. We begin with a gain spectrum, that is, gain versus the reduced mass energy ε for several values of the homogeneous linewidth factor γ. This leads to the limit of the δ-function lineshape function, which reveals important gain characteristics including the width of the gain region. We then show how the free-carrier theory justifies the linear-density gain model of Eq. (1.8) for bulk, while revealing its failure for quantum wells. Bulk and quantum-well gain spectra are compared for several different total carrier densities. The section closes with graphs that show how the emission and absorption parts of the gain contribute to the gain as functions of the total carrier density.

Figure 3-5 shows the gain spectrum of a bulk GaAs gain medium. The different curves are for different values of γ. The dashed curve, which is for $\gamma \rightarrow 0$, gives the inhomogeneously broadened limit, where the gain spectrum is the product of the relative density of states and the inversion, that is,

$$g = g_0 \sqrt{\varepsilon} [f_e(\varepsilon) + f_h(\varepsilon) - 1] \tag{66}$$

where the reduced mass energy $\varepsilon = \hbar\delta = \hbar(\omega - \nu)$ and Eq. (62) gives

$$g_0 = \frac{K|\mu(\epsilon)|^2}{4\epsilon_b \pi a_0^3 \epsilon_R^{3/2}} . \tag{67}$$

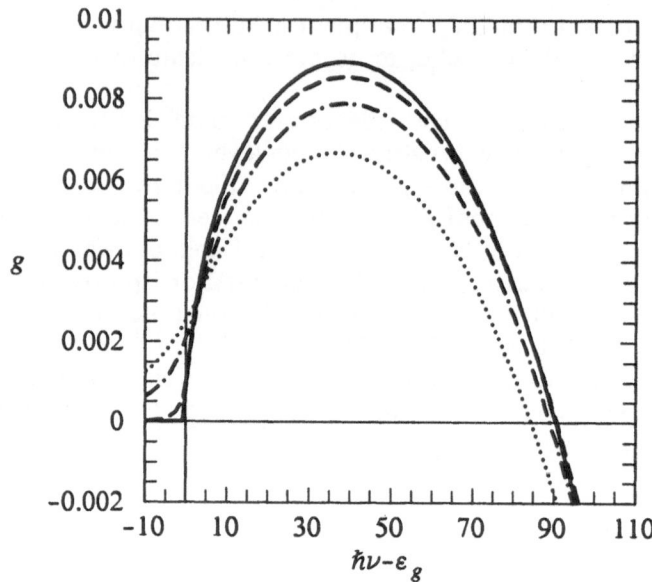

Fig. 3-5. Gain versus detuning $\hbar\delta = \hbar\nu - \varepsilon_{g0}$ as predicted by the free-carrier model Eq. (62). The curves are for $\gamma = 0$ (solid line), 1 meV (dashed), 5 meV (dot-dashed), and 10 meV (dotted). The $\gamma = 0$ case uses the δ-function line-shape formula (66).

Note that $\gamma = 1$ meV is very close to the inhomogeneously broadened limit given by $\gamma = 0$. There is evidence to believe that γ is typically about 10 meV, which gives a significantly different curve from that at the inhomogeneously broadened limit. In general, the peak gain decreases with increasing γ.

The use of a Lorentzian lineshape function overestimates the effects of homogeneous broadening because of its slowly decaying tails. This leads to some absorption at photon frequencies below the band gap. A more realistic lineshape function results from replacing the Lorentzian in the gain formula by a sech, that is

$$\frac{\gamma^2}{\gamma^2 + (\omega_{\mathbf{k}} - \nu)^2} \rightarrow \operatorname{sech}\left(\frac{\omega_{\mathbf{k}} - \nu}{\gamma}\right). \tag{68}$$

Note that the area under both of these functions is $\pi\gamma$. Figure 3-6 illustrates that the sech function decays faster (exponentially) than the Lorentzian. The physical reason for the exponential low energy tail of the lineshape is the *Urbach tail* of the electron-hole resonances in semiconductors. A comparison of the gain spectra computed with the different lineshapes is

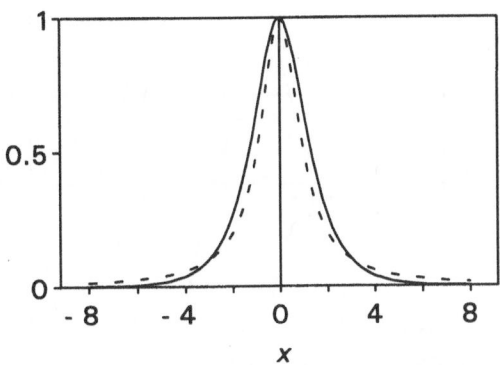

Fig. 3-6. The solid curve is a plot of sech(x), while the dashed curve is a plot of $1/(1 + x^2)$. The areas under both curves equal $\pi\gamma$.

shown in Fig. 3-7. We can see that the use of the sech lineshape function removes the problem of absorption below the bandgap energy. Furthermore, it predicts that transparency occurs at lower carrier densities than for the Lorentzian lineshape approximation. This is quite apparent in the plots of peak gain versus. carrier density in Fig. 3-8. The different curves correspond to different dephasing rates. One would not expect to use the same linewidth for the different lineshape functions, and Fig. 3-8 shows that in order to obtain similar predictions, γ_{sech} needs to be slightly larger than γ_{Lor}.

For the sech-lineshape approximation Eq. (53) no longer yields the correct result for the carrier-induced refractive index. To obtain the carrier-induced refractive index, we therefore use the Kramers-Kronig transformation,

$$\Delta(\delta n) = \frac{c}{\pi} P \int_0^\infty d\nu' \; \frac{\Delta g(\nu')}{\nu'^2 - \nu^2} \;, \qquad (69)$$

where $\Delta(\delta n)$ and Δg are the differences in refractive index and gain at two different carrier densities.

In order to have a rough approximation for the spectral width of the gain region we use the bracketed part of the δ-function lineshape formula (66) written in terms of the valence-band electron distribution as

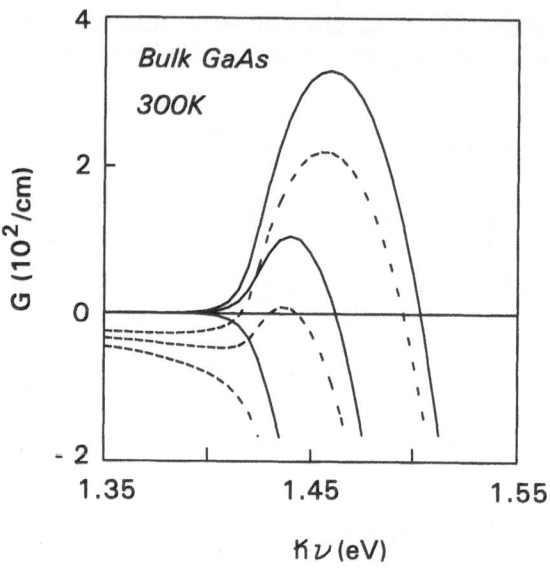

Fig. 3-7. The solid curves are the gain spectra calculated with a sech lineshape function, while the dashed curves are those calculated with a Lorentzian lineshape function. The carrier densities are $1 \times$, $2 \times$ and $3 \times 10^{18}/cm^3$ and $\gamma = 10^{13}/s$.

$$f_e(\varepsilon) + f_h(\varepsilon) - 1 = f_e(\varepsilon) - [1 - f_h(\varepsilon)] = f_e(\varepsilon) - f_v(\varepsilon) . \tag{70}$$

Here the valence-band electron distribution is given by

$$f_v(\varepsilon) = 1 - \frac{1}{e^{\beta(\varepsilon \, m_r/m_h - \mu_h)} + 1} = \frac{1}{1 + e^{-\beta(\varepsilon \, m_r/m_h - \mu_h)}}$$

$$= \frac{1}{e^{\beta(\varepsilon \, m_r/m_v + \mu_h)} + 1} . \tag{71}$$

The probability difference (70) is positive if $f_e(\varepsilon) > f_v(\varepsilon)$, i.e., if the conduction-band probability at energy ε is greater than the valence-band probability. The transparency point (crossover from gain to absorption) is given by the equal probability condition $f_e(\varepsilon) = f_v(\varepsilon)$, which occurs when the arguments of the two Fermi-Dirac exponential equal one another

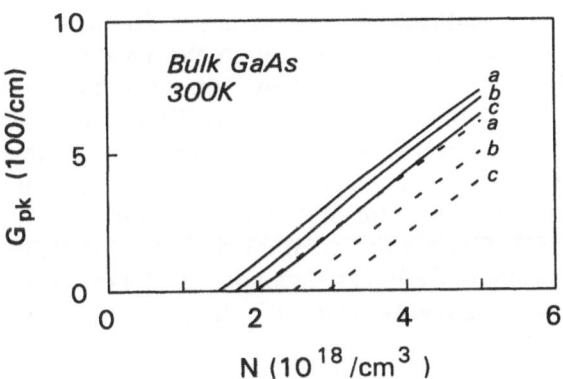

Fig. 3-8. Peak gain *vs* carrier density for sech (solid curves) and Lorentzian (dashed curves) lineshape functions. The different curves corresponds to different γ's: γ = (a) $10^{13}/s$, (b) $2\times10^{13}/s$ and (c) $3\times10^{13}/s$.

$$\frac{\varepsilon m_r}{m_e} - \mu_e = -\frac{\varepsilon m_r}{m_h} + \mu_h = 0 \,,$$

that is

$$\varepsilon = \mu_e + \mu_h \,. \tag{72}$$

The *total chemical potential*

$$\mu = \mu_e + \mu_h \tag{73}$$

is an important parameter in semiconductor laser theory since it defines the upper limit of the gain region with respect to the band-gap energy ε_{g0}, i.e., gain occurs in the spectral region

$$\varepsilon_{g0} < \hbar\omega < \varepsilon_{g0} + \mu \,. \tag{74}$$

This formula is correct also for the many-body case, if we replace

$$\varepsilon_{g0} \rightarrow \varepsilon_g = \varepsilon_{g0} + \delta\varepsilon_g \,,$$

where $\delta\varepsilon_g$ is the electronic bandgap renormalization discussed in Chap. 4.

This bandgap renormalization leads to a net frequency shift of the gain spectrum with carrier density. Although the free-carrier model does not predict a bandgap shift, it is useful in helping us understand the bandfilling effects of the semiconductor gain medium. For example, for bulk materials it gives a reasonably accurate description of the change in peak gain with carrier density, as long as we do not care at which energy that peak gain occurs. Figure 3-9a plots the peak gain as a function of carrier density for a bulk GaAs gain medium and various temperatures. These curves show an essentially linear dependence of peak gain on carrier density, which agrees reasonably well with the phenomenological gain expression

$$G = A_g(N - N_g) \tag{1.8}$$

for a bulk gain medium. However, we must still be careful when using Eq. (1.8), because of its limitations. One of them is that A_g and N_g, which are adjustable parameters in Eq. (1.8), have different values for different temperatures and effective masses. This is not a problem with the free-carrier model because it calculates the peak gain vs carrier density curves as functions of temperature and effective masses. Therefore, it can treat the variations of these curves under different experimental conditions in a self-consistent manner. For example, in Fig. 3-9a we use the free-carrier model to show that the gain degrades with increasing temperature. An interesting, although not very accurate derivation of Eq. (1.8) is to take γ sufficiently large that it can be factored outside the integral over ε. Using Eq. (59), we find Eq. (1.8) with the $-N_g$ term given by the -1 contribution in Eq. (62). Of course, γ is usually significantly smaller than μ, so this approach would not give a very accurate value for A_g or for N_g.

Another important limitation of the phenomenological gain expression (1.8) is that it applies only to a bulk gain medium. It is fortuitous that the product of the $\sqrt{\varepsilon}$ factor in the bulk density of states (58) with the probability inversion (70) has a nearly linear total-carrier-density dependence. Obviously, the constant 2D relative density of states (65) cannot give the same result. In fact, the peak gain vs carrier-density curve shown in Fig. 3-9b is calculated using Eq. (65) in the free-carrier gain formula. Notice the rollover in gain at high carrier density.

The difference in behavior between the free-carrier bulk gain and the idealized two-dimensional quantum-well gain is due to the different density of states and may be understood qualitatively by examining the Fermi distributions. In Fig. 3-10a, we plot electron and hole contributions to the integrand in the gain (62) along with the integrand versus the reduced-mass energy ε for three different carrier densities. We see that because of the relatively large hole mass m_h and consequently the large density of hole states (2.22), the hole states are only partially filled; in fact

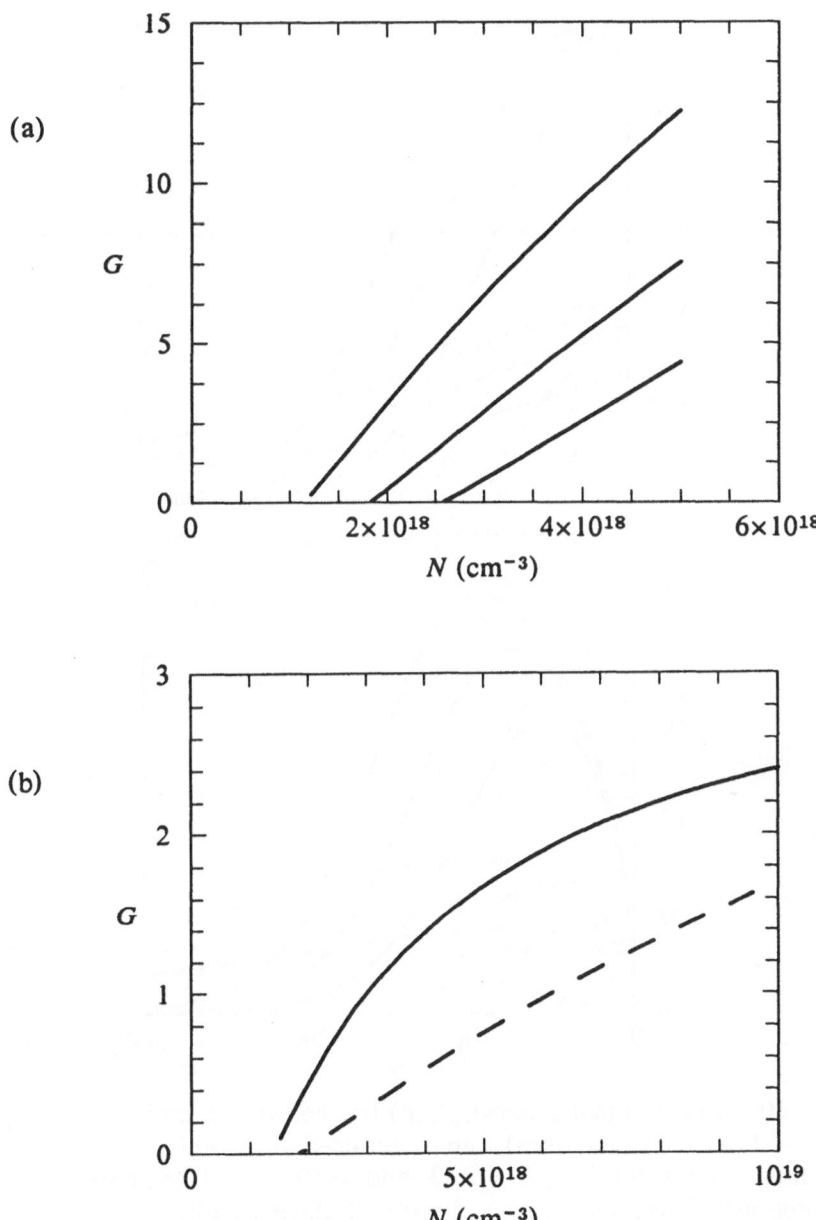

Fig. 3-9. (*a*) Peak gain (in 1000/*cm*) *vs* carrier density for a bulk GaAs gain medium. The different curves are for $T = 200\ K$ (top), $300\ K$ (middle), and $400\ K$ (bottom). (*b*) Peak gain *vs* carrier density for an idealized two-dimensional quantum-well gain medium at $T = 300\ K$. The dashed curve is for the bulk gain medium.

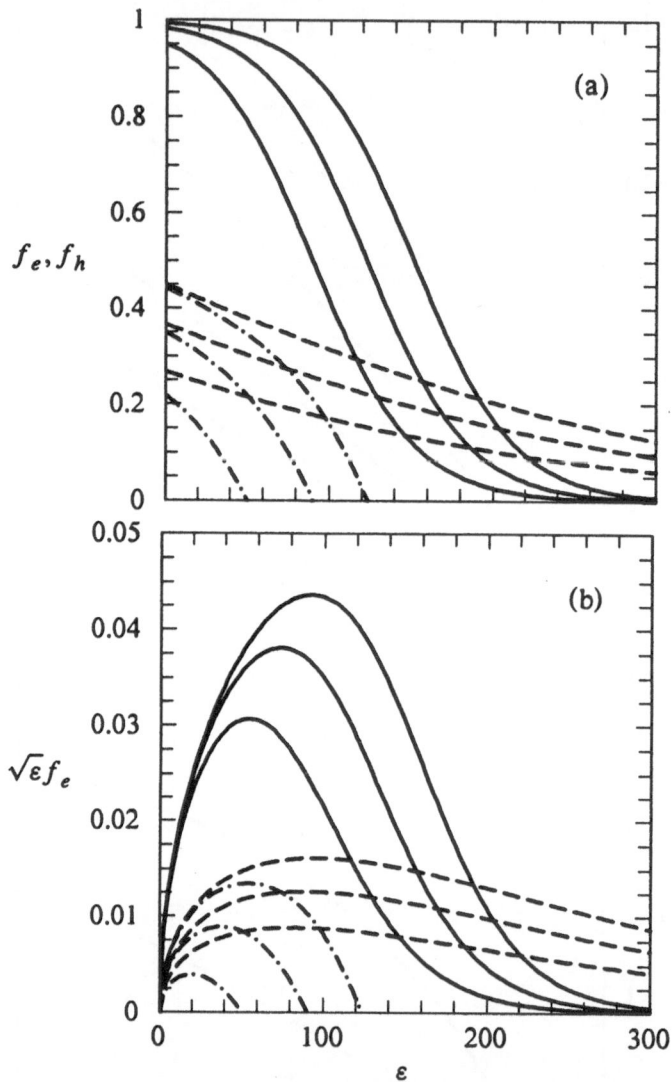

Fig. 3-10. (a) $f_e(\varepsilon)$ (solid lines), $f_h(\varepsilon)$ (dashed lines), and $f_e(\varepsilon)$ + $f_h(\varepsilon)$ - 1 (dot-dashed lines) versus reduced-mass energy ε for bulk total carrier densities $N = 2, 3,$ and 4×10^{18} cm^{-3}. (b) corresponding curves weighted by $\sqrt{\varepsilon}$ density of states factor.

they never reach the inversion value of one half. The addition of carriers then leads to a fairly uniform increase in $f_h(\varepsilon)$ versus ε. On the other hand, because of the relatively small m_e and small density of electron states, the lower energy electron states are almost completely filled, i.e., $f_e(\varepsilon \simeq 0) \simeq 1$. Consequently for $\varepsilon \simeq 0$, there is no more "room at the top" so that the addition of carriers cannot significantly increase the maximum value of $f_e(\varepsilon \simeq 0)$. Instead, the exclusion principle causes the additional electrons to preferentially occupy the vacant higher energy states.

In Fig. 3-10b, the corresponding curves with the bulk density of states factor $\sqrt{\varepsilon}$ are plotted. Two factors contribute to the increase in the $\sqrt{\varepsilon} f_e(\varepsilon)$ and $\sqrt{\varepsilon} f_h(\varepsilon)$ weighted distributions with carrier density. One is the increase in f_e and f_h and the other is the increase in the relative density of states $\sqrt{\varepsilon}$ as the higher energy states become occupied. Both are responsible for the increase in $\sqrt{\varepsilon} f_h$ with carrier density, whereas because of the exclusion principle only the latter plays a role in the case of $\sqrt{\varepsilon} f_e$.

The changes in the carrier energy distributions lead to changes in the inversion energy distribution as shown in the dot-dashed curves of Fig. 3-10b. Notice that the peak value varies essentially linearly with carrier density. Therefore, we expect the peak gain also to increase linearly with carrier density. For $\gamma \neq 0$, the gain peaks lie somewhat higher in energy than the peaks shown in Fig. 3-10b.

Figure 3-11 shows plots for a $2D$ quantum-well medium that correspond to the distributions in Fig. 3-10a. Since the $2D$ density of states is independent of the reduced-mass energy ε, the corresponding gain contributions are proportional to the values in Fig. 3-11. Hence for a zero linewidth, the peak gain occurs for $\varepsilon = 0$ and the occupation of higher energy states populated by increasing N does not increase the peak values for the carrier energy distributions. As a result, saturation effects are evident in Fig. 3-11; the peak values for $f_e(\varepsilon) + f_h(\varepsilon) - 1$ do not increase linearly with carrier density, which causes the peak gain in Fig. 3-7b to roll over. For a nonzero linewidth, the peak gain occurs somewhat above the band gap.

It is interesting to plot the individual gain and emission contributions given by Eqs. (36) and (37), respectively, to the gain factor

$$f_e(\varepsilon) + f_h(\varepsilon) - 1 = f_e f_h - (1 - f_e)(1 - f_h)$$
$$= [1 - e^{\beta(\varepsilon - \mu)}] f_e f_h . \tag{75}$$

As Fig. 3-12 shows, for low carrier densities the absorption factor $(1 - f_e)(1 - f_h)$ dominates, while at high densities, where μ is quite positive, the emission factor $f_e f_h$ dominates. Around the transparency region, both terms contribute. In operating laser conditions, the emission factor dominates the gain. More precisely, for the relatively light electron mass used in Fig. 3-11, $f_e(\varepsilon) \simeq 1$ for values of ε with appreciable gain. Hence for these values both $f_e(\varepsilon) + f_h(\varepsilon) - 1$ and $f_e(\varepsilon) f_h(\varepsilon)$ are approximately given by the hole probability $f_h(\varepsilon)$. For equal electron and hole masses, the

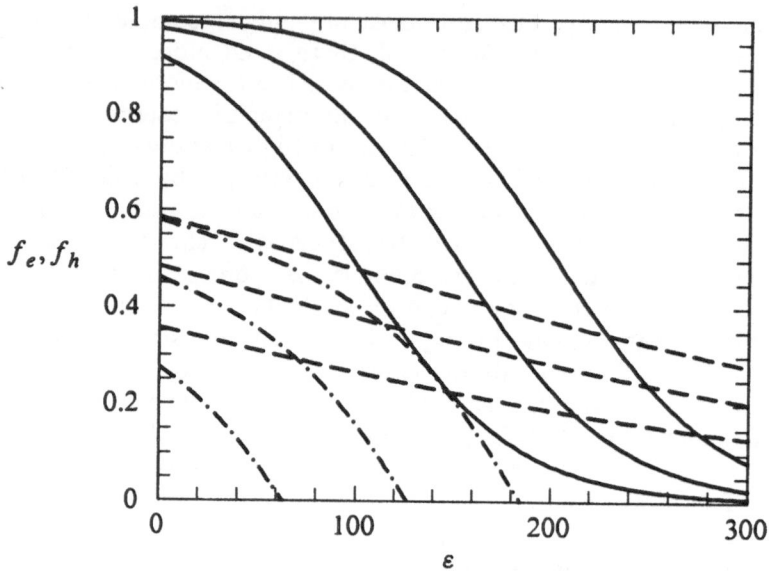

Fig. 3-11. $f_e(\varepsilon)$ (solid lines), $f_h(\varepsilon)$ (dashed lines), and $f_e(\varepsilon)$ + $f_h(\varepsilon) - 1$ (dot-dashed lines) versus reduced-mass energy ε for a quantum-well with width = 1.243×10^{-6} cm (the GaAs exciton Bohr radius), two-dimensional total carrier density $N = 2.486\times$, $3.73\times$, 4.97×10^{12} cm^{-2} (corresponds to $N = 2\times$, $3\times$, 4×10^{18} cm^{-3}) in order of increasing gain, $T = 300$ K, $m_e = 1.176 m_r$, and $m_h = 6.669 m_r$.

agreement is not as good as is illustrated by the dotted curves in Fig. 3-12. This is due to the fact that for reasonable gain values, there is "room at the top" for both the electrons and the holes for more of their kind.

3-4. Temperature Dependence of Gain

Referring to Fig. 3-9a, we see that the semiconductor gain decreases with increasing temperature. To understand this, we use a set of figures that are very similar to those in the previous section. Figure 3-13 plots the occupational probabilities of electron and hole states shown in Fig. 2-3 but here multiplied by the bulk relative density of states factor $\sqrt{\varepsilon}$. Notice that the peaks of the distributions decrease with increasing temperature, which leads to a decrease in the gain. Note also that some of the hole distributions have negative chemical potentials and look very much like the decaying tails of Maxwell-Boltzmann distributions. We see that both the

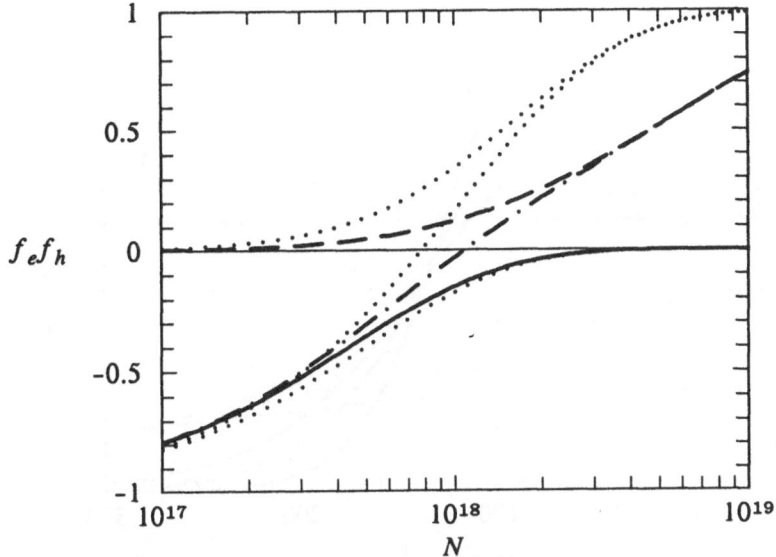

Fig. 3-12. Emission $f_e f_h$ (solid line), absorption $(1 - f_e)(1 - f_h)$ (dashed line), and their difference (75) (dot-dashed line) versus carrier density N (cm^{-3}) for a bulk medium, $T = 300 \ K$, and $\varepsilon = 0$ (band edge). The dotted curves give the corresponding values for equal electron and hole masses, $m_e = m_h = 2m$.

electron and hole distributions are sensitive to temperature change in the region of $\frac{1}{2}\mu$, which thanks to the $\sqrt{\varepsilon}$ factor is where the gain tends to peak. This leads to substantial temperature dependence in the gain, as contrasted with the idealized 2D quantum-well case discussed next.

Figure 3-14 shows similar plots for the quantum-well medium. Since at the gain peak $(\varepsilon = 0)$, only the hole distributions are changed significantly, the inversion is less affected by temperature than in the bulk case. As a result, the degradation of the idealized two-dimensional quantum-well gain with increasing temperature should be less than that of the bulk. This is indeed the case according to the lower set of curves in Fig. 3-15. These curves are calculated using Eqs. (52) and (65).

The lower group of quantum well curves in Fig. 3-15 suggest that a small density of states is advantageous for temperature insensitivity. A more Fermi-Dirac (or less Boltzmann) like energy distribution, with filled low energy states reduces the temperature dependence. Therefore, if we can alter the valence band curvature so that the hole effective mass, and consequently, the hole density of states is as small as that of the electrons, then gain degradation due the temperature increase should be reduced. To

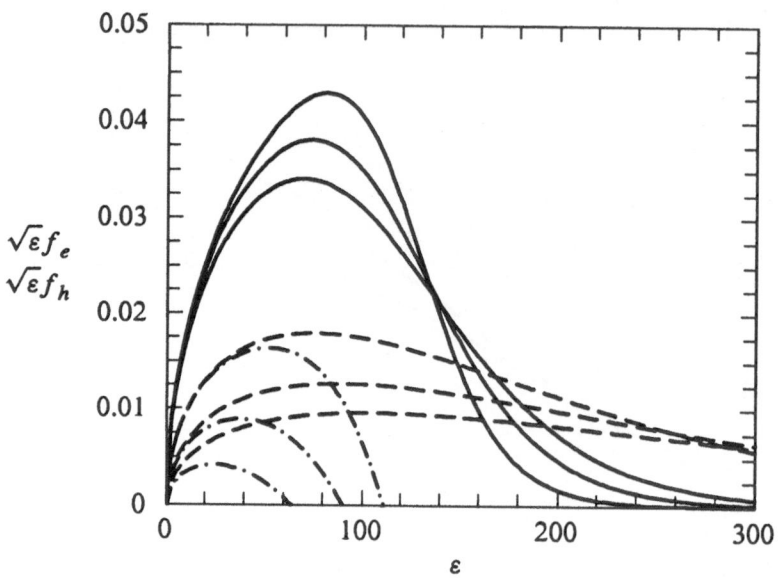

$\sqrt{\varepsilon}f_e$
$\sqrt{\varepsilon}f_h$

Fig. 3-13. Weighted bulk Fermi-Dirac distributions given by $\sqrt{\varepsilon}f_\alpha(\varepsilon)$ versus the reduced-mass energy $\varepsilon = \hbar^2k^2/2m$ for electrons ($\alpha = e$, solid lines) and holes ($\alpha = h$, dashed lines) for $T = 200$, 300 and 400 K (in order of decreasing peak values), and a total carrier density $N = 3\times10^{18}cm^{-3}$. Also included are the corresponding curves (dot-dashed lines) for the gain expression $\sqrt{\varepsilon}[f_e(\varepsilon) + f_h(\varepsilon) - 1]$. The effective masses are $m_e = 1.176m$ and $m_h = 6.669m$.

see if this is indeed the case, we compute the gain for an artificial quantum-well medium with equal hole and electron effective masses. The result is shown in the upper group of curves in Fig. 3-15. In practice, the reduction in the hole effective mass may be achieved with strained-layer quantum wells. Of course, the actual results are neither as good nor as straightforward as shown in Fig. 3-15, because the strain generally deforms the band curvature nonuniformly (nonparabolic bands) and alters the transition matrix elements. In fact, as we see in Chap. 6, the strained-layer quantum-well structures involve more than just a change in the effective masses, since the various valence bands get mixed together. Notice that a consequence of a small hole effective mass is an increase in gain rollover. This too can be understood in terms of the exclusion principle embodied in the Fermi-Dirac distributions, since for high N *both* the electron and the hole distributions fail to have room at the top for more carriers.

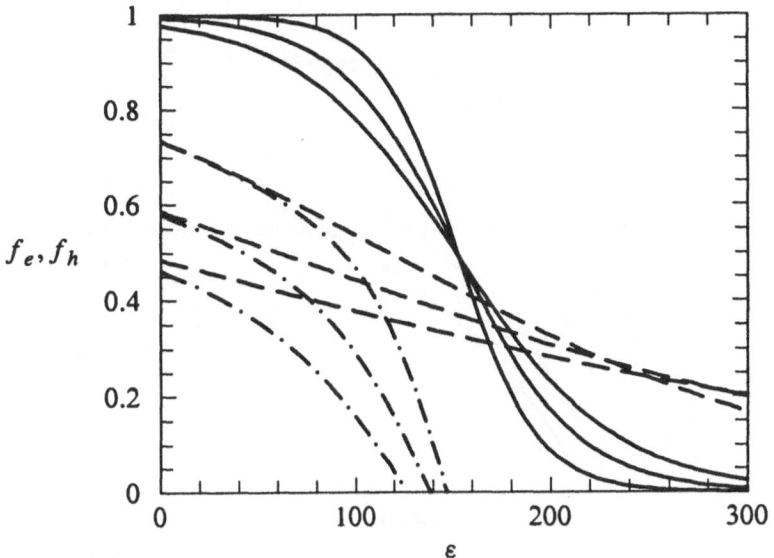

Fig. 3-14. Fermi-Dirac distributions given by $f_\alpha(\varepsilon)$ versus the reduced-mass energy ε for electrons ($\alpha = e$, solid lines) and holes ($\alpha = h$, dashed lines) for $T = 200\ K$, $300\ K$, and $400\ K$, $N = 3.73$ cm^{-2} (assuming an idealized 2D QW with width of $1.243\times10^{-6}\ cm$ this corresponds to $N = 3\times10^{18}\ cm^{-3}$). Also included are the corresponding curves (dot-dashed lines) for the quantum-well gain expression $f_e(\varepsilon) + f_h(\varepsilon) - 1$. The effective masses are $m_e = 1.176m$ and $m_h = 6.669m$.

3-5. Gain Saturation

In any oscillator, saturation by the oscillation intensity plays an important role. Without it, the oscillation amplitude would keep building up until something explodes! According to the linear-density gain model, the gain is given by

$$G = A_g(N - N_g) , \tag{1.8}$$

where A_g is a phenomenological gain coefficient, N is the total carrier density, and N_g is the "transparency" density, i.e., the carrier density at the crossover between absorption and gain. Equation (1.17) shows that the total carrier density difference $N - N_g$ saturates or "bleaches" as the light intensity I increases. Since the linear-density gain is proportional to $N -$

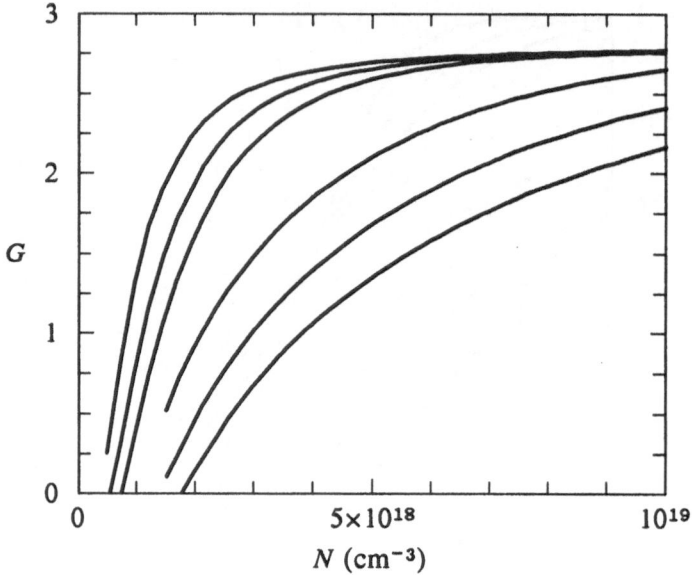

Fig. 3-15. Peak gain (in $1000/cm$) *vs* carrier density for idealized two-dimensional quantum well (lower three curves) and (*b*) quantum well with $m_e = m_h$ (model for strained quantum well, upper three curves). For each group, $T = 200\ K$, $300\ K$, and $400\ K$ in order of decreasing values.

N_g, G saturates the same way, namely

$$G = \frac{A_g(JT_1/ed - N_g)}{1 + IA_g T_1/\hbar\nu}. \tag{76}$$

This is illustrated in Fig. 3-16 for both absorbing (no injection current) and gain cases. We see the S-shaped saturation curves familiar in the saturation of homogeneously broadened two-level media. As for the two-level case, the simple theory predicts the same shape for absorption as for gain; only the sign is reversed. This theory approximates the peak gain for a given N value. One does not expect Eqs. (1.8) and (76) to be good approximations for absorbing media.

To get a better understanding of the nature of gain saturation, we consider the δ-function linewidth gain formula (66) for fixed detunings $\hbar\delta = \hbar\nu - \varepsilon_{g0}$ above the band gap for both gain and absorption media (with and without injected currents, respectively). Equation (66) is similar to Eq. (1.8) in that both consist of the difference between a term that increases as

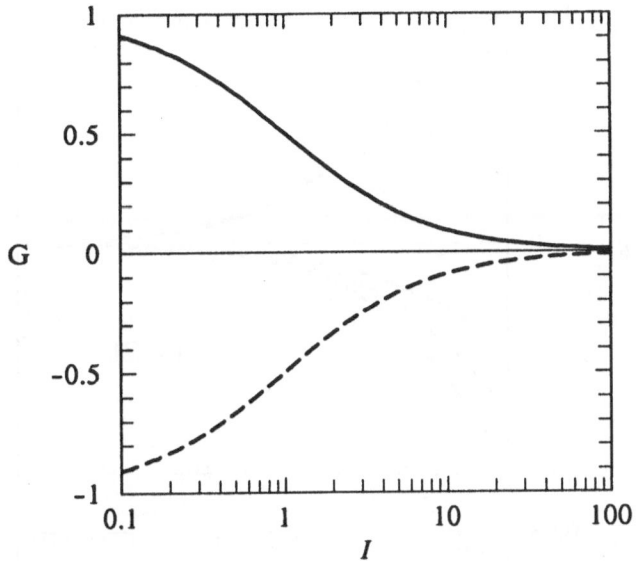

Fig. 3-16. Linear-density gain given by Eq. (76) versus applied light intensity I for gain (solid line) and absorbing (dashed line) media. The parameters are chosen such that the saturation intensity $h\nu/A_g T_1 = 1$, $A_g(JT_1/ed - N_g) = 1$ for the gain case, and $-A_g N_g = -1$ and $J = 0$ for the absorbing case.

N increases and a transparency value. However Eq. (66) has the advantages that it depends explicitly on the temperature, the carrier effective masses, the electric-dipole matrix element, and the field detuning $\hbar\delta$, and it takes the Fermi-Dirac character of the gain into account. It neglects many-body effects and spectral hole burning.

Figure 3-17 illustrates the dependence of g on I for a number of detunings $\hbar\delta$. The curves are generated by solving Eq. (54) for I in steady-state ($\dot{N} = 0$) as

$$I = \frac{\hbar\nu}{2g}\left[\Lambda - \frac{N}{T_1}\right], \tag{77}$$

and plotting g versus I as N is varied from $10^{17} cm^{-3}$ to $10^{19} cm^{-3}$ (2×10^{11} cm^{-2} to 2×10^{13} cm^{-2} for the 2-dimensional case). For $g < 0$ (below transparency, in the absorption region), Λ is set equal to 0, while for $g > 0$, Λ has a positive pump value.

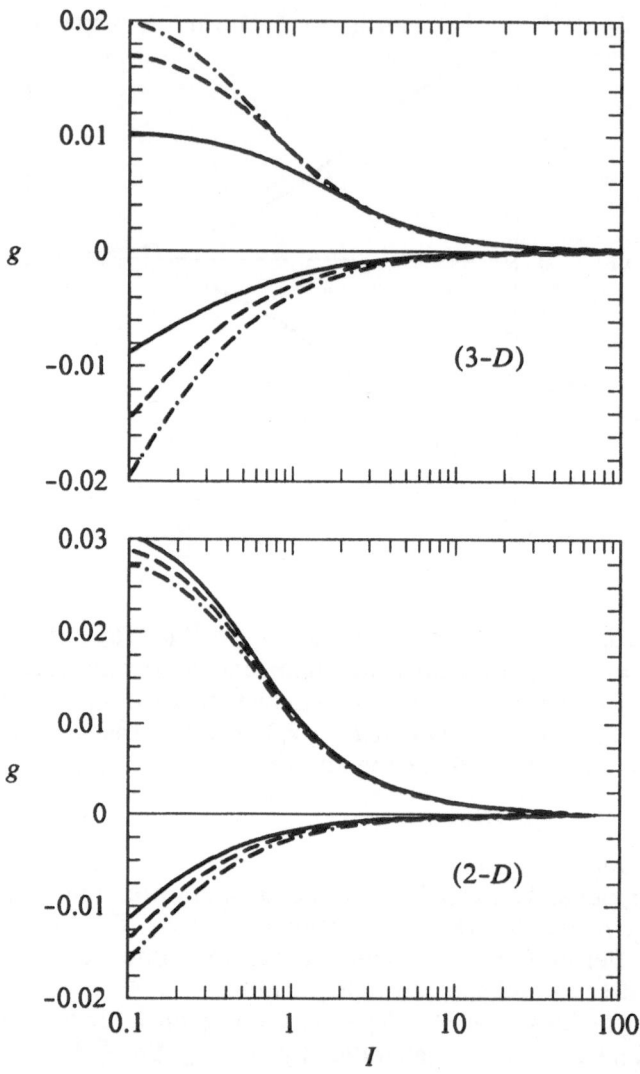

Fig. 3-17. (3-D) Weighted gain g of (66) versus intensity I for the detuning $h\nu - \varepsilon_{g0} = \hbar\delta = 10$ meV (solid lines), 30 meV (dashed lines), and 60 meV (dot-dashed lines). Positive g values are drawn for $\Lambda a_0^3 = .015$ carriers/s, $1/T_1 = .0008$ meV (see Eq. (77)), $g_0 = 1/2\pi^2\varepsilon_R^{3/2}$, $h\nu/2g_0 = 2\pi^2 a_0^3 \varepsilon_R^{3/2}$, $\varepsilon_R = 4.2$ meV, $m_e = 1.167m_r$, and $m_h = 6.669m_r$. Negative g values are drawn for $\Lambda = 0$. (2-D) Corresponding curves for 2-D gain $g = g_0[f_e(\hbar\delta) + f_h(\hbar\delta)-1]$, $g_0 = 1/2\pi\varepsilon_R$, $\Lambda a_0^2 = .015$, and $h\nu/2g_0 = 2\pi a_0^2 \varepsilon_R$.

For $I = 0$, the gain is that for $N = \Lambda T_1$, which is chosen to be quite large so that f_e and f_h have the maximum value of unity. As I increases, N decreases according to Eq. (77), but initially with little effect on g since the Fermi-Dirac distributions remain at their maximum value of 1. Since the holes are heavier, f_h starts to fall off first, later followed by f_e, thereby decreasing g. The Fermi-Dirac unity limit plays a relatively smaller role in the absorption case ($\Lambda = 0$, $g < 0$), since f_h is substantially less than one half over the entire intensity range, and f_e is less than a half for all but very large I.

We see from the curves and from Eq. (77) that for both two and three dimensions, increasing I indefinitly cannot make a transition between absorption and gain at the photon energy $\hbar\nu$. Instead, increasing I increases N for the absorption case and decreases N for the gain case, in either case driving $f_e(\hbar\delta) + f_h(\hbar\delta) - 1$ to zero, i.e., pulling the total chemical potential $\mu = \mu_e + \mu_h$ to $\hbar\delta = \hbar\nu - \varepsilon_{g0}$ and creating transparency at $\hbar\nu$. Since gain occurs for energies in the range $0 < \varepsilon < \mu$, this shows that in the absence of an injection current, a sufficiently strong pump wave of frequency ν can create a gain region below ν. This is due to the fact that carrier-carrier scattering redistributes the electron-hole pairs created by optical pumping above the carrier chemical potentials into the appropriate quasiequilibrium Fermi-Dirac distributions.

For the parameters chosen, the two-dimensional case needs a higher I to saturate the gain since the $\sqrt{\hbar\delta}$ factor is missing. This also leads to a reduced dependence on the detuning δ.

We can gain some further insight by expanding the susceptibility $\chi(N)$ in a first-order Taylor series about the zero-field value N_0. This N_0 can be generated by an injection current, optical pumping above the interaction region, or a combination of both. The first nonlinear term is called $\chi^{(3)}$, which, as we see shortly, is a strong function of N_0, field frequency, and temperature. Our derivation is valid also for the quasiequilibrium many-body theory and for all carrier densities and temperatures as long as the quasiequilibrium approximations are justified, i.e., as long as Fermi-Dirac distributions of the carriers exist in the laser. The main approximation is that the field intensity must be small enough to be treated by a third-order theory.

We write N as

$$N = N_0 + \Delta N \tag{78}$$

and expand the susceptibility $\chi(N)$ in the first-order Taylor series

$$\chi[N(\mathbf{r},t)] \simeq \chi(N_0) + \left.\frac{\partial\chi(N)}{\partial N}\right|_{N_0} \Delta N \ . \tag{79}$$

To find the ΔN resulting from weak field saturation, we expand the total carrier-density equation of motion to first order in $\Delta N \propto |\mathcal{E}|^2$ and take steady state ($\dot{N} = 0$). We write the equation of motion (54) for N as

$$\dot{N} = \Lambda - \Gamma(N) + \frac{\epsilon}{2\hbar}\chi''(N)|\mathcal{E}|^2 \ , \tag{80}$$

where Λ represents the optical pumping or carrier injection, $\Gamma(N)$ is a decay function including both radiative and nonradiative decay, ϵ is the host susceptibility, and $\mathcal{E}$ is the complex electric-field envelope. To lowest order and in steady state, we have

$$\Lambda = \Gamma(N_0) \ . \tag{81}$$

To first order in ΔN and $|\mathcal{E}|^2$, we have

$$0 = \Lambda - \Gamma(N_0) - \Delta N\Gamma_1 + \frac{\epsilon}{2\hbar}\chi''(N_0)|\mathcal{E}|^2 = -\Delta N\Gamma_1 + \frac{\epsilon}{2\hbar}\chi''(N_0)|\mathcal{E}|^2 \ ,$$

where the decay-rate coefficient

$$\Gamma_1 = \left.\frac{\partial\Gamma(N)}{\partial N}\right|_{N=N_0}. \tag{82}$$

This gives the weak-field intensity-induced carrier-density change

$$\Delta N = \frac{\epsilon}{2\hbar\Gamma_1}\chi''(N_0)|\mathcal{E}|^2 \ . \tag{83}$$

Substituting this change into Eq. (79), we have the approximate nonlinear susceptibility

$$\chi(N) \simeq \chi(N_0) + \frac{\epsilon\chi''(N_0)}{2\hbar\Gamma_1}\left.\frac{\partial\chi(N)}{\partial N}\right|_{N=N_0}|\mathcal{E}|^2 \ . \tag{84}$$

This gives the first two terms in the intensity expansion of the susceptibility as

$$\chi^{(1)}(N_0) = \chi(N_0) ,$$
(85)

$$\chi^{(3)}(N_0) = \frac{\epsilon \chi''(N_0)}{2\hbar\Gamma_1} \frac{\partial \chi(N)}{\partial N}\bigg|_{N=N_0} .$$
(86)

Figure 3-18a illustrates $\chi^{(3)}$ as a function of tuning above the band gap along with the gain $\propto -\chi''$ for three values of the total carrier density N for a bulk semiconductor according to the free-carrier model. In all cases, the fact that $\chi^{(3)}$ is proportional to $\chi''(N_0)$ forces the real and imaginary parts of $\chi^{(3)}$ to cross zero at the gain crossover. We see that within the gain region both $\mathrm{Re}\{\chi^{(3)}\}$ and $\mathrm{Im}\{\chi^{(3)}\}$ have similar shape to the gain $-\chi''$; however above the gain region the index turns around.

Figure 3-18b illustrates $\chi^{(3)}$ and the gain for three values of the temperature. As for Fig. 3-17a, the curves cross over at the gain crossover. Figures 3-19a and 3-19b are the same as Figs. 3-17a and 3-17b, respectively, except that a 2-D model is used instead of 3-D.

3-6. Carrier-Induced Refractive Index

For the free carrier theory, the background host variation represented by the -1 in Eq. (53) can be included in the host index n, that is, the carrier-induced refractive index reduces to

$$\delta n = - \frac{n}{2\epsilon_b \hbar\gamma V} \sum_{\mathbf{k}} |\mu_{\mathbf{k}}|^2 [f_{e\mathbf{k}} + f_{h\mathbf{k}}] \, \mathcal{L}(\omega_{\mathbf{k}} - \nu) \frac{\omega_{\mathbf{k}} - \nu}{\gamma} .$$
(87)

This approach avoids the problem that the $\Sigma_{\mathbf{k}}$ as given by Eq. (57) does not converge due to the long unphysical tails of the $\mathcal{L}(\omega_{\mathbf{k}} - \nu)(\omega_{\mathbf{k}} - \nu)/\gamma$ factor. Convergence is achieved with the dipole matrix element (see Sec. 6-3)

$$\mu_{\mathbf{k}} = \frac{\mu_{(\mathbf{k}=0)}}{1 + \epsilon/\epsilon_g} ,$$
(88)

but this alone still weights high energies too much. The problem stems

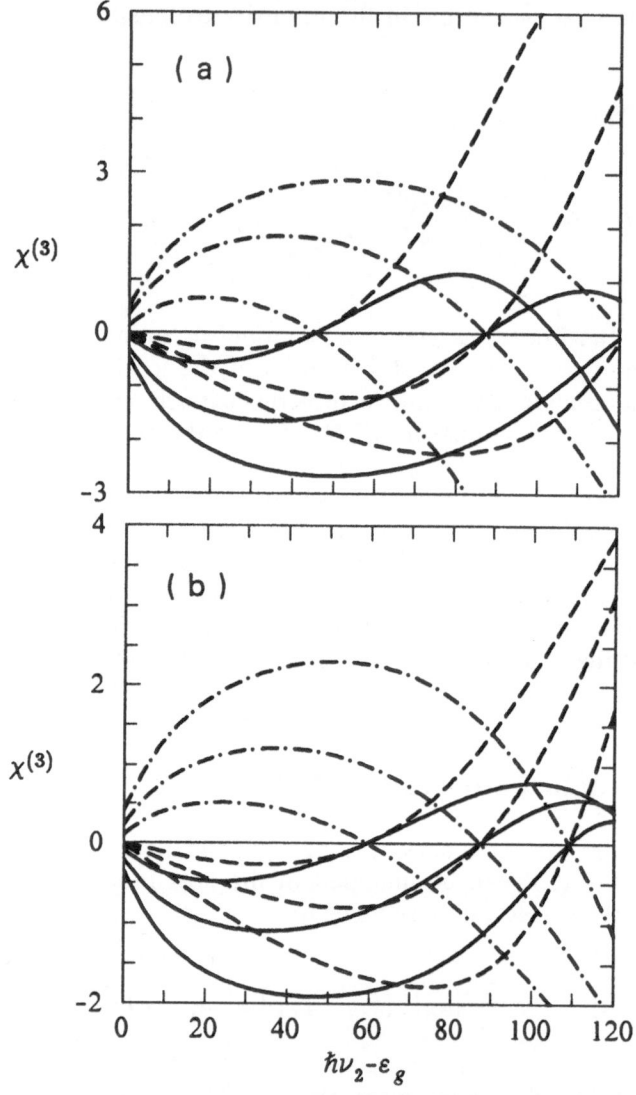

Fig. 3-18. (a) $\mathrm{Re}\{\chi^{(3)}\}$ (solid lines), $\mathrm{Im}\{\chi^{(3)}\}$ (dashed lines), and gain (dot-dashed lines) versus detuning of a weakly saturating field according to the free-carrier model for a bulk medium. The parameters chosen are $N = 2\times10^{18}$, 3×10^{18}, and 4×10^{18} cm^{-3} (in order of increasing gain crossover), $T = 300$ K, $m_e = 1.176m_r$, $m_h = 6.669m_r$, and $\gamma = 3$ meV. (b) Same as (a), but for $T = 200$ K, 300 K, and 400 K (in order of decreasing gain crossover) and $N = 3\times10^{18}$ cm^{-3}.

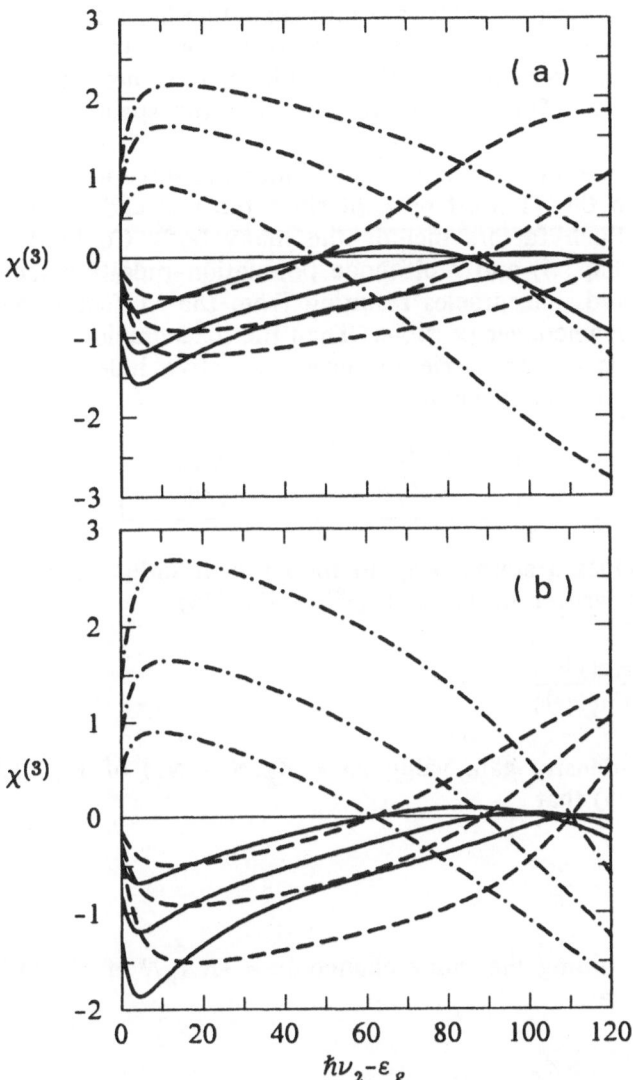

Fig. 3-19. (a) Re{$\chi^{(3)}$} (solid lines), Im{$\chi^{(3)}$} (dashed lines), and gain (dot-dashed lines) versus detuning of a weakly saturating field according to the free-carrier model for a 2-D medium. The parameters chosen are N = 2.486×10¹² (solid lines), 3.73×10¹² (dashed), 4.97×10¹² (dot-dashed) cm^{-3} (corresponds to N = 2×, 3×, 4×10¹⁸ cm^{-3}, respectively) (in order of increasing gain crossover), T = 300 K, m_e = 1.176m_r, m_h = 6.669m_r, and γ = 3 meV. (b) same as (a), but for T = 200 K, 300 K, and 400 K (in order of decreasing gain crossover)

both from the simple complex-Lorentzian lineshape factor and from the limitations of the effective-mass approximation. In real semiconductors the bands do not remain parabolic for higher k-values; they usually flatten out, leading to a finite bandwidth limiting the spectral range of optical transitions.

However if other complex factors enter the susceptibility summation in Eq. (50), then the -1 must be kept since the real and imaginary parts get mixed up. Such factors include the many-body Coulomb enhancement factor (see Chap. 4) and multimode population-pulsation factors (see Sec. 8-1). To avoid inaccuracies resulting from the -1 factor, one deals with index *changes* whenever possible. Then the inaccuracies cancel out.

Discussions of the carrier-induced refractive index often involve the linewidth enhancement factor

$$\alpha = \frac{\partial \chi'/\partial N}{\partial \chi''/\partial N} = -\frac{K}{n}\frac{\partial(\delta n)/\partial N}{\partial g/\partial N} \qquad (89)$$

since it provides a simple way to model such index contributions. This factor can be written in terms of $\chi^{(3)}$ of Eq. (86) as

$$\alpha = \frac{\text{Re}\{\chi^{(3)}\}}{\text{Im}\{\chi^{(3)}\}} . \qquad (90)$$

For a linear-density gain value $2g = A_g(N - N_g)$ of Eq. (1.8), we have using Eq. (2.75) that

$$\frac{\partial \chi''}{\partial N} = -\frac{A_g}{K} . \qquad (91)$$

Similarly combining the index change $\delta n = -RA_g N$ of Eq. (1.21) with Eq. (2.78), we have

$$\frac{\partial \chi'}{\partial N} = -\frac{RA_g}{K} . \qquad (92)$$

Dividing this by Eq. (91), we find that the antiguiding factor R is just the linewidth enhancement factor α

$$\alpha = R . \qquad (93)$$

If we lump the host index n into the N_g term, we can write the susceptibility as

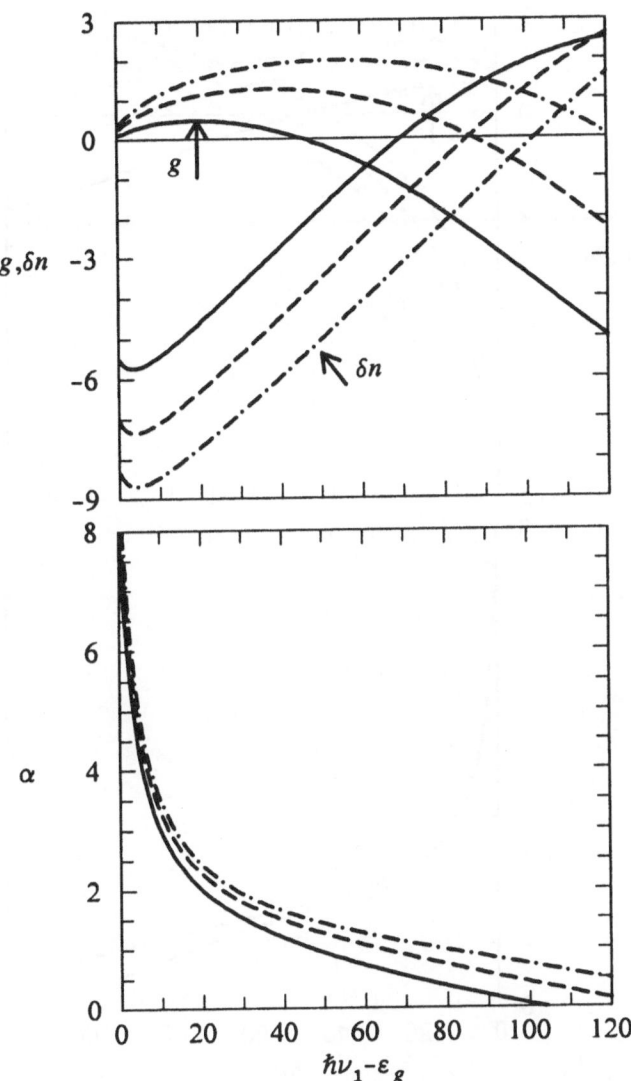

Fig. 3-20. Carrier-induced index change δn (87), gain g (52), and linewidth enhancement factor (89) versus detuning $\hbar\nu - \varepsilon_g$ above the band gap according to the free-carrier model for a bulk medium. $N = 2\times10^{18}$ (solid lines), 3×10^{18} (dashed), 4×10^{18} (dot-dashed) cm^{-3}, $T = 300\ K$, $m_e = 1.176m_r$, $m_h = 6.669m_r$, and $\gamma = 3\ meV$.

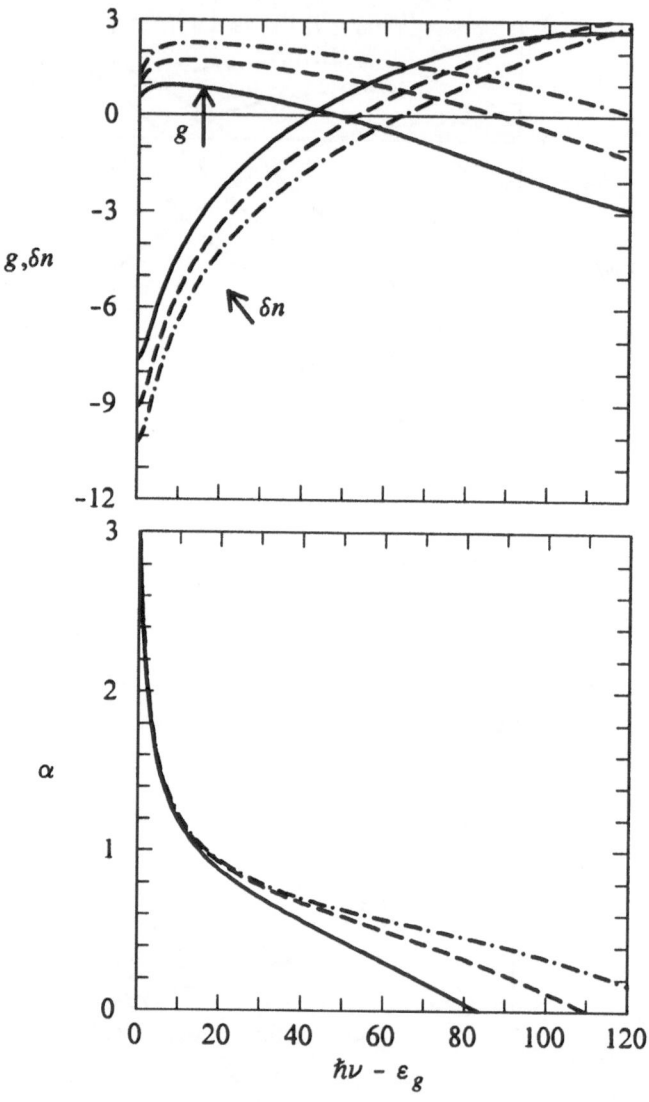

Fig. 3-21. Carrier-induced index change δn (87), gain g (52), and linewidth enhancement factor (89) versus detuning $\hbar\nu - \varepsilon_g$ above the band gap according to the free-carrier model for an idealized 2D gain medium. $N = 2.486 \times 10^{12}$ (solid lines), 3.73×10^{12} (dashed), 4.97×10^{12} (dot-dashed) cm^{-3} (corresponds to $2\times$, $3\times$ 4×10^{18} cm^{-3}, respectively), $T = 300 \ K$, $m_e = 1.176 \ m_r$, $m_h = 6.669 \ m_r$, and $\gamma = 3 \ meV$.

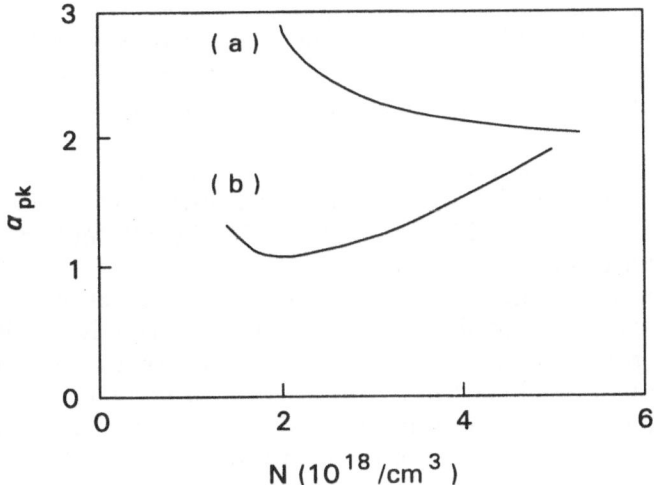

Fig. 3-22. α at the peak gain frequency *vs* carrier density for (*a*) bulk GaAs and (*b*) 5 *nm* $In_{.1}Ga_{.9}As$-GaAs quantum well.

$$\chi \simeq (i + \alpha)\chi'' . \tag{94}$$

These relationships are only valid as long as the susceptibility is a linear function of the total carrier density, which is typically only true for the gain peak of a bulk-gain medium. Nevertheless, the linewidth enhancement factor is an interesting quantity to study, since many problems, such as the laser linewidth itself, involve small changes in the susceptibility, for which α as a function of N, T, and tuning is an accurate measure.

We illustrate the carrier-induced refractive index δn, the gain g, and the linewidth enhancement factor α in Fig. 3-20 for a bulk medium and in Fig. 3-21 for a 2-*D* medium for three total carrier densities. The differences between the curves between Figs. 3-20 and 3-21 are due to the differences in the 3*D* bulk and the 2*D* quantum-well densities of states (in Chap. 5 we see other differences as well, due to many-body effects). For example, the relatively sharp increase in δn for quantum well is due to the abrupt (step function) rise in the 2*D* density of states.

As we mentioned before, we often do not need to compute the absolute phase shift since only the phase shift changes enter our expressions. For example, the linewidth or antiguiding parameter α of Eq. (89) is the change in the phase shift with respect to carrier density divided by the corresponding change in the gain. Figures 3-20 and 3-21 include α spectra computed using Eq. (89). The value of α at the peak gain is a good parameter for comparing the importance of the carrier-induced phase shift in determining the laser performance under different experimental conditions. The larger the value of α, the greater is the effect of the carrier-induced phase shift on the linewidth or the filamentation of a laser. Figure 3-22 depicts the dependence of $\alpha(\nu_{pk})$ on carrier density. Note that α is a function of carrier density, which contradicts a major assumption of the phenomenological model.

It is tempting to use Fig. 3-22 to come to the conclusion that the quantum-well laser has a smaller α. While this statement is usually true, the free-carrier results used in arriving at this conclusion do not describe the entire picture. Most importantly, they neglect many-body and bandstructure effects, which significantly affect the behavior of the gain medium. We elaborate on these features in the next three chapters.

Chapter 4
COULOMB EFFECTS

In the previous chapters, we discuss simple models for semiconductor gain, which occurs because of an inversion between the conduction and valence bands. To obtain a more realistic description of the gain medium, we need to include the Coulomb interaction between the charge carriers. The Coulomb potential is attractive between carriers in different bands (interband interaction) and repulsive for carriers in the same band (intraband interaction). Since Coulomb interaction processes always involve more than one carrier, the resulting effects are often called *many-body effects*, and quantum mechanical many-body techniques have to be used to analyze these phenomena.

As mentioned in Chap. 3, one of the most important consequences of the intraband Coulomb interaction is the rapid equilibration of the electrons and holes into quasiequilibrium Fermi-Dirac distributions. Under typical laser conditions the carrier-carrier scattering is also the dominant contributer to optical dephasing, which is the decay of the polarization of the medium. Another important many-body effect is plasma screening, which is the carrier-density-dependent weakening of the Coulomb interaction potential due to the presence of background charge carriers. These many-body Coulomb interactions significantly modify the gain and refractive index spectra. The spectral positions of the gain and index spectra are shifted through bandgap renormalization, and the shapes of the gain and index spectra are modified through interband Coulomb enhancement.

A schematic outline of a many-body theory is sketched in Fig. 4-1. Beginning with the many-body Hamiltonian for the interacting carriers, we derive equations of motion for the electron probability $n_{e\mathbf{k}}$ of Eq. (3.4), the hole probability $n_{h\mathbf{k}}$ (3.5), and the interband polarization $p_{\mathbf{k}}$ (3.3). The equations of motion couple these quantities to ensemble averages involving products of four particle operators, which, in turn, are coupled to ensemble averages of six operator products, and so on. This process produces a hierarchy of an infinite number of coupled differential equations involving ensemble averages of products of ever higher numbers of field operators. We can truncate the hierarchy by factorizing higher-order ensemble averages into products of the second-order averages, $n_{e\mathbf{k}}$, $n_{h\mathbf{k}}$ and $p_{\mathbf{k}}$. Factorizing the equations of motion for $n_{e\mathbf{k}}$, $n_{h\mathbf{k}}$ and $p_{\mathbf{k}}$ in this way, we find the

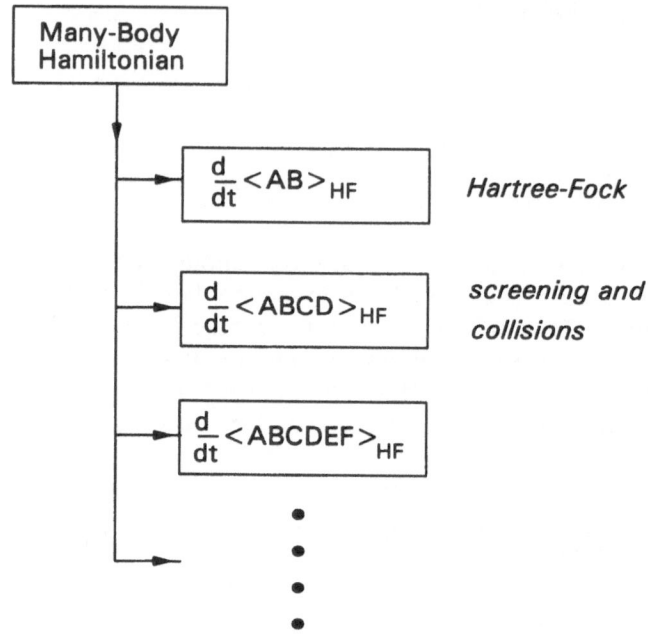

Fig. 4-1. A schematic outline of a many-body theory. *AB* represents a product of two creation or annihilation operators. The equations are split into the Hartree-Fock (HF) part and the higher-order correlations, $\langle ABCD \rangle$, $\langle ABCDEF \rangle$ and so on. The higher order correlations describe effects such as plasma screening and carrier collisions. The Hartree-Fock equations combined with the collision contributions give the semiconductor Bloch equations.

Hartree-Fock equations, so named because Hartree and Fock made this kind of approximation in studying the many-electron atom. Two important many-body effects appear in these Hartree-Fock equations, namely bandgap renormalization and Coulomb enhancement.

Corrections to the Hartree-Fock equations obtained by delaying the factorization procedure to the next level of the hierarchy give the collision and screening terms. At this stage, the derivation and its results are sufficiently complicated that we need to make additional approximations to obtain a usable laser theory. An important aspect of many-body theory is how to make these approximations consistently. One result of the derivation is the highly nonlinear Boltzmann equation for the intraband relaxation carrier densities.

The Hartree-Fock equations for $n_{e\mathbf{k}}$, $n_{h\mathbf{k}}$ and $p_{\mathbf{k}}$, combined with some form of the collision and screening contributions form a set of equations that play the same role as the optical Bloch equations in two-level systems. Therefore it has become customary to refer to them as the *semiconductor Bloch equations*. These equations reduce to atomic Bloch equations that have no decay when we drop the Coulomb interaction potential altogether, that is, in the limit of no carrier-carrier scattering, no plasma screening, no bandgap renormalization, and no Coulomb enhancement. In fact, since in the absence of carrier-carrier scattering, the probabilities ρ_{22} and ρ_{11} of Sec. 3-1 vanish, we see from Eq. (3.33) that the upper-level probability ρ_{33} is given simply by $f_{e\mathbf{k}}$, which equals $f_{h\mathbf{k}}$ for zero Coulomb interaction, and the lower-level probability ρ_{00} is given by $1 - f_{e\mathbf{k}}$. This collision-free limit is interesting, but it is irrelevant for semiconductor lasers, since various kinds of decay play major roles in these and most other lasers.

For the limiting case of low excitation conditions, the semiconductor Bloch equations contain the "hydrogen-like" Wannier equation, which describes the fundamental electron-hole pair or excitonic properties of a dielectric medium. From the semiconductor Bloch equations, one can derive the optical absorption and gain spectra at both low and high carrier densities, respectively. The excitonic states play an important role in the former, while the Coulomb effects may be viewed as a correction to the free-carrier theory in the latter.

The derivation of the plasma screening contributions using the fourth-order truncation discussed above is very complicated. Fortunately a somewhat phenomenological approach [e.g., see Harrison (1980), Ashcroft and Mermin (1976), and Haug and Koch (1993)] arrives at similar results. We use this simpler approach as outlined in Fig. 4-2, which also lets us skip the iterative step needed to incorporate plasma screening effects. We begin by presenting the many-body Hamiltonian in Sec. 4-1 and then derive the plasma screening contributions phenomenologically in Sec. 4-2. In Sec. 4-3, we use these results in modifying the many-body Hamiltonian to show the screening contributions explicitly, and we use the Hartree-Fock approximation to get the semiconductor Bloch equations.

The many-body effects of bandgap renormalization and Coulomb enhancement are discussed in Sec. 4-4 and 4-5, respectively. Section 4-5 treats only the low carrier-density limit to best illustrate the physical origin of the Coulomb enhancement phenomenon. Finally, Sec. 4-6 discusses the carrier-carrier collisions leading to the Boltzmann equation and the corresponding population and polarization relaxation in the semiconductor Bloch equations. In this section, we also include a brief outline of carrier-phonon collisions. Through this mechanism, carrier distributions dissipate energy and equilibrate with the lattice. The high carrier-density limit and corresponding approximations to the results of Sec. 4-5 are treated in Chap. 5. There we discuss many-body gain and carrier-induced refractive index for bulk semiconductors as well as how to treat many-body effects in quantum-well structures.

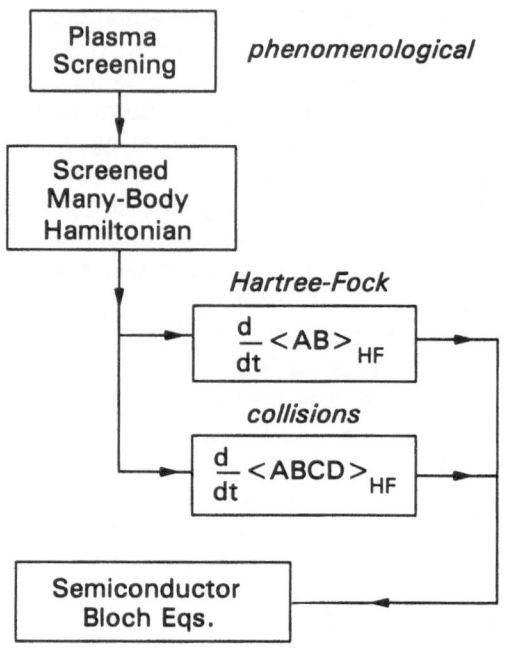

Fig. 4-2. We use an approach that takes advantage of a pheno-menological derivation of plasma screening, which is incorporated directly into the many-body Hamiltonian to give the screened Hartree-Fock equations. The remaining steps needed to arrive at the semiconductor Bloch equations are similar to those in Fig. 4-1.

4-1. Many-Body Hamiltonian

The Hamiltonian of an interacting electron system in a dielectric medium is

$$\mathcal{H} = \mathcal{H}_{kin} + \mathcal{H}_C , \tag{1}$$

where $\mathcal{H}_{kin}$ is the kinetic energy part of Eqs. (2.90) - (2.93), and $\mathcal{H}_C$ of Eq. (2.95) describes the Coulomb interaction among the electrons in the various energy bands. In order not to overburden the analysis with indices and band-structure details, we assume here the ideal two-band model of a semiconductor described by the total Hamiltonian

$$\mathcal{H} = \sum_{\mathbf{k}} \varepsilon_{c\mathbf{k}} a_{c\mathbf{k}}^{\dagger} a_{c\mathbf{k}} + \sum_{\mathbf{k}} \varepsilon_{v\mathbf{k}} a_{v\mathbf{k}}^{\dagger} a_{v\mathbf{k}}$$

$$+ \frac{1}{2} \sum_{\mathbf{k},\mathbf{k}',\mathbf{q} \neq 0} V_q \left[a_{c,\mathbf{k}+\mathbf{q}}^{\dagger} a_{c,\mathbf{k}'-\mathbf{q}}^{\dagger} a_{c\mathbf{k}'} a_{c\mathbf{k}} + a_{v,\mathbf{k}+\mathbf{q}}^{\dagger} a_{v,\mathbf{k}'-\mathbf{q}}^{\dagger} a_{v\mathbf{k}'} a_{v\mathbf{k}} \right.$$

$$\left. - 2 a_{c,\mathbf{k}+\mathbf{q}}^{\dagger} a_{v,\mathbf{k}'-\mathbf{q}}^{\dagger} a_{v\mathbf{k}'} a_{c\mathbf{k}} \right] , \tag{2}$$

where

$$V_q = \frac{1}{V} \int d^3r \, e^{-i\mathbf{q}\cdot\mathbf{r}} \, V(r) = \frac{4\pi e^2}{\epsilon_b V q^2} \tag{3}$$

is the Fourier transform of the Coulomb potential energy $V(r)$, and V (with no argument or subscript) is the volume of the semiconductor medium. As in Chap. 3, the momentum index $\mathbf{k}$ includes the spin index so that summation over $\mathbf{k}$ implies summing over the two possible spin orientations, and the subscript includes a comma only when ambiguity might otherwise arise.

The first two terms in the Hamiltonian (2) are the kinetic energies of the electrons in the conduction and valence bands. The next two terms describe the Coulomb interaction of the electrons within each band, i.e., the intraband Coulomb interaction, and the last term describes the interband Coulomb interaction between carriers in the valence and conduction band. Due to the Fermi properties of the creation and annihilation operators in the formalism of second quantization, all these Coulomb terms include both the direct and the exchange interactions, which are shown schematically in Fig. 2-9.

In the derivation of Eq. (2), we use the fact that the Coulomb scattering does not alter the spin orientation of an electron and that the $q=0$ contribution, which diverges, is cancelled by the $q=0$ terms from the electron-ion and ion-ion Coulomb potentials. We omit Coulomb terms that fail to conserve the number of electrons in each band, since such terms involve interband transitions, which are very unfavorable energetically.

For convenience, we transform the Hamiltonian into the electron-hole representation. For this purpose we use the electron and hole operators of Eqs. (2.86) and (2.84) in Eq. (2) and restore normal ordering of all creation and annihilation operators. This gives the Hamiltonian for interacting electrons and holes

$$\mathcal{H} = \sum_{\mathbf{k}} [(\varepsilon_{e\mathbf{k}} + \varepsilon_{g0}) \, a_{\mathbf{k}}^{\dagger} a_{\mathbf{k}} + \varepsilon_{h\mathbf{k}} \, b_{-\mathbf{k}}^{\dagger} b_{-\mathbf{k}}] + \frac{1}{2} \sum_{\mathbf{k},\mathbf{k}',\mathbf{q} \neq 0} V_q \, \Big[$$

$$a_{\mathbf{k}+\mathbf{q}}^{\dagger} a_{\mathbf{k}'-\mathbf{q}}^{\dagger} a_{\mathbf{k}'} a_{\mathbf{k}}$$

$$+ \, b_{\mathbf{k}+\mathbf{q}}^{\dagger} b_{\mathbf{k}'-\mathbf{q}}^{\dagger} b_{\mathbf{k}'} b_{\mathbf{k}} - 2 a_{\mathbf{k}+\mathbf{q}}^{\dagger} b_{\mathbf{k}'-\mathbf{q}}^{\dagger} b_{\mathbf{k}'} a_{\mathbf{k}} \Big] , \qquad (4)$$

where constant terms have been dropped because they only lead to a shift of the reference energy, which is irrelevant since only energy differences can be measured. The single particle energies in Eq. (4) are

$$\varepsilon_{e\mathbf{k}} = \frac{\hbar^2 k^2}{2m_e} , \qquad (5)$$

$$\varepsilon_{h\mathbf{k}} = -\varepsilon_{v\mathbf{k}} + \sum_{\mathbf{q} \neq 0} V_q = \frac{\hbar^2 k^2}{2m_h} , \qquad (6)$$

where the term containing V_q in $\varepsilon_{h\mathbf{k}}$ originates from the replacement of valence-band electron operators by hole operators in the interaction term

$$\frac{1}{2} \sum_{\mathbf{k},\mathbf{k}',\mathbf{q} \neq 0} V_q \, a_{v,\mathbf{k}+\mathbf{q}}^{\dagger} a_{v,\mathbf{k}'-\mathbf{q}}^{\dagger} a_{v\mathbf{k}'} a_{v\mathbf{k}}$$

of Eq. (4). Equation (6) differs from the free-carrier result [see Eq. (2.93) and following] in that the kinetic energy and therefore the hole effective mass include the Coulomb energy of the full valence band.

4-2. Plasma Screening

We begin our study of many-body effects with plasma screening, which is one of the most important consequences of the many-body Coulomb interaction. We follow an approach that uses arguments from classical electrodynamics and quantum mechanics. Given an electron at the origin of our coordinate system, we wish to know what effect this electron has on its surroundings. To find out, we introduce a test charge, i.e., a change sufficiently small as to cause negligible perturbation, at the location of interest. In a vacuum, the electrostatic potential due to the electron is

$\phi(r) = 4\pi e/r$. However in a semiconductor, there is a background dielectric constant ϵ_b, which is due to everything in the semiconductor in the absence of the carriers themselves, and there is the carrier distribution that is changed by the presence of our electron at the origin (see Fig. 4-3). The new carrier distribution, $\langle n_s(\mathbf{r}) \rangle$, in turn changes the electrostatic potential due to our electron. We denote the carrier density distribution as an expectation value since we plan to calculate it quantum mechanically.

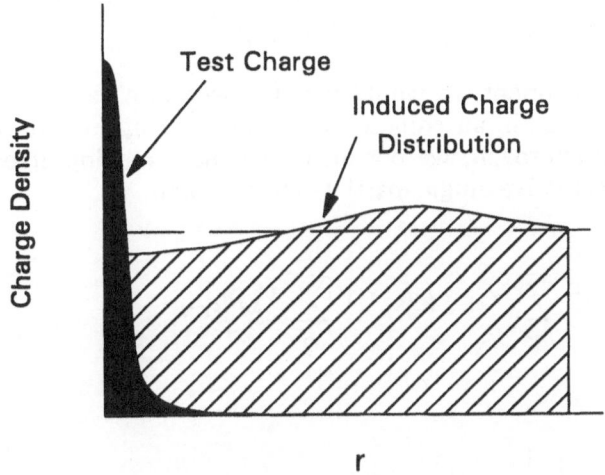

Fig. 4-3. An electron at the origin induces a change in the carrier distribution. The electron density is $n_e(\mathbf{r}) = \delta^3(\mathbf{r})$ and the new carrier distribution is $\langle n_s(\mathbf{r}) \rangle$, where the subscript s stands for screening, which is the net effect of the induced carrier distribution change.

To derive the induced carrier distribution, we first simplify the problem by assuming that the screening effects of an electron-hole plasma equal the sum of the effects resulting from the separate electron and hole plasmas. As such we neglect excitonic screening, which is not a bad approximation for the elevated carrier densities present in a semiconductor laser. Starting with the electron plasma, we note from Eqs. (2.97) and (2.98) that the corresponding quantum-mechanical operator for the screened electron charge distribution is

$$n_s(\mathbf{r}) = \frac{1}{V} \sum_{\mathbf{k},\mathbf{k'}} e^{i(\mathbf{k}-\mathbf{k'})\cdot\mathbf{r}} \, a_{\mathbf{k'}}^{\dagger} a_{\mathbf{k}} = \sum_{\mathbf{q}} n_{sq} \, e^{i\mathbf{q}\cdot\mathbf{r}} \,, \qquad (7)$$

where the Fourier transform of the charge operator is given by

$$n_{sq} = \frac{1}{V} \sum_{\mathbf{k}} a_{\mathbf{k}-\mathbf{q}}^{\dagger} a_{\mathbf{k}} \,. \qquad (8)$$

In a rigorous treatment we would use the electronic part of the many-body Hamiltonian to obtain an equation of motion for n_{sq}. With our more phenomenological approach, we postulate that the screening effects can be included in an effective single-particle Hamiltonian

$$\mathcal{H}_{eff} = \sum_{\mathbf{k}} \varepsilon_{e\mathbf{k}} \, a_{\mathbf{k}}^{\dagger} a_{\mathbf{k}} + V \sum_{\mathbf{q}} V_{sq} n_{s,-\mathbf{q}} \,, \qquad (9)$$

where

$$V_{sq} = \frac{1}{V} \int d^3 r \, V_s(r) \, e^{-i\mathbf{q}\cdot\mathbf{r}} \qquad (10)$$

with

$$V_s(r) = e\phi_s(r) \,, \qquad (11)$$

and $\phi_s(r)$ is the screened electrostatic potential.

With the effective Hamiltonian (9), we get the equation of motion

$$i\hbar \frac{d}{dt} a_{\mathbf{k}-\mathbf{q}}^{\dagger} a_{\mathbf{k}} = [a_{\mathbf{k}-\mathbf{q}}^{\dagger} a_{\mathbf{k}}, \mathcal{H}_{eff}]$$

$$= (\varepsilon_{\mathbf{k}} - \varepsilon_{\mathbf{k}-\mathbf{q}}) \, a_{\mathbf{k}-\mathbf{q}}^{\dagger} a_{\mathbf{k}} + \sum_{\mathbf{p}} V_{sp} (a_{\mathbf{k}-\mathbf{q}}^{\dagger} a_{\mathbf{k}+\mathbf{p}} - a_{\mathbf{k}-\mathbf{q}-\mathbf{p}}^{\dagger} a_{\mathbf{k}}) \,. \qquad (12)$$

Taking the expectation value and keeping only slowly varying terms, namely those with $\mathbf{p}=-\mathbf{q}$, we get

$$i\hbar \frac{d}{dt} \langle a^\dagger_{\mathbf{k-q}} a_{\mathbf{k}} \rangle = (\varepsilon_{\mathbf{k}} - \varepsilon_{\mathbf{k-q}}) \langle a^\dagger_{\mathbf{k-q}} a_{\mathbf{k}} \rangle + V_{sq} (n_{\mathbf{k-q}} - n_{\mathbf{k}}) . \quad (13)$$

We suppose that $\langle a^\dagger_{\mathbf{k-q}} a_{\mathbf{k}} \rangle$ has a solution of the form $e^{(\delta - i\omega)t}$, where the infinitesimal δ indicates that the perturbation has been switched on adiabatically, i.e., that we had a homogeneous plasma at $t = -\infty$. We further suppose that the induced charge distribution follows this response. This transforms Eq. (13) to

$$\langle a^\dagger_{\mathbf{k-q}} a_{\mathbf{k}} \rangle = V_{sq} \frac{n_{\mathbf{k-q}} - n_{\mathbf{k}}}{\hbar(\omega + i\delta) + \varepsilon_{\mathbf{k-q}} - \varepsilon_{\mathbf{k}}} \quad (14)$$

so that

$$\langle n_{sq} \rangle = \frac{V_{sq}}{V} \sum_{\mathbf{k}} \frac{n_{\mathbf{k-q}} - n_{\mathbf{k}}}{\hbar(\omega + i\delta) + \varepsilon_{\mathbf{k-q}} - \varepsilon_{\mathbf{k}}} . \quad (15)$$

The induced charge distribution is a source in Poisson's equation

$$\nabla^2 \phi_s(r) = -\frac{4\pi e}{\epsilon_b} [n_e(\mathbf{r}) + \langle n_s(\mathbf{r}) \rangle] . \quad (16)$$

The Fourier transform of this equation is

$$\phi_{sq} = \frac{4\pi e}{\epsilon_b q^2} \left[\frac{1}{V} + \langle n_{sq} \rangle \right] , \quad (17)$$

where for a point charge at the origin

$$n_{eq} = \frac{1}{V} \int d^3 r \, \delta^3(\mathbf{r}) \, e^{-i\mathbf{q}\cdot\mathbf{r}} = \frac{1}{V} . \quad (18)$$

Using $V_{sq} \equiv e\phi_{sq}$, we substitute Eq. (15) into (17) and solve for V_{sq} to find

$$V_{sq} = \frac{V_q}{1 - V_q \sum_{\mathbf{k}} \dfrac{n_{\mathbf{k-q}} - n_{\mathbf{k}}}{\hbar(\omega + i\delta) + \varepsilon_{\mathbf{k-q}} - \varepsilon_{\mathbf{k}}}} , \quad (19)$$

where the unscreened potential V_q is given by Eq. (3). Repeating the derivation for the hole plasma, and adding the electron and hole contributions, we find the screened Coulomb potential energy between carriers

$$V_{sq} = \frac{V_q}{\epsilon_q(\omega)} ,$$ (20)

where the longitudinal dielectric function $\epsilon_q(\omega)$ is given by

$$\epsilon_{\mathbf{q}}(\omega) = 1 - V_q \sum_{\substack{\mathbf{k} \\ \alpha=e,h}} \frac{n_{\alpha,\mathbf{k}-\mathbf{q}} - n_{\alpha\mathbf{k}}}{\hbar(\omega + i\delta) + \epsilon_{\alpha,\mathbf{k}-\mathbf{q}} - \epsilon_{\alpha\mathbf{k}}}$$

Lindhard formula
for the longitudinal dielectric function (21)

This formula describes a complex retarded dielectric function, i.e., the poles are in the lower complex frequency plane, and it includes spatial dispersion (q dependence) and spectral dispersion (ω dependence). In many practical situations, the Lindhard formula (21) is too complicated to use because of its continuum of poles. Fortunately, it is often sufficient to use a simplified version.

To see how to obtain a proper simplification, we look at the long wavelength ($q \to 0$) limit of the Lindhard formula. We assume a quasi-equilibrium system, where $n_{\alpha\mathbf{k}} = f_{\alpha\mathbf{k}}$ is the Fermi-Dirac distribution function. We expand $\epsilon_{\alpha,\mathbf{k}-\mathbf{q}}$ and $f_{\alpha,\mathbf{k}-\mathbf{q}}$ around $q = 0$ to find

$$\epsilon_{\alpha,\mathbf{k}-\mathbf{q}} - \epsilon_{\alpha\mathbf{k}} = \frac{\hbar^2}{2m_\alpha}(k^2 - 2\mathbf{k}\cdot\mathbf{q} + q^2) - \frac{\hbar^2 k^2}{2m_\alpha} \simeq - \sum_i \frac{\hbar^2 k_i q_i}{m_\alpha}$$ (22)

$$f_{\alpha,\mathbf{k}-\mathbf{q}} - f_{\alpha\mathbf{k}} = f_{\alpha\mathbf{k}} - \sum_j q_j \frac{\partial}{\partial k_j} f_{\alpha\mathbf{k}} + \cdots - f_{\alpha\mathbf{k}} \simeq - \sum_j q_j \frac{\partial f_{\alpha\mathbf{k}}}{\partial k_j} .$$ (23)

Inserting these expansions into Eq. (21), expanding the resulting denominator, noticing that $\Sigma_k \, \partial f_{\alpha\mathbf{k}}/\partial k = 0$, and integrating the remaining term by parts, we get the classical (or Drude) dielectric function

$$\epsilon_{q=0}(\omega) = 1 - \frac{\omega_{pl}^2}{\omega^2} , \tag{24}$$

where the square of the electron-hole plasma frequency is given by

$$\omega_{pl}^2 = \frac{4\pi N e^2}{\epsilon_b m_r} = 16\pi N a_0^3 \, (\epsilon_R / \hbar)^2 , \tag{25}$$

$N = V^{-1} \Sigma_k f_k$ is the electron or hole density, and m_r is the reduced electron-hole mass.

The classical result (24) is in many aspects too simplistic for a realistic description of many-body plasma screening effects in both active and passive semiconductors. Instead, the so-called *plasmon-pole approximation* replaces the continuum of electron-pair excitations, represented by the continuum of poles in the Lindhard formula, by a single effective plasmon pole. In this approximation, we replace the inverse Lindhard dielectric function by

$$\frac{1}{\epsilon_q(\omega)} = 1 + \frac{\omega_{pl}^2}{(\omega + i\gamma)^2 - \omega_q^2} , \tag{26}$$

which has the same structure as the long-wavelength plasma result (24), but instead of ω_{pl} in the denominator, it has the effective plasmon frequency ω_q defined by

$$\omega_q^2 = \omega_{pl}^2 (1 + q^2/\kappa^2) + C(\hbar q^2/4m_r)^2 . \tag{27}$$

Here C is a numerical constant usually taken between 1 and 4, and κ is the inverse static screening length

$$\kappa = \left[\frac{4\pi e^2}{\epsilon_b} \sum_{\alpha = e,h} \frac{\partial N}{\partial \mu} \right]^{1/2} . \tag{28}$$

Without discussing details of the derivation of Eq. (27), we just mention here that ω_q has been determined such that the dielectric function Eq. (21) fulfills certain sum rules [Mahan (1981), Lundquist (1967), Haug and Koch (1993)].

For many practical applications, one ignores the damped response of the screening represented by the γ in the dielectric function (26). In this "static" plasmon-pole approximation, the screened Coulomb potential (20) is given by

$$V_{sq} = V_q \left[1 - \frac{\omega_{pl}^2}{\omega_q^2} \right] , \tag{29}$$

where ω_q^2 is given by Eq. (27). This often allows analytic results, or at least much simpler numerical results for the effects of plasma screening.

4-3. Semiconductor Bloch Equations

Using the results of the previous section, we can phenomenologically include the effects of plasma screening by replacing the bare Coulomb potential V_q in Eq. (2) by the screened Coulomb potential V_{sq}, that is

$$\mathcal{H} = \sum_k [(\varepsilon_{ek} + \varepsilon_{g0}) a_k^\dagger a_k + \varepsilon_{hk}^s b_{-k}^\dagger b_{-k}]$$

$$+ \tfrac{1}{2} \sum_{\substack{k,k' \\ q \neq 0}} V_{sq} [a_{k+q}^\dagger a_{k'-q}^\dagger a_{k'} a_k + b_{k+q}^\dagger b_{k'-q}^\dagger b_{k'} b_k - 2 a_{k+q}^\dagger b_{k'-q}^\dagger b_{k'} a_k] . \tag{30}$$

Here

$$\varepsilon_{hk}^s = - \varepsilon_{vk} + \sum_{q \neq 0} V_{sq} \tag{31}$$

is the hole energy in the presence of the screened Coulomb potential. To express the hole-energy in terms of an effective mass, we have to remember that the hole effective mass is usually taken from low-excitation experiments, that is, for an unscreened Coulomb potential. Hence we rewrite Eq. (31) as

$$\varepsilon_{h\mathbf{k}}^s = -\varepsilon_{v\mathbf{k}} + \sum_{\mathbf{q}\neq 0} V_q + \sum_{\mathbf{q}\neq 0} [V_{sq} - V_q]$$

$$= \varepsilon_{h\mathbf{k}} + \sum_{\mathbf{q}\neq 0} [V_{sq} - V_q] = \frac{\hbar^2 k^2}{2m_h} + \Delta\varepsilon_{CH} , \qquad (32)$$

where

$$\Delta\varepsilon_{CH} \equiv \sum_{\mathbf{q}\neq 0} [V_{sq} - V_q] \qquad (33)$$

is called the *Debye shift*, or *Coulomb-hole (CH) self energy*. This term is independent of wave vector and is usually considered as one contribution to the bandgap shift.

Note that the Coulomb interaction Hamiltonian with the bare Coulomb potential, i.e., V_q, already contains the mechanism for plasma screening. Therefore, one should be concerned that the ad hoc replacement of V_q with V_{sq} in Eq. (4) might count screening effects twice. As becomes apparent during our derivation, we avoid such double counting by grouping the resulting terms so that each group can be associated with a physical effect. Terms that describe screening effects that are already taken into account by V_{sq} are then dropped.

Proceeding as in Sec. 3-1, we derive coupled equations of motion for the electron and hole populations $n_{e\mathbf{k}}$ and $n_{h\mathbf{k}}$ of Eqs. (3.4) and (3.5) as well as for the interband polarization $p_\mathbf{k}$ of Eq. (3.3). The derivation requires simple but lengthy operator commutations to reduce the commutators in the Heisenberg equations to

$$\dot{p}_\mathbf{k} = -i\omega'_\mathbf{k} p_\mathbf{k} - i\hbar^{-1}\mu_\mathbf{k} E(z,t)[n_{e\mathbf{k}} + n_{h\mathbf{k}} - 1]$$

$$+ \frac{i}{\hbar} \sum_{\mathbf{k}',\mathbf{q}\neq 0} V_{sq} \left[\langle a^\dagger_{\mathbf{k}'+\mathbf{q}} b_{-\mathbf{k}} a_{\mathbf{k}'} a_{\mathbf{k}+\mathbf{q}} \rangle + \langle b^\dagger_{\mathbf{k}'-\mathbf{q}} b_{\mathbf{k}'} a_{\mathbf{k}} b_{-\mathbf{k}-\mathbf{q}} \rangle \right.$$

$$\left. - \langle a^\dagger_{\mathbf{k}'+\mathbf{q}} b_{-\mathbf{k}+\mathbf{q}'} a_{\mathbf{k}'} a_{\mathbf{k}} \rangle - \langle b^\dagger_{\mathbf{k}'-\mathbf{q}} b_{-\mathbf{k}} b_{\mathbf{k}'} a_{\mathbf{k}-\mathbf{q}} \rangle + \langle b_{-\mathbf{k}+\mathbf{q}} a_{\mathbf{k}-\mathbf{q}} \rangle \delta_{\mathbf{k},\mathbf{k}'} \right] ,$$

$$(34)$$

$$\dot{n}_{e\mathbf{k}} = \frac{i}{\hbar} \left(\mu_{\mathbf{k}} p_{\mathbf{k}}^{*} - \mu_{\mathbf{k}}^{*} p_{\mathbf{k}} \right) E(z,t)$$

$$+ \frac{i}{\hbar} \sum_{\mathbf{k'},\mathbf{q} \neq 0} V_{s,q} \left[\langle a_{\mathbf{k}}^{\dagger} a_{\mathbf{k'}-\mathbf{q}}^{\dagger} a_{\mathbf{k}-\mathbf{q}} a_{\mathbf{k'}} \rangle - \langle a_{\mathbf{k}+\mathbf{q}}^{\dagger} a_{\mathbf{k'}-\mathbf{q}}^{\dagger} a_{\mathbf{k}} a_{\mathbf{k'}} \rangle \right.$$

$$\left. + \langle a_{\mathbf{k}}^{\dagger} a_{\mathbf{k}-\mathbf{q}} b_{\mathbf{k'}-\mathbf{q}}^{\dagger} b_{\mathbf{k'}} \rangle - \langle a_{\mathbf{k}+\mathbf{q}}^{\dagger} a_{\mathbf{k}} b_{\mathbf{k'}-\mathbf{q}}^{\dagger} b_{\mathbf{k'}} \rangle \right], \tag{35}$$

and

$$\dot{n}_{h\mathbf{k}} = \frac{i}{\hbar} \left(\mu_{\mathbf{k}} p_{\mathbf{k}}^{*} - \mu_{\mathbf{k}}^{*} p_{\mathbf{k}} \right) E(z,t)$$

$$+ \frac{i}{\hbar} \sum_{\mathbf{k'},\mathbf{q} \neq 0} V_{s,q} \left[\langle b_{-\mathbf{k}}^{\dagger} b_{\mathbf{k'}-\mathbf{q}}^{\dagger} b_{-\mathbf{k}-\mathbf{q}} b_{\mathbf{k'}} \rangle - \langle b_{-\mathbf{k}+\mathbf{q}}^{\dagger} b_{\mathbf{k'}-\mathbf{q}}^{\dagger} b_{-\mathbf{k}} b_{\mathbf{k'}} \rangle \right.$$

$$\left. + \langle a_{\mathbf{k'}+\mathbf{q}}^{\dagger} a_{\mathbf{k'}} b_{-\mathbf{k}}^{\dagger} b_{-\mathbf{k}+\mathbf{q}} \rangle - \langle a_{\mathbf{k'}+\mathbf{q}}^{\dagger} a_{\mathbf{k'}} b_{-\mathbf{k}-\mathbf{q}}^{\dagger} b_{\mathbf{k}} \rangle \right], \tag{36}$$

where the partly renormalized transition energy $\hbar\omega'$ is given by

$$\hbar\omega_{\mathbf{k}}' = \frac{\hbar^2 k^2}{2 m_r} + \epsilon_{g0} + \Delta\epsilon_{CH} . \tag{37}$$

Equations (34) - (36) show that the Coulomb interaction couples the two-operator dynamics to four-operator terms. One way to proceed is to factorize these terms into products of two-operator terms, yielding the Hartree-Fock limit of the equations. For a two-operator combination AB, we write

$$\frac{\partial}{\partial t} \langle AB \rangle = \frac{\partial}{\partial t} \langle AB \rangle_{HF} + \left[\frac{\partial}{\partial t} \langle AB \rangle - \frac{\partial}{\partial t} \langle AB \rangle_{HF} \right], \tag{38}$$

where HF indicates the Hartree-Fock contribution. The quantity inside the square bracket then contains both two and four-operator products, which we represent in general by $\langle ABCD \rangle$. According to the Heisenberg equation of motion (2.112) with the many-body Hamiltonian (30), $\langle ABCD \rangle$ obeys the equation of motion

$$\frac{\partial}{\partial t} \langle ABCD \rangle = \frac{\partial}{\partial t} \langle ABCD \rangle_{HF} + \left[\frac{\partial}{\partial t} \langle ABCD \rangle - \frac{\partial}{\partial t} \langle ABCD \rangle_{HF} \right], \qquad (39)$$

where $\partial \langle ABCD \rangle / \partial t$ contains expectation values of products of up to six operators. We can continue by deriving the equation of motion for

$$\langle ABCDEF \rangle \equiv \left[\frac{\partial}{\partial t} \langle ABCD \rangle - \frac{\partial}{\partial t} \langle ABCD \rangle_{HF} \right]$$

and so on. The result is a hierarchy of equations, where each succeeding equation describes a correlation among operators that is higher than the one before. In practice, we truncate the hierarchy by setting one of the square-bracketed quantities equal zero.

Returning to Eqs. (34) - (36), we first evaluate the Hartree-Fock contributions. To do so, we factorize all the expectation values of four-operator products into *all possible* operator combinations leading to products of densities and/or polarizations. For example, for $\langle a_{\mathbf{k}}^{\dagger} a_{\mathbf{k}'}^{\dagger} a_{\mathbf{p}} a_{\mathbf{p}'} \rangle$, we can have the two-operator combinations $\langle a_{\mathbf{k}}^{\dagger} a_{\mathbf{k}'}^{\dagger} \rangle \langle a_{\mathbf{p}} a_{\mathbf{p}'} \rangle$, $\langle a_{\mathbf{k}}^{\dagger} a_{\mathbf{p}} \rangle \langle a_{\mathbf{k}'}^{\dagger} a_{\mathbf{p}'} \rangle$, $\langle a_{\mathbf{k}}^{\dagger} a_{\mathbf{p}'} \rangle \langle a_{\mathbf{k}'}^{\dagger} a_{\mathbf{p}} \rangle$. Taking the anticommutation relations into account to get the proper signs between these combinations, we find

$$\langle a_{\mathbf{k}}^{\dagger} a_{\mathbf{k}'}^{\dagger} a_{\mathbf{p}} a_{\mathbf{p}'} \rangle \simeq \langle a_{\mathbf{k}}^{\dagger} a_{\mathbf{k}'}^{\dagger} \rangle \langle a_{\mathbf{p}} a_{\mathbf{p}'} \rangle - \langle a_{\mathbf{k}}^{\dagger} a_{\mathbf{p}} \rangle \langle a_{\mathbf{k}'}^{\dagger} a_{\mathbf{p}'} \rangle + \langle a_{\mathbf{k}}^{\dagger} a_{\mathbf{p}'} \rangle \langle a_{\mathbf{k}'}^{\dagger} a_{\mathbf{p}} \rangle$$

$$= 0 + [-\delta_{\mathbf{k},\mathbf{p}} \delta_{\mathbf{k}',\mathbf{p}'} + \delta_{\mathbf{k},\mathbf{p}'} \delta_{\mathbf{k}',\mathbf{p}}] n_{e\mathbf{k}} n_{e\mathbf{k}'} . \qquad (40)$$

Another example is

$$\langle a_{\mathbf{k}'+\mathbf{q}}^{\dagger} b_{\mathbf{q}-\mathbf{k}} a_{\mathbf{k}'} a_{\mathbf{k}} \rangle \simeq \delta_{\mathbf{k}',\mathbf{k}-\mathbf{q}} n_{e\mathbf{k}} P_{\mathbf{k}'} . \qquad (41)$$

Factorizing all the other four-operator products in this way and adding in collision contributions formally, we find the *Semiconductor Bloch Equations*

$$\dot{p}_{\mathbf{k}} = - i\omega_{\mathbf{k}}p_{\mathbf{k}} - i\Omega_{\mathbf{k}}(z,t)[n_{e\mathbf{k}} + n_{h\mathbf{k}} - 1] + \left.\frac{\partial p_{\mathbf{k}}}{\partial t}\right|_{col} \qquad (42)$$

$$\dot{n}_{e\mathbf{k}} = i[\Omega_{\mathbf{k}}(z,t)p_{\mathbf{k}}^* - \Omega_{\mathbf{k}}^*(z,t)p_{\mathbf{k}}] + \left.\frac{\partial n_{e\mathbf{k}}}{\partial t}\right|_{col} \qquad (43)$$

$$\dot{n}_{h\mathbf{k}} = i[\Omega_{\mathbf{k}}(z,t)p_{\mathbf{k}}^* - \Omega_{\mathbf{k}}^*(z,t)p_{\mathbf{k}}] + \left.\frac{\partial n_{h\mathbf{k}}}{\partial t}\right|_{col} \qquad (44)$$

Semiconductor Bloch Equations

The collision terms in these equations arise from many-body interactions that occur at a higher level in the hierarchy than the Hartree-Fock contributions. We derive these terms in Sec. 4-6. The Hartree-Fock contributions do contain two important many-body effects, namely a density-dependent contribution to the transition energy $\omega_{\mathbf{k}}$, and a renormalization of the electric-dipole interaction energy. Specifically, $\hbar\omega'_{\mathbf{k}}$ of Eq. (37) is replaced by the renormalized transition energy

$$\hbar\omega_{\mathbf{k}} = \frac{\hbar^2 k^2}{2m_r} + \varepsilon_{g0} + \Delta\varepsilon_{CH} - \sum_{\substack{\mathbf{k}' \\ \mathbf{k}' \neq \mathbf{k}}} V_{|\mathbf{k}-\mathbf{k}'|}\,(n_{e\mathbf{k}'} + n_{h\mathbf{k}'}) , \qquad (45)$$

and the Rabi frequency $\mu_{\mathbf{k}}E(z,t)/\hbar$ is renormalized to the value

$$\Omega_{\mathbf{k}}(z,t) = \frac{\mu_{\mathbf{k}}E(z,t)}{\hbar} + \frac{1}{\hbar}\sum_{\mathbf{k}' \neq \mathbf{k}} V_{|\mathbf{k}-\mathbf{k}'|}p_{\mathbf{k}'} . \qquad (46)$$

We discuss these two effects in the Secs. 4-4 and 4-5, respectively. First note that there is a large degree of symmetry in the semiconductor Bloch equations. They look like the two-level Bloch equations, with the exceptions that the transition energy and the electric-dipole interaction are renormalized, and the carrier probabilities $n_{\alpha\mathbf{k}}$ enter instead of the probability difference between upper and lower levels. The renormalizations are due to the many-body Coulomb interactions, and they couple equations for different **k** states, which is a significant complication. As discussed in the introduction to this chapter, if all Coulomb-potential contributions are dropped, i.e, V_q is set to 0, the semiconductor Bloch equations reduce to the undamped inhomogeneously broadened two-level Bloch equations. Of course, the limit $V_q = 0$ is unacceptable for semiconductors.

4-4. Bandgap Renormalization

In this section, we discuss the many-body effect known as bandgap renormalization, which is present in the full semiconductor Bloch equations. A contribution to this renormalization is the increasingly effective plasma screening that occurs with increasing carrier density. To see this, we note that at very low carrier densities, the lack of vacant valence band states together with the exclusion principle limit the ability of the valence electron distribution to change itself in order to effectively screen the Coulomb repulsion between a conduction electron and any one of the valence electrons. The corresponding abundance of vacant conduction band states are energetically inaccessible to a valence electron via the Coulomb interaction. At higher carrier densities, more vacant valence band states are available to allow the redistribution of charges for more effective screening. Since the screening of a repulsive interaction leads to a lowering of the conduction electron energy, the transition energy decreases with increasing density. This mechanism is described by the Debye shift, or Coulomb-hole (CH) self energy, $\Delta\varepsilon_{CH}$ given by Eq. (33). Recalling the static plasmon-pole approximation used in simplifying the Lindhard formula, we rewrite the Coulomb-hole self energy as

$$\Delta\varepsilon_{CH} = \sum_{\mathbf{q} \neq 0} [V_{sq} - V_q] = \sum_{\mathbf{q} \neq 0} V_q \left[\frac{1}{\epsilon_{\mathbf{q}}} - 1\right]$$

$$= - \sum_{\mathbf{q} \neq 0} V_q \frac{\omega_{pl}^2}{\omega_{pl}^2 \left(1 + \frac{q^2}{\kappa^2}\right) + \frac{C}{4}\left(\frac{\hbar q^2}{2m_r}\right)^2}. \tag{47}$$

Converting the sum to an integral, using formula (2.161) from Gradshteyn and Rhyzhik (1980), and expressing all parameters in terms of a_0 and ε_R [see Eqs. (2.7) and (2.8)], we write $\Delta\varepsilon_{CH}$ as

$$\Delta \varepsilon_{CH} = -2\varepsilon_R \frac{a_0 \kappa}{\sqrt{1 + \sqrt{C} a_0^2 \kappa^2 \varepsilon_R / \hbar \omega_{pl}}} . \qquad (48)$$

The second contribution to the carrier-density dependence of ω_k comes from the Hartree-Fock energy correction as described by Eq. (45). To highlight the physical origin of this energy correction, we define

$$\Delta \varepsilon_{SX,k} = - \sum_{k' \neq k} V_{s, |k-k'|} [n_{ek'} + n_{hk'}] , \qquad (49)$$

which is the screened-exchange (SX) shift. It is often a sufficiently good approximation to neglect the weak k-dependence in $\Delta \varepsilon_{SX,k}$ and just use a k-independent $\Delta \varepsilon_{SX}$. In this case, both contributions to the ω_k renormalization are independent of k and the renormalized bandgap is given simply by

$$\varepsilon_g = \varepsilon_{g0} + \Delta \varepsilon_{SX} + \Delta \varepsilon_{CH} . \qquad (50)$$

In Fig. 4-4 we plot the different contributions to the renormalized bandgap as function of scaled interparticle distance r_s defined by

$$\frac{4}{3} \pi r_s^3 = \frac{1}{N a_0^3} . \qquad (51)$$

4-5. Interband Coulomb Effects

The second many-body correction arising from the Hartree-Fock contributions is the renormalization of the electric-dipole interaction energy, which leads to the *excitonic* or *Coulomb enhancement* of the interband transition probability. In order to understand the physical origin of this effect, we examine the low density limit of the semiconductor Bloch equations.

In the low density limit, $n_e = n_h \simeq 0$, and the collision terms vanish because no scattering partners are available. Eq. (42) reduces to

$$\dot{p}_k = - i \omega_k p_k + i \Omega_k , \qquad (52)$$

which efficiently isolates the influence of the renormalized electric-dipole interaction frequency Ω_k. Choosing the plane-wave optical field

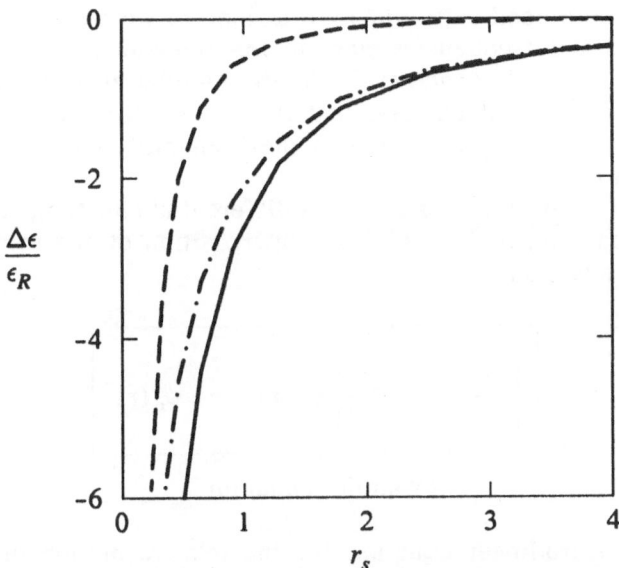

Fig. 4-4 Bandgap reduction as function of normalized interparticle distance r_s, Eq. (51), for the parameters of bulk GaAs at $T=300$K. The reduced gap is plotted as solid line, the screened exchange contribution as dashed line and the Coulomb-hole contribution as dash-dotted line, respectively.

$$E(\mathbf{R},t) = \frac{1}{2}E_0 e^{i(\mathbf{K}\cdot\mathbf{R} - \nu t)} + \text{c.c.} , \qquad (53)$$

where $\mathbf{R}$ is a center of mass coordinate, and making the rotating-wave approximation [see Sec. 3-2], we obtain

$$[\omega_k - \nu + i\gamma]\, p_{\mathbf{k}} = \Omega_{\mathbf{k}} , \qquad (54)$$

where γ is a small phenomenological damping coefficient that eventually is set to zero. Fourier transforming Eq. (54) back to coordinate space, we find

$$\left[-\frac{\hbar^2 \nabla_{\mathbf{r}}^2}{2m_r} - \frac{e^2}{\epsilon_b r} + \varepsilon_g - \hbar(\nu - i\gamma) \right] p(\mathbf{r}) = \mu E_0 e^{i(\mathbf{K}\cdot\mathbf{R}-\nu t)}\delta^3(\mathbf{r})\, V , \quad (55)$$

where we ignore the k-dependence of the interband dipole matrix element, which is usually a good approximation as long as we are only interested in small k-values and frequencies close to the fundamental absorption edge. The fact that it is the attractive Coulomb potential energy, $-e^2/(\epsilon_b r)$, that appears in our problem indicates that the renormalization of the electric-dipole interaction energy originates in the Coulomb interaction between an electron and a hole.

Equation (55) is an inhomogeneous differential equation, which may be solved by expanding $p(\mathbf{r})$ as a linear superposition of the solutions of the homogeneous equation

$$\left[-\frac{\hbar^2 \nabla_{\mathbf{r}}^2}{2m_r} - \frac{e^2}{\epsilon_b r} \right] \psi_n(\mathbf{r}) = \varepsilon_n \, \psi_n(\mathbf{r}) \tag{56}$$

Wannier equation

which is the Schrödinger equation for the relative motion of an electron and a hole interacting via the attractive Coulomb potential. In semiconductor physics this equation is known as the *Wannier equation*. As mentioned already in Chap. 2, there is a one-to-one correspondence between the electron-hole problem and the hydrogen atom if one replaces the proton by the valence-band hole. The solutions of the Wannier equation are therefore completely analogous to those of the hydrogen problem, which are discussed in most quantum mechanics textbooks. There are bound states called excitons, or more specifically *Wannier excitons*, and there are continuum states.

The bound and continuum eigenfunctions of the Wannier equation form an orthonormal basis set, so that we can write

$$p(\mathbf{r}) = \sum_n p_n \psi_n(\mathbf{r}) \,. \tag{57}$$

Substituting Eq. (57) into Eq. (55), multiplying by $\psi_m^*(\mathbf{r})$ and integrating over $\mathbf{r}$ yields

$$p_m = -\frac{\mu V \psi_m^*(\mathbf{r}=0)}{\hbar(\nu - i\gamma) - \varepsilon_g - \varepsilon_m} E_0 e^{i(\mathbf{K} \cdot \mathbf{R} - \nu t)} \,, \tag{58}$$

where we used the orthormality condition

$$\int d^3r \, \psi_m^*(\mathbf{r})\psi_n(\mathbf{r}) = \delta_{m,n} \; . \tag{59}$$

Inserting Eq. (58) into Eq. (57) gives

$$p(\mathbf{r}) = - \sum_n E_0 e^{i(\mathbf{K}\cdot\mathbf{R}-\nu t)} \frac{\mu V \psi_n^*(\mathbf{r}=0)}{\hbar(\nu-i\gamma)-\varepsilon_g-\varepsilon_n} \psi_n(\mathbf{r}) \; , \tag{60}$$

which has the Fourier transform

$$P_\mathbf{k} = - \sum_n E_0 e^{i(\mathbf{K}\cdot\mathbf{R}-\nu t)} \frac{\mu \psi_n^*(\mathbf{r}=0)}{\hbar(\nu-i\gamma)-\varepsilon_g-\varepsilon_n} \int d^3r \, \psi_n(\mathbf{r}) \, e^{i\mathbf{k}\cdot\mathbf{r}} \; . \tag{61}$$

Using an equation of the form of Eq. (2.117) for the polarization amplitude, we have

$$\mathscr{P}(\mathbf{R}) = - 2 e^{-i(\mathbf{K}\cdot\mathbf{R}-\nu t)} \sum_\mathbf{k} \mu^* P_\mathbf{k}$$

$$= - 2|\mu|^2 E_0 \sum_n \frac{|\psi_n(\mathbf{r}=0)|^2}{\hbar(\nu-i\gamma)-\varepsilon_g-\varepsilon_n} \; , \tag{62}$$

where $|\psi_n(\mathbf{r}=0)|^2$ is the probability of finding the electron–hole pair within the same atomic unit cell (zero spatial separation on our coarse grained length scale). The optical susceptibility $\chi(\nu)$ is then given by

$$\chi(\nu) = - \frac{2}{\epsilon}|\mu|^2 \sum_n \frac{|\psi_n(\mathbf{r}=0)|^2}{\hbar(\nu-i\gamma)-\varepsilon_g-\varepsilon_n} \; , \tag{63}$$

where we use a relation like Eq. (2.69). In cgs unity, the corresponding absorption coefficient $\alpha(\nu)$ is given by

$$\alpha(\nu) = \frac{4\pi\nu}{n_b c} \, \text{Im}\{\chi(\nu)\}$$

$$= \alpha_0 \left[\sum_{n=1}^{\infty} \frac{4\pi}{n^3} \delta(\Delta + n^{-2}) + \Theta(\Delta) \, \pi e \, \frac{\pi/\sqrt{\Delta}}{\sinh(\pi/\sqrt{\Delta})} \right], \qquad (64)$$

where $\Delta = (\hbar\nu - \varepsilon_g)/\varepsilon_R$, $\alpha_0 = 2 \, |\mu|^2/(\hbar n_b c a_0^3)$, and we have used the explicit form of the electron-hole pair eigenfunctions. Equation (64) is known as the *Elliott formula* and describes the bandgap absorption spectrum in an unexcited bulk semiconductor.

Equation (64) predicts that the absorption spectrum consists of a series of δ-functions at discrete energies. These resonances are the exciton peaks. The prefactor in front of the δ-functions in Eq. (64) shows that the exciton resonances have a rapidly decreasing oscillator strength $\propto n^{-3}$. The appearance of the exciton resonances in the absorption spectrum is a unique consequence of the electron-hole Coulomb attraction. The second term in Eq. (64), α_{cont}, describes the continuum absorption due to the ionized states. It can be written in terms of the free-carrier absorption

$$\alpha_{free}(\omega) = \alpha_0 \sqrt{\Delta}\,\Theta(\Delta) \,, \qquad (65)$$

as

$$\alpha_{cont} = \alpha_{free} \, \frac{\pi}{\sqrt{\Delta}} \, \frac{e^{\pi/\sqrt{\Delta}}}{\sinh(\pi/\sqrt{\Delta})} \,, \qquad (66)$$

where the correction to α_{free} is called the Sommerfeld or Coulomb enhancement factor. It is a simple exercise to verify that this factor approaches the value $2\pi/\sqrt{\Delta}$ for $\Delta \to 0$, which cancels the $\sqrt{\Delta}$ factor in the free-carrier absorption of Eq. (65) and yields a constant value at the bandgap. This is strikingly different from the square-root law of the free-carrier absorption.

If one takes into account the broadening of the exciton resonances caused by, for example, the scattering of electron-hole pairs with phonons, then only a few bound states can be spectrally resolved. An example of an absorption spectrum predicted by the Elliott formula is depicted in Fig. 4-5. In order to plot the spectrum in Fig. 4-5 we introduced a small amount of broadening. We see that the dominant feature is the $1s$-exciton absorption peak. The $2s$-exciton can also be resolved, but its height is only 1/8-th that of the $1s$-resonance. The other exciton states in GaAs materials usually appear only as a collection of unresolvable peaks just below the

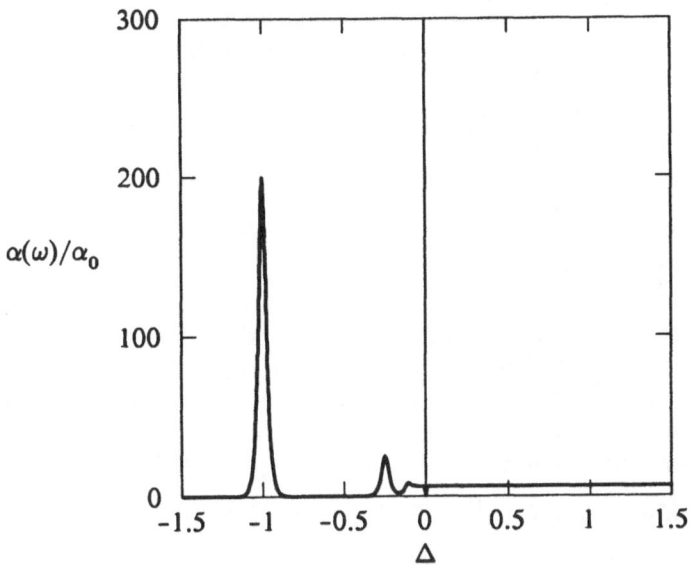

Fig. 4-5.   Optical spectrum at the absorption edge predicted by the Elliott formula.   The absorption coefficient $\alpha(\Delta)$ (in units of α_0) is plotted versus $\Delta = (\hbar\nu - \varepsilon_g)/\varepsilon_R$.   In order to show the absorption spectrum a small broadening of the exciton resonances has been introduced.   An absorption spectrum of this kind can be observed in high quality GaAs at low temperatures.

bandgap.   Note that the continuum absorption is almost constant in the spectral region shown.   These features have all been observed in spectra measured at very low temperatures and with extremely high-quality semiconductors.

Generally, the existence of resonances and the enhancement of the continuum optical spectrum can be traced back to the renormalization of the electric-dipole interaction energy.   This renormalization is caused by the attractive Coulomb interaction between electrons and holes and is responsible for the pronounced increase in the optical absorption around the absorption edge when compared to the free-carrier predictions.   The increased absorption, which is an example of the more general phenomenon of Coulomb enhancement, may be explained as follows.   Due to Coulomb attraction, an electron and a hole have a greater tendency to be in the vicinity of each other for a longer duration than would be the case if they were noninteracting particles.   This increases the interaction time, which in turn leads to a higher probability of an optical transition.

We note that this discussion has concentrated on unexcited or only weakly excited semiconductors. Gain occurs in highly excited media, where the excitons are all ionized, and the absorption becomes negative (gain) in the spectral range between the renormalized band edge and the chemical potential. Chapter 5 shows how Coulomb enhancement reshapes the gain spectra, a particularly important effect in quantum-well media.

4-6. Collision Processes

This final section of this chapter looks at relaxation processes for the carrier distributions and the interband polarization. These processes are not included in the Hartree-Fock approximation. Therefore to study them, we need to compare a higher-order equation of motion with the corresponding Hartree-Fock equation. For example, Eq. (35) for $\dot{n}_{e\mathbf{k}}$ contains the term $\langle a_{\mathbf{k}}^{\dagger} a_{\mathbf{k}'-\mathbf{q}}^{\dagger} a_{\mathbf{k}-\mathbf{q}} a_{\mathbf{k}'} \rangle$. In the Hartree-Fock approximation, this is given by

$$\langle a_{\mathbf{k}}^{\dagger} a_{\mathbf{k}'-\mathbf{q}}^{\dagger} a_{\mathbf{k}-\mathbf{q}} a_{\mathbf{k}'} \rangle_{HF} \simeq \langle a_{\mathbf{k}}^{\dagger} a_{\mathbf{k}} \rangle \langle a_{\mathbf{k}-\mathbf{q}}^{\dagger} a_{\mathbf{k}-\mathbf{q}} \rangle \, \delta_{\mathbf{k},\mathbf{k}'} . \tag{67}$$

To determine $\dot{n}_{e\mathbf{k}} - \dot{n}_{e\mathbf{k}}|_{HF}$, we need to evaluate

$$\delta \langle a_{\mathbf{k}}^{\dagger} a_{\mathbf{k}'-\mathbf{q}}^{\dagger} a_{\mathbf{k}-\mathbf{q}} a_{\mathbf{k}'} \rangle = \langle a_{\mathbf{k}}^{\dagger} a_{\mathbf{k}'-\mathbf{q}}^{\dagger} a_{\mathbf{k}-\mathbf{q}} a_{\mathbf{k}'} \rangle - \langle a_{\mathbf{k}}^{\dagger} a_{\mathbf{k}} \rangle \langle a_{\mathbf{k}-\mathbf{q}}^{\dagger} a_{\mathbf{k}-\mathbf{q}} \rangle \, \delta_{\mathbf{k},\mathbf{k}'} . \tag{68}$$

The time derivative of this equation is

$$\frac{\partial}{\partial t} \delta \langle a_{\mathbf{k}}^{\dagger} a_{\mathbf{k}'-\mathbf{q}}^{\dagger} a_{\mathbf{k}-\mathbf{q}} a_{\mathbf{k}'} \rangle = \langle \frac{\partial}{\partial t} a_{\mathbf{k}}^{\dagger} a_{\mathbf{k}'-\mathbf{q}}^{\dagger} a_{\mathbf{k}-\mathbf{q}} a_{\mathbf{k}'} \rangle$$

$$- \left[\langle \frac{\partial}{\partial t} a_{\mathbf{k}}^{\dagger} a_{\mathbf{k}} \rangle \langle a_{\mathbf{k}-\mathbf{q}}^{\dagger} a_{\mathbf{k}-\mathbf{q}} \rangle + \langle a_{\mathbf{k}}^{\dagger} a_{\mathbf{k}} \rangle \langle \frac{\partial}{\partial t} a_{\mathbf{k}-\mathbf{q}}^{\dagger} a_{\mathbf{k}-\mathbf{q}} \rangle \right] \delta_{\mathbf{k},\mathbf{k}'} . \tag{69}$$

Using the Heisenberg equation of motion with the full electron-hole Hamiltonian, we find

$$\frac{\partial}{\partial t}\delta\langle a_{\mathbf{k}}^{\dagger}a_{\mathbf{k}'-\mathbf{q}}^{\dagger}a_{\mathbf{k}-\mathbf{q}}a_{\mathbf{k}'}\rangle = \frac{i}{\hbar}\delta\langle a_{\mathbf{k}}^{\dagger}a_{\mathbf{k}'-\mathbf{q}}^{\dagger}a_{\mathbf{k}-\mathbf{q}}a_{\mathbf{k}'}\rangle \Delta\varepsilon_{e\mathbf{k}\mathbf{k}'\mathbf{q}}$$

$$+ \left.\frac{\partial}{\partial t}\delta\langle a_{\mathbf{k}}^{\dagger}a_{\mathbf{k}'-\mathbf{q}}^{\dagger}a_{\mathbf{k}-\mathbf{q}}a_{\mathbf{k}'}\rangle\right|_{Coul}, \qquad (70)$$

where

$$\Delta\varepsilon_{e\mathbf{k}\mathbf{k}'\mathbf{q}} = \varepsilon_{e\mathbf{k}} + \varepsilon_{e,\mathbf{k}'-\mathbf{q}} - \varepsilon_{e,\mathbf{k}-\mathbf{q}} - \varepsilon_{e\mathbf{k}'}. \qquad (71)$$

and

$$i\hbar\left.\frac{\partial}{\partial t}\delta\langle a_{\mathbf{k}}^{\dagger}a_{\mathbf{k}'-\mathbf{q}}^{\dagger}a_{\mathbf{k}-\mathbf{q}}a_{\mathbf{k}'}\rangle\right|_{Coul} = \langle [\mathscr{H}_{C}, a_{\mathbf{k}}^{\dagger}a_{\mathbf{k}'-\mathbf{q}}^{\dagger}a_{\mathbf{k}-\mathbf{q}}a_{\mathbf{k}'}]\rangle. \qquad (72)$$

Evaluation of this commutator leads to expressions containing products of up to six operators that are too lengthy to show here.

Formally integrating Eq. (70), we get

$$\delta\langle a_{\mathbf{k}}^{\dagger}a_{\mathbf{k}'-\mathbf{q}}^{\dagger}a_{\mathbf{k}-\mathbf{q}}a_{\mathbf{k}'}(t)\rangle = e^{(i\Delta\varepsilon_{e\mathbf{k}\mathbf{k}'\mathbf{q}}/\hbar - \gamma)t}$$

$$\times \int_{-\infty}^{t}dt' e^{-(i\Delta\varepsilon_{e\mathbf{k}\mathbf{k}'\mathbf{q}}/\hbar - \gamma)t'}\left.\frac{\partial}{\partial t}\delta\langle a_{\mathbf{k}}^{\dagger}a_{\mathbf{k}'-\mathbf{q}}^{\dagger}a_{\mathbf{k}-\mathbf{q}}a_{\mathbf{k}'}(t')\rangle\right|_{Coul}, \qquad (73)$$

where γ is a phenomenological decay constant added so that the integral vanishes at the lower boundary. Assuming that the Coulomb contribution is slowly varying compared to the exponential, we can move it outside the integral. The resulting integral can be readily evaluated to give

$$\delta\langle a_{\mathbf{k}}^{\dagger}a_{\mathbf{k}'-\mathbf{q}}^{\dagger}a_{\mathbf{k}-\mathbf{q}}a_{\mathbf{k}'}\rangle \simeq \frac{\frac{\partial}{\partial t}\delta\langle a_{\mathbf{k}}^{\dagger}a_{\mathbf{k}'-\mathbf{q}}^{\dagger}a_{\mathbf{k}-\mathbf{q}}a_{\mathbf{k}'}\rangle|_{Coul}}{i\Delta\varepsilon_{e\mathbf{k}\mathbf{k}'\mathbf{q}}/\hbar - \gamma}. \qquad (74)$$

A similar result is obtained for the other four-operator terms which appear in the equations of motion (34) - (36).

As a next step, we factorize all the six- and four-operator terms which occur in the Coulomb parts to obtain the simplest possible expression for the scattering terms. This gives

$$\frac{\partial n_{e\mathbf{k}}}{\partial t}\bigg|_{col} = - \sum_{\substack{\mathbf{k'},\mathbf{q} \\ \mathbf{q}\neq 0}} \frac{2\pi}{\hbar} V_q^2 \,\delta(\varepsilon_{e\mathbf{k}}+\varepsilon_{e,\mathbf{k'}+\mathbf{q}}-\varepsilon_{e\mathbf{k'}}-\varepsilon_{e,\mathbf{k}+\mathbf{q}})$$

$$\times \left[n_{e\mathbf{k}} n_{e,\mathbf{k'}+\mathbf{q}}(1-n_{e\mathbf{k'}})(1-n_{e,\mathbf{k}+\mathbf{q}}) - n_{e\mathbf{k'}} n_{e,\mathbf{k}+\mathbf{q}}(1-n_{e\mathbf{k}})(1-n_{e,\mathbf{k'}+\mathbf{q}}) \right]$$

$$- \sum_{\substack{\mathbf{k'},\mathbf{q} \\ \mathbf{q}\neq 0}} \frac{2\pi}{\hbar} V_q^2 \,\delta(\varepsilon_{e\mathbf{k}} + \varepsilon_{h\mathbf{k'}} - \varepsilon_{e,\mathbf{k}+\mathbf{q}} - \epsilon_{h,\mathbf{k'}-\mathbf{q}})$$

$$\times \left[n_{e\mathbf{k}} n_{h\mathbf{k'}}(1-n_{e,\mathbf{k}+\mathbf{q}})(1-n_{h,\mathbf{k'}-\mathbf{q}}) - n_{e,\mathbf{k}+\mathbf{q}} n_{h,\mathbf{k'}-\mathbf{q}}(1-n_{e\mathbf{k}})(1-n_{h\mathbf{k'}}) \right] , \tag{75}$$

where the Dirac delta functions come from letting $\gamma \rightarrow 0$. Equation (75) is the famous *Boltzmann equation* for electron-electron and electron-hole scattering. It has the form

$$\frac{\partial n_{e\mathbf{k}}}{\partial t}\bigg|_{col} = - n_{e\mathbf{k}} \,\Gamma_{e\mathbf{k}}^{out}\{n\} + (1 - n_{e\mathbf{k}}) \,\Gamma_{e\mathbf{k}}^{in}\{n\} , \tag{76}$$

where the rates $\Gamma_{e\mathbf{k}}^{out}\{n\}$ and $\Gamma_{e\mathbf{k}}^{in}\{n\}$ describe the effective scattering out of and into the state $\mathbf{k}$, respectively. The notation $\Gamma\{n\}$ symbolizes the functional dependence of Γ on the electron and hole distribution functions. The corresponding Boltzmann equation for the hole population $n_{h\mathbf{k}}$ is obtained from Eq. (75) by the interchange $e \longleftrightarrow h$. The terms appearing in the Boltzmann equation become obvious when we refer to Fig. 4-6.

It is convenient to write the total relaxation rate of $n_{\alpha\mathbf{k}}$ as a single decay rate

$$\gamma_{\alpha\mathbf{k}}\{n\} \equiv \Gamma_{e\mathbf{k}}^{out}\{n\} + \Gamma_{e\mathbf{k}}^{in}\{n\} , \tag{77}$$

in terms of which the carrier Boltzmann equations can be written as

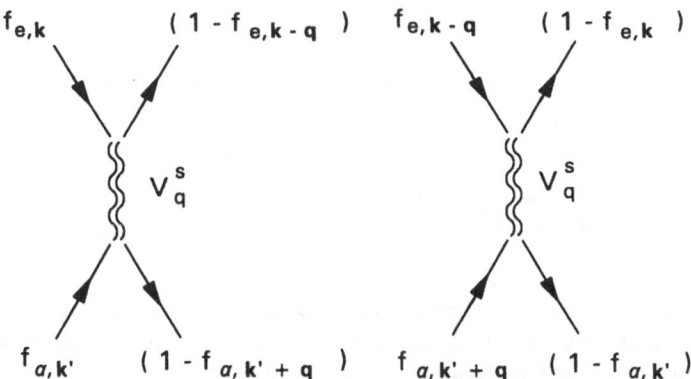

Fig. 4-6. Scattering into and out of the kth electronic state. A scattering partner is needed to conserve energy and momentum.

$$\left.\frac{\partial n_{\alpha k}}{\partial t}\right|_{col} = -\gamma_{\alpha k}\{n\}n_{\alpha k} + \Gamma_{\alpha k}^{in}\{n\} . \qquad (78)$$

Here we see explicitly that scattering both into and out of the state k are relaxation processes, although one increases the probability $n_{\alpha k}$, while the other decreases it.

It is straightforward to show that under quasiequilibrium conditions, i.e., $n_{\alpha k} = f_{\alpha k}$, the Boltzmann equation is identically zero. This implies that

$$f_{ek}\, \Gamma_{ek}^{out}\{f\} = (1 - f_{ek})\, \Gamma_{ek}^{in}\{f\} , \qquad (79)$$

for nonvanishing in- and out-scattering rates. Equation (79) describes a condition called *detailed balance*, for which the scattering into a state is balanced by the scattering out of that state. If the carrier distributions are sufficiently close to the quasiequilibrium Fermi-Dirac distributions, we can set $\Gamma_{\alpha k}^{in}\{n\} \simeq \Gamma_{\alpha k}^{in}\{f\}$ and $\Gamma_{\alpha k}^{out}\{n\} \simeq \Gamma_{\alpha k}^{out}\{f\}$ in Eq. (76). Further substituting $\Gamma_{ek}^{in}\{f\} = \gamma_{\alpha k}\{f\}f_{ek}$, which is a simple rearrangement of Eq. (79), we find

$$\left.\frac{\partial n_{\alpha k}}{\partial t}\right|_{col} \simeq - \gamma_{\alpha k}\{f\}[n_{\alpha k} - f_{\alpha k}] \, . \qquad (80)$$

This approximation fails (barely) to preserve the total carrier density N. To remedy this defect when studying interactions in the neighborhood of k_0, we choose $\gamma_{\alpha k_0}\{f\}$ instead of $\gamma_{\alpha k}\{f\}$, that is

$$\left.\frac{\partial n_{\alpha k}}{\partial t}\right|_{col} \simeq - \gamma_{\alpha k_0}\{f\}[n_{\alpha k} - f_{\alpha k}] \, . \qquad (81)$$

This does conserve the total carrier density since $\Sigma_k n_{\alpha k} = \Sigma_k f_{\alpha k} = VN$. Equation (81) is often used to simplify collision terms in the carrier distribution equations of motion [see Eq. (3.22)].

The scattering terms appearing in the polarization equation of motion may be obtained by taking the steps similar to those taken in going from Eq. (67) to Eq. (75). Keeping only contributions linear in p_k, Lindberg and Koch (1988) found a collision term of the general form

$$\left.\frac{\partial p_k}{\partial t}\right|_{col} = - A_k p_k + \sum_{q \neq 0} B_{k,q} p_{k+q} \, , \qquad (82)$$

where A_k and $B_{k,q}$ are complex functions that are proportional to the square of the Coulomb potential. The imaginary parts of these terms give rise to additional energy renormalizations and to potential corrections in the polarization equation. They are usually ignored when compared to the Hartree-Fork contributions, which depend linearly on the Coulomb potential. The terms with coefficients $Re\{B_{k,q}\}$ describe polarization transfer between the states k and $k + q$ due to the carrier-carrier collisions. These terms contribute to the *plasma screening* of the electron-hole Coulomb potential. We ignore these terms here, since we describe screening of the Coulomb potential phenomenologically, as discussed in Sec. 4-2. This leaves a term describing an exponential decay of the polarization

$$\left.\frac{\partial p_{\mathbf{k}}}{\partial t}\right|_{col} = -\gamma_{\mathbf{k}} p_{\mathbf{k}} \,, \tag{83}$$

with the rate

$$\gamma_{\mathbf{k}} = \mathrm{Re}(A_{\mathbf{k}}) = \tfrac{1}{2}[\Gamma_{e\mathbf{k}}^{out}\{n\} + \Gamma_{e\mathbf{k}}^{in}\{n\} + \Gamma_{h\mathbf{k}}^{out}\{n\} + \Gamma_{h\mathbf{k}}^{in}\{n\}]$$

$$= \tfrac{1}{2}[\gamma_{e\mathbf{k}}\{n\} + \gamma_{h\mathbf{k}}\{n\}] \,. \tag{84}$$

Following the vernacular of nuclear magnetic resonance of a spin-$\tfrac{1}{2}$ system, we introduce the induced-dipole lifetime

$$T_{2\mathbf{k}} = \frac{1}{\gamma_{\mathbf{k}}} \,, \tag{85}$$

which for the semiconductor case is both k-dependent and calculable. Carrier-carrier collisions that conserve k^2 are elastic, but in general carrier-carrier collisions involve energy exchange, a characteristic of T_1 (probability difference) relaxation times. Such "fast" T_1's are also met in molecular and dye lasers and result from collisions that transfer probabilities among vibrational bands. In those lasers, the quasiequilibrium level distributions are Maxwell-Boltzmann in character, due to their underlying one-electron nature. All such scattering events, elastic and inelastic, contribute to dipole dephasing.

Carrier-Phonon Scattering

Even though the carrier-carrier scattering process may dominate the fast carrier redistribution under typical laser densities, this process does not dissipate energy, since the kinetic energy is one of the conserved quantities in elastic carrier collisions. The other conserved quantities are total momentum and carrier number.

The most important source of energy dissipation, i.e., "carrier cooling" is caused by the coupling of the electronic system to the lattice. The dominant part of this carrier-phonon coupling can be modelled by the interaction Hamiltonian

$$\mathcal{H}_{e\text{-}p} = \sum_{\mathbf{k},\mathbf{q}} \hbar G_q a_{\mathbf{k}+\mathbf{q}}^{\dagger} a_{\mathbf{k}} (b_{\mathbf{q}} + b_{-\mathbf{q}}^{\dagger}) , \tag{86}$$

where $b_{\mathbf{q}}$ and $b_{\mathbf{q}}^{\dagger}$ are the annihilation and creation operators of longitudinal optical (LO) phonons, which are the quanta of the longitudinal polarization oscillations due to ionic displacements in a polar semiconductor. Equation (86) describes how an electron can be scattered in its band by emitting or absorbing one LO phonon.

The matrix element G_q for the linear interaction of the electrons with the lattice polarization is

$$G_q^2 = \frac{\omega_{LO} V_{sq}}{2\hbar} \left(\frac{1}{\epsilon_\infty} - \frac{1}{\epsilon_0} \right) , \tag{87}$$

where ω_{LO} is the LO-phonon frequency and ϵ_0 and ϵ_∞ are the low- and high frequency background dielectric constants of the medium. The coupling described by Eqs. (86) and (87) is usually called Fröhlich electron-LO phonon coupling. This coupling influences both, the carrier intraband relaxation and the electron-hole interband kinetics. For example, the electron-LO phonon coupling is responsible for the low-frequency lineshape (Urbach tail of the band-edge absorption) because it gives rise to (not spectrally resolved) phonon side bands in which not only a photon but also one or several thermal phonons are absorbed.

For our present discussion it is more important to study the electron intraband relaxation due to electron-LO phonon coupling. For this purpose we evaluate

$$\frac{dn_{e\mathbf{k}}}{dt} \bigg|_{e\text{-}p} = \frac{i}{\hbar} [\mathcal{H}_{e\text{-}p}, n_{e\mathbf{k}}] , \tag{88}$$

using second-order perturbation theory, i.e., we solve Eq. (88) by formal integration and iterate the result twice. In this way we obtain

$$\frac{dn_{e\mathbf{k}}}{dt}\bigg|_{e\text{-}p} = -2\pi \sum_{\mathbf{q},\pm} G_q^2 \, \delta\left(\Delta_{\mathbf{k},\mathbf{q}}^{\mp}\right)$$

$$\times \left[n_{e\mathbf{k}} \left[1 - n_{e,\mathbf{k}\text{-}\mathbf{q}}\right] (g_{\pm\mathbf{q}} + \tfrac{1}{2} \pm \tfrac{1}{2}) \right.$$

$$\left. - \left[1 - n_{e\mathbf{k}}\right] n_{e,\mathbf{k}\text{-}\mathbf{q}} \, (g_{\pm\mathbf{q}} + \tfrac{1}{2} \mp \tfrac{1}{2}) \right], \tag{89}$$

which is the Boltzmann collision integral for electron-LO phonon scattering. Following the notation in Haug and Koch (1993), we denote the frequency differences $\Delta^{\pm}$ as

$$\hbar\Delta_{\mathbf{k},\mathbf{q}}^{\pm} = \varepsilon_{\mathbf{k}} - \varepsilon_{\mathbf{k}\text{-}\mathbf{q}} \pm \hbar\omega_0 . \tag{90}$$

The different terms in Eq. (89) describe the transition rates in and out of state $\mathbf{k}$ under absorption or emission of LO-phonons. The first two terms describe the transition $\mathbf{k} \rightarrow \mathbf{k}\text{-}\mathbf{q}$ under emission (upper sign) or absorption (lower sign) of a phonon. The second two terms describe the transition $\mathbf{k}\text{-}\mathbf{q} \rightarrow \mathbf{k}$ under absorption or emission of a phonon. The function $g(\mathbf{q},t)$ is the phonon population function, which in general has to be computed self-consistently. However, it is often possible to simplify the problem by assuming that the phonons are in thermal equilibrium, so that

$$g_{\mathbf{q}} = \frac{1}{e^{\beta\hbar\omega_0} - 1} , \tag{91}$$

i.e., the phonon distribution is described by a thermal Bose function.

A Hierarchy of Relaxation Times

Numerical evaluations of the electron-phonon Boltzmann equation yield scattering times on the order of ps for typical semiconductor laser conditions. Hence in semiconductor lasers, the carrier-carrier scattering is the dominant contributor to the dipole dephasing rate constant $\gamma_{\mathbf{k}}$.

In describing the carrier probabilities $n_{\alpha\mathbf{k}}$ themselves, the carrier-carrier scattering rates $\gamma_{\alpha\mathbf{k}}\{n\}$ dominate the response on subpicosecond time scales. Superimposed on this fast response is the relatively slow response resulting from radiative and nonradiative recombination and pumping

processes along with the somewhat faster response that attempts to equilibrate the plasma and lattice temperatures via carrier-phonon scattering.
These processes change the quasiequilibrium Fermi-Dirac distributions to
which the carrier probabilities are driven by the carrier-carrier scattering.
Associated with these slower responses are T_1's on the order of nanoseconds for the total carrier density and picoseconds for the temperature
equilibration. Hence in general, semiconductor gain media involve a hierarchy of T_1 relaxation times. If we are interested in phenomena that vary
little in the carrier-carrier relaxation times, we may be able to adiabatically
eliminate the corresponding transients by assuming that the carrier distributions are described by Fermi-Dirac distributions. This is a quasiequilibrium approximation and the slow transient response may be describable
using a single long T_1.

However if we wish to study transient phenomena on a subpicosecond
time scale, we might wonder how to define a "fast" $T_{1\mathbf{k}}$ that corresponds to
$T_{2\mathbf{k}}$ of Eq. (85). Such a $T_{1\mathbf{k}}$ would be the lifetime of the inversion

$$d_k = n_{e\mathbf{k}} + n_{h\mathbf{k}} - 1 \tag{92}$$

as it shows up in the equation of motion

$$\dot{d}_{\mathbf{k}} = -\frac{d_{\mathbf{k}}}{T_{1\mathbf{k}}} + \text{other terms} . \tag{93}$$

Substituting Eqs. (78) and (92) into Eq. (93), we have

$$\left.\frac{\partial d_{\mathbf{k}}}{\partial t}\right|_{col} = -\gamma_{e\mathbf{k}}\{n\}n_{e\mathbf{k}} - \gamma_{h\mathbf{k}}\{n\}n_{h\mathbf{k}} + \Gamma_{e\mathbf{k}}^{in}\{n\} + \Gamma_{h\mathbf{k}}^{in}\{n\} . \tag{94}$$

To write $d_{\mathbf{k}}$ on the RHS of this equation, we introduce the probability
"sum"

$$s_{\mathbf{k}} = n_{e\mathbf{k}} + (1 - n_{h\mathbf{k}}) , \tag{95}$$

which complements the probability difference $d_{\mathbf{k}} = n_{e\mathbf{k}} - (1 - n_{h\mathbf{k}})$. In
terms of $d_{\mathbf{k}}$ and $s_{\mathbf{k}}$, we have that $n_{e\mathbf{k}} = \frac{1}{2}[d_{\mathbf{k}}+s_{\mathbf{k}}]$ and $n_{h\mathbf{k}} = \frac{1}{2}[d_{\mathbf{k}}-s_{\mathbf{k}}] + 1$.
Substituting these expressions into the equation of motion (94), we have

$$\dot{d}_{\mathbf{k}}\big|_{col} = -\tfrac{1}{2}[\gamma_{e\mathbf{k}}\{n\} + \gamma_{h\mathbf{k}}\{n\}]d_{\mathbf{k}} - \tfrac{1}{2}[\gamma_{e\mathbf{k}}\{n\} - \gamma_{h\mathbf{k}}\{n\}]s_{\mathbf{k}}$$

$$-\gamma_{h\mathbf{k}} + \Gamma^{in}_{e\mathbf{k}}\{n\} + \Gamma^{in}_{h\mathbf{k}}\{n\} . \qquad (96)$$

The simplest thing to note is that if $\gamma_{e\mathbf{k}}\{n\} = \gamma_{h\mathbf{k}}\{n\}$, then by Eqs. (84), (85), and (93), we have $T_{1\mathbf{k}} = 1/\gamma_{\mathbf{k}} = T_{2\mathbf{k}}$.

Unless the electron and hole effective masses are equal, $\gamma_{e\mathbf{k}}\{n\} \neq \gamma_{h\mathbf{k}}\{n\}$, and we would not expect to be able to describe the fast response with a single $T_{1\mathbf{k}}$. A similar situation is met in the case of the two-level atom problem with different upper- and lower-level decay constants. There it is fairly well known that the T_1 given by

$$T_1 = \frac{1}{2}\left[\frac{1}{\gamma_a} + \frac{1}{\gamma_b}\right] \qquad (97)$$

describes steady-state saturation correctly, although it fails to account for the transient response in general. Following a procedure similar to that used to derive Eq. (97), we consider the equation of motion for the sum term $s_{\mathbf{k}}$. Using Eqs. (78) and (95) and dropping the $\{n\}$ for typographical simplicity, we have

$$\dot{s}_{\mathbf{k}}\big|_{col} = -\gamma_{e\mathbf{k}}n_{e\mathbf{k}} + \Gamma^{in}_{e\mathbf{k}} + \gamma_{h\mathbf{k}}n_{h\mathbf{k}} - \Gamma^{in}_{h\mathbf{k}}$$

$$= -\tfrac{1}{2}\gamma_{e\mathbf{k}}[d_{\mathbf{k}}+s_{\mathbf{k}}] + \tfrac{1}{2}\gamma_{h\mathbf{k}}[d_{\mathbf{k}}-s_{\mathbf{k}} + 2] + \Gamma^{in}_{e\mathbf{k}} - \Gamma^{in}_{h\mathbf{k}}$$

$$= -\gamma_{\mathbf{k}}s_{\mathbf{k}} - \tfrac{1}{2}[\gamma_{e\mathbf{k}} - \gamma_{h\mathbf{k}}]d_{\mathbf{k}} + \gamma_{h\mathbf{k}} + \Gamma^{in}_{e\mathbf{k}} - \Gamma^{in}_{h\mathbf{k}} . \qquad (98)$$

In steady state ($\dot{s}_{\mathbf{k}}\big|_{col} = 0$), this gives

$$s_{\mathbf{k}} = -\frac{\tfrac{1}{2}[\gamma_{e\mathbf{k}} - \gamma_{h\mathbf{k}}]d_{\mathbf{k}} + \gamma_{h\mathbf{k}} + \Gamma^{in}_{e\mathbf{k}} - \Gamma^{in}_{h\mathbf{k}}}{\gamma_{\mathbf{k}}} . \qquad (99)$$

As a rough approximation to the near steady-state transient behavior of $d_{\mathbf{k}}$, we substitute Eq. (99) into Eq. (96) and simplify to find

$$\dot{d}_{\mathbf{k}}\big|_{col} = -\frac{d_{\mathbf{k}}}{T_{1\mathbf{k}}} + (\text{functions of } \gamma\text{'s}) , \qquad (100)$$

where the carrier-carrier-scattering probability-difference decay time $T_{1\mathbf{k}}$

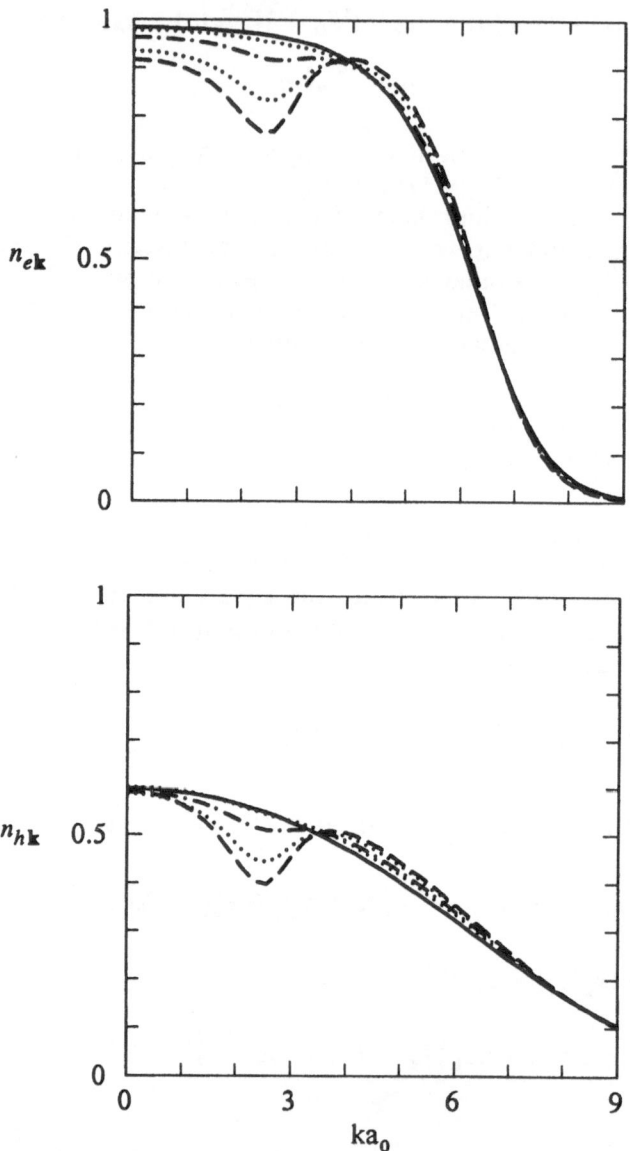

Fig. 4-7. Relaxation of disturbed Fermi distribution functions for electrons (top) and holes (bottom) at a density $N = 3 \times 10^{18}$ cm^{-3} and temperature $T \simeq 300\ K$ obtained by numerically solving the Boltzmann equation using the dynamically screened Coulomb potential in RPA approximation.   The times are: $t = 0$ (long dashed), 21 fs (dotted), 75 fs (dash-dotted), 147 fs (dotted), 796 fs (solid). [From Binder $et\ al.$ (1992)].

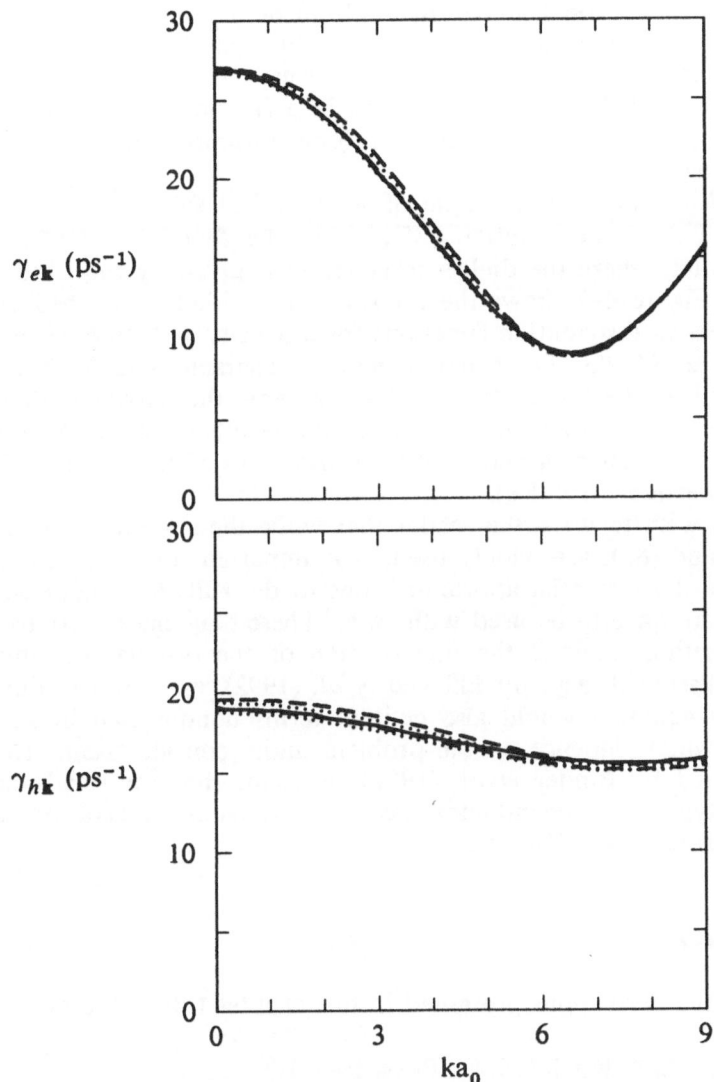

Fig. 4-8. Carrier-carrier scattering rates extracted from Fig. 4-7. (top) electrons, (bottom) holes. [From Binder *et al.* (1992)]

is given approximately by

$$T_{1\mathbf{k}} = \frac{1}{2}\left[\frac{1}{\gamma_{e\mathbf{k}}} + \frac{1}{\gamma_{h\mathbf{k}}}\right],$$

(101)

see also Eq. (3.49). As noted above, in the limit of $\gamma_{ek} = \gamma_{hk} = \gamma_k$, $T_{1k} = T_{2k}$. Equation (101) is similar to the two-level-atom Eq. (97), which is also only valid near steady state. The "steady state" considered here, however, is reached to all intents and purposes in a fraction of a picosecond. Hence Eq. (101) could be useful in describing saturation involving hole burning by a single-mode field.

In Fig. 4-7 we show an example of a numerical solution of the full carrier-carrier Boltzmann equation (75), using the dynamically screened Coulomb potential, where the dielectric function is given by the Lindhard formula (21). Figure 4-7 shows the relaxation of initially disturbed electron and hole Fermi distribution functions for a density $n = 3 \times 10^{18}$ cm^{-3} and temperature $T \simeq 300$ K. These parameters are characteristic for a room temperature bulk GaAs laser. In Fig. 4-8 we show the carrier scattering rates Γ defined in Eq. (76) corresponding to the results shown in Fig. 4-7. We see that typical scattering and dephasing times are of the order of 50 - 100 fs (1 femtosecond = 10^{-15} s).

Finally, we wish to warn the reader that while the simplified collision equations (81) and (83), are widely used, it is important to remember that they are results of substantial approximations to the full Boltzmann equation and therefore have to be used with care. These diagonal equations involve approximations beyond the linearization of the original Boltzmann equation. As discussed, e.g., by ElSayed et al. (1992), a rigorously linearized Boltzmann equation would also contain terms nondiagonal in k, the importance of which depends on the problem under consideration. However, as discussed by Binder et al. (1992), it seems that for steady-state operation of typical semiconductor lasers, a diagonal relaxation rate approximation is usually sufficient.

REFERENCES

Major parts of the derivations presented in this chapter follow the paper:

Lindberg, M. and S.W. Koch (1988), Phys. Rev. **B38**, 3342.

For more details on the semiconductor Bloch equations and for further references see:

Koch, S.W., N. Peyghambarian, and M. Lindberg (1988), J. Phys. **C21**, 5229.

Haug, H. and S.W. Koch (1989), Phys. Rev. **A39**, 1887.

Haug, H. (1988), Ed., *Optical Nonlinearities and Instabilities in Semiconductors*, Academic, New York (1988)

Stahl, A. and I. Balslev (1987), *Electrodynamics of the Semiconductor Band Edge*, Springer Tracts in Modern Physics 110, Springer Verlag, Berlin.

Haug, H. and S.W. Koch (1993), *Quantum Theory of the Optical and Electronic Properties of Semiconductors*, 2nd Ed., World Scientific, Singapore

More details of the presented analysis of the carrier Boltzmann equation can be found in:

Binder, R., D. Scott, A.E. Paul, M. Lindberg, K. Henneberger, and S.W. Koch (1992), Phys. Rev. **B45**, 1107.

The linearized Boltzmann equation is discussed by:

ElSayed, K,. T. Wicht, H. Haug, and L. Banyai (1992), Z. Physik **B86**, 345.

Discussions of the two-level Bloch equations can be found in:

Allen, L. and J. H. Eberly (1975), *Optical Resonances and Two-Level Atoms*, John Wiley, New York; reprinted (1987) with corrections by Dover, New York.

Meystre, P. and M. Sargent III (1991), *Elements of Quantum Optics*, 2nd Ed., Springer-Verlag, Heidelberg.

Sargent III, M., M.O. Scully, and W.E. Lamb (1977), *Laser Physics*, Addison-Wesley, Reading, MA.

The classical theory of plasma screening is discussed in:

Ashcroft, N.W. and N.D. Mermin (1976), *Solid State Theory*, Saunders College, Philadelphia.

Harrison, W. A. (1980), *Solid State Theory*, Dover Publ. New York

Haug, H. and S.W. Koch (1993), Op. Cit.

General many-body theory and sum rules are discussed in:

Lundquist, B.I. (1967), Phys. Konden. Mat. **6**, 193 and 206.

Mahan, G.D. (1981), *Many Particle Physics*, Plenum Press, New York.

For the modifications of the plasmon-pole approximation in an electron-hole plasma see:

Haug, H. and S. Schmitt-Rink (1984), Op. Cit.

Zimmermann, R. (1988), *Many-Particle Theory of Highly Excited Semi-conductors*, Teubner, Berlin.

We have used the integral tables in

Gradshteyn, I.S. and I.M. Rhyzhik (1980), *Tables of Integals, Series and Products*, Academic Press, New York.

Chapter 5
MANY-BODY GAIN

In Chap. 4, we discuss the basic many-body theory of the semiconductor gain medium and derive the semiconductor Bloch equations, which contain the many-body modifications due to the Coulomb interaction between the carriers. The many-body effects manifest themselves as plasma screening, bandgap renormalization, interband Coulomb attraction, and carrier-carrier scattering.

In this chapter, we use the semiconductor Bloch equations and make the quasiequilibrium approximation to derive expressions for the gain and the carrier-induced phase-shift. Our derivation makes a Padé approximation that is valid at the high carrier densities typically found in semiconductor lasers. Both bulk and quantum-well gain media are treated. To demonstrate the evaluation of the many-body gain and phase-shift formulas, we rewrite them in computer-oriented ways and present examples that reveal the different contributions to the semiconductor gain and phase-shift spectra. In some cases, the many-body effects help explain laboratory observations of semiconductor laser behavior. An example is the presence of gain below the unpumped semiconductor bandgap. Once considered to be an anomaly, it is now well known that this gain shift is due to bandgap renormalization.

Section 4-5 shows that the interband Coulomb interaction dominates the semiconductor absorption spectra for low electron-hole-pair densities. Even though the strength of the Coulomb enhancement is reduced by screening under typical lasing conditions, it still has significant effects on the spectral position, shape, and magnitude of the gain and phase-shift spectra. Because Coulomb enhancement calculations are somewhat more complicated than those for bandgap renormalization, there is a tendency to ignore them. Unfortunately, such a theory, which is actually the free-carrier theory with an ad hoc inclusion of bandgap renormalization, is frequently referred to in the literature as a "many-body theory". We show in this chapter that such an approach can lead to predictions that are in worse agreement with a real many-body theory than the predictions of a free-carrier model. The errors result from the fact that it is inconsistent to account for Fermionic exchange effects and Coulomb repulsion, which give rise to bandgap renormalization, and at the same time ignore the Coulomb attraction, which gives rise to Coulomb enhancement.

Assuming a simple two-band semiconductor, we demonstrate in this chapter that the many-body effects are most important for gain-medium properties that involve carrier-density-dependent changes in gain or index. The linewidth enhancement factor α is such a quantity. Sections 5-2 and 5-3 show that the many-body theory predicts structure in the α spectra that is absent in the free-carrier results. Furthermore, we demonstrate that the many-body effects are enhanced in two dimensions, making them even more important in a quantum-well medium than in a bulk medium.

5-1. Padé Approximation

To obtain the gain and carrier-induced refractive index (phase shift), we begin with the semiconductor Bloch equation (4.42) for the induced dipole $p_\mathbf{k}$, here including the dipole decay rate (4.83) given by the relaxation rate approximation of Sec. 4-6,

$$\dot{p}_\mathbf{k} = - (i\omega_\mathbf{k} + \gamma_\mathbf{k})p_\mathbf{k} - i\Omega_\mathbf{k}(z,t)[n_{e\mathbf{k}} + n_{h\mathbf{k}} - 1] . \tag{1}$$

Formally integrating from $-\infty$ to t, we have

$$p_\mathbf{k}(t) = -i\int_{-\infty}^{t} dt'\, e^{(i\omega_\mathbf{k} + \gamma_\mathbf{k})(t'-t)}\Omega_\mathbf{k}(z,t')[n_{e\mathbf{k}}(t') + n_{h\mathbf{k}}(t') - 1] . \tag{2}$$

As in Sec. 3-2 for the free-carrier theory, we make the rate equation approximation, which assumes that the carrier probabilities and the electric-field envelope vary little in the time $T_2 \equiv 1/\gamma_\mathbf{k}$. Using the field (3.42), we have

$$p_\mathbf{k}(t) = - \frac{i}{\hbar}[n_{e\mathbf{k}}(t) + n_{h\mathbf{k}}(t) - 1]\left[\tfrac{1}{2}\mu_\mathbf{k} E(z)\frac{e^{i[Kz - \nu t - \phi(z)]}}{i(\omega_\mathbf{k} - \nu) + \gamma_\mathbf{k}} \right.$$

$$\left. + \sum_{\mathbf{k}'\neq\mathbf{k}} V_{s,|\mathbf{k}-\mathbf{k}'|}\int_{-\infty}^{t} dt'\, e^{i(\omega_\mathbf{k} + \gamma_\mathbf{k})(t'-t)}p_{\mathbf{k}'}(t') \right] . \tag{3}$$

We solve Eq. (3) by iteration in powers of the screened Coulomb interaction energy $V_{s,|\mathbf{k}-\mathbf{k}'|}$. To lowest order, i.e., setting $V_{s,|\mathbf{k}-\mathbf{k}'|} = 0$ in Eq. (3), we find the free-carrier result (3.44)

$$p_{\mathbf{k}}^{(0)}(t) = -\frac{i\mu_{\mathbf{k}}}{2\hbar}E(z)e^{i[Kz - \nu t - \phi(z)]}\frac{n_{e\mathbf{k}}(t) + n_{h\mathbf{k}}(t) - 1}{i(\omega_{\mathbf{k}}-\nu) + \gamma_{\mathbf{k}}}$$

$$= \frac{1}{2}E(z)e^{i[Kz - \nu t - \phi(z)]}\chi_{\mathbf{k}}^{(0)}(t) \ , \tag{4}$$

where for later convenience we introduce the k-dependent susceptibility function

$$\chi_{\mathbf{k}}^{(0)}(t) = -\frac{i\mu_{\mathbf{k}}}{\hbar}\frac{n_{e\mathbf{k}}(t) + n_{h\mathbf{k}}(t) - 1}{i(\omega_{\mathbf{k}}-\nu) + \gamma_{\mathbf{k}}} . \tag{5}$$

Substituting the lowest-order result into Eq. (3) and noting that $\chi_{\mathbf{k}}^{(0)}(t)$ varies little in the time T_2, we find the first-order contribution

$$p_{\mathbf{k}}^{(1)}(t) = -\frac{i}{2\hbar}E(z)e^{i[Kz-\nu t-\phi(z)]}\frac{n_{e\mathbf{k}}(t) + n_{h\mathbf{k}}(t) - 1}{i(\omega_{\mathbf{k}}-\nu) + \gamma_{\mathbf{k}}}\sum_{\mathbf{k'}\neq\mathbf{k}}V_{s,|\mathbf{k}-\mathbf{k'}|}\chi_{\mathbf{k}}^{(0)}(t)$$

$$= \frac{1}{2}E(z)e^{i[Kz-\nu t-\phi(z)]}\chi_{\mathbf{k}}^{(0)}(t)q(\mathbf{k},t) \ , \tag{6}$$

where the complex dimensionless factor

$$q(\mathbf{k},t) = \frac{1}{\mu_{\mathbf{k}}}\sum_{\mathbf{k'}\neq\mathbf{k}}V_{s,|\mathbf{k}-\mathbf{k'}|}\chi_{\mathbf{k}}^{(0)}(t) . \tag{7}$$

In principle, we could continue to iterate Eq. (3) in this way until reaching any desired accuracy. However, this process does not converge rapidly and offers no substantial CPU-time improvement over the direct numerical solution of Eq. (3) obtained by discretizing the integral and using a matrix inversion [Haug and Koch (1993)]. On the other hand, we obtain a remarkably accurate result for gain media by treating $p_{\mathbf{k}}^{(0)}$ and $p_{\mathbf{k}}^{(1)}$ as the first two terms of a geometrical series, which we then "resum" [Haug and Koch (1989)]. This approximation is the simplest kind of Padé approximation [see, e.g., Gaves-Morris (1973)]. Accordingly adding Eqs. (4) and (6) and resumming, we find

$$p_{\mathbf{k}}(t) \simeq p_{\mathbf{k}}^{(0)} + p_{\mathbf{k}}^{(1)} = \frac{1}{2}E(z)e^{i[Kz - \nu t - \phi(z)]}\chi_{\mathbf{k}}^{(0)}(t)[1 + q(\mathbf{k},t)]$$

$$\simeq \frac{1}{2}E(z)e^{i[Kz - \nu t - \phi(z)]}\frac{\chi_{\mathbf{k}}^{(0)}(t)}{1 - q(\mathbf{k},t)} . \tag{8}$$

Substituting this equation into Eq. (2.117), using the constitutive relation (2.69), and the self-consistency equations (2.72) and (2.73), we find the gain and carrier-induced phase-shift

$$g - i\frac{d\phi}{dz} = \frac{iK}{2}\chi = \frac{K}{2\epsilon\hbar V}\sum_{\mathbf{k}}\frac{|\mu_{\mathbf{k}}|^2}{1 - q(\mathbf{k})}\frac{n_{e\mathbf{k}} + n_{h\mathbf{k}} - 1}{\gamma_{\mathbf{k}} + i(\omega_{\mathbf{k}}-\nu)} \qquad (9)$$

where the carrier-induced phase-shift is related to the carrier-induced refractive index by Eq. (2.77) and K is the magnitude of the laser-field wave vector. Equation (9) is quite convenient for numerical evaluation since it expresses the many-body susceptibility in terms of the free-carrier result (3.50), which is given by Eq. (9) with $q = 0$. The many-body effects can be identified explicitly as 1) a carrier density dependence of the transition energy, $\omega_{\mathbf{k}}(N)$, and 2) the factor $1/(1-q(\mathbf{k}))$, which represents the Coulomb enhancement in our Padé approximation.

5-2. Bulk Semiconductors

Using Eq. (4.29) for V_{sq} with (4.27) for ω_q^2, we can write the screened Coulomb potential for a bulk semiconductor medium as

$$V_{sq} = \frac{8\pi\epsilon_R a_0}{Vq^2}\left[\frac{\epsilon_q}{\epsilon_\kappa} + \frac{C}{4}\frac{\epsilon_q^2}{\epsilon_{pl}^2}\right]\left[1 + \frac{\epsilon_q}{\epsilon_\kappa} + \frac{C}{4}\frac{\epsilon_q^2}{\epsilon_{pl}^2}\right]^{-1}, \qquad (10)$$

where a_0 and ϵ_R are the exciton Bohr radius and the Rydberg energy, $\epsilon_q = \hbar^2 q^2/2m_r$, $\epsilon_\kappa = \hbar^2\kappa^2/2m_r$, $\epsilon_{pl} = \hbar\omega_{pl}$, κ is the inverse screening length, and ω_{pl} is the electron-hole plasma frequency. The vector $\mathbf{q}$ equals $\mathbf{k} - \mathbf{k}'$ so that $q^2 = k^2 + k'^2 - 2kk'\cos\theta$, where θ is the angle between $\mathbf{k}$ and $\mathbf{k}'$. Substituting Eq. (10) into Eq. (7) and evaluating the summation as an integral, we get

$$q(k) = -\frac{i\epsilon_R^2 a_0^3}{\pi\hbar}\int_0^\infty dk'\frac{k'^2}{\sqrt{\epsilon_k\epsilon_{k'}}}\frac{1 + \epsilon_k/\epsilon_g}{1 + \epsilon_{k'}/\epsilon_g}\frac{f_{ek'}+f_{hk'}-1}{\gamma + i(\omega_{k'}-\nu)}\Theta(k,k'), \qquad (11)$$

where we drop the k-dependence of $\gamma_{\mathbf{k}}$, in anticipation of evaluating it at the wavenumber of the relevant laser mode. Furthermore, we assume quasiequilibrium conditions so that $n_{\alpha k} \simeq f_{\alpha k}$, where the Fermi-Dirac distributions $f_{\alpha k}$ are given by Eq. (2.12). The angular integration function

$$\Theta(k,k') = \int_0^\pi d\theta \sin\theta \frac{1 + \dfrac{C}{4}\dfrac{\varepsilon_\kappa \varepsilon_q}{\varepsilon_{pl}^2}}{1 + \dfrac{\varepsilon_q}{\varepsilon_\kappa} + \dfrac{C}{4}\dfrac{\varepsilon_q^2}{\varepsilon_{pl}^2}} = \Theta_1(k,k') + \Theta_2(k,k') \qquad (12)$$

can be evaluated analytically. For the first term, we get

$$\Theta_1(k,k') = \frac{1}{2}\ln\left[\frac{1 + \dfrac{\varepsilon_{k_+}}{\varepsilon_\kappa} + \dfrac{C}{4}\dfrac{\varepsilon_{k_+}^2}{\varepsilon_{pl}^2}}{1 + \dfrac{\varepsilon_{k_-}}{\varepsilon_\kappa} + \dfrac{C}{4}\dfrac{\varepsilon_{k_-}^2}{\varepsilon_{pl}^2}}\right] , \qquad (13)$$

where $k_\pm = |\mathbf{k} \pm \mathbf{k'}|$. The second term is given by

$$\Theta_2(k,k') = \frac{1}{\sqrt{D}}\left\{ \text{atan}\left[\frac{1}{\sqrt{D}}\left(\frac{C\varepsilon_{k_+}\varepsilon_\kappa}{2\varepsilon_{pl}^2}+1\right)\right] - \text{atan}\left[\frac{1}{\sqrt{D}}\left(\frac{C\varepsilon_{k_-}\varepsilon_\kappa}{2\varepsilon_{pl}^2}+1\right)\right]\right\},$$

$$\qquad (14)$$

for $D \equiv C\varepsilon_\kappa^2/\varepsilon_{pl}^2 - 1 > 0$ and

$$\Theta_2(k,k') = \frac{1}{2\sqrt{-D}}\ln\left[\frac{\dfrac{C\varepsilon_{k_+}\varepsilon_\kappa}{2\varepsilon_{pl}^2} + 1 + \sqrt{-D}}{\dfrac{C\varepsilon_{k_-}\varepsilon_\kappa}{2\varepsilon_{pl}^2} + 1 + \sqrt{-D}} \frac{\dfrac{C\varepsilon_{k_-}\varepsilon_\kappa}{2\varepsilon_{pl}^2} + 1 - \sqrt{-D}}{\dfrac{C\varepsilon_{k_+}\varepsilon_\kappa}{2\varepsilon_{pl}^2} + 1 - \sqrt{-D}}\right] \qquad (15)$$

for $D \leq 0$. The k'-integration in Eq. (11) has to be performed numerically.

The Coulomb-Hole self-energy contribution to the bandgap renormalization is given by Eq. (4.48) and the screened-exchange contribution (4.49) may be approximated by the integral

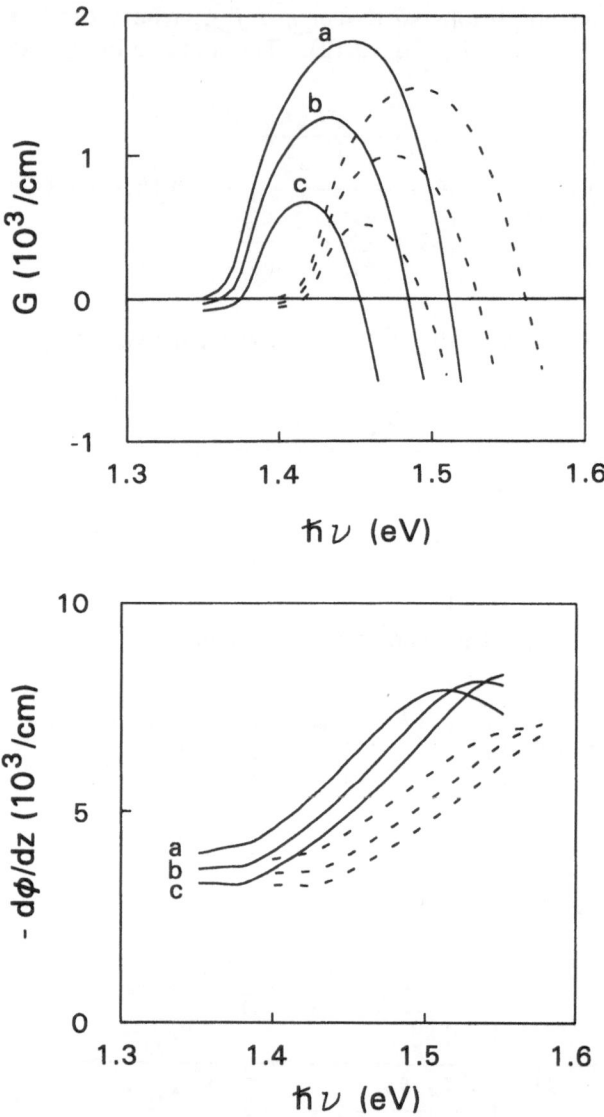

Fig. 5-1. (top) Intensity gain $G = 2\,g$ and (bottom) carrier-in-duced phase shift versus photon energy $\hbar\nu$ according to the many-body theory (solid lines) and free-carrier theory (dashed lines). The carrier densities are $N = 3\times10^{18}\ cm^{-3}$, $4\times10^{18}\ cm^{-3}$, and $5\times10^{18}\ cm^{-3}$ in order of increasing maxima. We use the parameters for bulk GaAs at room temperature.

$$\Delta\varepsilon_{SXk} = -\frac{4}{\pi}\varepsilon_R a_0 \int_0^\infty dk' \frac{\dfrac{\varepsilon_{k'}}{\varepsilon_\kappa} + \dfrac{C}{4}\dfrac{\varepsilon_{k'}{}^2}{\varepsilon_{pl}{}^2}}{1 + \dfrac{\varepsilon_{k'}}{\varepsilon_\kappa} + \dfrac{C}{4}\dfrac{\varepsilon_{k'}{}^2}{\varepsilon_{pl}{}^2}} [f_{ek'} + f_{hk'}]. \tag{16}$$

Using Eqs. (13) - (15) and (4.49) in Eq. (11), and the result in Eq. (9), we calculate the gain and carrier-induced phase-shift. For bulk semiconductors, we convert the summation over states in Eq. (9) to a three-dimensional integral and find

$$g - i\frac{d\phi}{dz} = \frac{\nu|\mu|^2}{2\varepsilon_0 nc\hbar\pi^2} \int_0^\infty \frac{dk\,k^2}{(1 + \varepsilon_k/\varepsilon_g)^2} \frac{f_{ek} + f_{hk} - 1}{\gamma + i(\omega_k - \nu)} \frac{1}{1 - q(k)}, \tag{17}$$

where we again used the quasiequilibrium approximation. Figure 5-1 plots gain and carrier-induced phase shift spectra calculated using Eq. (17) for an undoped bulk GaAs medium and different carrier densities. In the free-carrier results (dashed curves), band filling is the only cause for the density dependence of gain and phase shift, while in the many-body results, bandgap renormalization and Coulomb enhancement also contribute. An obvious difference between the two theories is the overall frequency shift due to the bandgap renormalization. The many-body spectra for the different carrier densities are also frequency shifted relative to one another since the bandgap renormalization is a carrier density dependent function. Note that Fig. 5-1 (top) shows the existence of gain at frequencies below the unexcited semiconductor bandgap, which for GaAs is $1.42eV$. This feature is a consequence of the bandgap renormalization, which decreases the bandgap when the carrier density is increased in GaAs-type materials. The spectral region of optical gain is basically bounded by the renormalized bandgap from below and by the total chemical potential from above

$$\varepsilon_g' \le \hbar\omega_{gain} \le \varepsilon_g' + \mu_e + \mu_h, \tag{18}$$

which replaces ε_{g0} in Eq. (3.74) by ε_g'.

Coulomb enhancement has the effect of reshaping and increasing the magnitude of the gain and absorption spectra. It is most noticeable at low carrier densities, especially when it is possible to resolve the exciton absorption peaks. At the elevated densities needed for gain, Coulomb enhancement effects are not as drastic because plasma screening mitigates the electron-hole Coulomb attraction. Figure 5-2 shows the effects of Coulomb enhancement at a carrier density of 4×10^{18} cm^{-3}. The dashed curves are obtained using the renormalized bandgap in the free-carrier gain formula. Even though this ad hoc inclusion of bandgap renormaliza-

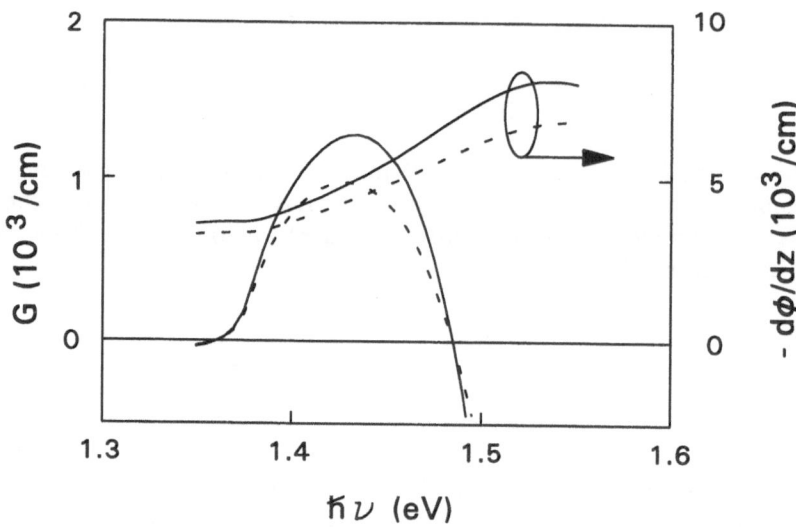

Fig. 5-2. Intensity gain G and carrier-induced phase shift versus photon energy $\hbar\nu$ predicted by many-body theory (solid line) and by free-carrier theory (dashed line) with ad hoc bandgap renormalization. The carrier density is $N = 4\times10^{18}\ cm^{-3}$.

tion into the free-carrier theory is sometimes used in the literature, it is actually inconsistent because it neglects the Coulomb attraction between an electron and a hole, while it takes into account the exchange interaction and Coulomb repulsion between two electrons or two holes. To be consistent, Coulomb attraction and repulsion have to be treated at the same level of approximation. When using Eq. (9), we need to keep in mind that the factor $1/[1-q(k)]$ is only valid at high carrier densities. It ignores the existence of the bound electron-hole states (excitons), and is therefore not correct for densities below the Mott density.

Figure 5-3 shows the peak gains (G_{peak}) as predicted by the many-body theory and the free-carrier theory for a bulk GaAs. The difference between the two curves is due to Coulomb enhancement, which causes an approximately 20% increase in the peak gain. Figure 5-3 also shows the dependence of the peak gain frequency ν_{peak} on carrier density. Both band filling and Coulomb enhancement lead to a blue shift of the peak-gain frequency, whereas bandgap renormalization leads to a red shift. In a bulk semiconductor with GaAs-type effective masses, band-filling effects dominate the density-dependent peak-gain shift. The difference between the solid and dot-dashed ν_{peak} curves in Fig. 5-3 is due to the reshaping of the gain spectra by Coulomb enhancement. Note that according to the many-body theory, the peak gain for a carrier density between 2×10^{18}

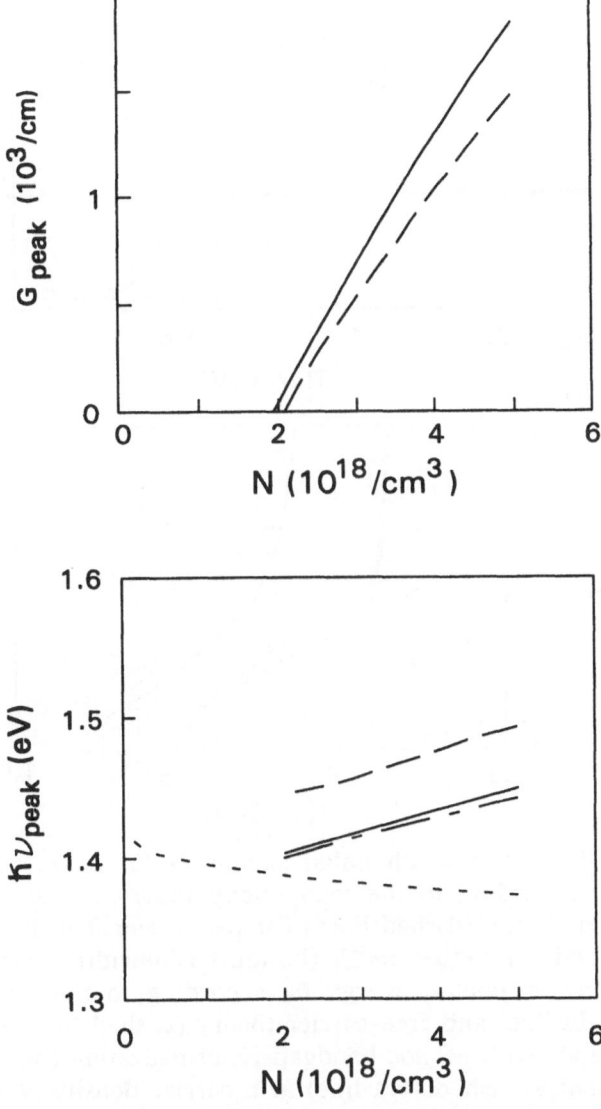

Fig. 5-3. (top) Peak gain G_{peak} and (bottom) peak-gain energy $\hbar\nu_{peak}$ versus carrier density N according to the many-body theory (solid lines), free-carrier theory (dashed lines), and free-carrier theory with a renormalized bandgap (dot-dashed lines). For the peak gain, the two free-carrier models coincide. The renormalized bandgap is indicated by the dotted curve.

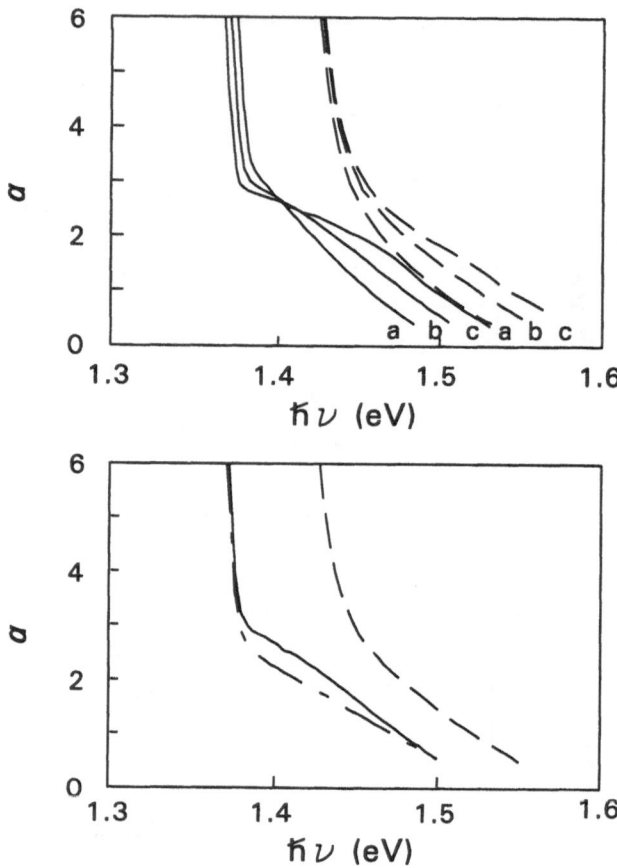

Fig. 5-4. (top) Linewidth enhancement factor α versus photon energy $\hbar\nu$ according to the many-body theory (solid lines) and free-carrier theory (dashed lines) for the carrier densities $N = (a)$ 3, (b) 4, and (c) 5×10^{18} cm^{-3}. (bottom) Linewidth enhancement factor α versus photon energy $\hbar\nu$ according to the many-body theory (solid line) and free-carrier theory (dashed line) and free-carrier theory with ad hoc bandgap renormalization (no Coulomb enhancement - dash-dotted line) at a carrier density $N = 4 \times 10^{18}$ cm^{-3}.

cm^{-3} and slightly above 3×10^{18} cm^{-3} occurs below the unrenormalized bandgap, around the frequency of the exciton resonance of the unexcited material.

As discussed in Sec. 3-6, the linewidth enhancement or antiguiding factor α is a useful quantity in semiconductor laser theory. Here we present some numerical results for the α factor, mainly to emphasize the importance of the carrier-induced refractive-index (phase shift) changes. Examples of the computed α spectra for bulk GaAs are shown in Fig. 5-4. Note the appearance of a saddle in the many-body results at high carrier densities. This saddle is observed in experiments in bulk and quantum-well gain media. Our theory associates the existence of this saddle with the many-body interactions.

Figure 5-4 (bottom) shows the different many-body contributions to the α spectra. Both bandgap renormalization and Coulomb enhancement have noticeable effects. A quantitative comparison between the experimental and theoretical spectra requires a more accurate description of the bandstructure than the two band model. Therefore we postpone such a comparison to Chap. 6, where we include realistic bandstructures.

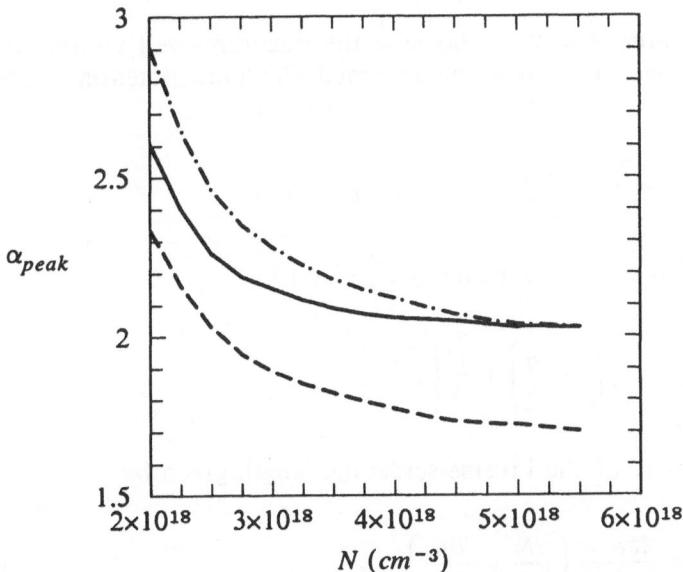

Fig. 5-5. Linewidth enhancement factor α at the peak gain frequency versus carrier density. The solid curve is the many-body result, the dash-dotted curve is the free-carrier result and the dashed curve is obtained using the renormalized bandgap in the free-carrier theory.

In Fig. 5-5, we show the value of α at the peak gain frequency as a function of carrier density. We see that the inclusion of the bandgap renormalization in the free-carrier theory actually yields worse agreement with the many-body results than the unmodified free-carrier theory. By ignoring the Coulomb enhancement, we significantly underestimate the values of α at the peak gain.

5-3. Quantum Wells

To treat the many-body Coulomb interactions in a quantum well, we follow a procedure that is similar to our analysis of the bulk medium. Typically well widths for lasers are less than 15nm, which is sufficiently narrow to approximate the carriers as a two-dimensional plasma. Then $r \simeq x^2 + y^2$ and z is direction perpendicular to the well. The two-dimensional Fourier transform of the Coulomb potential is

$$V_q = \frac{2\pi e^2}{\epsilon_b \, qA},\tag{19}$$

where the area $A = V/w$ and w is the quantum-well width. A derivation similar to Sec. 4-2 gives the screened Coulomb potential (4.29) with the two-dimensional plasma frequency ω_{pl} given by

$$\omega_{pl}^2 = \frac{2\pi e^2 N w q}{\epsilon_b m_r} = 8\pi N w a_0{}^3 \epsilon_R{}^2 q/\hbar^2 \, ,\tag{20}$$

the effective plasmon frequency ω_q given by

$$\omega_q^2 = \omega_{pl}^2 \left(1 + \frac{q}{\kappa} \right) + \frac{C}{4} \left[\frac{\hbar q^2}{2m_r} \right]^2 ,\tag{21}$$

and the square of the inverse screening length given by

$$\kappa = \frac{2\pi e^2 w}{\epsilon_b} \left(\frac{\partial N}{\partial \mu_e} + \frac{\partial N}{\partial \mu_h} \right) . \; \cdot\tag{22}$$

In the two-band and quasiequilibrium approximations near the band edge, Eq. (22) yields

$$\kappa = \frac{2}{a_0} \left[\frac{m_e}{m_r} f_{e0} + \frac{m_h}{m_r} f_{h0} \right] ,\tag{23}$$

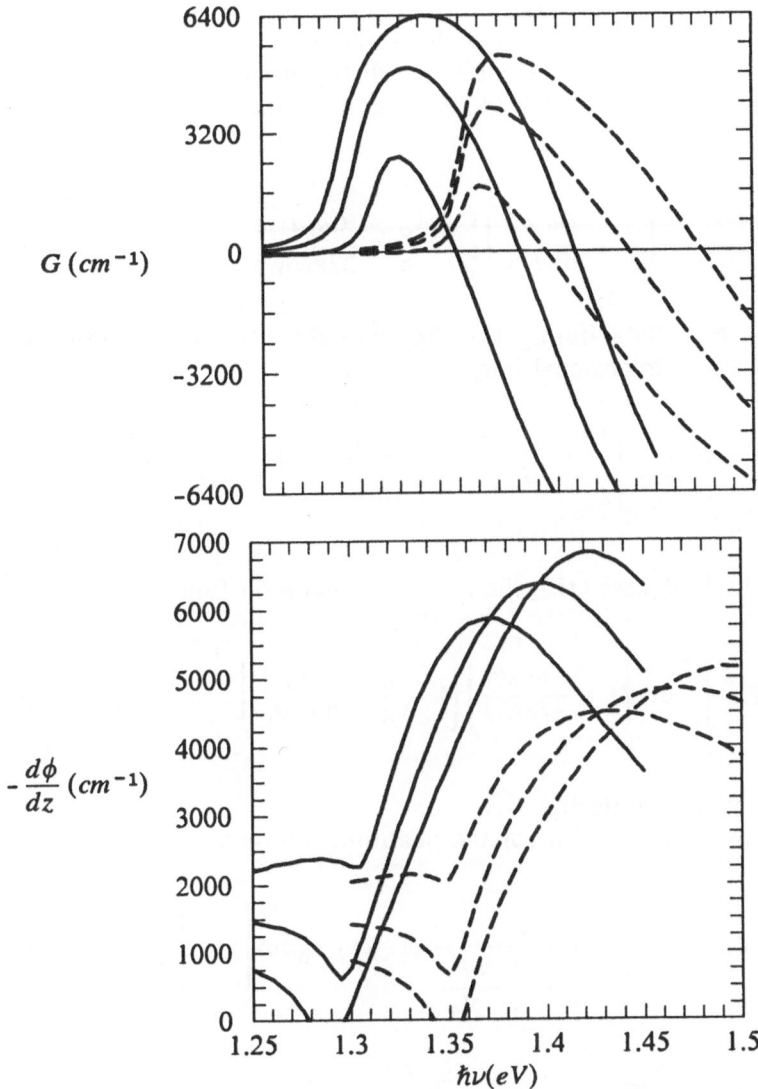

Fig. 5-6. (top) quantum-well gain and (bottom) carrier-induced phase shift spectra for carrier densities $N = 2 \times 10^{18}$ cm^{-3}, 3×10^{18} cm^{-3}, and 4×10^{18} cm^{-3} in order of increasing peak gain and peak index according to the many-body theory (solid lines), and free-carrier theory (dashed lines). We use the parameters for a 5nm $In_{.1} Ga_{.9} As/GaAs$ strained quantum-well structure.

where $f_{\alpha 0}$ denotes the Fermi–Dirac distribution at $k = 0$. This analytic result can be verified easily by using the explicit $2D$ expression (2.33) for the chemical potential and the Fermi–Dirac distribution for $k = 0$.

From Eqs. (19) – (23) and (4.29), we find the screened Coulomb potential

$$V_{sq} = \frac{4\pi\epsilon_R a_0}{Aq}\left[\frac{q}{\kappa} + \frac{Ca_0 q^3}{32\pi Nw}\right]\left[1 + \frac{q}{\kappa} + \frac{Ca_0 q^3}{32\pi Nw}\right]^{-1}, \qquad (24)$$

where $q = |\mathbf{k} - \mathbf{k}'|$. Substituting Eq. (24) into Eq. (7) and evaluating the summation as a two-dimensional integral, we obtain

$$q(k) = -\frac{ia_0\epsilon_R}{\pi\hbar\kappa\mu_k}\int_0^\infty dk'\, k'\, \mu_{k'}\, \frac{f_{ek'} + f_{hk'} - 1}{\gamma + i(\omega_{k'}-\nu)}\,\Theta(k,k'), \qquad (25)$$

where, unlike the bulk case (12), the angular integration function

$$\Theta(k,k') = \int_0^{2\pi} d\theta \left[1 + \frac{C\kappa a_0 q^2}{32\pi Nw}\right]\left[1 + \frac{q}{\kappa} + \frac{Ca_0 q^3}{32\pi Nw}\right]^{-1} \qquad (26)$$

cannot be evaluated analytically.

Equations (4.47) and (4.49) for the quantum well give

$$\Delta\varepsilon_{CH} = -2\varepsilon_R a_0 \int_0^\infty \frac{dq}{1 + \dfrac{q}{\kappa} + \dfrac{Ca_0 q^3}{32\pi Nw}} \simeq -2\varepsilon_R a_0 \kappa \ln\left[1 + \sqrt{\frac{32\pi Nw}{C\kappa^3 a_0}}\right] \qquad (27)$$

$$\Delta\varepsilon_{SX} = -\frac{2\varepsilon_R a_0}{\kappa}\int_0^\infty dk\, k\, \frac{1 + \dfrac{C\kappa a_0 k^2}{32\pi Nw}}{1 + \dfrac{k}{\kappa} + \dfrac{Ca_0 k^3}{32\pi Nw}}\,[f_{ek} + f_{hk}]. \qquad (28)$$

Equations (25), (27) and (28) are then used in Eq. (9) to calculate the gain and carrier-induced phase shift. Converting the summation over states in Eq. (9) to a two-dimensional integral, we find

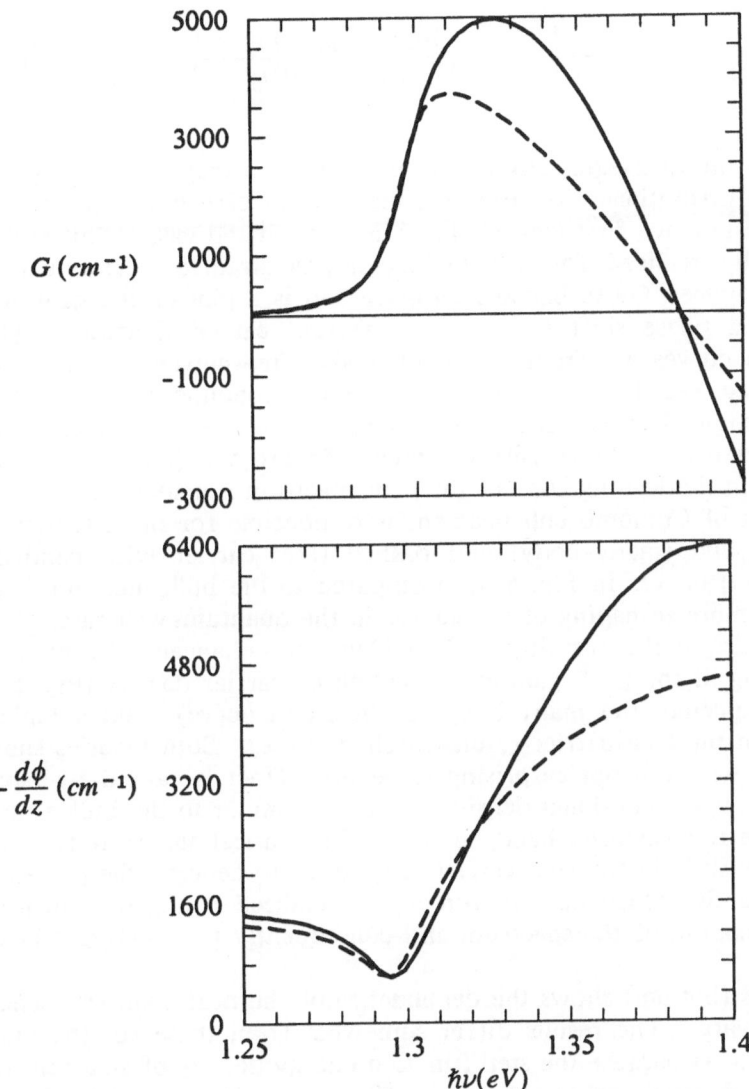

Fig. 5-7. Coulomb enhancement effects on gain and carrier-induced phase shift spectra in a quantum-well medium according to the many-body theory (solid lines) and the free-carrier theory (dashed lines) with an ad hoc inclusion of bandgap renormalization. The curves are for a carrier density of 3×10^{18} cm^{-3}.

$$g - i\frac{d\phi}{dz} = \frac{\nu}{2\pi\epsilon_0 nc\hbar w} \int_0^\infty dk\,k\; \frac{|\mu_k|^2}{1 - q(\mathbf{k})}\; \frac{f_{ek} + f_{hk} - 1}{\gamma + i(\omega_{\mathbf{k}} - \nu)}.$$ (29)

Here we used the quasiequilibrium approximation, so that f_{ek} and f_{hk} are Fermi-Dirac distributions. To study the many-body effects in a quantum-well gain medium, we first look at the simple two-band case. This situation is actually realized for a 5nm $In_{.1}Ga_{.9}As$ strained quantum well sandwiched between GaAs barriers. Figure 5-6 is a plot of the gain and carrier-induced phase shift spectra for different carrier densities. The solid (dashed) curves are from the many-body (free-carrier) theory. As discussed in Sec. 3-6, band-edge effects are more pronounced in the quantum-well medium because of the step function instead of square root energy dependence of the density of states. Figure 5-6 (top) shows that the sharpness in the leading edge of the gain spectrum is smoothed out.

The effect of Coulomb enhancement is responsible for the differences between the solid (many-body) and dashed (free-carrier with bandgap renormalization) curves in Fig. 5-7. Compared to the bulk medium (Fig. 5-2), there is more reshaping of the curves in the quantum-well case. One gets a better idea of the magnitude of the Coulomb enhancement contribution by looking at the peak gain as a function of carrier density [Fig. 5-8 (top)]. As expected, the many-body results (solid curve) yield a higher peak gain than the free-carrier results (dashed curve). Both theories show gain rollover, which is not surprising since this effect is caused by band filling with a two-dimensional density of states. Similar to the bulk situation, using the renormalized bandgap in the free-carrier theory only leads to a frequency shift in the free-carrier spectrum. In general, the presence of other subbands, which have different renormalized bandgaps, can lead to further reshaping of the spectrum and consequently to a different peak gain.

Figure 5-8 (bottom) shows the dependence of the peak gain frequency on carrier density. The results differ somewhat from those for the bulk medium (Fig. 5-3) because the step function energy density of states in the quantum-well medium reduces the band-filling contribution to the peak-gain frequency shift, so that the many-body interactions play a greater role. Consequently, the many-body (solid curve) and free-carrier (dashed curve) predictions of $d\nu_{pk}/dN$ differ more than in the bulk case. These differences are not bigger since the effects of Coulomb enhancement and bandgap renormalization are opposed to each other. This is best demonstrated by noting that without Coulomb enhancement [dot-dashed curve in Fig. 5-8 (bottom)], the peak-gain frequency shifts red with increasing carrier density.

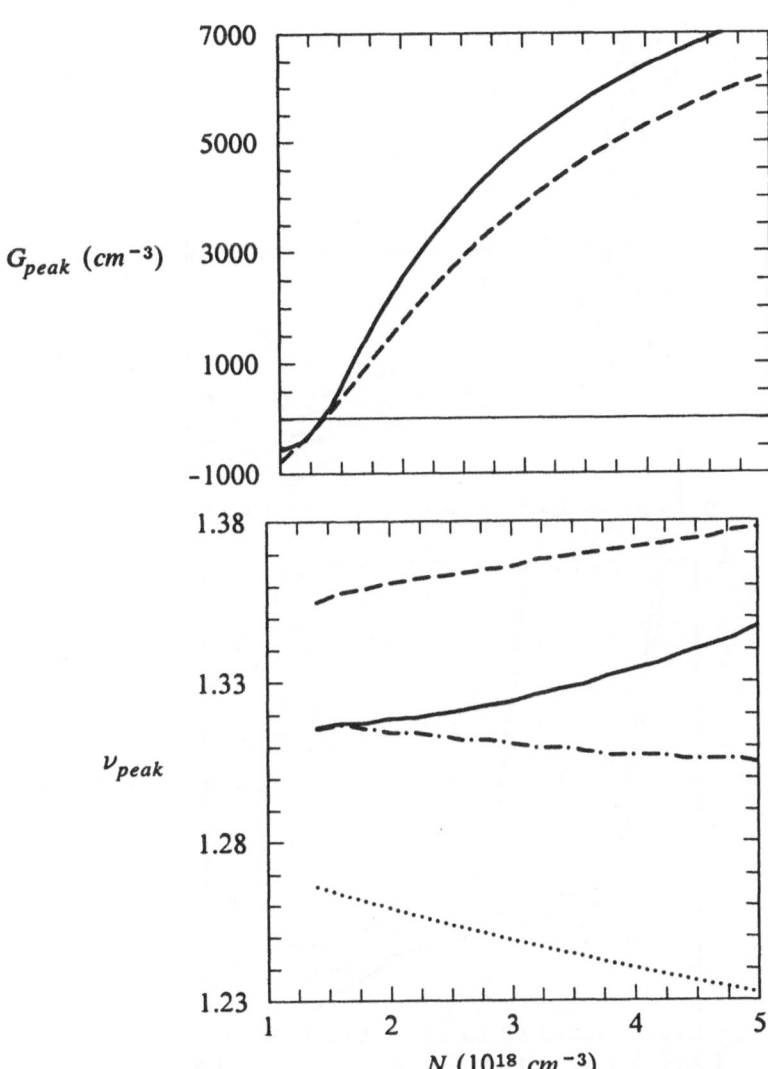

Fig. 5-8. (top) many-body (solid curve) and free-carrier (dashed curve) predictions of quantum-well peak gain versus carrier density. (bottom) Peak-gain frequency versus carrier density for the many-body (solid line), the free-carrier theory (dashed line), and the free-carrier theory with renormalized bandgap (dot-dashed curve). The dotted curve is the renormalized bandgap.

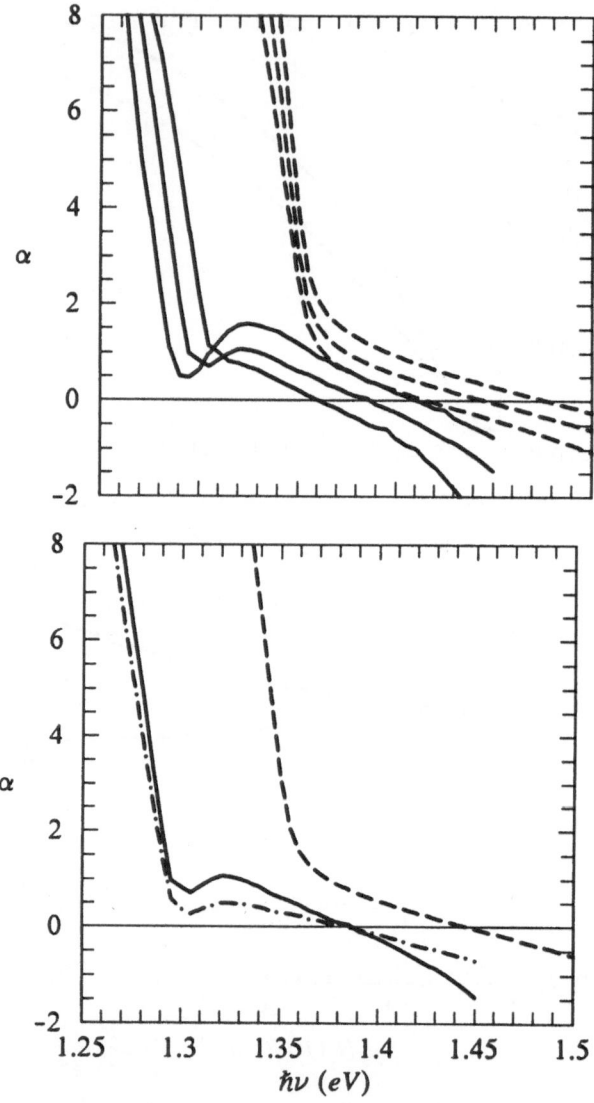

Fig. 5-9. (top) quantum-well linewidth-enhancement factor α versus photon energy for carrier densities 2, 3, and 4×10^{18} cm^{-3}. (bottom) shows the different many-body contributions for a carrier density of 3×10^{18} cm^{-3}. The solid curves are the many-body results, the dashed curves are the free-carrier results and the dotted curve is obtained using the renormalized bandgap in the free-carrier theory.

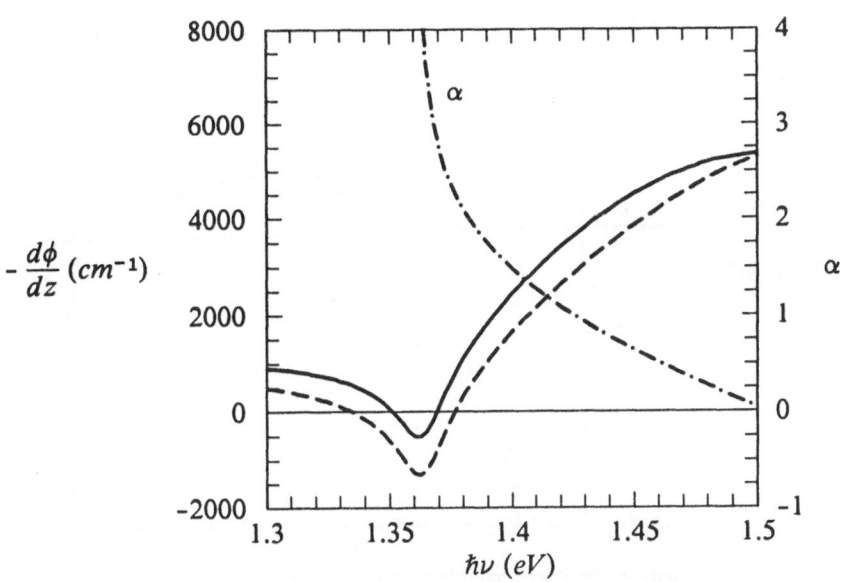

Fig. 5-10. Phase shift spectra for the carrier densities 4×10^{18} cm^{-3} (solid curve) and 5×10^{18} cm^{-3} (dashed curve) and the corresponding linewidth-enhancement factor α spectrum for the free-carrier theory.

As alluded to earlier, the more pronounced features in the quantum-well gain and phase-shift spectra lead to noticeable differences in the α spectra for bulk and quantum-well materials. The solid curves in Fig. 5-9 (top) show substantial structure in the quantum-well α spectra, especially at high carrier densities. Comparison with the free-carrier spectra (dashed curves) reveals that the structures are completely due to the many-body interactions. Figure 5-9 (bottom) shows that both bandgap renormalization and Coulomb enhancement contribute to the α structure. The physical mechanisms responsible for the strong frequency dependence of α is analyzed in Fig. 5-10. Figure 5-10 shows the phase change $d\phi/dz$ for two carrier densities as predicted by the free-carrier theory. The dashed curve is for a higher density than the solid curve. The difference between the two curves gives the numerator of α, whose spectrum is shown in the right graph. Bandgap renormalization shifts the two curves by different amounts. The higher density (dashed) curve is shifted more and Fig. 5-11 (top) shows that this leads to an overall smaller separation between the two curves. The effects of the irregularities in the differences of $d\phi/dz$ lead to the pronounced structure in the α spectrum. Crossings between the $d\phi/dz$ curves result in zero and negative values for α. Coulomb enhancement tends to separate the curves, giving higher values for α, but the structure in

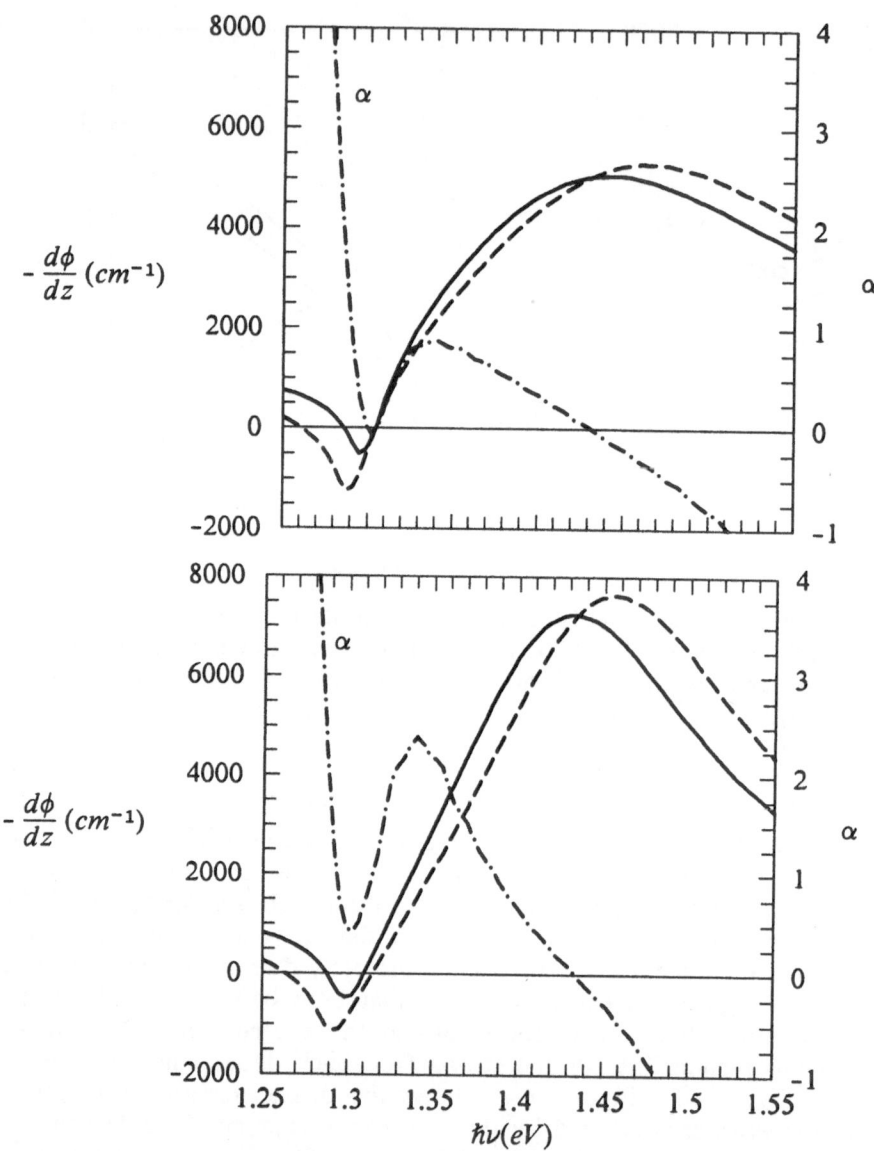

Fig. 5-11. Phase shift spectra for the carrier densities $4 \times 10^{18}\ cm^{-3}$ (solid curve) and $5 \times 10^{18}\ cm^{-3}$ (dashed curve) and the corresponding linewidth-enhancement factor α spectrum for the free-carrier theory with ad hoc bandgap renormalization (top) and for the full many-body theory (bottom).

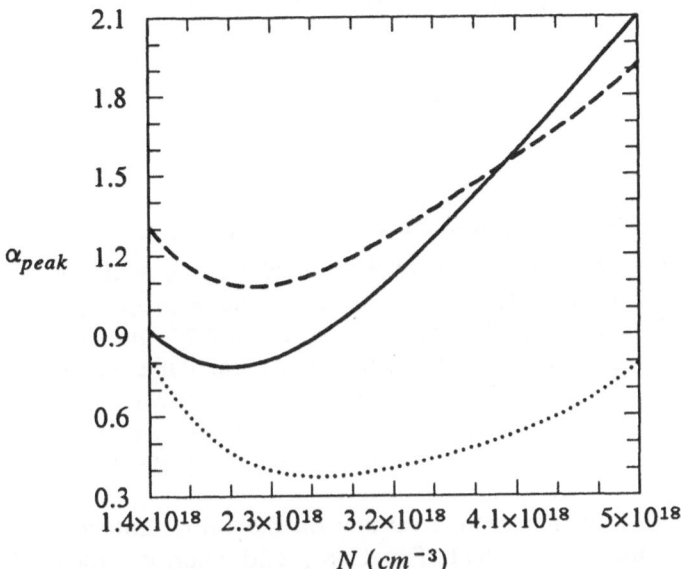

Fig. 5-12. Linewidth enhancement factor α at the peak gain frequency versus carrier density. The solid curve is the many-body result, the long-dashed curve is the free-carrier result, and the dot-dashed curve is obtained using the renormalized bandgap in the free-carrier theory.

the spectrum remains [Fig. 5-11 (bottom)]. Finally, Fig. 5-12 depicts the dependence of α at the peak gain frequency on carrier density. As in the bulk case, the ad hoc inclusion of bandgap renormalization to the free-carrier theory results in worse agreement with the many-body results.

REFERENCES

See also the references to Chap. 4.

Gaves-Morris, P.R. (1973), Ed., *Padé Approximants and Their Application*, Academic Press, N.Y.

Haug, H. and S.W. Koch (1989), Phys. Rev. **A39**, 1887.

Haug, H. and S.W. Koch (1993), *Quantum Theory of the Optical and Electronic Properties of Semiconductors*, 2nd Edition, World Scientific, Singapore.

Chapter 6
BAND MIXING AND STRAIN IN QUANTUM WELLS

The ability to change the energy-level structure of the gain medium through material and structure design is a unique property of semiconductor lasers. To take full advantage of this capability, one needs to be able to predict the bandstructure that results from a particular laser heterostructure arrangement. In this chapter, we show a procedure for performing bandstructure calculations that are relevant to the laser physicist. In order to be concise, the details of the derivation of many of the formulas are left to Apps. $B - D$.

Section 6-1 begins with a review of the important concepts of lattice periodicity, unit cells, Bloch functions, and energy bands. Section 6-2 describes the zone center ($k = 0$) states that are relevant for laser transitions that take place in a III-V compound like GaAs. To calculate the bandstructure that evolves from these states, we use the Kane or $\mathbf{k \cdot p}$ theory, which is described in Sec. 6-3. Kane theory allows one to compute the bandstructure in the neigborhood of zone center perturbatively, using the energy eigenvalues and the basic symmetries of the zone-center states. Specializing to the top valence bands in typical III-V compounds, Sec. 6-4 introduces the Luttinger theory, which is most convenient to use when valence-band mixing effects become important. Since we are interested in the bandstructure modifications under quantum-confinement conditions, such as quantum wells or superlattices, we include the confinement geometry in the Luttinger Hamiltonian in Sec. 6-5. As we show, quantum confinement can modify valence-band-structure significantly, mixing especially the top two bulk semiconductor valence bands, namely, the heavy-hole and light-hole bands. In order to deal with semiconductor heterostructures that are not grown under lattice-matched conditions, Sec. 6-6 introduces the concept of elastically strained systems and includes these strain effects in our valence-band-structure calculations. Section 6-7 summarizes the results of the earlier sections by presenting a cook book description of a bandstructure calculation. The remaining sections apply the theory and techniques developed in Secs. 6-1 through 6-7 to study some common laser heterostructures. The GaAs-AlGaAs quantum well is treated in Sec. 6-8, where we demonstrate quantum-well width effects. Section 6-9 studies the strained quantum-well structure, InGaAs-AlGaAs. This study reveals the increase in the average highest valence band curvature with compressive strain. We also illustrate consequences of the strain-induced bandstructure changes, such as the reduced carrier density for transparency and polarization discrimination. Section 6-10 describes the strained quantum-well structure, InGaAs-InP, which has the property of being tensile-strained, unstrained or compressive-strained, depending on the InGaAs composition. In this section, we demonstrate the effects of strain on the threshold carrier density, the differential gain, and the linewidth enhancement factor. Finally, Sec. 6-11 describes the InGaP-InAlGaP quantum well, which has gain at visible wavelengths. This section shows the dependence of the lasing wavelength on the heterostructure dimension and composition, and the dependence of the laser threshold on the lasing wavelength.

6-1. Bloch Theorem

An electron in a crystal sees a periodic potential due to the ions present at each lattice site. This potential is modified by delocalized electrons that are originally bound to each atom making up the crystal. The net result may be approximated by a periodic effective potential V_0, such that an electronic energy eigenstate $|\phi_{nk}\rangle$ in the solid obeys the time-independent Schrödinger equation

$$\left[\frac{p^2}{2m_0} + V_0 \right] |\phi_{nk}\rangle = \varepsilon_{nk} |\phi_{nk}\rangle , \tag{1}$$

where m_0 is the mass of the free electron, n is the band index, and k is the electron wavevector. Translational symmetry in the lattice dictates that the energy eigenfunctions obey the *Bloch theorem*

$$\boxed{\langle \mathbf{r} + \mathbf{R} | \phi_{nk} \rangle = e^{i\mathbf{k}\cdot\mathbf{R}} \langle \mathbf{r} | \phi_{nk} \rangle} \tag{2}$$

Bloch theorem

This condition is satisfied when

$$\langle \mathbf{r} | \phi_{nk} \rangle = e^{i\mathbf{k}\cdot\mathbf{r}} \langle \mathbf{r} | n\ \mathbf{k} \rangle , \tag{3}$$

$$\langle \mathbf{r} + \mathbf{R} | n\ \mathbf{k} \rangle = \langle \mathbf{r} | n\ \mathbf{k} \rangle , \tag{4}$$

where $\langle \mathbf{r} | n\ \mathbf{k} \rangle$ is the *lattice periodic function*. Substituting Eq. (3) into Eq.

(1), we find the eigenvalue equation for the lattice periodic part of the wavefunction

$$\left[\frac{p^2}{2m_0} + V_0 + \frac{\hbar}{m_0}\,\mathbf{k}\cdot\mathbf{p}\right]|n\,\mathbf{k}\rangle = \left[\varepsilon_{n\mathbf{k}} - \frac{\hbar^2 k^2}{2m_0}\right]|n\,\mathbf{k}\rangle . \tag{5}$$

The eigenstates are orthonormal

$$\langle \phi_{m\mathbf{k}} | \phi_{n\mathbf{q}} \rangle = \delta_{m,n}\,\delta_{\mathbf{k},\mathbf{q}} , \tag{6}$$

or equivalently,

$$\langle m\,\mathbf{k} | n\,\mathbf{k} \rangle = \delta_{m,n} . \tag{7}$$

The first-principles calculation of energy bands and eigenfunctions in a solid is a specialized area in the field of condensed matter physics. First one must develop an accurate model for the effective potential V_0. Then one solves Eq. (1), or equivalently Eq. (5), with the appropriate boundary conditions (2) or (4). The solution requires complicated numerical computation schemes, and a first-principles solution of all the $\mathbf{k}$ states involved in a laser transition can quickly become a prohibitively lengthy process. $\mathbf{k}\cdot\mathbf{p}$ theory provides a shortcut through this process. This theory allows one to compute the energy eigenstates and the wavefunctions in the vicinity of any given k, and in particular in the vicinity of $k = 0$, which is the region most relevant for optical transitions in lasers. To use the method, we need to know the $k = 0$ eigenstates, which are discussed in the next section.

6-2. Electronic States at $\mathbf{k} = 0$

For a III-V compound like GaAs, the conduction and valence band states at $k = 0$ are shown in Fig. 6-1. Without spin-orbit coupling, the $m_s = \pm\frac{1}{2}$ states are uncoupled and degenerate. For typographical simplicity, we suppress the spin indices in working with $|l\,m_l\rangle$ and include a factor of 2 in the appropriate sums over states. The conduction-band $k = 0$ state has s-like symmetry, with zero orbital angular momentum. We designate this state by $|S\rangle = |0\,0\rangle$. The $k = 0$ states in the top valence band have p-like symmetry, which may be represented by the $l = 1$ states, $|1\,\pm1\rangle$ and $|1\,0\rangle$.

In the next section, where we deal with the momentum-operator matrix elements, it is more convenient to work with the following combinations of orbital angular momentum eigenstates $|l\,m_l\rangle$

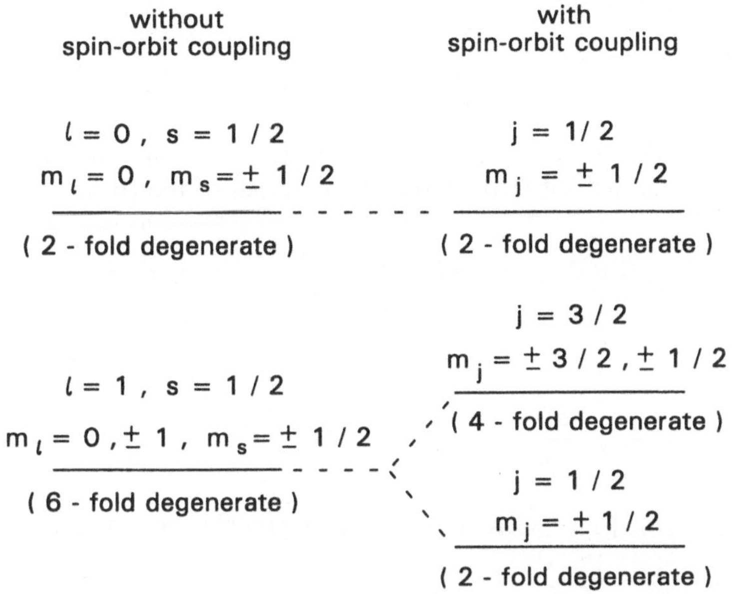

| without spin-orbit coupling | with spin-orbit coupling |

$$l = 0 , \; s = 1/2$$
$$m_l = 0 , \; m_s = \pm \, 1/2$$
$$\overline{}$$

(2 - fold degenerate)

$$j = 1/2$$
$$m_j = \pm \, 1/2$$
$$\overline{}$$

(2 - fold degenerate)

$$j = 3/2$$
$$m_j = \pm \, 3/2 , \pm \, 1/2$$

(4 - fold degenerate)

$$l = 1 , \; s = 1/2$$
$$m_l = 0 , \pm \, 1 , \; m_s = \pm \, 1/2$$

(6 - fold degenerate)

$$j = 1/2$$
$$m_j = \pm \, 1/2$$

(2 - fold degenerate)

Fig. 6-1. The electron eigenstates at $k = 0$ that play a role in optical transitions. In the absence of spin–orbit coupling, the eigenstates are $|l \; s \; m_l \; m_s\rangle$, where l, m_l and s, m_s are the total and z-component quantum numbers for the orbital and spin angular momenta, respectively. With spin–orbit coupling, the eigenstates become $|l \; s; \; j \; m_j\rangle$, where j and m_j are the quantum numbers for the total (orbital plus spin) angular momentum.

$$|X\rangle = \frac{1}{\sqrt{2}} [|1 \; -1\rangle + |1 \; 1\rangle] \,, \tag{8a}$$

$$|Y\rangle = \frac{i}{\sqrt{2}} [|1 \; -1\rangle - |1 \; 1\rangle] \,, \tag{8b}$$

$$|Z\rangle = |1 \; 0\rangle \,. \tag{8c}$$

In the coordinate representation, the corresponding eigenfunctions have the symmetry

$$\langle \mathbf{r}|X\rangle \propto x f(r) \,, \tag{9a}$$

$$\langle \mathbf{r}|Y \rangle \propto yf(r) , \tag{9b}$$

$$\langle \mathbf{r}|Z \rangle \propto zf(r) , \tag{9c}$$

where $f(r)$ is a spherically symmetric function.

The coupling between the electron spin and orbital angular momenta contributes a spin-orbit term in the Hamiltonian of the form

$$\frac{1}{2m_0^2 c^2} (\mathbf{S} \times \nabla V) \cdot \mathbf{p} = \xi\, \mathbf{S} \cdot \mathbf{L} , \tag{10}$$

where

$$\xi = \frac{1}{2m_0^2 c^2 r} \frac{dV}{dr}$$

is the spin-orbit function [see, e.g., Schiff (1968)]. To obtain Eq. (10) we use

$$(\mathbf{S} \times \nabla V) \cdot \mathbf{p} = (\nabla V \times \mathbf{p}) \cdot \mathbf{S} . \tag{11}$$

For a central potential

$$\nabla V(\mathbf{r}) = \frac{dV}{dr} \frac{\mathbf{r}}{r} , \tag{12}$$

so that

$$(\nabla V \times \mathbf{p}) \cdot \mathbf{S} = \frac{1}{r} \frac{dV}{dr} (\mathbf{r} \times \mathbf{p}) \cdot \mathbf{S} = \frac{1}{r} \frac{dV}{dr} \mathbf{L} \cdot \mathbf{S} . \tag{13}$$

We assume that the spin-orbit interaction is small compared to the electrostatic interaction (*Russell-Saunders case* or *LS coupling scheme*). Furthermore, since

$$\mathbf{J}^2 = (\mathbf{L} + \mathbf{S})^2 = \mathbf{L}^2 + \mathbf{S}^2 + 2\mathbf{L} \cdot \mathbf{S} , \tag{14}$$

the new energy eigenstates are also eigenstates of the total angular momentum $\mathbf{J}$. These states are denoted as $|l\, s;\, j\, m_j \rangle$, and they can be expressed in terms of the old ones by

$$|l\,s;\,j\,m_j\rangle = \sum_{m_l,\,m_s} |l\,s\,m_l\,m_s\rangle\langle l\,s\,m_l\,m_s|l\,s;\,j\,m_j\rangle\,, \tag{15}$$

where $\langle l\,s\,m_l\,m_s|l\,s;\,j\,m_j\rangle$ are the Clebsch-Gordan coefficients. For the conduction bands, the new states are simply

$$|0\,\tfrac{1}{2};\,\tfrac{1}{2}\,\tfrac{1}{2}\rangle = |0\,\tfrac{1}{2}\,0\,\tfrac{1}{2}\rangle \tag{16a}$$

$$|0\,\tfrac{1}{2};\,\tfrac{1}{2}\,-\tfrac{1}{2}\rangle = |0\,\tfrac{1}{2}\,0\,-\tfrac{1}{2}\rangle\,. \tag{16b}$$

For the valence bands, $l = 1$ and $s = \tfrac{1}{2}$ give $j = \tfrac{3}{2}$ and $\tfrac{1}{2}$. According to Eqs. (10) and (14), the spin-orbit interaction energetically shifts the $j = \tfrac{1}{2}$ states below the $j = \tfrac{3}{2}$ states. The amount of the shift is $\Delta \simeq 9\langle\xi\rangle/8$, where $\langle\xi\rangle$ is the expectation value of the spin-orbit function taken with the solutions of the problem without spin-orbit coupling. $\langle\xi\rangle$ is usually treated as a parameter obtained from experiment. Numerical values of the spin-orbit energy Δ are 0.11 eV in InP or 0.34 eV in GaAs, respectively. As a consequence of this large energy splitting usually only the $j = \tfrac{3}{2}$ states are directly involved in optical transitions. These eigenstates can be expressed in terms of the states of Eq. (8). We find

$$|1\,\tfrac{1}{2};\,\tfrac{3}{2}\,\tfrac{3}{2}\rangle = -\,|1\,\tfrac{1}{2}\,1\,\tfrac{1}{2}\rangle\,, \tag{17a}$$

$$|1\,\tfrac{1}{2};\,\tfrac{3}{2}\,\tfrac{1}{2}\rangle = -\,\sqrt{\tfrac{1}{3}}\,|1\,\tfrac{1}{2}\,1\,-\tfrac{1}{2}\rangle + \sqrt{\tfrac{2}{3}}\,|1\,\tfrac{1}{2}\,0\,\tfrac{1}{2}\rangle\,, \tag{17b}$$

$$|1\,\tfrac{1}{2};\,\tfrac{3}{2}\,-\tfrac{1}{2}\rangle = \sqrt{\tfrac{2}{3}}\,|1\,\tfrac{1}{2}\,0\,-\tfrac{1}{2}\rangle + \sqrt{\tfrac{1}{3}}\,|1\,\tfrac{1}{2}\,-1\,\tfrac{1}{2}\rangle\,, \tag{17c}$$

$$|1\,\tfrac{1}{2};\,\tfrac{3}{2}\,-\tfrac{3}{2}\rangle = |1\,\tfrac{1}{2}\,-1\,-\tfrac{1}{2}\rangle\,, \tag{17d}$$

$$|1\,\tfrac{1}{2};\,\tfrac{1}{2}\,\tfrac{1}{2}\rangle = \sqrt{\tfrac{2}{3}}\,|1\,\tfrac{1}{2}\,1\,-\tfrac{1}{2}\rangle + \sqrt{\tfrac{1}{3}}\,|1\,\tfrac{1}{2}\,0\,\tfrac{1}{2}\rangle \tag{17e}$$

$$|1\,\tfrac{1}{2};\,\tfrac{1}{2}\,-\tfrac{1}{2}\rangle = \frac{1}{\sqrt{3}}\,|1\,\tfrac{1}{2}\,0\,-\tfrac{1}{2}\rangle - \sqrt{\tfrac{2}{3}}\,|1\,\tfrac{1}{2}\,-1\,\tfrac{1}{2}\rangle \tag{17f}$$

6-3. k·p Theory

Assuming that we know the band energies and eigenstates at the momentum value k_0, we show in this section how the k·p theory allows one to compute the states in the vicinity of k_0. Note that $|\phi_{nk}\rangle$ cannot be expanded in terms of $|\phi_{mk_0}\rangle$, because these functions are orthogonal to one another whenever $k_0 \neq k$. However, since the states for each k are complete, we may write

$$|n\ k\rangle = \sum_m c_{nmk}\ |m\ k_0\rangle . \tag{18}$$

If $\langle r|n\ k_0\rangle$ is periodic with respect to translation by a lattice vector, then $\langle r|n\ k\rangle$ is also periodic, so that the Bloch theorem is satisfied. The problem is then to use Eq. (5) to obtain the expansion coefficient c_{nmk}. For this purpose we first rewrite Eq. (5) as

$$(\mathcal{H}_0 + \lambda\mathcal{H}_1)|n\ k\rangle = W_{nk}|n\ k\rangle , \tag{19}$$

where

$$\mathcal{H}_0 = \frac{p^2}{2m_0} + V_0 , \tag{20}$$

$$\mathcal{H}_1 = \frac{\hbar}{m_0}k\cdot p . \tag{21}$$

$$W_{nk} = \varepsilon_{nk} - \frac{\hbar^2 k^2}{2m_0} , \tag{22}$$

We treat $\mathcal{H}_1$ as a perturbation and use λ to keep track of the order of the terms in the perturbation expansion.

In the following we discuss the analysis for the example of $k_0 = 0$, i.e., for the states around the center of the Brillouin zone (Γ point). These states are the most relevant ones for the description of optical transitions near the semiconductor absorption edge. The zone-center eigenstates and energies

$$|m\ k=0\rangle \equiv |m\rangle \quad \text{and} \quad \epsilon_{m,k=0} \equiv \epsilon_m , \tag{23}$$

are solutions of the eigenvector equation

$$\mathcal{H}_0|m\rangle = \varepsilon_m |m\rangle \ . \tag{24}$$

At first we ignore the spin-orbit coupling so that the states $|m\rangle$ are given by $|l \ m_l\rangle$. For the conduction band, we have the nondegenerate eigenstate $|S\rangle = |0 \ 0\rangle$, and for the valence band, we have the degenerate eigenstates $|1 \pm 1\rangle$ and $|1 \ 0\rangle$, or $|X\rangle$, $|Y\rangle$, and $|Z\rangle$.

The details of the perturbation treatment are given in App. *B*. The result for bands with only spin-degenerate zone center eigenstates, such as the conduction band, is

$$\varepsilon_{n\mathbf{k}} = \varepsilon_n + \frac{\hbar^2 k^2}{2m_n} , \tag{25}$$

where the *effective mass* is given by Eq. (B.21) as

$$\frac{1}{m_n} \equiv \frac{1}{m_{n,eff}} = \frac{1}{m_0}\left(1 + \frac{2}{m_0}\sum_{m \neq n} \frac{|\langle n|p_x|m\rangle|^2}{\varepsilon_n - \varepsilon_m}\right). \tag{26}$$

Values of the effective masses for the lowest conduction bands in InAs, GaAs and InP are $0.027m_0$, $0.066m_0$ and $0.073m_0$, respectively.

For the bands with degenerate zone center eigenstates, for example the $l = 1$ valence band states, we need to follow a different approach. Suppose the first N eigenstates $|1\rangle$, $|2\rangle$, $|3\rangle$, $\cdots$, $|N\rangle$ of $\mathcal{H}_0$ are degenerate, so that

$$\varepsilon_1 = \varepsilon_2 = \varepsilon_3 = \cdots = \varepsilon_N \ . \tag{27}$$

We need to find N new orthonormal eigenstates $|n\rangle'$ that are nondegenerate at $k \neq 0$. To do so, we write the new eigenstates as linear superpositions of the old ones, that is

$$|n\rangle' = \sum_{m=1}^{N} |m\rangle \langle m|n\rangle' \tag{28}$$

where $1 \leq n \leq N$ and $\langle m|n\rangle' = 0$ for $m > N$. In App. *B*, we show that this leads to a set of N coupled equations that can be solved for $\varepsilon_{n\mathbf{k}}$ and $\langle m|n\rangle'$. In matrix formalism, we have

$$\underline{\mathcal{H}}\, \mathbf{A} = \varepsilon_{n\mathbf{k}}\mathbf{A} \, , \tag{29}$$

where $\underline{\mathcal{H}}$ is an $N \times N$ matrix with the matrix elements

$$\mathcal{H}_{ij} = \left[\varepsilon_1 + \frac{\hbar^2 k^2}{2m_0}\right]\delta_{i,j} + \frac{\hbar^2}{m_0^{\,2}} \sum_m{}' \frac{\mathbf{k}\cdot\mathbf{p}_{im}\ \mathbf{k}\cdot\mathbf{p}_{mj}}{\varepsilon_1 - \varepsilon_m} \, , \tag{30}$$

and $\Sigma_m{}'$ denotes the exclusion of the $l = 1$ to N states from the summation. In other words, the summation involves only the remote bands. The elements of the vector $\mathbf{A}$ are the probability amplitudes

$$A_j = \langle j\,|\,n\rangle' \, . \tag{31}$$

For non-trivial solutions,

$$\det(\underline{\mathcal{H}} - \varepsilon_{n\mathbf{k}}\,\underline{I}) = 0 \, , \tag{32}$$

where $\underline{I}$ is the identity matrix. The solutions of the resulting secular equation yield $\varepsilon_{n\mathbf{k}}$ for $n = 1$ to N. The transformation matrix $\underline{S}$ that diagonalizes $\underline{\mathcal{H}}$ gives the new eigenstates at $k = 0$ according to

$$\underline{S}^{-1}\underline{\mathcal{H}}\underline{S} = \underline{E} \, , \tag{33}$$

or

$$S_{mn} = \langle m\,|\,n\rangle' \, , \tag{34}$$

and the elements of the matrix $\underline{E}$ are

$$\varepsilon_{nm} = \varepsilon_{n\mathbf{k}}\delta_{n,m} \, . \tag{35}$$

Using the results of the $\mathbf{k}\cdot\mathbf{p}$ theory, we can derive Eq. (3.88) for the dipole matrix element $\mu_{\mathbf{k}}$. To this end, we investigate the matrix element of the space operator between two bands λ and λ'

$$r_{\lambda'\lambda}(\mathbf{k}',\mathbf{k}) = \langle \phi_{\lambda'\mathbf{k}'}\,|\,r\,|\,\phi_{\lambda\mathbf{k}}\rangle \, . \tag{36}$$

To evaluate Eq. (36) we consider only interband transitions, $\lambda \neq \lambda'$. Furthermore, we notice that

$$\langle \phi_{\lambda'\mathbf{k}'}\,|\,r\,|\,\phi_{\lambda\mathbf{k}}\rangle\,(\varepsilon_{\lambda\mathbf{k}} - \varepsilon_{\lambda'\mathbf{k}'}) = \langle \phi_{\lambda'\mathbf{k}'}\,|[r, \mathcal{H}_0]|\,\phi_{\lambda\mathbf{k}}\rangle \, , \tag{37}$$

since

$$\mathcal{H}_0|\lambda \ \mathbf{k}\rangle = \varepsilon_{\lambda\mathbf{k}}|\lambda \ \mathbf{k}\rangle \ . \tag{38}$$

Using this and the fact that

$$[\mathbf{r}, \mathcal{H}_0] = \frac{i\hbar}{m_0} \ \mathbf{p} \ , \tag{39}$$

where $\mathbf{p}$ is the momentum operator, we obtain

$$\mathbf{r}_{\lambda'\lambda}(\mathbf{k}',\mathbf{k}) = \frac{i\hbar}{m_0(\varepsilon_{\lambda\mathbf{k}} - \varepsilon_{\lambda'\mathbf{k}'})} \ \langle \phi_{\lambda'\mathbf{k}'}|\mathbf{p}|\phi_{\lambda\mathbf{k}}\rangle \ \ . \tag{40}$$

The momentum operator is diagonal in the coordinate representation

$$\langle \mathbf{r}'|\mathbf{p}|\mathbf{r}\rangle = \delta(\mathbf{r}-\mathbf{r}') \ \frac{\hbar}{i} \ \nabla \ , \tag{41}$$

so that

$$\langle \phi_{\lambda'\mathbf{k}'}|\mathbf{p}|\phi_{\lambda\mathbf{k}}\rangle = \int_{L^3} d^3r \ \phi^*_{\lambda'\mathbf{k}'}(\mathbf{r}) \ \mathbf{p} \ \phi_{\lambda\mathbf{k}}(\mathbf{r}) \ . \tag{42}$$

Using the leading order of the k·p result given by Eq. (B.10), we get

$$\langle \mathbf{r}|\phi_{\lambda\mathbf{k}}\rangle \simeq \frac{e^{i\mathbf{k}\cdot\mathbf{r}}}{\sqrt{V}} \ \langle \mathbf{r}|\lambda\rangle \ , \tag{43}$$

so that

$$\langle \phi_{\lambda'\mathbf{k}'}|\mathbf{p}|\phi_{\lambda\mathbf{k}}\rangle \simeq \frac{1}{V} \int_V d^3r \ e^{-i(\mathbf{k}'-\mathbf{k})\cdot\mathbf{r}} \langle \lambda'|\mathbf{r}\rangle \ (\hbar\mathbf{k} + \mathbf{p}) \ \langle \mathbf{r}|\lambda\rangle \ . \tag{44}$$

To continue with our evaluation, we split the integral over the crystal volume into the sum over the unit cells and the integral within a unit-cell

$$\frac{1}{V} \int_V \rightarrow \frac{1}{N} \sum_\nu \frac{1}{v} \int_v , \tag{45}$$

where $V = Nv$. Correspondingly, the space vector is written as

$$\mathbf{r} \rightarrow \mathbf{R}_{\nu} + \mathbf{r} ,$$

where $\mathbf{r}$ varies within one unit cell and $\mathbf{R}_{\nu}$ is the position vector of unit cell ν. Inserting these expressions into Eq. (44), we get

$$\langle \phi_{\lambda' \mathbf{k}'} | \mathbf{p} | \phi_{\lambda \mathbf{k}} \rangle = \sum_{\nu} \frac{e^{-i(\mathbf{k}'-\mathbf{k})\cdot \mathbf{R}_{\nu}}}{N}$$

$$\times \int_{\nu} d^3 r \, \frac{\exp(-i(\mathbf{k}'-\mathbf{k})\cdot \mathbf{r})}{\nu} \langle \lambda' | \mathbf{r} \rangle \, (\hbar \mathbf{k} + \mathbf{p}) \langle \mathbf{r} | \lambda \rangle . \quad (46)$$

The unit-cell integral yields the same result for all unit cells and can be moved out of the summation over the unit cells, which then yields $\delta_{\mathbf{k},\mathbf{k}'}$ and

$$\langle \phi_{\lambda' \mathbf{k}'} | \mathbf{p} | \phi_{\lambda \mathbf{k}} \rangle = \delta_{\mathbf{k},\mathbf{k}'} \langle \lambda' | \mathbf{p} | \lambda \rangle = \delta_{\mathbf{k},\mathbf{k}'} \, \mathbf{p}_{\lambda' \lambda}(0) . \quad (47)$$

Here we denote

$$\langle \lambda' | \mathbf{p} | \lambda \rangle = \frac{1}{\nu} \int_{\nu} d^3 r \, \langle \lambda' | \mathbf{r} \rangle \, (\hbar \mathbf{k} + \mathbf{p}) \langle \mathbf{r} | \lambda \rangle . \quad (48)$$

The term $\propto \hbar \mathbf{k}$ disappears in going from Eq. (46) to (47) because of the orthogonality of the lattice periodic functions and the $\lambda \neq \lambda'$ requirement.

Collecting all contributions to the dipole matrix element, we get

$$e\mathbf{r}_{\lambda' \lambda}(\mathbf{k}',\mathbf{k}) = \mu_{\lambda' \lambda}(\mathbf{k},\mathbf{k}') = \frac{i e \hbar \mathbf{p}_{\lambda' \lambda}(0)}{m_0 (\varepsilon_{\lambda' \mathbf{k}} - \varepsilon_{\lambda \mathbf{k}})} \delta_{\mathbf{k},\mathbf{k}'} \quad (49)$$

or

$$\mu_{\lambda' \lambda}(\mathbf{k}',\mathbf{k}) = \delta_{\mathbf{k},\mathbf{k}'} \, \mu_{\lambda' \lambda}(0) \frac{\varepsilon_{\lambda'} - \varepsilon_{\lambda}}{\varepsilon_{\lambda' \mathbf{k}} - \varepsilon_{\lambda \mathbf{k}}} \equiv \mu_{\mathbf{k}} . \quad (50)$$

For the case of two parabolic bands with effective masses m_{λ} and $m_{\lambda'}$ and dispersions

$$\varepsilon_{\lambda' \mathbf{k}} = \varepsilon_g + \frac{\hbar^2 k^2}{2m_{\lambda'}} \quad \text{and} \quad \varepsilon_{\lambda \mathbf{k}} = \frac{\hbar^2 k^2}{2m_{\lambda}} , \quad (51)$$

we recover the Kane matrix element $\mu_{\mathbf{k}}$ of Eq. (3.88).

6-4. Luttinger Hamiltonian

Specializing our discussion to the two top valence bands, we first ignore the spin of the electron so that the $k = 0$ eigenstates have $l = 1$ and are three-fold degenerate. Leaving the details of the calculations to App. B, we just note here that by ordering the valence band states such that $|j\rangle$ for $j = 1$, 2 and 3 is $|X\rangle$, $|Y\rangle$ and $|Z\rangle$, we transform the Hamiltonian matrix (30) into

$$\mathcal{H} = \begin{pmatrix} \varepsilon_1 + Ak_x^2 + B(k_y^2 + k_z^2) & Ck_x k_y & Ck_x k_z \\ Ck_x k_y & \varepsilon_1 + Ak_y^2 + B(k_x^2 + k_z^2) & Ck_y k_z \\ Ck_x k_z & Ck_y k_z & \varepsilon_1 + Ak_z^2 + B(k_x^2 + k_y^2) \end{pmatrix}, \quad (52)$$

where A, B, and C are given by Eqs. (B.36), (B.37). and (B.39), respectively.

In order to include spin-orbit coupling, we have to consider the basis set made up of $|j\ m_j\rangle$. Concentrating on the top two valence bands, we have to deal only with the four eigenstates $|1\rangle = |\frac{3}{2}\ \frac{3}{2}\rangle$, $|2\rangle = |\frac{3}{2}\ -\frac{1}{2}\rangle$, $|3\rangle = |\frac{3}{2}\ \frac{1}{2}\rangle$, and $|4\rangle = |\frac{3}{2}\ -\frac{3}{2}\rangle$. As shown in App. B, this leads to the *Luttinger Hamiltonian* for the hole states

$$\mathcal{H} = \begin{pmatrix} \mathcal{H}_{hh} & -c & -b & 0 \\ -c^* & \mathcal{H}_{lh} & 0 & b \\ -b^* & 0 & \mathcal{H}_{lh} & -c \\ 0 & b^* & -c^* & \mathcal{H}_{hh} \end{pmatrix} \quad (53)$$

Luttinger Hamiltonian

where

$$\mathcal{H}_{hh} = \frac{\hbar^2 k_z^2}{2m_0}(\gamma_1 - 2\gamma_2) + \frac{\hbar^2(k_x^2 + k_y^2)}{2m_0}(\gamma_1 + \gamma_2), \quad (54)$$

$$\mathcal{H}_{lh} = \frac{\hbar^2 k_z^2}{2m_0} (\gamma_1 + 2\gamma_2) + \frac{\hbar^2(k_x^2 + k_y^2)}{2m_0} (\gamma_1 - \gamma_2) , \tag{55}$$

$$c = \frac{\sqrt{3}\hbar^2}{2m_0} [\gamma_2(k_x^2 - k_y^2) - 2i\gamma_3 k_x k_y] , \tag{56}$$

$$b = \frac{\sqrt{3}\hbar^2}{m_0} \gamma_3 k_z (k_x - ik_y) . \tag{57}$$

The *Luttinger parameters* are defined as

$$\gamma_1 = - 2m_0(A + 2B)/3\hbar^2 , \tag{58}$$

$$\gamma_2 = - m_0(A - B)/3\hbar^2 , \tag{59}$$

$$\gamma_3 = - m_0 C/3\hbar^2 . \tag{60}$$

These parameters are determined experimentally using various techniques sensitive to the bandstructure. The Luttinger parameters for many semiconductor materials are listed in books like Landolt-Börnstein. Typical values for some III-V semiconductors are

	γ_1	γ_2	γ_3	
GaAs	6.85	2.1	2.9	
InAs	19.67	8.37	9.29	(61)
InP	6.35	2.08	2.76	

This list shows that

$$\gamma_1 > \gamma_2 \simeq \gamma_3 , \tag{62}$$

which is correct for almost all (cubic) semiconductors.

In the vicinity of $k = 0$, one may introduce the so-called *axial approximation*, where the Luttinger parameters γ_2 and γ_3 in the function c of Eq. (56) are replaced by an effective Luttinger parameter

$$\bar{\gamma} = \tfrac{1}{2}(\gamma_2 + \gamma_3) . \tag{63}$$

With the axial approximation, the Luttinger Hamiltonian can be transformed into a block diagonal form by a unitary transformation [see Eqs. (B.56) - (B.65)], so that

$$\mathcal{H}' = \begin{pmatrix} \mathcal{H}^U & 0 \\ 0 & \mathcal{H}^L \end{pmatrix} \qquad (64)$$

block-diagonal Luttinger Hamiltonian

where

$$\mathcal{H}^U = \begin{pmatrix} \mathcal{H}_{hh} & R \\ R^* & \mathcal{H}_{lh} \end{pmatrix}, \qquad (65)$$

$$\mathcal{H}^L = \begin{pmatrix} \mathcal{H}_{lh} & R \\ R^* & \mathcal{H}_{hh} \end{pmatrix}, \qquad (66)$$

$$R = |c| - i|b| . \qquad (67)$$

The new basis states $|1\rangle$, $|2\rangle$, $|3\rangle$, and $|4\rangle$ are given by Eq. (B.65).

The axial approximation removes the small anisotropy of the band-structure for different directions in the k_x, k_y plane, resulting in cylindrical symmetry around the k_z axis [Altarelli (1985)]. To completely diagonalize the block-diagonal Luttinger Hamiltonian (64), we only need to diagonalize each block individually. This gives the eigenvalues

$$\begin{aligned} \varepsilon_{lh,\mathbf{k}} &= \tfrac{1}{2}(\mathcal{H}_{hh} + \mathcal{H}_{lh}) + \sqrt{\tfrac{1}{4}(\mathcal{H}_{hh} - \mathcal{H}_{lh})^2 + |c|^2 + |b|^2} \\ &= \hbar^2 k^2 (\gamma_1 + 2\gamma_2)/2m_0 , \end{aligned} \qquad (68)$$

$$\begin{aligned} \varepsilon_{hh,\mathbf{k}} &= \tfrac{1}{2}(\mathcal{H}_{hh} + \mathcal{H}_{lh}) - \sqrt{\tfrac{1}{4}(\mathcal{H}_{hh} - \mathcal{H}_{lh})^2 + |c|^2 + |b|^2} \\ &= \hbar^2 k^2 (\gamma_1 - 2\gamma_2)/2m_0 , \end{aligned} \qquad (69)$$

where we use $\bar{\gamma}$ and γ_2 interchangably, as we do from now on in this book. Since the upper and lower blocks give the same results, the bandstructure is composed of two twice-degenerate isotropic valence bands with effective masses $m_0/(\gamma_1 \pm 2\gamma_2)$. Using the Luttinger parameters for GaAs, we obtain

$$m_{hh} \equiv m_0/(\gamma_1 - 2\gamma_2) \simeq 0.38\, m_0 \qquad (70)$$

$$m_{lh} \equiv m_0/(\gamma_1 + 2\gamma_2) \simeq 0.09\, m_0 , \qquad (71)$$

which are the heavy-hole and light-hole effective masses, respectively. Figure 6-2 shows the GaAs bandstructure of interest for optical transitions.

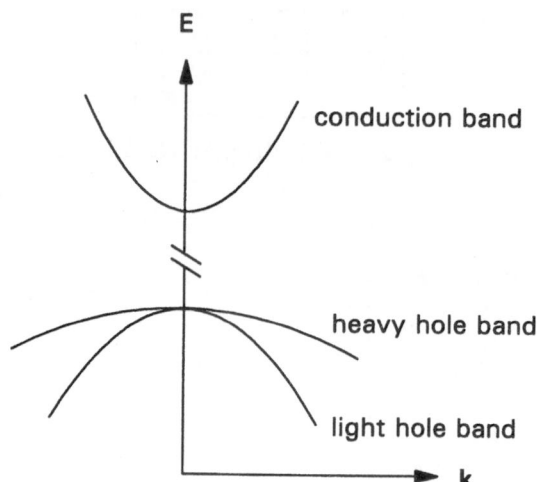

Fig. 6-2. The GaAs energy bands involved in optical transitions. The effective masses for the conduction, heavy-hole and light-hole bands are $.067m_0$, $.377m_0$ and $.09m_0$, respectively.

6-5. Quantum Wells

For quantum-well structures the time-independent Schrödinger equation becomes

$$\left[\frac{p^2}{2m_0} + V_0 + V_{con}\right]|\phi_\lambda^{QW}\rangle = \varepsilon_\lambda |\phi_\lambda^{QW}\rangle , \qquad (72)$$

where V_{con} is the potential due to the epitaxially grown heterostructure, and λ represents the combination of quantum numbers to be specified later for identifying the quantum-well states. We assume that the potential V_{con} is sufficiently small and varies sufficiently little within a unit cell, so that the new eigenstates may be approximated as the linear superposition of the bulk material eigenstates

$$|\phi_\lambda^{QW}\rangle = \sum_{n,\mathbf{k}} |\phi_{n\mathbf{k}}\rangle\langle\phi_{n\mathbf{k}}|\phi_\lambda^{QW}\rangle ,\tag{73}$$

where $|\phi_{n\mathbf{k}}\rangle$ is the bulk eigenstate satisfying Eq. (1). This approach is called the *envelope function approximation* because in the coordinate representation Eq. (73) becomes

$$\langle\mathbf{r}|\phi_\lambda^{QW}\rangle = \sum_m W_{\lambda m}(\mathbf{r})\langle\mathbf{r}|m\rangle ,\tag{74}$$

where $W_{\lambda m}(\mathbf{r})$ varies slowly compared to $\langle\mathbf{r}|m\rangle$ and therefore plays the role of an envelope function. We introduce the Fourier transform

$$W_{\lambda m}(\mathbf{r}) = \sum_{\mathbf{k}} e^{i\mathbf{k}\cdot\mathbf{r}} W_{\lambda m\mathbf{k}} ,\tag{75}$$

and show in App. C, Eq. (C.6), that

$$W_{\lambda m\mathbf{k}} = \sum_n \langle m\,|n\,\mathbf{k}\rangle\langle\phi_{n\mathbf{k}}|\phi_\lambda^{QW}\rangle \tag{76}$$

is the sum of products of probability amplitudes. Substituting Eq. (74) into (72), multiplying the result by $\exp(-i\mathbf{k}'\cdot\mathbf{r})\langle j\,|\mathbf{r}\rangle$, and integrating over the volume of the crystal, we find

$$\left[\frac{\hbar^2 k^2}{2m_0} + \varepsilon_n\right] W_{\lambda n\mathbf{k}} + \frac{\hbar}{m_0}\sum_m \mathbf{k}'\cdot\mathbf{p}_{nm} W_{\lambda m\mathbf{k}} + \sum_{\mathbf{k}} V_{con,\mathbf{k}-\mathbf{k}'} W_{\lambda n\mathbf{k}'}$$
$$= \varepsilon_\lambda W_{\lambda n\mathbf{k}} ,\tag{77}$$

which shows that the confinement potential does not mix states from different bands. In deriving Eq. (77) we make use of the fact that the quantum-well structure gives rise to two length scales: one for coarse spatial variations that are of the order of the heterostructure and the other for fine spatial variations that are of the order of the lattice unit cell. It is sometimes useful to differentiate between these two length scales, and we do so in the following by using upper-case letters (e.g., $\mathbf{R}$, Z, and $\mathbf{R}_\perp$) for the coarse variations and lower-case letters (e.g., $\mathbf{r}$, x, y, and z) for the fine variations.

To solve the set of equations given by Eq. (77), we again use perturbation theory. First we consider the bands that are nondegenerate at $k = 0$, for example the s-like conduction band when spin is ignored. Then a calculation similar to the one in Sec. B-2 for the bulk material shows that

$$\left[\varepsilon_n + \frac{\hbar^2 k^2}{2m_n}\right] W_{\lambda n \mathbf{k}} + \sum_{\mathbf{k}'} V_{con,\mathbf{k}-\mathbf{k}'} W_{\lambda n \mathbf{k}'} = \varepsilon_\lambda W_{\lambda n \mathbf{k}} , \qquad (78)$$

where the effective mass m_n for the nth band of the bulk material is defined in Eq. (26). The Fourier transform of Eq. (78) is

$$\left[-\frac{\hbar^2}{2m_n}\nabla^2 + V_{con}(Z)\right] W_{\lambda n}(\mathbf{R}) = \varepsilon_{\lambda n} W_{\lambda n}(\mathbf{R}) , \qquad (79)$$

where Z is in the direction of epitaxial growth. This equation is separable so that

$$W_{\lambda n}(\mathbf{R}) = A_{n_z n}(Z) B_{\mathbf{k}_\perp n}(\mathbf{R}_\perp) , \qquad (80)$$

where $\lambda \rightarrow n_z \mathbf{k}_\perp$ and $\mathbf{R}_\perp$ is a vector which varies in the plane of the heterostructure. The envelope functions obey

$$\left[-\frac{d}{dZ}\frac{\hbar^2}{2m_n}\frac{d}{dZ} + V_{con}(Z)\right] A_{n_z n}(Z) = \varepsilon_{n_z n} A_{n_z n}(Z) , \qquad (81)$$

i.e., the motion in the direction perpendicular to the well is that of a particle with mass m_n in a one-dimensional square potential well. Since the effective mass in the Z-direction of a quantum-well heterostructure depends on space, we must symmetrize the second derivative as shown in Eq. (81) to avoid implicitly dealing with a non-Hermitian Hamiltonian. For the case of a single quantum well, Eq. (81) reduces to Eq. (2.37). Furthermore,

$$-\frac{\hbar^2}{2m_n}\nabla_\perp^2 B_{\mathbf{k}_\perp n}(\mathbf{R}_\perp) = \varepsilon_{\mathbf{k}_\perp n} B_{\mathbf{k}_\perp n}(\mathbf{R}_\perp) , \qquad (82)$$

which describes free-particle motion in the plane of the well. The total energy is

$$\varepsilon_{n_z \mathbf{k}_\perp n} = \varepsilon_n + \varepsilon_{n_z n} + \varepsilon_{\mathbf{k}_\perp n} , \tag{83}$$

where

$$\varepsilon_{\mathbf{k}_\perp n} = \hbar^2 k_\perp^2 / 2m_n \tag{84}$$

and the solution for $\varepsilon_{n_z n}$ is discussed in Sec. 2-4. Hence the energy of the quantum-confined electron has contributions from the bulk structure, the quantum confinement and the free motion in x-y plane. Instead of an infinite number of states in every direction, the z-direction has only a finite number of states, which are the bound solutions of Eq. (81).

The effects of quantum confinement are much more interesting for bands that are degenerate at $k = 0$, such as the top two valence bands in GaAs and similar materials. We show in App. C that second-order degenerate perturbation theory yields the equations

$$\left[-\frac{\partial}{\partial Z} \frac{\hbar^2}{2m_{hhZ}} \frac{\partial}{\partial Z} - \frac{\hbar^2}{2m_{hh\perp}} \left(\frac{\partial^2}{\partial X^2} + \frac{\partial^2}{\partial Y^2} \right) + V_{con}(Z) \right] W_{\lambda n}(\mathbf{R})$$

$$+ \frac{\sqrt{3}\hbar^2}{2m_0} k_\perp \left[\gamma_2 k_\perp - 2\gamma_3 \frac{d}{dZ} \right] W_{\lambda m}(\mathbf{R}) = \varepsilon_{\lambda 1} W_{\lambda n}(\mathbf{R}) \tag{85a}$$

for the combinations $n, m = 1, 2$ and 4, 3 and

$$\left[-\frac{\partial}{\partial Z} \frac{\hbar^2}{2m_{lhZ}} \frac{\partial}{\partial Z} - \frac{\hbar^2}{2m_{lh\perp}} \left(\frac{\partial^2}{\partial X^2} + \frac{\partial^2}{\partial Y^2} \right) + V_{con}(Z) \right] W_{\lambda n}(\mathbf{R})$$

$$+ \frac{\sqrt{3}\hbar^2}{2m_0} k_\perp \left[\gamma_2 k_\perp - 2\gamma_3 \frac{d}{dZ} \right] W_{\lambda m}(\mathbf{R}) = \varepsilon_{\lambda 2} W_{\lambda n}(\mathbf{R}) , \tag{85b}$$

for $n, m = 2, 1$ and 3, 4. Similar to the situation for the bulk material, the solutions to Eqs. (85) are doubly degenerate and we only need to consider either $n = 1,2$ or 3,4. We identify $n = 1$ and 4 as hh (for heavy hole) and $n = 2$ and 3 as lh (for light hole), so that

$$m_{hhz} = m_0 / (\gamma_1 - 2\gamma_2) , \tag{86a}$$

$$m_{lhz} = m_0 / (\gamma_1 + 2\gamma_2) , \tag{86b}$$

$$m_{hh\perp} = m_0/(\gamma_1 + \gamma_2) , \tag{86c}$$

$$m_{lh\perp} = m_0/(\gamma_1 - \gamma_2) . \tag{86d}$$

As in the nondegenerate case, the quantum-well states are products of two-dimensional (X,Y) free-particle eigenstates with the one-dimensional (Z) square-well eigenstates of Eq. (80). The quantum numbers are $\lambda \rightarrow (n_z, \mathbf{k}_\perp)$, where n_z is between 1 and $N_{Z,hh}$ for the heavy holes and between 1 and $N_{Z,lh}$ for the light holes. A difference between the present situation and the case of isotropic bulk semiconductors is that the effective mass in the Z-direction is different from that in the transverse direction for both heavy and light holes. In fact, according to Eq. (86), the states with the heavier effective mass in the Z-direction also have the lighter effective mass in the transverse direction. This is commonly referred to as *mass reversal*. The convention is to use the terms *light* and *heavy holes* according to the respective mass in Z-direction.

To solve the set of coupled equations described by Eq. (85), we first rewrite them in a matrix formulation

$$\underline{\mathcal{H}} \, \mathbf{W}_{nk_\perp} = \varepsilon_{nk_\perp} \mathbf{W}_{nk_\perp}, \tag{87}$$

where n is the band index and $\underline{\mathcal{H}}$ is a $N_z \times N_z$ matrix with $N_z = N_{z,hh} + N_{z,lh}$. The diagonal matrix elements are

$$\mathcal{H}_{ii} = -\frac{d}{dZ} \frac{\hbar^2}{2m_{nZ}} \frac{d}{dZ} + \frac{\hbar^2 k_\perp^2}{2m_{n\perp}} \tag{88}$$

and the off-diagonal elements are

$$\mathcal{H}_{ij} = \frac{\sqrt{3}\hbar^2}{2m_0} k_\perp \left[\gamma_2 k_\perp - 2\gamma_3 \frac{d}{dZ} \right] . \tag{89}$$

The computation of the matrix elements is shown in more detail in Sec. 6-7.

To illustrate the effect of the off-diagonal elements, we consider the case of the single quantum well discussed in Sec. 2-4. Using the eigenfunctions, Eq. (2.50), for A we obtain the bandstructure shown in Fig. 6-3. The dashed lines are the top heavy- and light-hole bands obtained by ignoring the off-diagonal elements $\mathcal{H}_{ij}$ of Eq. (89). Without this band-mixing the hole-bands intersect at a finite k-value. This unphysical band-crossing is removed through the off-diagonal coupling, which is included in the full results obtained by solving Eq. (87) (solid lines in Fig. 6-3).

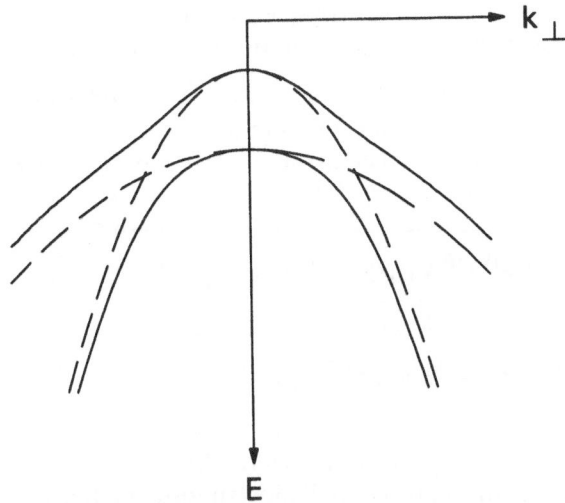

Fig. 6-3. Computed top two valence bands for 5 *nm* GaAs sandwiched between Al$_{.08}$Ga$_{.92}$As. The effect of band mixing is depicted by the difference between the solid curves (full result) and dashed curves (without bandmixing).

When defining the matrix elements of $\mathcal{H}$, we order the eigenstates so that $m = hh$ and $n_z = n$ for $n \leq N_{z,hh}$, and $m = lh$ and $n_z = n - N_{z,hh}$ for $n > N_{z,hh}$. Using the same notation, the new eigenfunctions for the holes are

$$\langle \mathbf{r} | \phi^h_{n\mathbf{k}_\perp} \rangle = e^{i\mathbf{k}_\perp \cdot \mathbf{R}_\perp} \sum_{m=1}^{2} \sum_{n_m=1}^{N_m} A_{n_m m}(Z) \langle \mathbf{r} | m \rangle \, , \qquad (90a)$$

for the upper block and

$$\langle \mathbf{r} | \phi^h_{n,\mathbf{k}_\perp} \rangle = e^{i\mathbf{k}_\perp \cdot \mathbf{R}_\perp} \sum_{m=3}^{4} \sum_{n_m=1}^{N_m} A_{n_m m}(Z) \langle \mathbf{r} | m \rangle \, , \qquad (90b)$$

for the lower block of the Luttinger Hamiltonian (64). The functions $A_{n_m m}$ are products of probability amplitudes and square well eigenfunctions. Solutions of Eq. (87) for $\varepsilon_{nk_\perp}$ and $W_{nk_\perp}$, respectively $A_{n_m m}$, then gives us the quantum-well bandstructure and eigenstates.

For later reference we also list here the conduction-band wavefunctions

$$\langle \mathbf{r} | \phi^e_{l\mathbf{k}_\perp} \rangle = e^{i\mathbf{k}_\perp \cdot \mathbf{R}_\perp} C_l(Z) \langle \mathbf{r} | S \uparrow \rangle \ , \tag{91a}$$

$$\langle \mathbf{r} | \phi^e_{l,\mathbf{k}_\perp} \rangle = e^{i\mathbf{k}_\perp \cdot \mathbf{R}_\perp} C_l(Z) \langle \mathbf{r} | S \downarrow \rangle \ , \tag{91b}$$

for the states with spin $\frac{1}{2}$ and $-\frac{1}{2}$, respectively.

In order to use the computed bandstructure in the calculation of gain and refractive index, we have to replace the kinetic energy terms in the microscopic expressions by the respective single-particle energies computed using Eq. (87). In addition, we have to take into consideration that the dipole matrix elements also change. We therefore compute the dipole matrix elements for light which propagates along the x-y plane of the quantum well, as is usually the case in edge emitting semiconductor lasers. Here we can have two possible polarization directions of the light: in the case of a TM mode the polarization is parallel to the z-direction, and for a TE mode the polarization is in the x-y plane, where we assume it is along the x-direction.

For the TM mode we therefore have to evaluate the matrix element

$$\mu_{TM} = \langle \phi^e_{l\mathbf{k}_\perp} | ez | \phi^h_{n\mathbf{k}_\perp} \rangle \tag{92}$$

and for the TE mode

$$\mu_{TE} = \langle \phi^e_{l\mathbf{k}_\perp} | ex | \phi^h_{n\mathbf{k}_\perp} \rangle \ . \tag{93}$$

To compute the quantum-well gain and refractive index, we always need the absolute square of the respective matrix element. In App. C we show that the results are

$$
|\mu_{TM}|^2 = \frac{1}{3}\left|\sum_{n_m=1}^{N_m}\langle A_{n_m\,m}|C_l\rangle\right|^2 |\langle S\uparrow|ez|Z\uparrow\rangle|^2
\tag{94}
$$

for the conduction band with spin ↑, and $m = 2, 3$, since the TM mode couples to the light-hole band only. Correspondingly, we get for the TE mode

$$
|\mu_{TE}|^2 = \frac{|\langle S\uparrow|ex|X\uparrow\rangle|^2}{4}\left[\left|\sum_{n_1=1}^{N_1}\langle A_{n_1 1}|C_l\rangle\right|^2 + \frac{1}{3}\left|\sum_{n_2=1}^{N_2}\langle A_{n_2 2}|C_l\rangle\right|^2\right.
$$
$$
\left. + \frac{2}{\sqrt{3}}\left(\sum_{n_1=1}^{N_1}\langle C_l|A_{n_1 1}\rangle\right)\left(\sum_{n_2=1}^{N_2}\langle A_{n_2 2}|C_l\rangle\right)\cos(2\phi)\right],
\tag{95}
$$

where ϕ is the angle between **k** and **x**.

6-6. Strained Quantum Wells

Semiconductor heterostructures can be grown epitaxially with two materials that are not perfectly lattice matched, provided this mismatch is not too large. Too large a mismatch may prevent epitaxial growth altogether or lead to fractures, island formation, and other undesirable defects. However under proper conditions, one may obtain a stable structure in which the materials are under elastic strain. One situation of practical interest is the case of relatively thick barriers and thin quantum wells grown on a substrate of barrier material. Ideally, in this case the well material grows with the lattice constant a_b of the barrier material. Hence the well material is under strain, which is compressive (tensile) if the lattice constant of the barrier is smaller (larger) than that of the well material. Strained-layer quantum wells are interesting for applications in semiconductor lasers, because they allow 1) a wider range of material combinations, and 2) a certain amount of bandstructure and hence gain engineering. The simplest example is a frequency shift of the gain spectrum, but gain increases are also possible.

To appreciate the bandstructure engineering aspect, we discuss in this section the simplest modifications of quantum-well bandstructures caused by elastic strain. For this purpose, we modify the analysis of the previous sections of this chapter to include the most important strain effects. In particular, we repeat the $\mathbf{k \cdot p}$ analysis, here with a perturbation due to strain as well as that due to the $\mathbf{k \cdot p}$ term. This theory is originally due to Pikus and Bir (1960), and we more or less follow their original work.

Rather than present the general theory of bandstructures in strained semiconductors, we restrict our discussion to the ideal case mentioned above of a single quantum well with cubic symmetry and bulk lattice constant a_w, which grows under elastic strain in the x-y plane and assumes the lattice constant a_b of the barrier material. Under these conditions the strain tensor is

$$e = \begin{pmatrix} e_{xx} & 0 & 0 \\ 0 & e_{yy} & 0 \\ 0 & 0 & e_{zz} \end{pmatrix},$$

(96)

where the in-plane components are given by the lattice mismatch parameters

$$e_{xx} = e_{yy} = (a_b - a_w)/a_b \equiv e_0 .$$

(97)

The fact that there is no net force acting perpendicular to the quantum-well plane leads to

$$e_{zz} = - 2e_0 C_{12}/C_{11} ,$$

(98)

where the quantities C_{ij} are the *elastic moduli* or *elastic stiffness constants*. Because of cubic symmetry, only C_{11}, C_{12}, and C_{44} of the possible twenty-one C_{ij} have nonzero values. These values can be found, e.g., in Landolt-Börnstein (1982). The fact that the off-diagonal elements of the strain tensor are zero is also a consequence of the cubic symmetry.

The Schrödinger equation still holds in the strained material, but the coordinates are those of the strained system. Since the epitaxially grown layers can tolerate only small amounts of strain, the linearized expression

$$r'_\alpha = r_\alpha + \Sigma_\beta e_{\alpha\beta} r_\beta$$

(99)

can be use to relate any component of the vector $\mathbf{r'}$ in the strained system to the corresponding vector $\mathbf{r}$ in the unstrained system. Using Eq. (99), we find the Schrödinger equation for the strained system in the coordinate system of the unstrained system to be

$$\left[- \frac{\hbar^2}{2m_0} \nabla^2 + V_0(\mathbf{r}) + \frac{\hbar}{m_0} \mathbf{k}\cdot\mathbf{p} + \sum_{\alpha\beta} \mathcal{S}^{\alpha\beta} e_{\alpha\beta} \right] \langle \mathbf{r}|n\mathbf{k}\rangle$$

$$= \left[\varepsilon_{n\mathbf{k}} - \frac{\hbar^2 k^2}{2m_0} \right] \langle \mathbf{r}|n\mathbf{k}\rangle \;, \tag{100}$$

where the effect of strain leads to the additional term proportional to

$$\mathcal{S}^{\alpha\beta} = \frac{\hbar}{m_0} k_\alpha p_\beta + \frac{\hbar^2}{m_0} e_{\alpha\beta} \frac{\partial^2}{\partial r_\alpha \partial r_\beta} + V_{\alpha\beta} \;. \tag{101}$$

Now we repeat the $\mathbf{k}\cdot\mathbf{p}$ analysis, here including a perturbation due to strain in addition to that of the $\mathbf{k}\cdot\mathbf{p}$ term. We treat only the hole band-structure since in the lowest-order approximation the strain effects on the conduction band amount to a rigid shift. In our derivation, we find a one-to-one correspondence between the terms proportional to $k_\alpha k_\beta$ and to $e_{\alpha\beta}$. Hence we can generalize the Luttinger Hamiltonian simply by adding the proper $e_{\alpha\beta}$ terms to the $k_\alpha k_\beta$ elements. As in the unstrained case, where we use empirical Luttinger parameters, we describe the effects of strain using two new empirical parameters, the so-called *hydrostatic* and *shear-deformation potentials*. These terms yield the strained Luttinger Hamiltonian

$$\mathcal{H}_{total} = \mathcal{H} + \mathcal{H}_{strain} \;, \tag{102}$$

where is $\mathcal{H}_{strain}$ is a diagonal matrix

$$\mathcal{H}_{strain} = \begin{pmatrix} -\delta\varepsilon_H -\frac{1}{2}\delta\varepsilon_S & 0 & 0 & 0 \\ 0 & -\delta\varepsilon_H +\frac{1}{2}\delta\varepsilon_S & 0 & 0 \\ 0 & 0 & -\delta\varepsilon_H +\frac{1}{2}\delta\varepsilon_S & 0 \\ 0 & 0 & 0 & -\delta\varepsilon_H - \frac{1}{2}\delta\varepsilon_S \end{pmatrix}, \tag{103}$$

with

$$\delta\varepsilon_H = 2a_1 e_0 (C_{11} - C_{12})/C_{11} \;, \tag{104}$$

$$\delta\epsilon_S = 2a_2 e_0 (C_{11} + 2C_{12})/C_{11} , \tag{105}$$

where a_1 and a_2 are the hydrostatic and shear deformation potentials, respectively, and play the same role as the Luttinger parameters do in unstrained configurations.

To compute the strain-induced energy changes, we diagonalize the total Hamiltonian (102) including the strain terms to obtain

$$\epsilon_{1/2} = \frac{\mathcal{H}_{hh} + \mathcal{H}_{lh}}{2} + \delta\epsilon_H \pm \sqrt{\tfrac{1}{4}(\mathcal{H}_{hh} - \mathcal{H}_{lh} - \delta\epsilon_S)^2 + |c|^2 + |b|^2}$$

$$= \frac{\hbar^2 k^2}{2m_0}(\gamma_1 \pm 2\gamma_2) + \delta\epsilon_H \mp \frac{\delta\epsilon_S}{2} + O(k_\alpha k_\beta, \alpha \neq \beta) . \tag{106}$$

This result shows that $\delta\epsilon_H$ shifts all valence bands by same amount. Hence it can be considered a strain-induced bandgap shift. The term $\tfrac{1}{2}\delta\epsilon_S$, however, enters the heavy- and light-hole energies with opposite sign. Using the definitions (70) and (71), we obtain from Eq. (106)

$$\epsilon_{hh} = \frac{\hbar^2 k^2}{2m_{hh}} - \delta\epsilon_H - \tfrac{1}{2}\delta\epsilon_S \tag{107}$$

$$\epsilon_{lh} = \frac{\hbar^2 k^2}{2m_{lh}} - \delta\epsilon_H + \tfrac{1}{2}\delta\epsilon_S , \tag{108}$$

showing the occurrence of a strain-induced heavy-hole light-hole splitting by the amount $\delta\epsilon_S$.

Since in our approximations the strain contributions to Eq. (102) are k-independent and additive, we can use the Luttinger Hamiltonian for quantum wells by replacing k_z by $-i\partial/\partial z$, and by symmetrizing the second-order derivatives. Adding the strain Hamiltonian and using the proper boundary conditions (App. D), we have all the ingredients needed to compute valence-band structures in strained-layer quantum wells.

6-7. Bandstructure Calculation

In this section, we describe the steps for calculating a bandstructure. Table 6-1 lists the relevant bulk material parameters for some compounds of interest to semiconductor laser and additional parameters needed for strained structures are given in Table 6-2.

	γ_1	γ_2	γ_3	$m_c(m_0)$	ϵ_0	ϵ_∞	$\varepsilon_g(eV)$
GaAs	6.85	2.1	2.9	0.0665	13.71	10.9	1.423
InAs	19.67	8.37	9.29	0.027	15.15	12.25	0.35
InP	6.35	2.08	2.76	0.064	12.61	9.61	1.35
GaP	4.2	0.98	1.66	0.15	11.1	9.07	2.74
AlP	3.47	0.06	1.15	0.22	9.8	7.54	3.58

Table 6-1. Material parameters for common semiconductor laser compounds. Listed are the Luttinger parameters γ_i, the effective electron mass m_c, the low- and high-frequency dielectric constants, and the room-temperature bandgap energies, respectively.

	$a(\text{Å})$	C_{11}	C_{12}	C_{44}	$a_1(eV)$	$a_2(eV)$
GaAs	5.6533	11.88	5.38	5.94	-7.1	-1.7
AlAs	5.660	12.5	5.3	5.4	-5.64	-1.5
InAs	6.0583	8.33	4.53	3.96	-5.9	-1.8
InP	5.8687	10.22	5.76	4.6	-6.35	-2.0
GaP	5.451	14.1	6.2	7.0	-9.3	-1.5
AlP	5.451	13.2	6.3	6.2	-5.54	-1.6

Table 6-2. Strain-related material parameters. Listed are the lattice constant a, the elastic stiffness constants C_{ij}, and the hydrostatic and deformation potentials a_1 and a_2, respectively.

In these tables, the electron effective mass m_c is given with respect to the free electron mass, the energies are in eV, the bandgaps are for 300K, and the elastic stiffness constants C_{ij} are in units of $10^{11} dyn/cm^2$. The values for the parameters are usually determined by fits to experimental data and some discrepancies exist among the different sources, especially in the parameters associated with strain. Some of the sources used are listed in the references at the end of this chapter. The values shown in the tables are the ones used in the calculations in the following sections. For the compounds, AlGaAs, InGaAs and GaInP, the necessary data for the band-structure calculations are obtained using properly weighted averages of the parameters in the tables. For example, for $In_x Ga_{1-x} As$

$$\bar{a} = a_{InAs} x + a_{GaAs}(1 - x) . \tag{109}$$

If strain is present, the next step is to use Eqs. (104) and (105) to cal-culate the strain-induced bandgap shifts for the heavy and light holes. Given a band-offset ratio, i.e., the fraction of the difference in the quan-tum well and barrier bandgaps to be associated with the conduction and valence bands, we solve Schrödinger equation for the square-well eigen-states for the electron, the heavy hole, and the light hole. The conduction band problem is then essentially solved since the bands are parabolic with curvature given by the electron effective mass and band edges given by the square-well energies. The valence bands are determined by diagonalizing the Luttinger Hamiltonian as described in Sec. 5. The dimension of the Luttinger Hamiltonian depends on the total number of bound square-well eigenstates for the holes. For example, let us assume that there are four such states. Taking spin into account, there are eight hole states, which give an 8×8 Luttinger Hamiltonian. Since the Luttinger Hamiltonian is block diagonal, we end up diagonalizing a 4×4 matrix. Each diagonal matrix element is the sum of a square-well energy and the kinetic energy of the heavy or light hole. The off-diagonal matrix elements are given by Eq. (89). Only the heavy and light-hole states are coupled. Using the square-well eigenfunctions, Eqs. (2.46) and (2.48), which fall off exponen-tially outside the well, $\propto exp(-K_i z)$, and oscillate sinusoidally inside the well, $\propto sin(k_i z)$, we find the matrix elements coupling even (cosine) heavy and light holes to be

$$\langle n\, hh | \mathcal{H} | m\, lh \rangle = \frac{\sqrt{3}\, \hbar^2 k_\perp^2\, \gamma_2}{2 m_0} B_{hhn} B_{lhm}$$

$$\times \left[\frac{sin(\tfrac{1}{2} k_+ w)}{k_+} + \frac{sin(\tfrac{1}{2} k_- w)}{k_-} + \frac{2 cos(\tfrac{1}{2} k_{hhn} w) cos(\tfrac{1}{2} k_{lhm} w)}{K_{hhn} + K_{lhm}} \right], \tag{110}$$

where

$$k_\pm = k_{hhn} \pm k_{lhm} \tag{111}$$

and the B_{hhn} and B_{lhm} are the normalization constants of the quantum-well wavefunctions. Similarly the matrix elements coupling the odd (sine) heavy and light holes are

$$\langle n\ hh|\mathscr{H}|m\ lh\rangle = \frac{\sqrt{3}\ \hbar^2 k_\perp^2\ \gamma_2}{2m_0}A_{hhn}A_{lhm}$$

$$\times\left[\frac{\frac{1}{2}\sin(k_-w)}{k_-} - \frac{\sin(\frac{1}{2}k_+w)}{k_+} + \frac{2\sin(\frac{1}{2}k_{hhn}w)\sin(\frac{1}{2}k_{lhm}w)}{K_{hhn} + K_{lhm}}\right],\qquad(112)$$

those for the odd heavy and even light holes are

$$\langle n\ hh|\mathscr{H}|m\ lh\rangle = \frac{\sqrt{3}\ \hbar^2 k_\perp\gamma_3}{m_0}A_{hhn}B_{lhm}$$

$$\times\left[\frac{\sin(\frac{1}{2}k_-w)}{k_-/k_{lhm}} - \frac{\sin(\frac{1}{2}k_+w)}{k_+/k_{lhm}} + 2\frac{\cos(\frac{1}{2}k_{hhn}w)\cos(\frac{1}{2}k_{lhm}w)}{(K_{hhn} + K_{lhm})/K_{lhm}}\right],\qquad(113)$$

and those for the even heavy and odd light holes are

$$\langle n\ hh|\mathscr{H}|m\ lh\rangle = -\frac{\sqrt{3}\hbar^2 k_\perp\gamma_3}{m_0}B_{hhn}A_{lhm}$$

$$\times\left[\frac{\sin(\frac{1}{2}k_-w)}{k_-/k_{lhm}} + \frac{\sin(\frac{1}{2}k_+w)}{k_+/k_{lhm}} - \frac{2\cos(\frac{1}{2}k_{hhn}w)\sin(\frac{1}{2}k_{lhm}w)}{(K_{hhn} + K_{lhm})/K_{lhm}}\right].\qquad(114)$$

Diagonalizing the Luttinger Hamiltonian gives the valence band energies and eigenfunctions for a value of k. Repeating the process for a range of k values gives the valence bandstructure and its associated eigenfunctions. We show the results of doing so for several common laser structures in the following sections.

6-8. GaAs-AlGaAs Quantum Wells

A widely used heterostructure in present semiconductor lasers is a GaAs quantum well between AlGaAs barriers. This structure lases typically between 780 to 870nm. The growth technology is well established and there is considerable data on device reliability. Devices made with this structure include the first laser arrays.

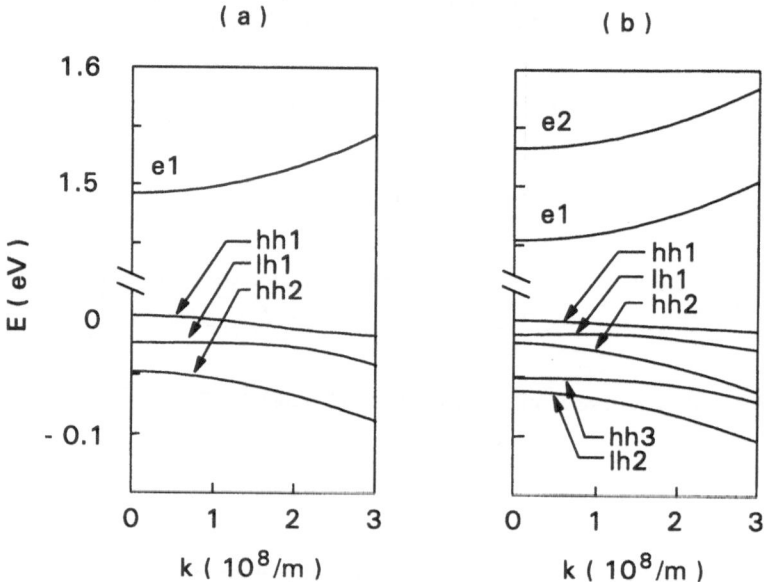

Fig. 6-4.   Bandstructures of (a) 5nm and (b) 10nm GaAs-$Al_{0.2}Ga_{0.8}As$ quantum wells.   All examples in the remainder of this chapter are for a temperature of 300K.

The relevant bulk material parameters are given in Table 6-1. For the bandgap of $Al_x Ga_{1-x} As$, one may use the empirical formula

$$\varepsilon_g = 1.519 + 1.247x - .0005405T^2/(T + 204) , \tag{115}$$

for $x < .45$, or

$$\varepsilon_g = 1.519 + 1.247x + 1.147(x - 0.45)^2 - .0005405T^2/(T + 210) , \tag{116}$$

for $x \geq .45$. The band offset ratio is usually taken to be 0.67. Figure 6-4 shows two examples of GaAs-AlGaAs quantum-well bandstructures. For narrow well widths, the bandstructure is relatively simple, consisting of few bands. For example, Fig. 6-4*a* shows a 5*nm* GaAs quantum well with only one conduction and three valence bands. We label each band according to its state at the zone center, $k = 0$. The conduction bands are labeled en, where $n = 1$ and 2 are the square-well bound state quantum numbers. Similarly the valence bands are labeled, hh1, lh1 and hh2, where hh and lh stand for heavy hole and light hole, respectively. We emphasize that away from the zone center, the valence bands become mixtures of heavy and light hole states with possibly different square-well quantum numbers. This is evident by the changes in the band curvatures at $k > 0$. What our notation does is to indicate the dominant state in a band, in the neighborhood of zone center. With a wider quantum well the bandstructure becomes more complicated because of the increase in the number of bands. Figure 6-4*b* shows that with a 10*nm* quantum well, the number of conduction and valence bands increases to two and five, respectively, with the valence bands originating from three heavy-hole and two light-hole states at $k = 0$.

The bandstructure calculation also gives the electron and hole eigenfunctions, which we use in Eqs. (94) and (95) to compute the dipole transition matrix elements. Figure 6-5 shows the results for the 5*nm* quantum well. The curves exhibit strong k dependence. In terms of the gain, the important values of the dipole matrix elements are those around zone center because that is where the inversion is highest. Since there is usually little band mixing there, one can obtain an idea of the significance of the individual transitions to the dipole matrix elements through symmetry arguments. In Appendix C we show that because of the symmetry of the electron and hole lattice periodic functions, the TM dipole matrix element only couples the electron state to the light hole state, while the TE dipole matrix element gets contributions from both electron to heavy hole and electron to light hole transitions. This property shows up in Fig. 6-5*b* where the TM e1-hh1 dipole matrix element is negligible for small k. Its value increases at higher k because of the mixing of heavy hole and light hole states in the hh1 band. Appendix C also shows that the dipole matrix elements involve inner products of envelope functions. Since the envelope functions are the orthonormal eigenfunctions of a square potential well, a dipole matrix element coupling states with different square well quantum numbers is negligible around zone center. This explains the smaller values for the e1-hh2 curves in the figure. Again, because of state mixing these matrix elements grow with increasing k.

In the earlier chapters, our discussions are limited to two-band bulk and quantum-well structures because we were unable to compute exact bandstructures. This being no longer true, we can now perform comparisons of behaviors of realistic gain structures. An example is the comparison of quantum-well structures with different well widths. We have already showed that the bandstructure can vary noticeably with quantum-well

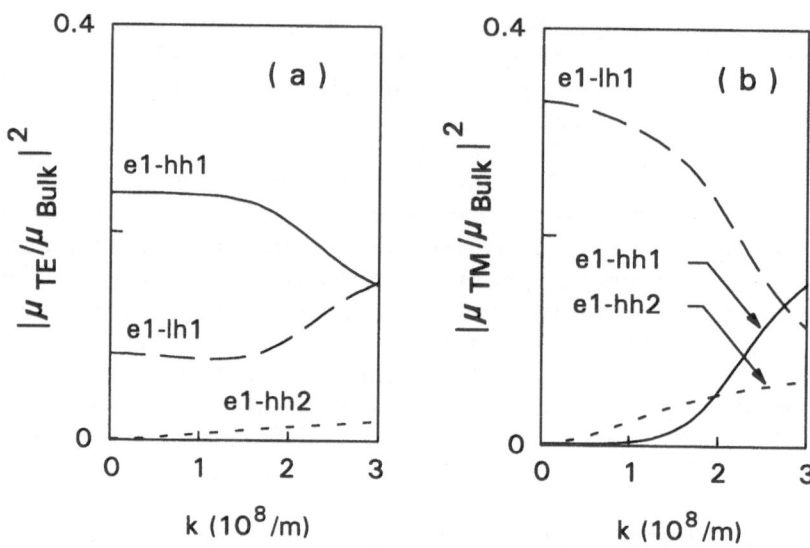

Fig. 6-5. Absolute square of the dipole matrix elements versus k for the bandstructure shown in Fig. 6-4a. The transitions are e1-hh1 (solid line), e1-lh1 (long dashed line) and e1-hh2 (short dashed line). μ_{Bulk} is the bulk GaAs dipole matrix element for the electron to heavy hole transition.

width. Comparison of Figs. 6-6a and 6-6b shows the corresponding changes in the gain spectra. These spectra are computed using the many-body gain theory. First we examine the $5nm$ GaAs-Al$_{0.2}$Ga$_{0.8}$As gain spectra. Note that the TM gain spectra are shifted toward higher frequencies and a higher carrier density is necessary to reach transparency. These features occur because the TM transitions to heavy hole states are forbidden, as discussed in the previous paragraph. Figure 6-6b shows that a wider well width gives broader gain spectra. There is also a greater change in the shape of the gain spectrum with carrier density. At a carrier density of $N = 4\times10^{18}/cm^3$, the TE and TM $10nm$ quantum-well gain spectra have simple shapes, because the small carrier density can only create appreciable electron and hole populations in the e1, hh1 and lh1 bands. At the peak gain frequency, the main contributing transitions involve the recombination of an electron from e1 and a hole from hh1 and lh1. The transitions terminating in the different valence bands are essentially indistinguishable because the energy differences between the these bands are not much greater than the transition linewidth $\hbar\gamma$. When the carrier density is suffi-

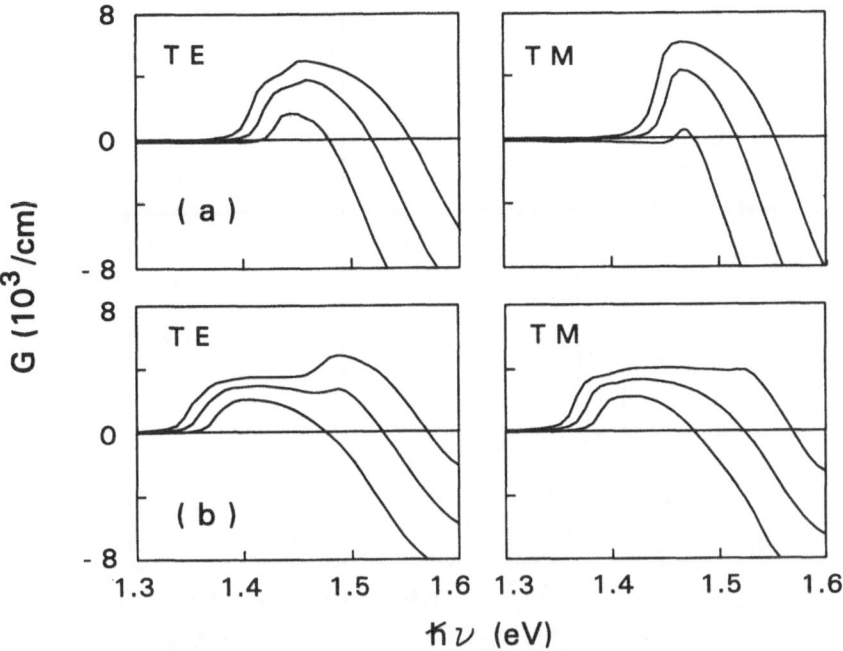

Fig. 6-6. TE and TM gain spectra for (a) $5nm$ and (b) $10nm$ GaAs-Al$_{0.2}$Ga$_{0.8}$As quantum wells for carrier densities, $N = 4\times$, $6\times$ and $8\times10^{18}/cm^3$. Unless stated otherwise, all our gain calculations are for a transition linewidth of $10^{13}/s$.

ciently increased so that an inversion is established between the conduction band, e2, and the valence bands, hh2 and lh2, then a secondary peak appears in the gain curve, as seen in the TE and TM gain spectra at $N = 6\times10^{18}/cm^3$ and $8\times10^{18}/cm^3$, respectively. These peak are due to the combined contributions of transitions originating from e1 and e2. They may eventually overtake the original peaks, as shown in the TE gain spectrum for $N = 8\times10^{18}/cm^3$.

The increased bandstructure complexity in wider quantum wells also results in additional structures in the α spectra. Figure 6-7 shows large variations in the linewidth enhancement factor α of the $10nm$ quantum well, with values ranging from close to zero to over 3, in the frequency region where gain exists. As shown in Fig. 5-9, the dispersion is due to many-body Coulomb effects. That α has pronounced minima at certain frequencies has important implications on the fundamental linewidth of quantum-well lasers. As α approaches zero, the refractive index is increasingly decoupled from changes in gain (e.g., due to spontaneous emission), resulting in a reduction in the spontaneous emission contribution to the laser linewidth. As is discussed in detail in Chap. 10, the coupling between

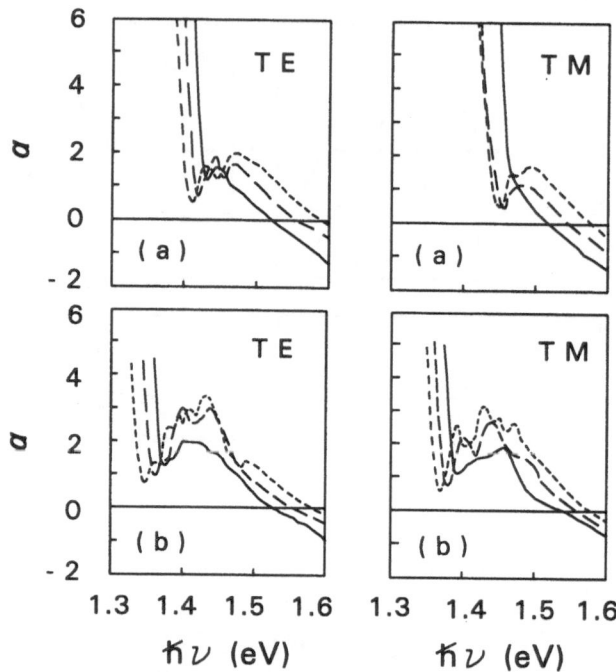

Fig. 6-7. Spectra of linewidth enhancement factor spectra for (a) 5*nm* and (b) 10*nm* GaAs-Al$_{0.2}$Ga$_{0.8}$As. The solid, long-dashed and short-dashed curves are for carrier densities, N = 4×, 6×, 8×10^{18}/cm^3, respectively.

refractive index and gain also leads to filamentation or self-focusing, which limits the scalability of semiconductor lasers to higher output power. Here small α values allow broad-area semiconductor lasers to operate more efficiently far above threshold. Whether we can make use of these advantages depends on α at the peak gain frequency, which is where the laser operates. Figure 6-8 plots the TE linewidth enhancement factor at the peak gain frequency α_{pk}, versus carrier density for the 5*nm* and 10*nm* quantum wells. While α_{pk} for the 5*nm* quantum well is relatively constant, that for the 10*nm* quantum well varies noticeably. It even has a discontinuity that is due to the abrupt shift in the peak-gain frequency (see TE gain spectra for N = 6× and 8×10^{18}/cm^3 in Fig. 6-5b). Note that α_{pk} becomes significantly smaller after the discontinuity, which suggests that we may get a more stable laser by designing it to operate at a higher threshold.

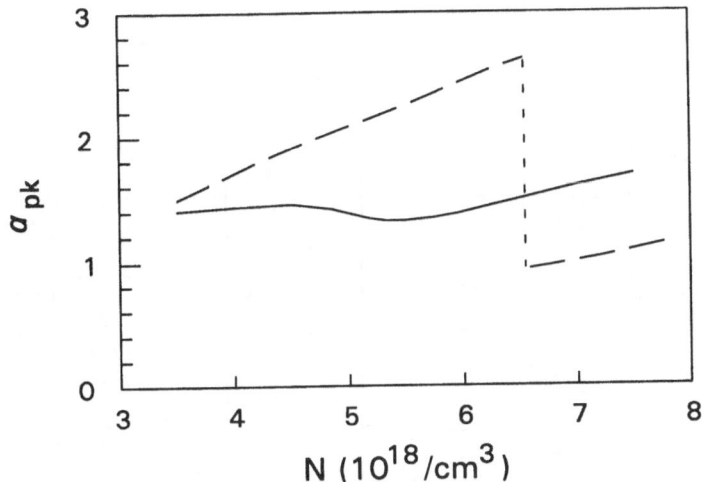

Fig. 6-8. TE linewidth enhancement factor at the peak gain frequency for a $5nm$ (solid curve) and a $10nm$ (dashed curve) GaAs-Al$_{0.2}$Ga$_{0.8}$As quantum well.

6-9. InGaAs-AlGaAs Strained Quantum Wells

As described in Sec. 6-6, quantum wells may be grown with materials having different lattice constants. The thin quantum well material deforms so as to be lattice matched to the barriers. This deformation creates a strain in the quantum well. The capability to grow strained quantum wells gives the semiconductor laser access to a wider range of lasing wavelengths. An extensively investigated strained quantum well structure is InGaAs between AlGaAs barriers. Lasers fabricated with this structure typically lase between $910nm$ and $980nm$, with wavelengths as long as $1.07\mu m$ having been reported. This wavelength range is of interest because it overlaps with the absorption spectrum of Erbium-doped fibers.

In addition to the lasing wavelength, strain also changes the shape of the valence bands. To see this for the InGaAs-AlGaAs structure, we first refer to Table 6-2, which shows that the lattice constant for InGaAs is larger than that for AlGaAs. As a result, the quantum well is under compressive strain. To compute the bandstructure, we use Eqs. (115) and (116) for the bandgap of AlGaAs and the following empirical formulas for the bandgap of InGaAs. For In$_x$Ga$_{1-x}$As at 77K

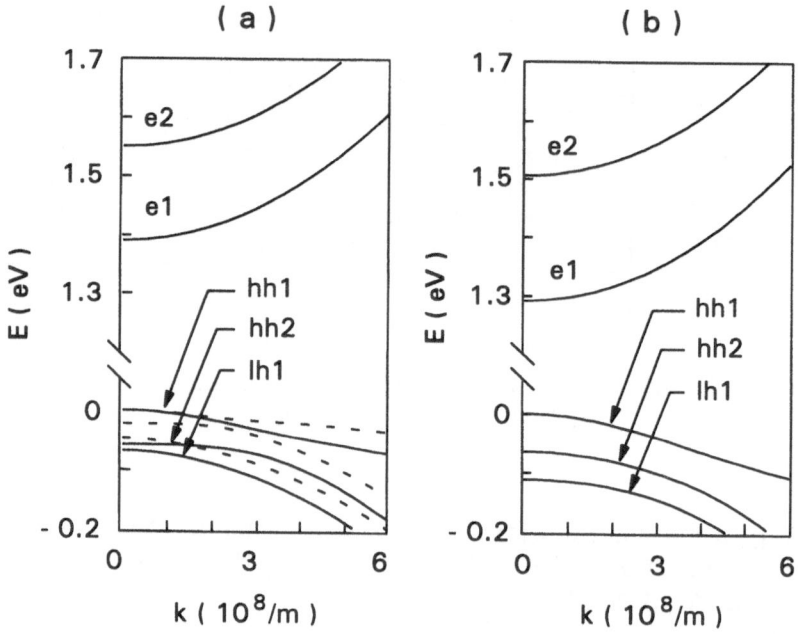

Fig. 6-9.   Bandstructures for $5nm$ $In_x Ga_{1-x} As$-$Al_{0.2}Ga_{0.8}As$ strained quantum wells with (a) $x = 0.1$ and (b) $x = 0.2$.   For comparison, the dashed curves depict the valence bands for $5nm$ $GaAs$-$Al_{0.2}Ga_{0.8}As$.

$$\varepsilon_g = 1.508 - 1.47x + 0.375x^2 , \tag{117}$$

and at 300K

$$\varepsilon_g = 1.43 - 1.53x + 0.45x^2 , \tag{118}$$

where the bandgaps at the other temperatures are assumed to be those computed by linear interpolation or extrapolation.   Figure 6-9 shows the calculated bandstructures for InGaAs-AlGaAs with indium concentrations of 0.1 and 0.2. According to Eqs. (107) and (108), compressive strain causes the heavy and light hole square-well states to shift so as to increase the energy separation between the hh1 and lh1 bands at $k = 0$. This moves the crossing of the heavy and light hole bands to higher k values (refer to Fig. 6-3).   Consequently, the highest valence band, hh1, maintains its smaller curvature over a larger region, as may be seen by comparing the solid and dashed curves in Fig. 6-9a. Figure 6-9b shows that further in-

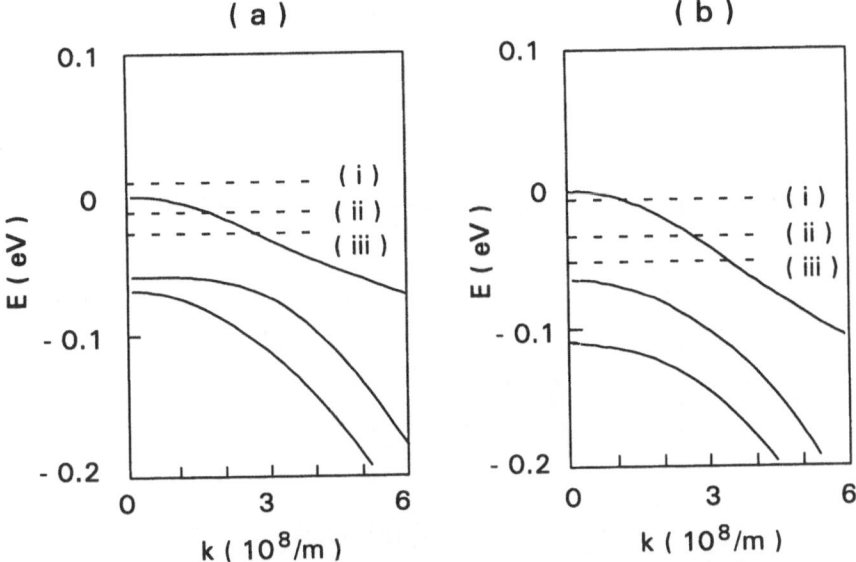

Fig. 6-10.   Hole chemical potentials (dashed lines) for 5nm In$_x$Ga$_{1-x}$As-Al$_{0.2}$Ga$_{0.8}$As, where (a) $x = 0.1$, (b) $x = 0.2$. The densities are N = (i) 2×, (ii) 4×, and (iii) 6×10^{18}/cm^3. For the unstrained GaAs-AlGaAs, the hole chemical potentials for the same densities all lie inside the bandgap.

creasing the indium concentration to 0.2 increases the average hh1 band curvature.

A consequence of the changes in the bandstructure due to compressive strain is a reduction in the carrier density at transparency. As discussed in the earlier chapters, increasing a band curvature (or equivalently for a parabolic band, reducing its effective mass) eases the population of its states. Figure 6-10 shows that for a given carrier density, the hole chemical potential lies further inside the valence bands for higher indium concentration. Since a bottleneck in the creation of gain in a semiconductor medium is in the population of the hole states, the introduction of strain, which increases the highest energy valence band curvature, helps alleviate the problem. This is seen in Fig. 6-11, where we plot the peak gain versus carrier density for InGaAs-AlGaAs with different amount of strain.

Another result of compressive strain is polarization discrimination. Compressively strained structures can have significantly smaller TM gain than TE gain (compare Figs. 6-12a and 6-12b). This is another conse-

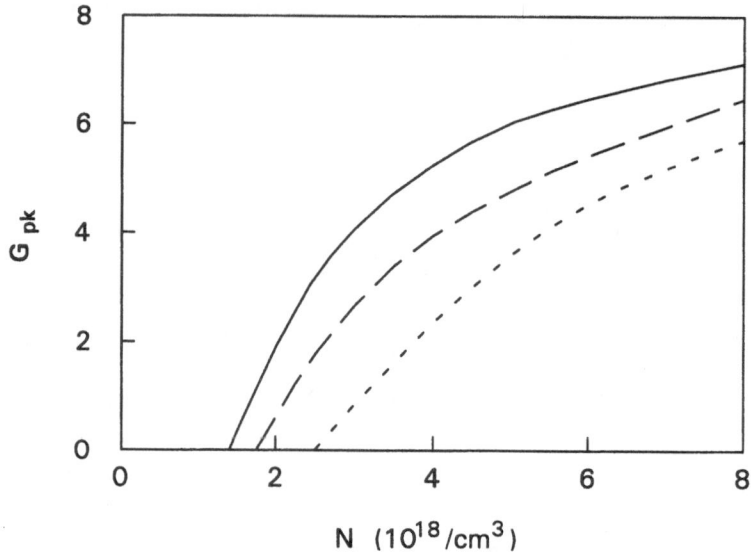

Fig. 6-11. Plots of peak gain versus carrier density for $5nm$ $In_x Ga_{1-x} As/Al_{0.2}Ga_{0.8}As$, with $x = 0$ (short-dashed curve), 0.1 (long-dashed curve) and 0.2 (solid curve). They reveal reductions in carrier density at transparency with increasing strain. The curves are calculated using the many-body gain formula.

quence of the TM dipole transition occurring only between the electron and light-hole states. The increased splitting between the hh1 and lh1 bands means that the lh1 band, which provides most of the TM gain, is harder to populate. Furthermore, because of the greater hh1 and lh1 band separation, the TE and TM gain spectra are further apart in frequency than is the case for the unstrained structure (compare Figs. 6-6 and 6-12).

Figure 6-13a compares the linewidth enhancement factor at the peak gain frequency α_{pk} and the peak gain G_{pk} for the $5nm$ $In_{0.2}Ga_{0.8}As$-$Al_{0.2}Ga_{0.8}As$ and the $5nm$ $GaAs$-$Al_{0.2}Ga_{0.8}As$ structures. For low values of peak gain, α_{pk} in the strained quantum well is below that of the unstrained one. However α_{pk} gradually increases with G_{pk}, so that for a peak gain of $\simeq 7000$ cm^{-1}, both strained and unstrained media have the same α_{pk}. Beyond this point, a further increase in peak gain results in a much larger variation in α_{pk} for the strained quantum well. The influence of the well width is shown in Fig. 6-13b, where the results for the TE mode in a $10nm$ $In_{0.2}Ga_{0.8}As$-$Al_{0.2}Ga_{0.8}As$ strained quantum well are plotted. Similar to the unstrained case shown in Fig. 6-8, the onset of the e2-hh2 transition

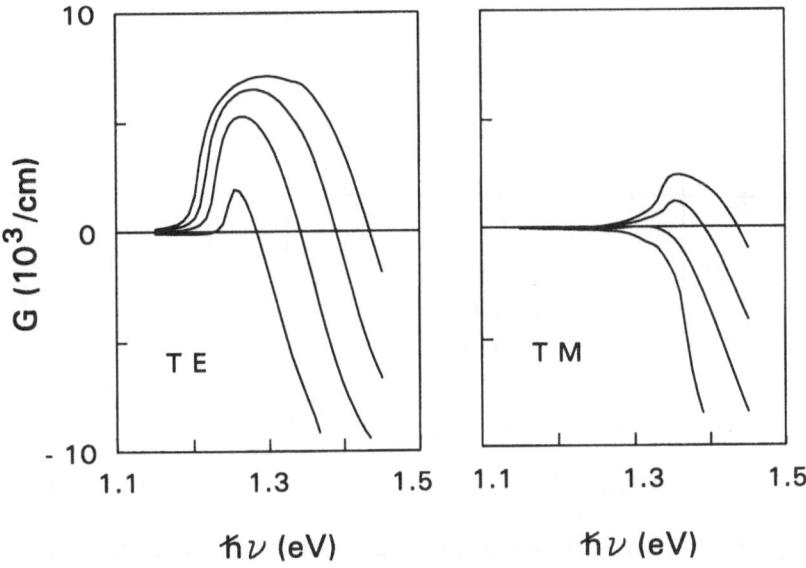

Fig. 6-12. (left, *a*) TE and (right, *b*) TM gain spectra for *5nm* In$_{0.2}$Ga$_{0.8}$As-Al$_{0.2}$Ga$_{0.8}$As showing much lower TM gain for the same carrier density. The carrier densities are N = 2×, 4×, 6×, 8×10^{18}/*cm*^{3}.

leads to a large drop in α_{pk}. So far the calculations presented assume a Lorentzian line shape with a constant phenomenological linewidth, γ = 10^{13}/s. Actually, the semiconductor gain lineshape is not truly Lorentzian, nor is the linewidth constant. A genuine line-shape theory is somewhat involved. However we can estimate the influence of the linewidth on our results by computing α_{pk} for different linewidths. Comparing the results for γ = 10^{13}/s and 4×10^{13}/s, we found that the quantitative results are changed somewhat, but the qualitative results are not strongly linewidth dependent. In particular, the abrupt drop in α_{pk} persists even for the broad linewidth case.

Figure 6-14 shows a comparison with experiment. Experimental data on the α spectrum for a laser operating with a *6nm* In$_{0.2}$Ga$_{0.8}$As-GaAs strained quantum-well structure show a saddle between 1.28*eV* and 1.29*eV*. We are able to reproduce this feature in the computed α spectra at a carrier density of $N \simeq 2.5 \times 10^{18}$/*cm*^{3}. Our theory associates the saddle to many-body Coulomb effects. In fact, this saddle has the same origin as the dispersion in the α spectra of GaAs-AlGaAs quantum wells shown in Fig.

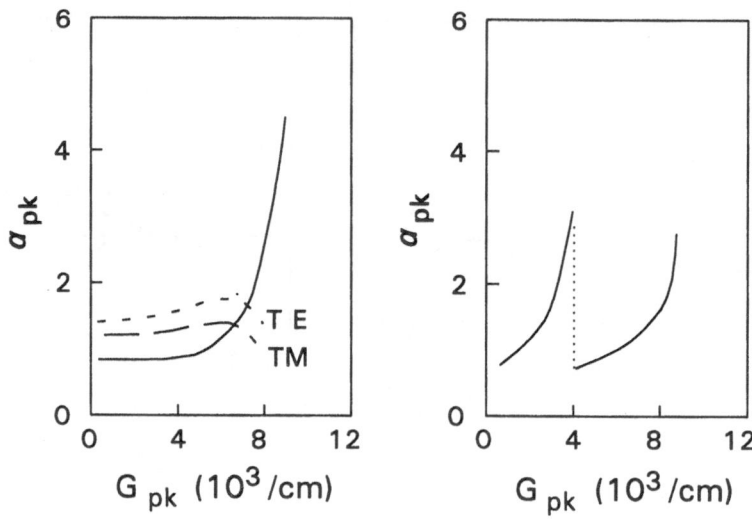

Fig. 6-13. (left) Linewidth enhancement factor at peak gain as a function of peak gain for the TE mode in $5nm$ $In_{0.2}Ga_{0.8}As$-$Al_{0.2}Ga_{0.8}As$ (solid curve); the TM gain is negligible in this case. Short-dashed and long-dashed curves, respectively, show the results for the TE and TM modes in the GaAs-$Al_{0.2}Ga_{0.8}As$ quantum well. (right) Linewidth enhancement factor at peak gain versus peak gain for $10nm$ $In_{0.2}Ga_{0.8}As$-$Al_{0.2}Ga_{0.8}As$.

6-7. In contrast, a free-carrier theory (dashed curves) yields an α that monotonously decreases with frequency.

The InGaAs-GaAs structure introduced in the previous paragraph brings out an interesting feature of InGaAs strained quantum wells. If the barriers are GaAs instead of AlGaAs, then the confinement potential becomes too weak to bound the light hole, regardless of well width. Consequently, only the heavy hole bands exist. There is little band mixing and the bands are all parabolic close to the zone center with the same effective mass as shown in Fig. 6-15a. Figure 6-15b shows that the idealized two-band system may be achieved with a $5nm$ $In_{0.1}Ga_{0.9}As$-GaAs structure.

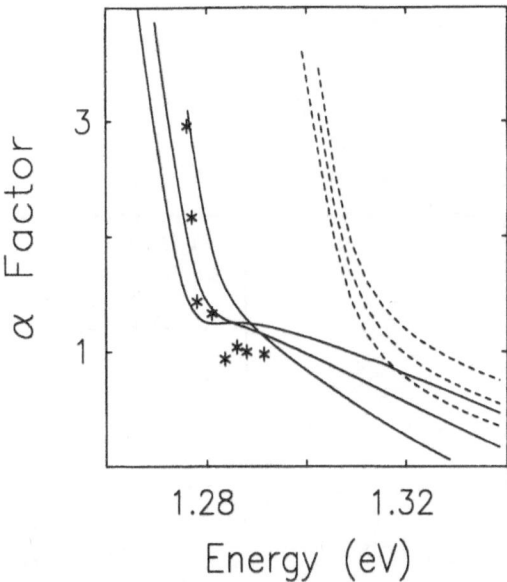

Fig. 6-14. Comparison between experimentally measured values of α (stars) [Dutta *et al.* (1990)] and our calculations for an $6nm$ $In_{0.2}Ga_{0.8}As$-GaAs strained quantum well. The many-body results are shown as solid curves (from right to left, $N = 1.5\times$, $2\times$, $2.5\times10^{18}/cm^3$, and the free-carrier results are plotted as dashed curves (from right to left, $N = 2.5\times$, $2\times$, $1.5\times10^{18}/cm^3$.)

6-10. InGaAs-InP

Heterostructures with $In_x Ga_{1-x} As$ quantum wells between InP barriers have the interesting property of being either strained or unstrained. The structure is unstrained for $x \simeq 0.53$, under tensile or compressive strain for smaller or larger indium concentration, respectively. Figure 6-16 shows the bandstructures of $5nm$ $In_x Ga_{1-x} As$-InP for the three cases. These bandstructures are computed using the parameters given in Tables 6-1 and 6-2. The bandgap of InGaAs is given in Eqs. (117) - (118) and the bandgap of InP is

$$\varepsilon_g = 1.42667 - .000326436T , \qquad (119)$$

which is determined using the linear fit to the data in Landolt-Börnstein. For $x = 0.33$, there are three valence bands originating from one light and two heavy hole states at $k = 0$, with the light hole state being associated

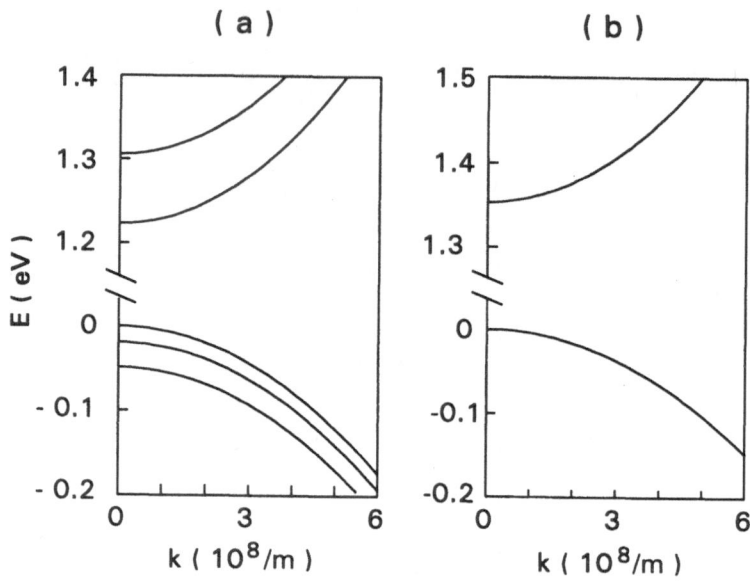

Fig. 6-15. (*a*) With GaAs barriers only the (parabolic) heavy hole and the conduction bands exist. The bandstructure is computed for a 5*nm* $In_{0.2}Ga_{0.8}As$-GaAs structure. (*b*) With a 5*nm* $In_{0.1}Ga_{0.9}As$-GaAs structure, one realizes the idealized two-band system.

with the upper most valence band. With $x = 0.53$, the number of conduction and valence bands increases to two and four, respectively, because of the deeper well. Note that now the upper most valence band is predominately heavy hole in character and therefore has a higher average band curvature than is the case with $x = 0.33$. For $x = 0.73$, the light hole bandgap is shifted further to the point where two heavy hole states lie above it. This further decreases the mixing of heavy and light hole states in the upper most valence band, resulting in an even higher average hh1 band curvature. Not to be forgotten is the overall bandgap change with strain (from $0.73eV$ with $x = 0.73$ to $1.01eV$ with $x = .33$), which broadens the range of accessible wavelength for lasing. Lasers fabricated with InGaAs-InP quantum wells typically lase between $1.45\mu m$ to $1.62\mu m$, which makes them of interest in optical-fiber communications.

Figure 6-17 plots the gain spectra for the different structures shown in Fig. 6-16. The curves are computed using the many-body theory and the carrier density is varied to give a peak gain of approximately $4000/cm$ for all the structures. The explanation for the gain spectrum variation in rela-

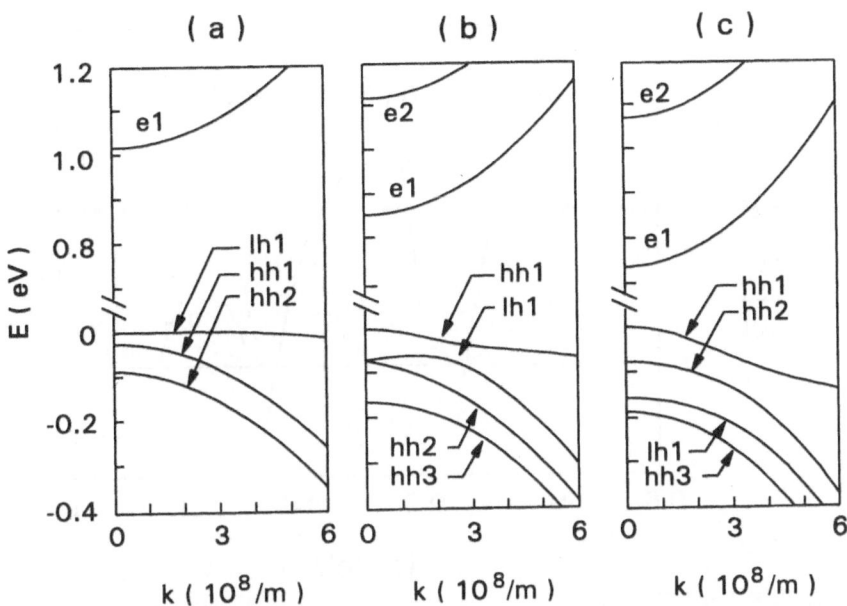

Fig. 6-16.   Bandstructures for $5nm$ $In_x Ga_{1-x}$ As-InP with (a) x = 0.33, (b) x = 0.53 and (c) x = 0.73, showing the increase in the average hh1 band curvature with increasing indium concentration.

tion to the corresponding bandstructures is similar to that for the InGaAs-AlGaAs structures, especially for the x = 0.73 structure, which is under compressive strain. The gain spectrum is dominated by the e1-hh1 transition, since these bands have the highest electron and hole populations at the chosen density of N = 1.6×10^{18}/cm^3. There is no TM gain because the lh1 band, which is its primarily contributer, is too far removed from the band edge to be populated at this density. For the unstrained structure with x = 0.53, the highest valence band is still hh1. Therefore the e1-hh1 transition still dominates. However the energy separation between hh1 and lh1 is reduced. This has two effects. One is a higher hole population in lh1 and the other is a stronger heavy hole-light hole coupling, which increases the light hole content in hh1, and conversely the heavy hole content in lh1. This is responsible for the reduction in the difference in peak gain values between the TE and TM modes, as shown in the figure. For the indium concentration giving tensile strain, we have a reversed situation. The highest valence band is lh1, which is mostly light hole in character around the zone center and therefore contributes more to the TM than to the TE mode. As we can see, the TM mode dominates. For indium concentration x ≥ 0.53 the TE gain spectra are red shifted with respect to

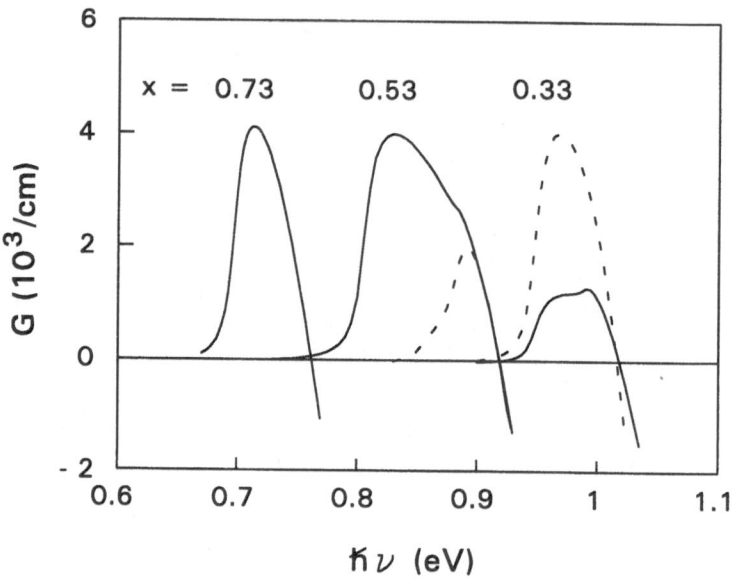

Fig. 6-17. $In_x Ga_{1-x}$ As-InP gain spectra for TE mode (solid curve) and TM mode (dashed line) for the indium concentrations $x = 0.33$, 0.53 and 0.73. The densities, $N = 4.5 \times 10^{18}/cm^3$ for $x = 0.33$, $N = 4 \times 10^{18}/cm^3$ for $x = 0.53$ and $N = 1.6 \times 10^{18}/cm^3$ for $x = 0.73$, are chosen to give a peak gain of $\simeq 4000/cm$.

the TM gain spectra because the top valence band is predominately heavy hole like around zone center. On the other hand, for indium concentration, $x < 0.53$, the top valence band is predominately light hole like around the zone center and therefore the TE gain spectra are blue shifted with respect to the TM gain spectra. There is, of course, an overall shift of the gain spectra to higher frequencies with decreasing indium concentration in InGaAs, due to the corresponding increase in the alloy bandgap.

More details on the effects of strain are best described with the following figures. Figure 6-18 is a plot of the threshold carrier density, N_{th} versus indium concentration in InGaAs for different threshold gains, G_{th}. The $G_{th} = 0$ curve gives the transparency carrier density. The indium concentrations considered cover the estimated range with critical thicknesses greater than $5nm$. As expected from previous results, the threshold carrier density is reduced by compressive strain. There can also be a slight drop in threshold carrier density for a TM mode with tensile strain. The curves show that strain can be used for polarization selection. Tension favors the TM mode while compression favors the TE mode.

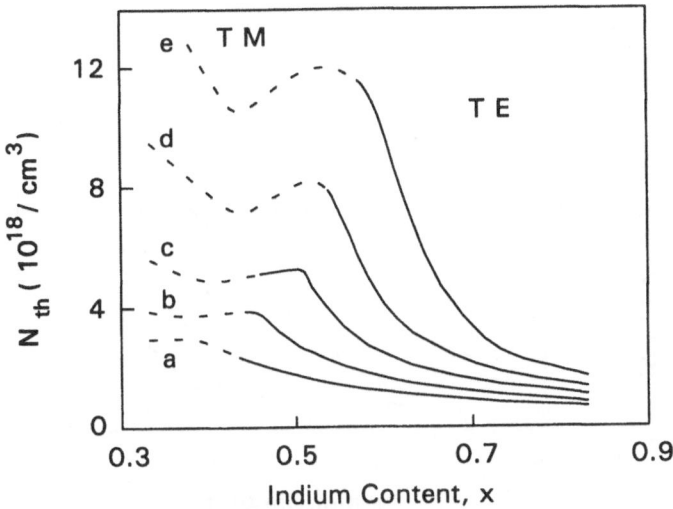

Fig. 6-18. Threshold carrier density versus indium content x for $5nm$ $In_x Ga_{1-x}$ As-InP for threshold gain G_{th} = (a) 0, (b) 2000/cm, (c) 4000/cm, (d) 6000/cm, and (e) 8000/cm. The dashed curves indicate where the TM modes have higher gains and the solid curves indicate where the TE modes have higher gains.

Using the real part of the derivative of the many-body gain formula with respect to carrier density, we compute the differential gain G'. When using the many-body gain formula, one must be careful because unlike the free-carrier model, this derivative involves more than the Fermi-Dirac distributions. There are contributions from bandgap renormalization and Coulomb enhancement, in addition to band filling. Figure 6-19 is a plot of G' as a function of indium concentration in InGaAs, for different threshold gains. The values of G' are those for the peak gain frequencies. We see that the differential gain does indeed increase with compressive strain. There are also discontinuities in the curves. One cause of the discontinuities is the transition from TE to TM mode. Another cause is the jump in the peak gain frequency when the second conduction band becomes appreciably populated. For a $5nm$ quantum well width, the second conduction band is only present for $x < .69$. For wider well widths, a second conduction band is possible at higher compressive strain. Also for wider well widths, more than two conduction subbands may be present. Finally, the decrease in G' with increasing gain is due to gain rollover, which is a

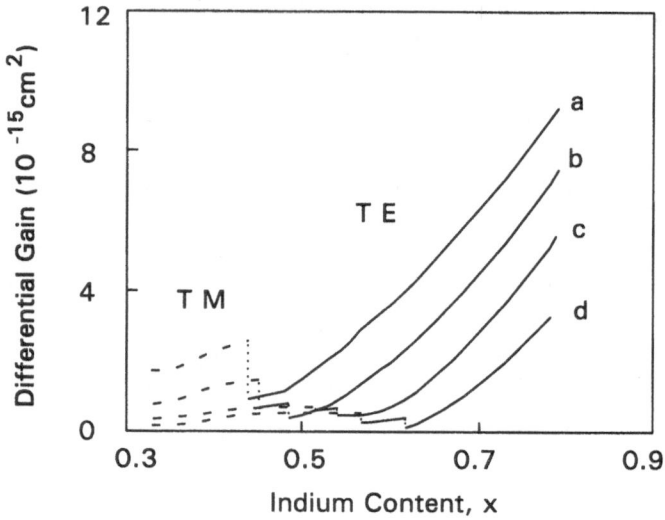

Fig. 6-19. Differential gain at peak gain versus indium concentration x for $5nm$ $In_x Ga_{1-x} As$-InP for threshold gain G_{th} = (a)
$2000/cm$, (b) $4000/cm$, (c) $6000/cm$, and (d) $8000/cm$. The solid
lines indicate where the lasing modes have TE polarization and
the long dashed curves indicate where the lasing modes have TM
polarization. The short dashed lines are used to connect the discontinuous sections of a constant threshold gain curve.

consequence of the two-dimensional density of states in quantum-well
structures.

    Figure 6-20 shows the refractive index derivative, $d(\delta n)/dN$ versus
indium concentration in InGaAs. The values are those for the peak gain
frequency and they are computed by taking the derivative of the imaginary
part of Eq. (5.29) with respect to the carrier density. We find noticeable
contributions from the many-body corrections. One many-body effect is
the positive $d(\delta n)/dN$ at high thresholds. The many-body effects are even
more significant at frequencies away from the gain peak.

    Figure 6-21 is a plot of α computed at the peak gain frequency versus
indium concentration in InGaAs for different threshold gains. Consistent
with existing results from experiments, α_{pk} exhibits a complicated dependence on gain (or carrier density) and strain, as evident by the various
crossings among the curves. On the other hand, there are also common
features among strained quantum-well structures of different materials and

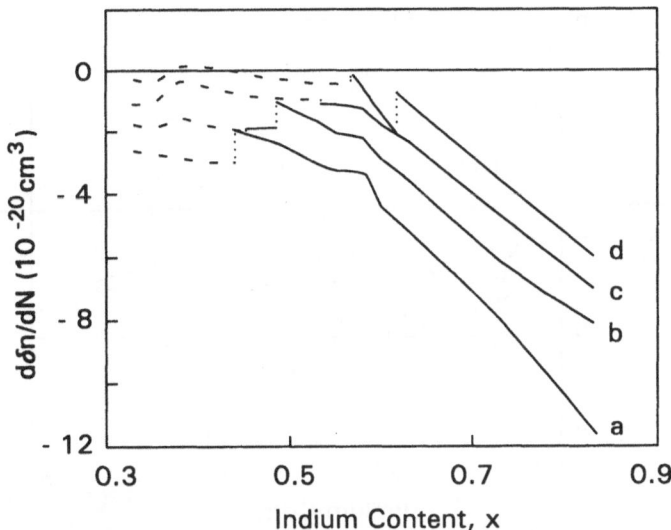

Fig. 6-20. Carrier density derivative of the carrier-induced refractive index at peak gain versus indium concentration x for $5nm$ $In_x Ga_{1-x}$ As-InP for threshold gain of (a) 2000/cm, (b) 4000/cm, (c) 6000/cm, and (d) 8000/cm. The notation used is similar to the one in Fig. 6-19.

dimensions. One is that the larger the threshold gain, the greater is the dependence of α on strain. Another is that the number of pronounced peaks in an α_{pk} versus strain equals one minus the number of conduction bands. The existence of higher conduction bands is inhibited in narrow wells and with high compressive strain. On the $d\alpha_{pk}/dx < 0$ side of the peaks, the increase in α_{pk} with decreasing strain is caused by bandstructure effects. On the $d\alpha_{pk}/dx > 0$ side, the relatively sharper (and sometimes discontinuous) drop in α_{pk} with decreasing strain is due to the abrupt jump in the peak gain frequency. These frequency shifts are typically greater than $100meV$ and they occur when higher conduction subbands become appreciably populated. The many-body effects are also significant on the $d\alpha_{pk}/dx > 0$ side. Here, the effects of bandgap renormalization, which tends to decrease α_{pk}, are comparable to those due to band filling, which tend to increase α_{pk}. In fact, under certain conditions the many-body effects can dominate and lead to negative values for α_{pk}, as shown in the figures.

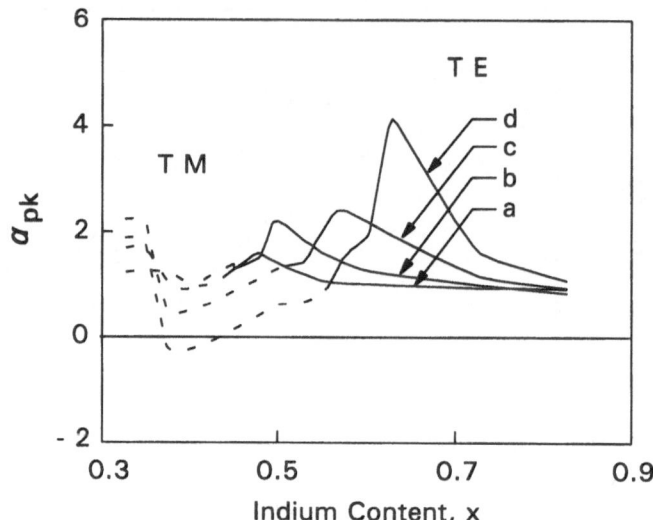

Fig. 6-21. Linewidth enhancement factor α at peak gain versus indium content x for $5nm$ $In_x Ga_{1-x}$ As-InP for threshold gain G_{th} = (a) $2000/cm$, (b) $4000/cm$, (c) $6000/cm$, and (d) $8000/cm$. The notation used is similar to that in Fig. 6-19.

6-11. InGaP-InAlGaP

Heterostructures made of InGaP quantum wells and InAlGaP barriers are of great interest because they provide gain in the visible wavelength region below $700nm$. Recently electrically injected vertical cavity surface emitting lasers (VCSELs) using these heterostructures demonstrated operation at wavelengths between 639 and 661 nm. The bulk material parameters for InP, GaP and AlP are given in Tables 6-1 and 6-2. Parameter values for InGaP and InAlGaP are taken to be the properly weighed averages of those of InP, GaP and AlP. The bandgap of $In_{1-x} Ga_x P$ at room temperature as given by Adachi (1982) is

$$\varepsilon_g = 1.35 + 0.643x + 0.786x^2 . \tag{120}$$

However the results of VCSEL experiments are found to fit better to the formula [Stringfellow *et al.* (1972)],

$$\varepsilon_g = 1.421 + 0.73x + 0.7x^2 . \tag{121}$$

The bandgap for $In_{1-x}(Al_yGa_{1-y})_xP$ is even more uncertain. A reasonable fit to available data appears to be

$$\varepsilon_g = E_g(In_{1-x}Ga_xP) + 0.6y ,\tag{122}$$

for $y \leq 0.6$ and

$$\varepsilon_g = E_g(In_{1-x}Ga_xP) + 0.36 ,\tag{123}$$

otherwise.

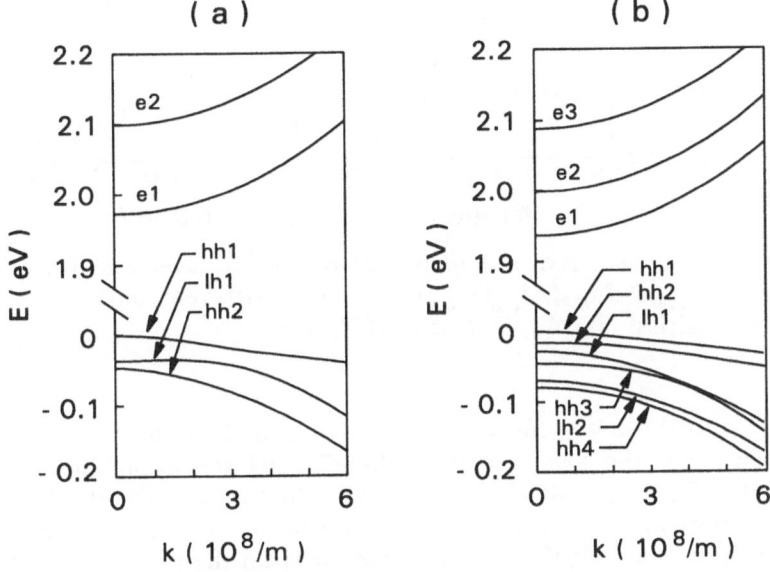

Fig. 6-22. Bandstructures for (a) 5nm and (b) 10nm $In_{0.54}Ga_{0.46}P$-$In_{0.5}(Al_{0.4}Ga_{0.6})_{0.5}P$.

Reported values for the band-offset ratio ranges from $\simeq 0.39$ to $\simeq 0.67$. Using Tables 6-1 and 6-2, Eqs. (121) and (122), and the more recently reported band-offset ratio of 0.67 [Dawson and Duggan (1993)], we computed the bandstructures of 5nm and 10nm $In_{0.54}Ga_{0.46}P$-$In_{0.5}(Al_{0.4}Ga_{0.6})_{0.5}P$ quantum wells. These structures have been successfully incorporated into lasers. The results are shown Fig. 6-22. In both cases the quantum wells are under compressive strain, while the $In_{0.5}Ga_{0.5}P$-$In_{0.5}(Al_{0.4}Ga_{0.6})_{0.5}P$ structure is unstrained.

Figure 6-23 shows the corresponding TE and TM gain spectra. In Fig. 6-23a, the slight bump in the high frequency end of the TE gain spectrum

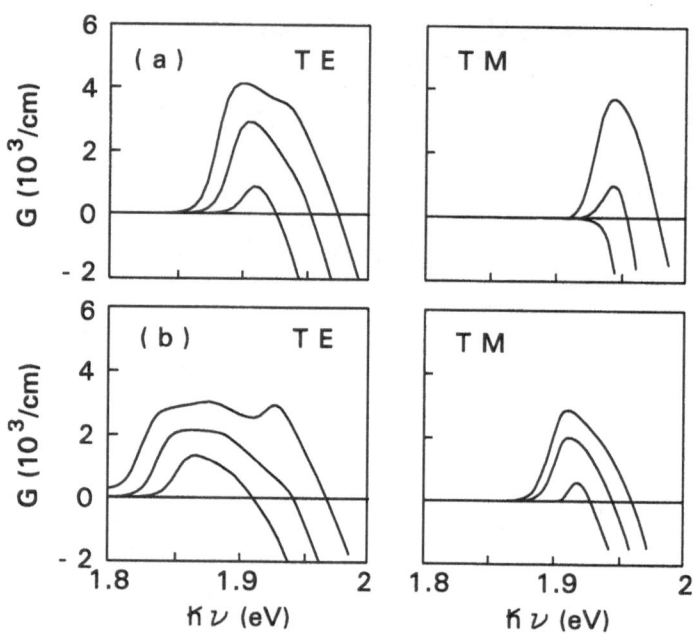

Fig. 6-23. TE and TM gain spectra for (a) 5nm and (b) 10nm $In_{0.54}Ga_{0.46}P$-$In_{0.5}(Al_{0.4}Ga_{0.6})_{0.5}P$. The different curves are for the densities, $N = 4\times$, $6\times$ and $8\times10^{18}/cm^3$.

for $N = 8\times10^{18}/cm^3$ is due to the onset of contributions from the $e2 \rightarrow hh2$ transition. For the 5nm structure, the $e2 \rightarrow hh2$ transition does not play a role in the TE gain until $N > 8\times10^{18}/cm^3$. In contrast, for the 10nm structure, the contributions from the $n = 2$ transitions, $e2 \rightarrow hh2$ and $e2 \rightarrow lh2$, are significant at $N = 8\times10^{18}/cm^3$, as is evident from the pronounced secondary peak in the gain spectrum. However the $n = 3$ transition $e3 \rightarrow hh3$ is insignificant until the densities are too high to be of practical interest. Figure 6-23 also shows the gain for TM modes. The blue shift in the TM spectra and the higher carrier density needed to reach transparency are typical of compressive strained systems where the highest valence band is predominately heavy-hole like around the zone center. Note that since the $e2 \rightarrow hh2$ transition does not contribute to the TM dipole matrix element, there is only one peak in the TM gain spectrum for the 5nm structure, regardless of carrier density. However because of the $e2 \rightarrow lh2$ transition, a second gain peak occurs in the 10nm structure at sufficiently high carrier densities.

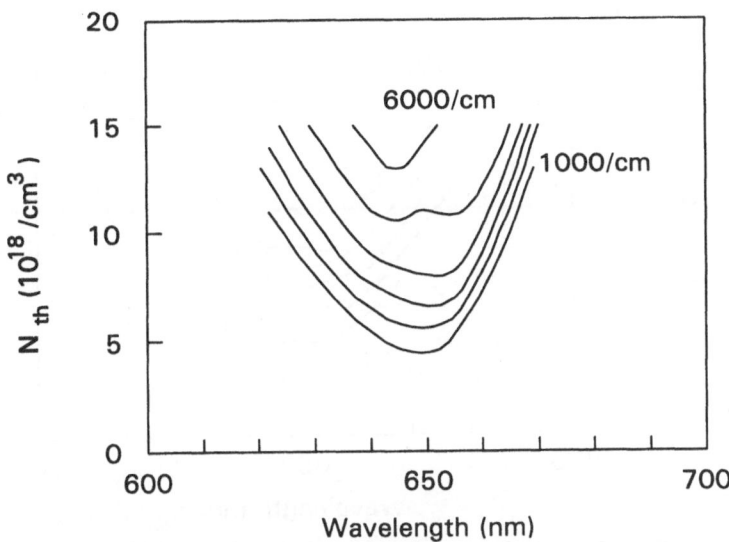

Fig. 6-24. Threshold carrier density as a function of lasing wavelength for $5nm$ $In_{0.54}Ga_{0.46}P$-$In_{0.5}(Al_{0.4}Ga_{0.6})_{0.5}P$. The different curves are for different threshold gains, $1000/cm \leq G_{th} \leq 6000/cm$ in $1000/cm$ increments. The threshold gain is defined by Eq. (1.7) and it is the gain inside the active region (not the modal gain). We assume a TE polarization.

In terms of determining the operating laser wavelength for a given heterostructure, a more useful plot is shown in Fig. 6-24. From the gain spectra, we extracted the carrier density as a function of wavelength for given gain. For an edge emitter, which typically operates with a long and lossy resonator, the resonator resonances are essentially continuous in the wavelength scale used in Fig. 6-24. Lasing occurs at the minima of the curves, which means that we expect an edge emitter fabricated with a $5nm$ $In_{0.54}Ga_{0.46}P$-$In_{0.5}(Al_{0.4}Ga_{0.6})_{0.5}P$ heterostructure to have a laser wavelength of around $650nm$. For higher threshold lasers with $G_{th} \geq 5000/cm$, the wavelength may decrease to $640nm$. Bandstructure, band filling, and many-body carrier-carrier interactions play important roles in determining the positions of the minima, and the shifts in the positions of the minima with respect to threshold gain. The many-body effects enter via bandgap renormalization, which shifts the entire gain spectrum, and Coulomb enhancement, which reshapes the gain spectrum. Therefore only the many-body gain theory is valid for this analysis.

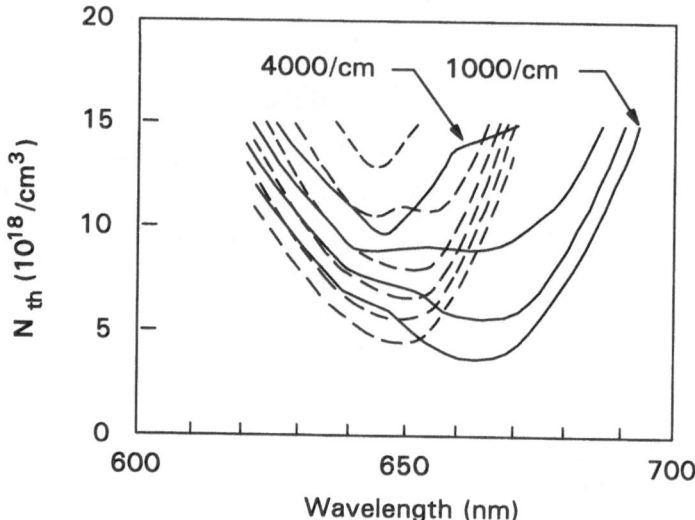

Fig. 6-25.   Threshold carrier density as a function of lasing wavelength for $10nm$ $In_{0.54}Ga_{0.46}P-In_{0.5}(Al_{0.4}Ga_{0.6})_{0.5}P$ (solid curves).   We superimpose the curves for the $5nm$ structure for comparison (dashed curves).   The solid curves are for the threshold gains, $G_{th} = 1000/cm$, to $4000/cm$ in $1000/cm$ increments. The dashed curves are for the threshold gains, $G_{th} = 1000/cm$ to $6000/cm$ in $1000/cm$ increments.   We assume a TE polarization.

The information contained in Fig. 6-24 is even more relevant for the design of VCSELs.   The short, high-Q resonator of a VCSEL has well resolved resonances that are spaced far apart.   Typically there is only one VCSEL resonance within the high reflectivity region, which means that it is important to design the optical resonator so that one of its resonances falls at the peak of the gain spectrum.   The minima of the curves in Fig. 6-24 show the optimum wavelengths for a resonator resonance.   In addition, the curves tell us the price we pay, in terms of increased threshold carrier density, when the resonator resonance deviates from the curve minimum.   Note that because the curves are asymmetric about their minima, it is better to design a resonator with a resonance that tends toward the short-wavelength side of a curve minimum.

The steepness of the curves shown in Fig. 6-24 suggests a relatively strict requirement for resonator-gain matching.   The requirement is somewhat relaxed for the $10nm$ $In_{0.54}Ga_{0.46}P-In_{0.5}(Al_{0.4}Ga_{0.6})_{0.5}P$ structure, especially for lasers with a threshold gain in the neighborhood of $3000/cm$

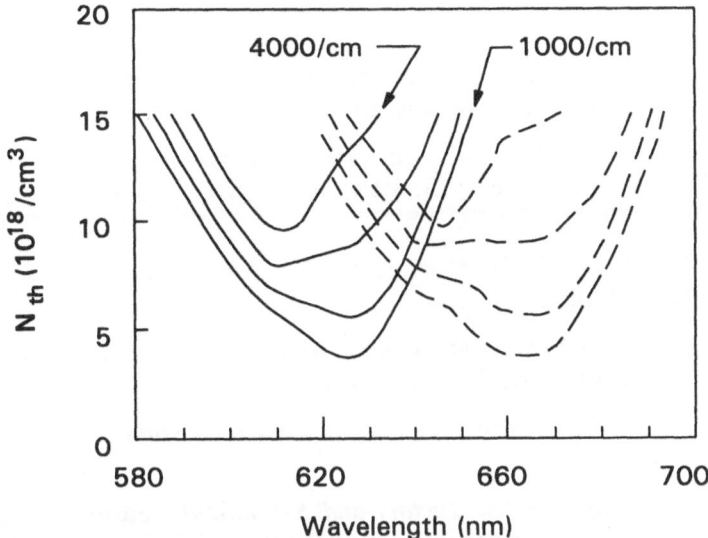

Fig. 6-26. Carrier density necessary to reach threshold as a function of lasing wavelength for $10nm$ $In_{0.54}(Al_{0.2}Ga_{0.8})_{0.46}P$-$In_{0.5}(Al_{0.4}Ga_{0.6})_{0.5}P$ (solid curves). We superimpose the curves for the $10nm$ $In_{0.54}Ga_{0.46}P$-$In_{0.5}(Al_{0.4}Ga_{0.6})_{0.5}P$ structure for comparison (dashed curves). Both sets of curves are for the threshold gains, $G_{th} = 1000/cm$, to $4000/cm$ in $1000/cm$ increments and we assume a TE polarization.

(Fig. 6-25). Here the threshold carrier density remains relatively constant for a wavelength region between $640nm$ and $665nm$. However lasers with the $10nm$ structure operates optimally at the longer wavelength of $660nm$. If one is willing to pay the price of a higher threshold, then a lasing wavelength of $645nm$ is possible by operating where transitions originating from e2 are significant.

Shorter wavelength operation is can be realized with the addition of aluminum in the quantum well. Figure 6-26 shows that lasing at $620nm$ is possible with a $10nm$ $In_{0.54}(Al_{0.2}Ga_{0.8})_{0.46}P$-$In_{0.5}(Al_{0.4}Ga_{0.6})_{0.5}P$ structure. In fact, if we are willing to operate with a high gain threshold of $G_{th} \geq 3000/cm$, we may even be able to reduce the lasing wavelength to $610nm$. Doing so increases the threshold carrier density by more than a factor of two.

REFERENCES

Much of the material discussed in this chapter can be found in many solid state physics textbooks and review articles, e.g.,

Altarelli, M. (1985), p.12 in *Heterojunctions and Semiconductor Superlattices*, Eds. G. Allan, G. Bastard, N. Boccara, M. Lannoo and M. Voos, Springer Verlag, Berlin.

Ashcroft, N. W. and N.D. Mermin (1976), *Solid State Physics*, Saunders College (HRW), Philadelphia.

Bastard, G. (1988), *Wave Mechanics Applied to Semiconductor Hetero-structures*, Les Editions de Physique, Paris.

Callaway, J. (1974), *Quantum Theory of the Solid State*, Part *A*, Academic Press, New York.

Kane, E. O. (1966), *Semiconductors and Semimetals*, edited by R. K. Willardson and A. C. Beer, Academic, New York, *p.* 75.

Kittel, C. (1971), *Introduction to Solid State Physics*, Wiley & Sons, New York; Kittel, C. (1967) *Quantum Theory of Solids*, Wiley & Sons, New York.

The block diagonalization of the Luttinger Hamiltonian has been done by

Broido, D.A. and L.J. Sham (1985), Phys. Rev. **B31**, 888.

The Hamiltonian for strained semiconductors has been derived by

Bir, G.L. and G.E. Pikus (1974), *Symmetry and Strain-Induced Effects in Semiconductors*, Wiley & Sons, New York.

Pikus, G.E. and G.L. Bir (1960), Sov. Phys. - Solid State 1, 1502 [Fiz. Tverd. Tela (Leningrad) 1, 1642 (1959)];

For papers and reviews dealing with bandstructure calculations and optical properties of strained superlattices see, e.g.,

Ahn, D. and S.L. Chuang (1988), IEEE J. Quantum Electron. 24, 2400;

Chuang, S.L. (1991), Phys. Rev. **B43**, 9649;

Dawson, M.D., and G. Duggan (1993), Phys. Rev. **B47**;

Duggan, G. (1990), SPIE **1283**, 206;

Marzin, J.Y. (1986), *Heterojunctions and Semiconductor Superlattices*, eds. G. Allan, G. Bastard, and M. Voos, Springer, Berlin, *p*. 161.

The spin-orbit coupling scheme is discussed, e.g., in

Schiff, L. (1968), *Quantum Mechanics*, McGraw-Hill, New York. Chap. 12.

The experimental results in Fig. 6-14 are from

Dutta, N.K., J. Wynn, D. L. Sivco, and Y. Cho (1990), Appl. Phys. Lett. **56**, 2293.

For references on the composition dependence of the $In_{1-x} Ga_x P$ bandgap see, e.g.,

Adachi, S. (1982), J. Appl. Phys. **53**, 8775.

Stringfellow, G. B., P. F. Lindquist, R. A. Burmeister (1972), J. Electron. Mater. **1**, 437.

Some results of the many-body calculations are presented in

Chow, W. W., M. Pereira, and S. W. Koch (1992), Appl. Phys. Lett. **61**, 758.

Pereira, M., S. W. Koch, and W. W. Chow (1993), Journ. Opt. Soc. **B10**, 765.

A large number of material parameters for many semiconductors can be found in

Landolt-Börnstein (1982), *Numerical Data and Functional Relationships in Science and Technology*, *ed*. K.H. Hellwege, Vol. 17 Semiconductors, edited by O. Madelung, M. Schulz and H. Weiss, Springer Verlag, Berlin.

Chapter 7
SEMICLASSICAL LASER THEORY

In this chapter, we use the polarization of the semiconductor medium derived in the preceding chapters as a source in Maxwell's equations for the laser electromagnetic field. The discussion is primarily oriented toward the prediction of plane-wave, single-mode laser output as a function of parameters such as temperature, injection current, cavity losses, effective masses, and carrier decay rates. More complicated field configurations involving multiple modes are discussed in later chapters. The discussion is couched in terms of a general susceptibility of the medium and applies equally well to the quasiequilibrium many-body treatment of the semiconductor active medium given in Chaps. 4 through 6 as it does to the simple free-carrier model of Chap. 3.

Section 7-1 modifies the approach discussed in Sec. 2-5 to derive slowly-varying Maxwell equations for cavity modes. This modal approach is ideal for high-Q cavities and it describes some features of the internal low-Q cavities given directly by the facet reflectivities. It differs from the approach of Sec. 2-5 in that the time dependence of the field amplitudes and phases are retained instead of the longitudinal spatial dependence. This is called the *uniform amplitude approximation*. More generally, one can consider both dependencies along with transverse variations.

Section 7-2 combines the single-mode self-consistency equations with the microscopic polarization (Secs. 3-2, 5-2 and 5-3) to find the amplitude and frequency-determining equations and their steady-state solutions. The output characteristics of bulk and quantum-well semiconductor lasers are discussed.

Section 7-3 considers the stability of the single-mode solution and the relaxation oscillations that occur as the diode approaches steady-state operation. Spontaneous emission continually perturbs this steady-state solution, thereby driving the relaxation oscillations and producing sidebands in the laser output spectrum. Section 7-4 considers injection locking of the laser frequency. This is a problem that can also lead to interesting instabilities, as is illustrated for the example of a single mode VCSEL.

Section 7-5 discusses the problem a laser operating with two coupled resonators. The longitudinal field coupling is realized in several existing semiconductor laser experiments. They include coupled cavity lasers, external resonator lasers, and lasers subjected to feedback from external mirrors. We introduce the concept of supermodes of a composite-cavity laser to treat this problem. The coupling of the composite-cavity modes through the gain medium is discussed. This coupling results in interesting mode hopping and dynamical switching phenomena. Section 7-6 discusses arrays of semiconductor lasers. The individual lasers are coupled primarily by evanescent waves. An important question concerning laser arrays is, how dissimilar can the lasers be in a phase locked array? We show how the composite cavity mode treatment may be used to determine lockbands.

7-1. Multimode Maxwell Equations

As discussed in Chap. 1, the monolithic edge emitting semiconductor diode has a built-in low-Q cavity, within which substantial spatial variations occur across the beam profile and along the longitudinal direction. Such variations can be much less important for a diode within an external mirror cavity or for a VCSEL. As discussed in Sec. 1-9, a very simple laser theory uses the round-trip oscillation condition (1.6) along with the linear-density gain formula (1.8) and the steady-state solution of the total carrier density equation of motion (1.16). This approach assumes that the total carrier density does not depend on the laser longitudinal coordinate (z). As such it implicitly assumes that the field amplitude and phase are uniform (do not vary) along the z axis.

In this section we generalize this uniform amplitude and phase approximation to include variations in time. Coupled with appropriate formulas for the gain and total carrier density, this allows us to predict not only the laser output intensity, but also whether the solution is stable with respect to intensity fluctuations, and how the output approaches the steady-state oscillating point. In particular, we see in Sec. 7-3 that this leads to relaxation oscillations, which are the hallmark of an instability phenomenon that plagues some laser-diode applications and delights physicists fascinated by dynamical chaos. The stability question is pursued further in Chap. 8, which discusses when other laser modes might build up.

Although this chapter is primarily concerned with single-mode operation, it is no more difficult to calculate the multimode modal Maxwell's equations than the single-mode equations and we need the multimode generality in later chapters, notably Chap. 8. Consequently we suppose that the electromagnetic field in the laser cavity can be represented by a scalar electric field $E(z,t)$ written as the multimode superposition of plane-wave modes

$$E(z,t) = \frac{1}{2} \sum_{m} E_m(t) \exp[-i(\nu_m t + \phi_m)] U_m(z) + \text{c.c.} \tag{1}$$

Here ν_m is the frequency of the mth mode, and $E_m(t)$ and $\phi_m(t)$ are the mth mode amplitude and phase, respectively, which are assumed to vary little in an optical period. The functions $U_m(z)$ specify the mode variations along the laser axis and consist of standing waves

$$U_m(z) = \sin(K_m z) \tag{2}$$

for a two-mirror laser and running waves

$$U_m(z) = \exp(iK_m z) \tag{3}$$

for a unidirectional ring laser (see Fig. 7-1). The boundary conditions for the standing wave cavity are

$$U_m(0) = U_m(L) = 0 , \tag{4}$$

which is satisfied when

$$K_m = \frac{m\pi}{L} . \tag{5}$$

The mode index m is an integer in the neighborhood of 10^3. For the ring resonator, both the eigenfunctions and their first derivations have to be continuous through the ring, so that

$$U_m(0) = U_m(L) , \tag{6}$$

$$\frac{d}{dz} U_m(0) = \frac{d}{dz} U_m(L) , \tag{7}$$

which is satisfied when

$$K_m = \frac{2\pi m}{L} . \tag{8}$$

The laser frequency is $\nu \simeq K_m c/n$, where c is the speed of light in vacuum and n is the refractive index of the medium, which is approximately 3.5

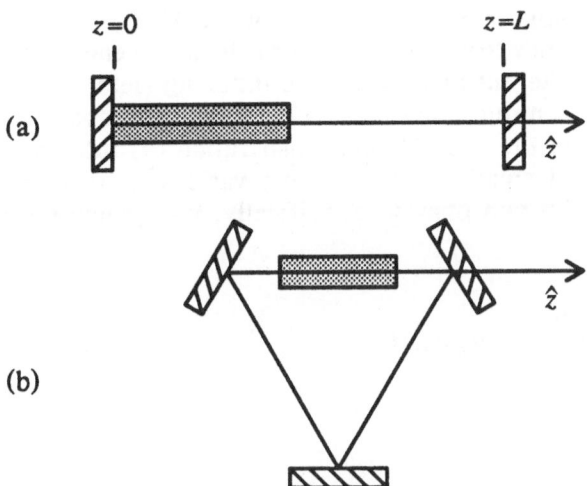

$z=0$ $z=L$

(a)

$\hat{z}$

(b)

$\hat{z}$

Fig. 7-1. (*a*) Diagram of laser showing reflectors in plane perpendicular to laser (*z*) axis and active medium between external-
mirror reflectors. The active-medium facets are assumed to be
antireflection coated to help suppress multiple cavity effects. (*b*)
Corresponding ring laser. Although both running waves oscillate
in some ring lasers, e.g., in ring-laser gyros, we suppose that only
one direction oscillates, a simplification that avoids complications
such as spatial hole burning.

for GaAs. Since the frequency remains the same outside the laser medium,
the free space wave vector becomes $K_{0m} = K_m /n$.
    The unidirectional ring laser configuration is wittingly or unwittingly a
very popular configuration, since the longitudinal spatial dependence
effectively cancels out of the problem. Hence theories that ignore this
dependence are for all intents and purposes assuming the unidirectional
ring configuration.
    The multimode field of Eq. (1) induces a polarization of the medium
of the form

$$P(z,t) = \frac{1}{2} \sum_{m} \mathscr{P}_m(t) \exp[-i(\nu_m t + \phi_m)] U_m(z) + \text{c.c.} \qquad (9)$$

in which the complex polarization component $\mathscr{P}_m(t)$ also varies little in an optical period. They are complex inasmuch as in general the induced polarization has a different phase from the inducing field.

As for the running-wave field (2.67) and polarization (2.68) in Sec. 2-5, we substitute the field (1) and polarization (9) into the wave equation (2.66) and neglect small terms involving various derivatives of the slowly varying amplitudes and phases. Specifically, we assume the validity of the inequalities

$$\left| \frac{\partial E_m(t)}{\partial t} \right| \ll \nu_m E_m(t) \tag{10}$$

$$\left| \frac{\partial \phi_m(t)}{\partial t} \right| \ll \nu_m \tag{11}$$

$$\left| \frac{\partial \mathscr{P}_m(t)}{\partial t} \right| \ll \nu_m |\mathscr{P}_m(t)| , \tag{12}$$

which allow us to neglect the small contributions $\ddot{E}_m$, $\ddot{\phi}_m$, $\dot{E}_m \dot{\phi}_m$, and $\dot{\mathscr{P}}_m$. This procedure is the time-variation version of the derivation of Eq. (2.71) from Eq. (2.70). Here in addition to setting the real and imaginary parts separately equal to zero, we need to project onto the orthogonal spatial modes $U_m(z)$. Since the problem is linear and real, we can derive relations using Eqs. (1) and (9) without the complex conjugates; the relations must then also be true for the full quantities. Accordingly substituting Eqs. (1) and (9) without the complex conjugates into the wave equation (2.66), projecting onto $U_m(z)$, multiplying through by $\exp[i(\nu_m t + \phi_m(z))]$, dropping second derivatives of slowly varying quantities and $\dot{E}_m \dot{\phi}_m$, we find

$$K_m^2 E_m - i\mu_0 \epsilon (\nu_m + \dot{\phi}_m)[2\dot{E}_m - i(\nu_m + \dot{\phi}_m)E_m]$$

$$= \mu_0 \nu_m^2 \mathscr{P}_m + \text{some negligible terms involving } \mathscr{P}_m ,$$

where ϵ ($= \epsilon_0 n^2$) is the permittivity of the semiconductor with no carriers. Noting that $\mu_0 \epsilon_0 = c^{-2}$, where c is the speed of light in vacuum, we have

$$[\Omega_m^2 - (\nu_m + \dot{\phi}_m)^2]E_m - 2i\nu_m \dot{E}_m = \frac{\nu_m^2}{\epsilon} \mathscr{P}_m ,$$

where the passive cavity frequency Ω_m is defined in terms of K_m by

$$\Omega_m = (c\,K_m)/n \;. \tag{13}$$

Making an approximation reminiscent of the rotating-wave approximation

$$\Omega_m^2 - (\nu_m + \dot{\phi}_m)^2 = (\Omega_m + \nu_m + \dot{\phi}_m)(\Omega_m - \nu_m - \dot{\phi}_m)$$

$$\simeq 2\nu_m(\Omega_m - \nu_m - \dot{\phi}_m)\;, \tag{14}$$

and setting real and imaginary parts separately equal to zero, we find the *self-consistency equations*

$$\dot{E}_m = -\frac{\nu}{2Q_m}E_m - \frac{\nu}{2\epsilon}\mathrm{Im}\{\mathscr{P}_m\} \tag{15}$$

$$\nu_m + \dot{\phi}_m = \Omega_m - \frac{\nu}{2\epsilon}\mathrm{Re}\{\mathscr{P}_m\}/E_m \tag{16}$$

so named because the field parameters ultimately appearing in the formulas for the $\mathscr{P}_m$ are taken to be the very same as the parameters in Eq. (1) (see Fig. 2-8). In Eq. (15), we have included a phenomenological decay term $-(\nu/2Q_m)E_m$, where Q_m is the cavity *quality factor* for nth mode.

In general, the Q of an oscillator is defined by

$$Q = \frac{\text{energy stored}}{\text{energy lost per radian}}\;. \tag{17}$$

Hence Q/ν is the energy stored divided by the energy lost per second. The $\frac{1}{2}$ enters since Eq. (15) is an amplitude equation of motion, rather than an intensity or energy equation of motion. We also use the generic frequency ν, which could be ν_m or the frequency of a different oscillating mode, instead of ν_m explicitly since it simplifies multimode coefficients and leads to negligible difference. One is comparing the difference $|\nu - \nu_m|$ to ν_m itself, which is rarely more than one part in a million. If it is, one has to be more precise, here using ν_m.

The cavity Q can be calculated from an analysis based on the roundtrip discussion of Sec. 1-3. In the absence of gain, the intensity after a roundtrip in a two-mirror cavity is given by

$$I(2L) = R_1 R_2 e^{-2\alpha_{\mathrm{abs}}L}\,I(0) = I(0)\exp\{-2L\,[\alpha_{\mathrm{abs}} - ln(R_1 R_2)/2L]\}\;. \tag{18}$$

Hence the loss per roundtrip is given by

$$\frac{\text{loss}}{\text{roundtrip}} = 2L\,\alpha_{\text{abs}} - ln(R_1 R_2) .$$

According to Eq. (17), an oscillator of circular frequency ν has a loss/time given by ν/Q. Noting that the roundtrip time $\tau = 2nL/c$, we see that

$$\frac{\nu}{Q} = \frac{\text{loss}}{\text{roundtrip}} \frac{\text{roundtrip}}{\text{time}} = [2L\,\alpha_{\text{abs}} - ln(R_1 R_2)]\frac{c}{2nL}$$

$$= \frac{c}{n}\alpha_{\text{abs}} - \frac{c}{2nL}ln(R_1 R_2) . \tag{19}$$

In the limit that $R_1 = R_2 = 1 - T$ with $T \ll 1$ and negligible α_{abs}, we have

$$\frac{\nu}{Q} \simeq -\frac{c}{nL}ln(1 - T) \simeq \frac{cT}{nL} . \tag{20}$$

The physical interpretation of the self-consistency equations (15) and (16) is similar to that for the z-dependent versions of Eqs. (2.72) and (2.73). Equation (15) is a "Beer's law" that describes how absorptive and amplifying contributions combine to influence the mode amplitude. Equation (16) can be interpreted in terms of a medium-induced change in the refractive index. In these equations, time variations are of interest rather than spatial variations. This allows us to study the buildup of laser modes, their approach to steady-state operating points, and their stability. One noticeable difference between the two equation pairs is the frequency shift of the oscillating frequency ν_m from the passive cavity frequency Ω_m caused by the carrier-induced change in the refractive index. In Eq. (2.73), a phase change results that changes the wavelength of the propagating wave. In Eq. (16), the modal wavelength is fixed by the roundtrip constructive interference condition, so that the modal frequency must accommodate any index change.

7-2. Single-Mode Semiconductor Laser Theory

The polarization of the semiconductor medium is given both in modal form by Eq. (9) and as the expectation value of the polarization operator (2.115). In terms of the p_k of Eq. (3.3), the latter reads as

$$P(z,t) = \frac{1}{V}\sum_{k} \mu_k^* p_k(z,t) + \text{c.c.,} \tag{21}$$

where we allow p_k to have both time and spatial dependence. Projecting Eqs. (9) and (21) onto the mode factor $U_m(z)$, we find the slowly-varying complex polarization

$$\mathscr{P}_m(t) = e^{i(\nu_m t + \phi_m)} \frac{2}{\mathscr{M}} \int_0^L dz\, U_m^*(z) \sum_k \mu_k^* p_k(z,t)\,, \qquad (22)$$

where $\mathscr{M}$ is the mode normalization factor

$$\mathscr{M} = \int_0^L dz\, |U_m(z)|^2 = \begin{cases} \frac{1}{2}L\,, & \text{two-mirror cavity} \\ L\,, & \text{unidirectional ring} \end{cases} \qquad (23)$$

which is independent of the mode index m. It is convenient to write this component in terms of a k-dependent dimensionless susceptibility function

$$p_k(z,t) = \chi_k E(z,t)\,, \qquad (24)$$

in terms of which, Eq. (22) reads as

$$\mathscr{P}_m(t) = e^{i(\nu_m t + \phi_m)} \frac{2}{\mathscr{M}} \int_0^L dz\, U_m^*(z)\, E(z,t) \sum_k \mu_k^* \chi_k\,. \qquad (25)$$

In general χ_k depends on z and slowly on time through its dependence on the total carrier density N. Chapters 4 through 6 derive the many-body gain for a variety of semiconductor configurations assuming that N is given. For a laser, we need to find the value of N consistent with the laser output intensity. To this end, we add the pump and recombination contributions from Eq. (3.15) to the many-body electric-dipole contributions in Eqs. (4.43) and (4.44). This gives the k-dependent carrier probability equation of motion

$$\dot{n}_{\alpha k} = \Lambda_{\alpha k} - B_k n_{ek} n_{hk} - \gamma_{nr} n_{\alpha k} + [i\Omega_k(z,t)p_k^* + \text{c.c.}] + \left.\frac{\partial n_{\alpha k}}{\partial t}\right|_{col}, \qquad (26)$$

where the Rabi interaction energy $\Omega_k(z,t)$ is given by Eq. (4.46). Summing this over k, we find that the total carrier density obeys the equation of motion

$$\dot{N} = \Lambda - \Gamma(N) + \frac{1}{V} \sum_{\mathbf{k}} [i\Omega_{\mathbf{k}}(z,t)p_{\mathbf{k}}^{*} + \text{c.c.}]$$

$$= \Lambda - \Gamma(N) + \frac{1}{\hbar V} \sum_{\mathbf{k}} [i\mu_{\mathbf{k}} E(z,t)p_{\mathbf{k}}^{*} + \text{c.c.}]$$

$$+ \frac{1}{\hbar V} \sum_{\mathbf{k},\mathbf{k}'} V_{|\mathbf{k}-\mathbf{k}'|} i[p_{\mathbf{k}'} p_{\mathbf{k}}^{*} - p_{\mathbf{k}'}^{*} p_{\mathbf{k}}]$$

$$= \Lambda - \Gamma(N) + \frac{1}{\hbar V} \sum_{\mathbf{k}} [i\mu_{\mathbf{k}} E(z,t)p_{\mathbf{k}}^{*} + \text{c.c.}] , \qquad (27)$$

where

$$\Gamma(N) = \gamma_{nr} N + \frac{1}{V} \sum_{\mathbf{k}} B_{\mathbf{k}} n_{e\mathbf{k}} n_{h\mathbf{k}} \qquad (28)$$

and the third step follows by interchanging $\mathbf{k}$ and $\mathbf{k}'$ in the second summation. Physically the explicit cancellation of the Coulomb term in Eq. (27) reflects the fact that the carrier-carrier scattering does not change the total carrier density. Hence the many-body renormalization of the field only contributes to $\dot{N}$ through the polarization component $p_{\mathbf{k}}$. Substituting Eq. (24), we have

$$\dot{N} = \Lambda - \Gamma(N) + \frac{2|E(z,t)|^2}{\hbar V} \sum_{\mathbf{k}} \text{Im}\{\mu_{\mathbf{k}}^{*} \chi_{\mathbf{k}}\} . \qquad (29)$$

For the standing-wave case (2), $|U_m(z)|^2 = \sin^2(K_m z)$, while for the running-wave case (3), $|U_m(z)|^2 = 1$. Hence for a standing wave according to Eq. (29), N varies periodically in space with a period equal to half a wavelength, an important phenomenon known as *longitudinal spatial hole burning*. No such variation occurs for a running wave. Spatial hole burning is known to modify mode coupling in the multimode operation of liquid and solid-state lasers and even to a limited degree in gas lasers. If the carriers diffuse through several wavelengths before recombining, one could argue that they would see an average field intensity, for which we replace $\sin^2(K_n z) = \frac{1}{2} - \frac{1}{2}\cos(2K_n z)$ by the average value $\frac{1}{2}$. This is a remarkably good approximation for some gas-laser configurations. Chapter 8 considers spatial hole burning in greater detail.

For a simple result, suppose $E(z,t)$ of (1) is a single-mode running-wave field characterized by the mode function (3), that is

$$E(z,t) = \frac{1}{2} E_m(t) \exp[i(K_m z - \nu_m t - \phi_m)] + \text{c.c.} \tag{30}$$

Then in the rotating-wave approximation, the z integral in Eq. (22) is trivial and the slowly-varying polarization component (22) reduces to

$$\mathscr{P}_m(t) = E_m(t) \frac{1}{V} \sum_{\mathbf{k}} \mu_{\mathbf{k}}^* \chi_{\mathbf{k}}(t)$$

$$= \epsilon \chi_m E_m , \tag{31}$$

where $\chi_m = \chi_m' + i\chi_m''$ is the complex mode susceptibility (note that the susceptibility functions χ_m and $\chi_{\mathbf{k}}$ are related but different). Substituting Eqs. (30) and (31) into Eq. (29), we have

$$\dot{N} = \Lambda - \Gamma(N) + \frac{\epsilon}{2\hbar} \chi_m'' E_m^2 . \tag{32}$$

This result is valid for the full quasiequilibrium many-body χ_m and a similar expression is valid for the plane-wave saturation that we encounter in multiwave mixing of Sec. 8-1.

Substituting Eq. (31) into the self-consistency equations (15) and (16), we have the single-mode amplitude- and frequency-determining equations

$$\dot{E}_m = -\frac{\nu}{2Q_m} E_m - \frac{\nu}{2} \chi_m'' E_m , \tag{33}$$

$$\nu_m + \dot{\phi}_m = \Omega_m - \frac{\nu}{2} \chi_m' , \tag{34}$$

where for the many-body χ of Eq. (5.9), we have the mode susceptibility

$$\chi_m(t) = -\frac{i}{\hbar \gamma \epsilon V} \sum_{\mathbf{k}} \frac{|\mu_{\mathbf{k}}|^2}{1 - q(k)} \frac{f_{e\mathbf{k}} + f_{h\mathbf{k}} - 1}{\gamma + i(\omega_{\mathbf{k}} - \nu_m)} . \tag{35}$$

This is the same as $\chi(z)$ of Eq. (3.50) except that $\chi_m(t)$ depends slowly on the time t (through N) and frequency ν_m rather than the position z and frequency ν, the factor $[1 - q(\mathbf{k})]^{-1}$ models the Coulomb-enhancement in the Padé approximation of Sec. 5-1, and $\omega_{\mathbf{k}}$ includes bandgap renormalization. The free-carrier result uses an $\omega_{\mathbf{k}}$ without the renormalization contributions and sets $q(\mathbf{k}) = 0$. As usual to avoid continually dividing by n^2, we define the susceptibility χ_m by $\mathscr{P}_m = \epsilon \chi_m E_m$, instead of by the usual MKS definition $\mathscr{P}_m = \epsilon_0 \chi_m E_m$. The Fermi-Dirac distributions are determined by the temperature T and the carrier density N determined by Eq. (32). Note that χ_m for other semiconductor configurations, such as strained-layer quantum wells can be used in place of Eq. (35). Our basic approximation is that the semiconductor is in quasiequilibrium, both with respect to the carriers and with respect to the lattice. This approximation is generally good for single-mode operation, and can be valid for some kinds of multimode operation as well (see Chap. 8).

Steady-state laser operation occurs when $\dot{E}_m = 0$. Above the laser threshold, this implies the steady-state oscillation condition

$$\boxed{\chi_m'' = -\frac{1}{Q_m}} \,, \tag{36}$$

i.e., the *saturated gain equals the cavity losses*. According to Eqs. (36) and (35), the probability difference $f_{e\mathbf{k}} + f_{h\mathbf{k}} - 1$ and hence the total carrier density N must remain at their respective laser-threshold values, a feature known as *gain clamping*. In particular, Eq. (36) determines the steady-state single-mode laser carrier density N and the threshold pump value Λ_{th} as functions of the cavity loss rate constant $\nu/2Q_m$ and temperature T. As long as quasiequilibrium conditions are satisfied the intensity determines the total carrier density independently of the momentum $\mathbf{k}$. Therefore, we say that the semiconductor laser medium *saturates homogeneously*, in spite of the linear-gain profile being inhomogeneously broadened.

In terms of the "photon-number" density $n(t)$ defined by (note that this is a classical quantity in spite of the name we give it)

$$n(t) = \frac{\epsilon E_m{}^2}{\hbar\nu} \,, \tag{37}$$

the carrier-density equation of motion (32) becomes

$$\dot{N} = \Lambda - \Lambda_{th} + \tfrac{1}{2}\nu\chi_m'' n \,, \tag{38}$$

where the threshold pump value $\Lambda_{th} = \Gamma(N)$ (that for $n = 0$) is given by Eq. (28). Equation (28) must be evaluated numerically using the N that satisfies the gain-clamping formula (36). Multiplying the field amplitude Eq. (33) by $2\epsilon E_m/\hbar\nu$, we have the equation of motion

$$\dot{n} = -\nu n(\chi_m'' + 1/Q_m) .$$

(39)

The steady-state photon number density is determined by the condition $\dot{n} = \dot{N} = 0$. Accordingly using Eqs. (36) and (38), we have

$$\boxed{n = \frac{\Lambda - \Lambda_{\text{th}}}{\nu/2Q_m} = (\mathcal{N} - 1)\frac{\Lambda_{\text{th}}}{\nu/2Q_m}} ,$$

(40)

where the relative excitation $\mathcal{N}$ is defined by the ratio

$$\mathcal{N} = \frac{\Lambda}{\Lambda_{\text{th}}} .$$

(41)

Equation (40) shows that the laser intensity increases as a linear function of pump rate with a slope proportional to $\Lambda_{\text{th}}Q_m\nu$ as shown in Fig. 7-2a. The curve shown in this figure is equivalent to the L-I curves used for characterizing semiconductor lasers. For the L-I curves one plots the light output power, which is proportional to n, versus the injection current, which is proportional to Λ. The steady-state gain-clamping condition holds the carrier density at the threshold value as shown in Fig. 7-2b. Appendix A shows that the two-level single-mode unidirectional intensity also increases linearly with pump rate.

Figure 7-3 compares the n vs Λ curves for bulk and quantum-well semiconductor lasers. The different curves are for different threshold gains, due to, for example, different resonator lengths or facet coatings. Note that the bulk Λ_{th} increases uniformly with g_{th}, which is not the case for the quantum-well laser. This difference is due to gain rollover in the quantum-well structure and it brings up an important consideration in the design of the quantum-well lasers. Depending on the quantum-well width and composition, it is possible that the threshold gain is too high for threshold to ever be reached.

Another interesting comparison is in the temperature dependence of bulk, quantum-well and strained quantum-well lasers. Figure 7-4 shows the n vs Λ curves for the different types of lasers for a variety of temperatures. The difference in behavior may be understood by referring to the discussion in Section 3-4.

According to Eq. (34), the shift in the laser oscillation frequency ν_m from the passive cavity value Ω_m is given by $-\frac{1}{2}\nu\chi_m'$. The oscillation frequency ν_m lies in the gain region, usually in the vicinity of the peak gain. Examining the real part of Eq. (35) in the free-carrier approximation [set

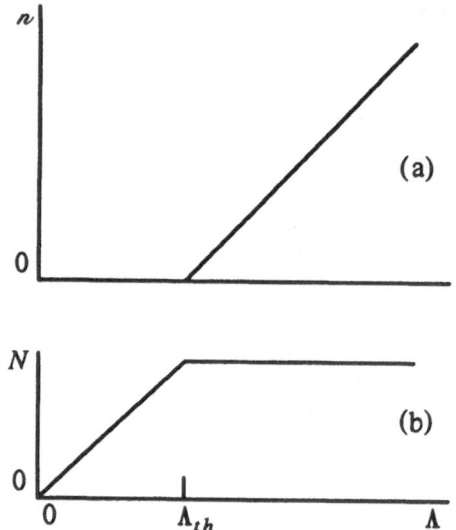

Fig. 7-2. (a) Photon number density n of Eq. (40) versus injection-current pump Λ showing linear dependence of n above the threshold pump value Λ_{th}. (b) Total carrier density N versus Λ showing how N is clamped at threshold value by the steady-state saturated-gain equals loss condition (36).

$q(\mathbf{k}) = 0$ and use the unrenormalized $\omega_{\mathbf{k}}$]

$$\chi'_m(t) = -\frac{1}{\hbar\gamma\epsilon V} \sum_{\mathbf{k}} |\mu_{\mathbf{k}}|^2[f_{e\mathbf{k}} + f_{h\mathbf{k}} - 1]\mathcal{L}(\omega_{\mathbf{k}}-\nu_m)(\omega_{\mathbf{k}} - \nu_m)/\gamma] , \quad (42)$$

we note that $\hbar\omega_{\mathbf{k}}$ values below $\hbar\nu_m$ give positive contributions to χ'_m (a negative frequency shift to ν_m); $\hbar\omega_{\mathbf{k}}$ values greater than $\hbar\nu_m$ but less than the total chemical potential $\mu = \mu_e + \mu_h$ give negative contributions to χ'_m (a positive frequency shift), and $\hbar\omega_{\mathbf{k}}$ values above μ (in the absorption region) give positive contributions (a negative frequency shift). The sum of all these contributions tends to be negative (a positive frequency shift), due to the large contribution from the absorptive region lying above the gain region. Equation (2.78) shows that a negative χ'_m corresponds to a negative index change δn.

We have ignored the transverse spatial dependence of real laser beams by using a plane-wave theory. Thinking for a moment about a beam with

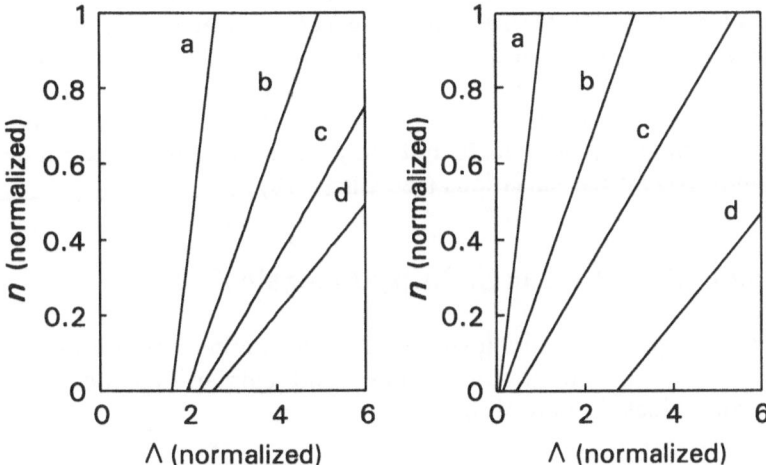

Fig. 7-3. Photon number density versus pump for bulk (left figure) and quantum-well (right figure) GaAs lasers. The different curves correspond to the modal threshold gains (*a*) 1 ×, (*b*) 3 ×, (*c*) 5 ×, and (*d*) 7 × $10^{11}/s$.

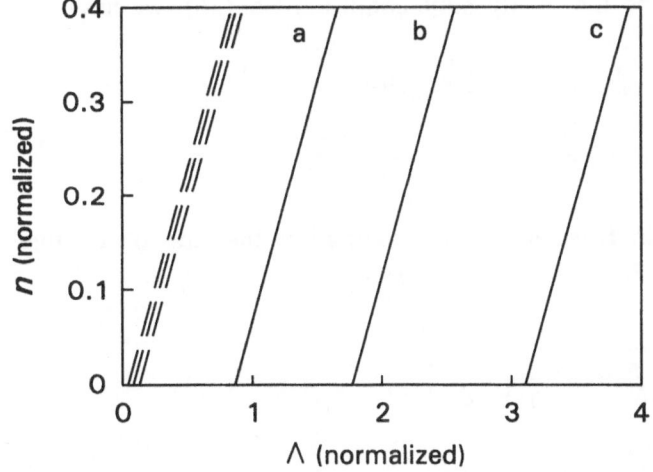

Fig. 7-4. Photon number density versus pump for bulk GaAs (solid lines) and $5nm \; In_{.1}Ga_{.9}As$-GaAs quantum-well (dashed lines) lasers. The different curves are for the temperatures (*a*) 200K, (*b*) 300K, and (*c*) 400K and the threshold gain $2 \times 10^{11}/s$.

a roughly Gaussian transverse dependence, we note that regions of larger intensity correspond to smaller total carrier density and hence smaller values of the distributions $f_{\alpha \mathbf{k}}$. This implies that the carrier-induced refractive index depends on the transverse variation of the laser-beam intensity. This leads to focussing of the beam if $d\delta n/dI > 0$ and defocussing if $d\delta n/dI < 0$. Defocussing and focussing are commonly referred to as *antiguiding* and *filamentation*, respectively. These effects are ignored in the present plane-wave theory, but they have consequences for a models that include transverse variations (see Chap. 10).

7-3. Single-Mode Linear-Stability Analysis

To determine if the single-mode photon number density of Eq. (40) is stable, we expand the photon-number and carrier densities about their steady-state values n_s and N_s as

$$n(t) = n_s + \delta n(t) \tag{43}$$

$$N(t) = N_s + \delta N(t) . \tag{44}$$

This is a very simple theory, since it ignores the fact that the laser field has a phase that can also change. A more general approach is given in Chap. 8. Substituting Eqs. (43) and (44) into the equations of motion (38) and (39) and using the steady-state condition Eq. (36), we have

$$\frac{d}{dt}\delta n(t) = -\nu n_s \frac{\partial \chi''_m}{\partial N}\delta N(t) \tag{45}$$

$$\frac{d}{dt}\delta N(t) = -\Gamma_1'\delta N(t) - \frac{\nu}{2Q}\delta n(t) , \tag{46}$$

where in the free-carrier approximation the rate of change of the gain with respect to the carrier density is

$$\frac{\partial \chi''_m}{\partial N} = -\frac{1}{\hbar \gamma \epsilon V}\sum_k |\mu_\mathbf{k}|^2 \mathscr{L}_m (g_{e\mathbf{k}} + g_{h\mathbf{k}}) , \tag{47}$$

the derivatives $g_{\alpha \mathbf{k}}$ of the Fermi-Dirac distributions are given by

$$g_{\alpha \mathbf{k}} = \frac{\partial f_{\alpha \mathbf{k}}}{\partial N} = \frac{\partial f_{\alpha \mathbf{k}}}{\partial \mu_\alpha}\frac{\partial \mu_\alpha}{\partial N} = \frac{f_{\alpha \mathbf{k}}(1-f_{\alpha \mathbf{k}})}{V^{-1}\Sigma_{\mathbf{k}'}f_{\alpha \mathbf{k}'}(1-f_{\alpha \mathbf{k}'})} , \tag{48}$$

the zero-field relaxation rate of the carrier density

$$\Gamma_1 = \frac{\partial \Lambda_{th}}{\partial N} = \gamma_{nr} + \frac{1}{V} \sum_k B_{\mathbf{k}} (f_{e\mathbf{k}} g_{h\mathbf{k}} + f_{h\mathbf{k}} g_{e\mathbf{k}}) , \qquad (49)$$

and the power-broadened relaxation rate

$$\Gamma_1' = \Gamma_1 - \frac{\nu}{2} \frac{\partial \chi_m''}{\partial N} n_s . \qquad (50)$$

It is convenient to write Eqs. (45) and (46) as the single matrix equation

$$\frac{d}{dt} \begin{pmatrix} \delta n(t) \\ \delta N(t) \end{pmatrix} = \begin{pmatrix} 0 & -\nu n_s \partial \chi_m'' / \partial N \\ -\nu/2Q & -\Gamma_1' \end{pmatrix} \begin{pmatrix} \delta n(t) \\ \delta N(t) \end{pmatrix} . \qquad (51)$$

The coupled set of equations (38) and (39) are stable if Eq. (51) predicts that the deviations δN and δn decay to zero in time. This happens if the eigenvalues of the matrix, call it A, on the RHS of Eq. (51) both have negative real parts. The eigenvalue equation is given by setting $\det(A - \Lambda I) = 0$, which gives

$$\Lambda(\Gamma_1' + \Lambda) - \frac{\nu^2 n_s}{2Q} \frac{\partial \chi_m''}{\partial N} = 0 . \qquad (52)$$

This has the solutions

$$\Lambda_\pm = -\frac{\Gamma_1'}{2} \pm \sqrt{\frac{\Gamma_1'^2}{4} + \frac{\partial \chi_m''}{\partial N} \frac{\nu^2 n_s}{2Q}} . \qquad (53)$$

Since $\partial \chi_m'' / \partial N$ is negative, both roots have negative real parts implying that single-mode operation is stable. The solutions are oscillatory yielding linear relaxation oscillations provided

$$-\frac{\partial \chi_m''}{\partial N} \frac{n_s \nu^2}{2Q} > \frac{\Gamma_1'^2}{4} . \qquad (54)$$

In general the LHS of this inequality dominates, giving the relaxation oscillation frequency

$$\Omega_R \simeq \sqrt{\frac{\frac{1}{2}\partial\chi_m''}{\partial N}n_s\nu^2\chi_m''} \quad.$$

(55)

Nonlinear relaxation oscillations can be studied by direct integration of the coupled equations of motion (38) and (39) as illustrated in Fig. 7-5.

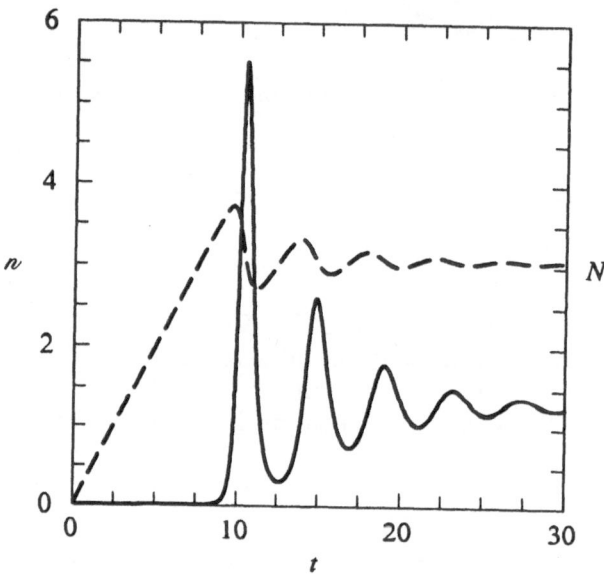

Fig. 7-5. Photon-number density n (solid line) and carrier density N (carriers per a_0^3, dashed line) versus time for a linear-density gain model with $A_g = 3$, $\nu/2Q = 3$, $N_g = 2.05944$, $\gamma_{nr} = .004$, and $\Lambda = .4$.

7-4. Injection Locking

A method to stabilize a laser is to dampen its fluctuations with an injected field from a stable laser. To see how injection locking occurs, we use an approach similar to that for optical bistability in passive cavities. Consider the cavity configuration in Fig. 7-6. We write the total field just inside the cavity as

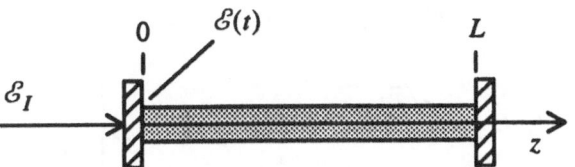

Fig. 7-6. Two-mirror laser cavity with a complex internal field amplitude $\mathcal{E}(t)$ and an injected signal amplitude $\mathcal{E}_I$.

$$\mathcal{E}(t+\tau) = \sqrt{T}\,\mathcal{E}_I + f(\mathcal{E}(t)) , \qquad (56)$$

where $\mathcal{E}(t)$ is a complex electric field at the time t and $f(\mathcal{E})$ represents the amplification and losses that $\mathcal{E}$ experiences in a two-mirror cavity round-trip of time $\tau = 2Ln/c$. Here we take the carrier frequencies associated with $\mathcal{E}(t)$ and $\mathcal{E}_I$ to be identical, relegating any frequency difference to the phase of $\mathcal{E}(t)$. We can use this equation to derive a coarse-grained time rate of change of $\mathcal{E}(t)$ by writing

$$\frac{d\mathcal{E}(t)}{dt} \simeq \frac{\mathcal{E}(t+\tau) - \mathcal{E}(t)}{\tau} . \qquad (57)$$

Such an approach is unable to deal with some interesting difference-equation instabilities, but is nevertheless very useful in studying locking effects. Combining Eqs. (56) and (57), we have

$$\frac{d\mathcal{E}(t)}{dt} = \frac{\sqrt{T}}{\tau}\mathcal{E}_I + \frac{f(\mathcal{E}(t)) - \mathcal{E}(t)}{\tau} = \frac{\sqrt{T}}{\tau}\mathcal{E}_I + \frac{d\mathcal{E}(t)}{dt}\bigg|_{laser}, \qquad (58)$$

where $[d\mathcal{E}(t)/dt]|_{laser}$ is the time rate of change in the absence of the injected signal.

Converting this to the amplitude and phase notation using $\mathcal{E} = E_m(t)\exp[-i\phi_m(t)]$ and $\mathcal{E}_I = E_I(t)\exp[-i\phi_I]$, we have

$$\frac{d\mathcal{E}(t)}{dt} = (\dot{E}_m - i\dot{\phi}_m E_m)e^{-i\phi_m} = \frac{\sqrt{T}}{\tau}E_I e^{-i\phi_I} + \frac{d\mathcal{E}(t)}{dt}\bigg|_{laser} .$$

Multiplying through by $\exp(i\phi_m)$, we have

$$\dot{E}_m - i\dot{\phi}_m E_m = \frac{\sqrt{T}}{\tau} E_I e^{i\Psi} + [\dot{E}_m - i\dot{\phi}_m E_m]\Big|_{laser},$$

where the relative phase angle

$$\Psi(t) = \phi_m(t) - \phi_I . \tag{59}$$

Equating the real and imaginary parts separately to zero and using the amplitude- and frequency-determining equations (33) and (34) for the laser contribution, we find

$$\dot{E}_m = -\frac{\nu}{2Q_m} E_m - \frac{\nu}{2} X''_m E_m + \frac{\sqrt{T}}{\tau} E_I \cos\Psi \tag{60}$$

$$\nu_m + \dot{\phi}_m = \Omega_m - \frac{\nu}{2} X'_m - \frac{\sqrt{T}}{\tau} \frac{E_I}{E_m} \sin\Psi , \tag{61}$$

where ν_m is the carrier frequency of the injected signal and $\phi_m(t)$ contains any deviation of the cavity-field frequency.

Equation (61) is often written in the generic form

$$\boxed{\dot{\Psi} = a + b\sin\Psi} , \tag{62}$$

where for the injected-signal problem the coefficients are given by

$$a = \Omega_m - \nu_m - \frac{\nu}{2} X'_m \tag{63}$$

$$b = -\frac{\sqrt{T}}{\tau} \frac{E_I}{E_m} = -\frac{\sqrt{T} c}{2nL} \frac{E_I}{E_m} \simeq -\frac{\nu}{2Q_m} \frac{E_I}{\sqrt{T} E_m} , \tag{64}$$

where we use Eq. (20) in the final approximation. Equation (62) is called the *mode-locking* or *frequency-locking equation*, since it locks the phase of a mode to that of another signal. This equation applies to a wide variety of lasers and classical oscillators, such as the locking of a pair of nearby clock pendula or digital clock circuits, the frequency locking of a pair of tuning forks, and the injection locking of a Van der Pol vacuum-tube triode oscillator. In the laser arena [see Chap. 8], in addition to injection locking, there is the locking of longitudinal modes into a periodic array, the locking of the oppositely directed running waves in a ring laser (bête noire of ring-laser gyros), the locking of orthogonal circular polarizations in a Zeeman laser, and the locking of adjacent lasers in a transverse laser array.

Before considering the specific features of the injected-signal problem, we describe the general features of Eq. (62) under the assumption that Eqs. (38), (60), and (62) decouple, that is, that changes in $\Psi(t)$ affect $E_m(t)$ and hence $N(t)$ sufficiently little that we can treat the coefficients a and b as constants independent of $\Psi(t)$. This approximation is clearly not always valid, but it leads to simple solutions that are quite useful and can also serve as a basis for understanding more complicated behavior.

First suppose that

$$|a| < |b| . \tag{65}$$

Then according to Eq. (62), there are two values of Ψ for which $\sin\Psi = -a/b$ yielding $\dot\Psi = 0$, namely

$$\Psi_s = \begin{cases} -\sin^{-1}(a/b) \\ \pi + \sin^{-1}(a/b) \end{cases}. \tag{66}$$

To check their stability, we substitute

$$\Psi(t) = \Psi_s + \delta\Psi(t) \tag{67}$$

into Eq. (62) and find

$$\frac{d}{dt}\delta\Psi(t) = b\cos(\Psi_s)\delta\Psi(t) , \tag{68}$$

which shows that Ψ_s is stable provided

$$b\cos\Psi_s < 0 . \tag{69}$$

Noting that $b^2 = b^2(\cos^2\Psi_s + \sin^2\Psi_s) = b^2\cos^2\Psi_s + a^2$, we have

$$b\cos\Psi_s = \pm \sqrt{b^2 - a^2} , \tag{70}$$

so the stable value corresponds to $-\sqrt{b^2-a^2}$. Specifically if $b > 0$, then the stable value is $\Psi_s = \pi + \sin^{-1}(a/b)$, while if $b < 0$, $\Psi_s = -\sin^{-1}(a/b)$.

If $|a| \gg |b|$, the relative phase angle Ψ changes essentially linearly in time at the rate $d\Psi/dt = a$. As $|a|$ decreases toward $|b|$, the b term in Eq. (62) subtracts from the a term in one half of the cycle and adds in the other half. This leads to a "slipping" behavior, as Ψ "slips" past the point where the b term tries to cancel the a term. When $|a/b|$ decreases to unity, Ψ gets to the slipping point and sticks. For $|a| > |b|$, it is useful to calculate the average frequency defined by the reciprocal of the $\Psi(t)$ period

$$\Delta\nu = \frac{2\pi}{t(\Psi=2\pi) - t(\Psi=0)} \, ,$$

From Eq. (62), we have

$$\frac{2\pi}{\Delta\nu} = \int_{t(0)}^{t(2\pi)} dt = \int_0^{2\pi} \frac{d\Psi}{a + b\sin\Psi} = \frac{2\pi}{(a^2 - b^2)^{1/2}} \, , \tag{71}$$

which gives

$$\Delta\nu = \sqrt{a^2 - b^2} \, . \tag{72}$$

For the injection-locked laser, we note that $b \propto 1/L$, which is large for tiny cavities, such as the VCSEL (Vertical-Cavity Surface-Emitting Laser). Specifically for a 1-μm cavity, an index $n = 3.5$, a mirror transmission $T = .01$, and $E_I/E_m = 0.1$, we have

$$b = -\frac{\sqrt{T}\, c}{2nL} \frac{E_I}{E_m} \simeq -.433\times10^{12} \, \frac{rad}{s} \, , \tag{73}$$

i.e., about 70 GHz, which is on the order of the cavity bandwidth. This means that according to the present model, a relatively small injected signal can lock the VCSEL over the oscillation region.

One may inquire about the linear stability of the coupled set of three equations (38), (60), and (62). For this purpose, it is convenient to write Eq. (60) in terms of the locking coefficient b as

$$\dot{E}_m = -\left[\frac{\nu}{2Q_m} + \frac{\nu}{2}\chi_m'' + b\cos\Psi\right]E_m \, . \tag{74}$$

Note that the steady-state condition (36) for the laser with no injected signal is *not* valid here; instead the steady-state solution to Eq. (74) is

$$\frac{\nu}{2Q_m} + \frac{\nu}{2}\chi_s'' = -b_s\cos\Psi_s = \pm\sqrt{b_s^2 - a_s^2} \, . \tag{75}$$

Near the edges of the locking region, Eq. (75) reduces to Eq. (36), but for small a_s, significant deviations occur.

Substituting the deviations (44), (67) and

$$E_m(t) = E_s + \delta E(t) , \tag{76}$$

into the equations of motion (74), (62), and (32) and using the steady-state value $a_s = -b_s \sin\Psi_s$, we have

$$\frac{d}{dt}\delta E(t) = b_s \cos\Psi_s \, \delta E(t) - a_s E_s \delta\Psi - \frac{\nu}{2}E_s \frac{\partial \chi_s''}{\partial N}\delta N(t) \tag{77}$$

$$\frac{d}{dt}\delta\Psi(t) = \frac{a_s}{E_s}\delta E(t) + b_s \cos\Psi_s \, \delta\Psi(t) - \frac{\nu}{2}\frac{\partial\chi_s'}{\partial N}\delta N(t) , \tag{78}$$

$$\frac{d}{dt}\delta N(t) = \frac{\nu n_s \chi_s''}{E_s}\delta E(t) - \Gamma_1'\delta N(t) . \tag{79}$$

Here we use χ_s'' to denote the steady-state value of χ_m'' and we neglect the small N dependence of b that results from changes in the index of refraction.

Equations (77) through (79) can be written in matrix form as (for typographical simplicity, we omit the subscript s on a_s and b_s)

$$\frac{d}{dt}\begin{bmatrix} \delta E(t) \\ \delta\Psi(t) \\ \delta N(t) \end{bmatrix} = \begin{bmatrix} b\cos\Psi_s & -aE_s & -\frac{1}{2}\nu E_s \partial\chi_s''/\partial N \\ a/E_s & b\cos\Psi_s & -\frac{1}{2}\nu\partial\chi_s'/\partial N \\ \nu n_s \chi_s''/E_s & 0 & -\Gamma_1' \end{bmatrix}\begin{bmatrix} \delta E(t) \\ \delta\Psi(t) \\ \delta N(t) \end{bmatrix} . \tag{80}$$

Expanding the determinant across the bottom row, we find that the eigenvalues of this matrix are determined by

$$\tfrac{1}{2}n_s\nu^2\chi_s''\left[a\frac{\partial\chi_s'}{\partial N} + \frac{\partial\chi_s''}{\partial N}(b\cos\Psi_s-\lambda)\right] - (\Gamma_1' + \lambda)[\lambda^2 - 2\lambda b\cos\Psi_s + b^2] = 0 ,$$

which gives the eigenvalue equation

$$\lambda^3 + (\Gamma_1' - 2b\cos\Psi_s)\lambda^2 + [\Omega_R^2 + b^2 - 2\Gamma_1'b\cos\Psi_s]\lambda$$

$$- \tfrac{1}{2}n_s\nu^2\chi_s''a\frac{\partial\chi_s'}{\partial N} - \Omega_R^2 b\cos\Psi_s + \Gamma_1'b^2$$

$$= \lambda^3 + c_1\lambda^2 + c_2\lambda + c_3 = 0 , \tag{81}$$

in which we define

$$\Omega_R^2 = \tfrac{1}{2} n_s \nu^2 \chi_s'' \frac{\partial \chi_s''}{\partial N} . \tag{82}$$

In the absence of an injected signal, this reduces to the square of Eq. (55). Note that for $\dot{\Psi} = 0$, the limit $b \to 0$ implies $a \to 0$, so that in this limit Eq. (81) with (36) reduces to Eq. (52) as it should.

A set of stationary solutions to (38), (60), and (62) is stable if and only if the real parts of the eigenvalues given by Eq. (81) are all negative. The Hurwitz criterion states that the roots of this polynomial have negative real parts if and only if the coefficients c_n and the quantity $H_2 = c_1 c_2 - c_3$ are positive. The condition $c_1 > 0$ implies

$$\Gamma_1' - 2b\cos\Psi_s = \Gamma_1' \pm 2\sqrt{b^2 - a^2} > 0 , \tag{83}$$

which is a weaker condition than Eq. (69) and might allow bistable operation, i.e., either sign of $\pm\sqrt{b^2 - a^2}$. The condition $c_2 > 0$ gives the similar inequality

$$\Omega_R^2 + b^2 - 2\Gamma_1' b\cos\Psi_s > 0 . \tag{84}$$

In terms of the *linewidth enhancement factor*

$$\alpha = \frac{\partial \chi'/\partial N}{\partial \chi''/\partial N} , \tag{85}$$

the condition $c_3 > 0$ gives

$$c_3 = -\Omega_R^2(a\alpha + b\cos\Psi_s) + \Gamma_1' b^2 > 0 . \tag{86}$$

This inequality favors $a < 0$ since at the gain peak in bulk media, the linewidth enhancement factor α is positive and typically about equal to 2. In fact, taking the derivative of the free-carrier Eq. (42) with respect to N, we find

$$\frac{\partial \chi_m'}{\partial N} = -\frac{1}{\hbar \gamma \epsilon V} \sum_{\mathbf{k}} |\mu_{\mathbf{k}}|^2 [g_{e\mathbf{k}} + g_{h\mathbf{k}}] \mathscr{L}(\omega_{\mathbf{k}} - \nu_m) \frac{\omega_{\mathbf{k}} - \nu_m}{\gamma} . \tag{87}$$

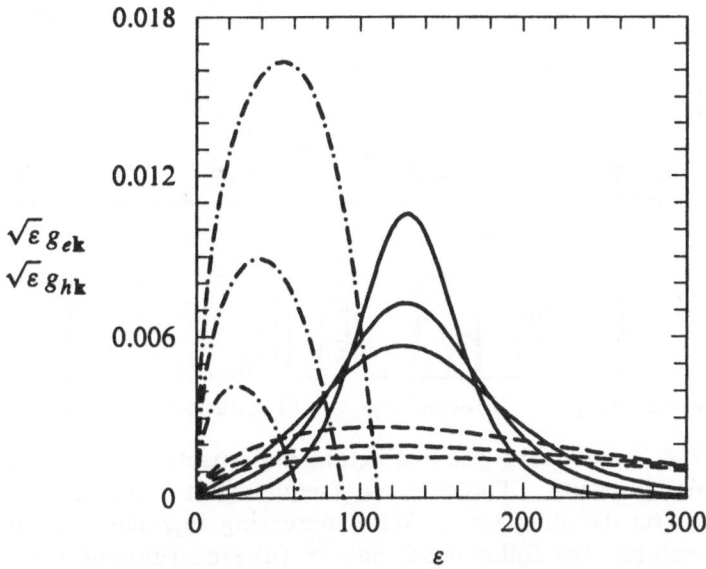

Fig. 7-7. Weighted derivatives $g_{\alpha k}$ given by Eq. (48) of the qua-siequilibrium Fermi-Dirac distributions $f_{\alpha k}$ versus reduced-mass energy ε for electrons ($\alpha = e$, solid lines) and holes ($\alpha = h$, dashed lines) for $T = 200$, 300 and $400\ K$ (in order of decreasing peak values), and a total carrier density $N = 3 \times 10^{18} cm^{-3}$. Also included are the corresponding curves (dot-dashed lines) for the gain expression $\sqrt{\varepsilon}[f_e(\varepsilon) + f_h(\varepsilon) - 1]$. The values of these and other parameters are the same as those for the weighted Fermi-Dirac distributions in Fig. 3-6. Note that the $g_{\alpha k}$ peak above the gain region.

For different effective masses of electrons and holes, $\sqrt{\varepsilon}[g_{ek} + g_{hk}]$ peaks above the gain region, as illustrated in Fig. 7-7. Since this occurs for $\omega_k > \nu_m$, we see that $\partial \chi'_m / \partial N$ of Eq. (87) is negative. Furthermore $\chi''_s < 0$, since we are analyzing a gain medium. Hence the sign of the lead term in Eq. (87) is that of the detuning factor a. Similarly Eq. (47) shows that $\partial \chi''_m / \partial N$ is negative, so that the linewidth enhancement factor α is positive in the gain region.

Steady-state predictions of the theory can be found along these lines. Configurations that have no steady state are characterized by limit cycles or chaos. The semiconductor laser in the presence of an injected field pro-vides us with a practical system to study such instabilities. This is shown in Fig. 7-8, where with changing injected intensity, a semiconductor laser can be made to exhibit very different dynamical behavior. For each in-

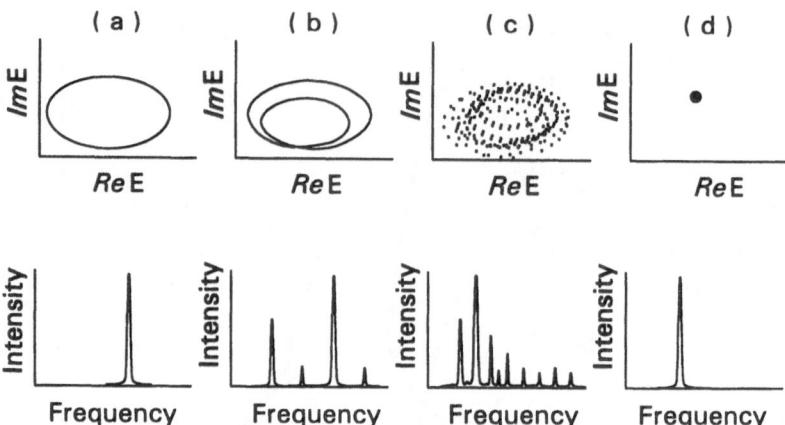

Fig. 7-8. Laser field phase diagram and spectrum for increasing injected intensity. The injected frequency is detuned from that of the free-running laser. With increasing injected intensity the laser exhibits the following behavior: (a) free-running, (b) bifurcation, (c) chaos and (d) injection locked.

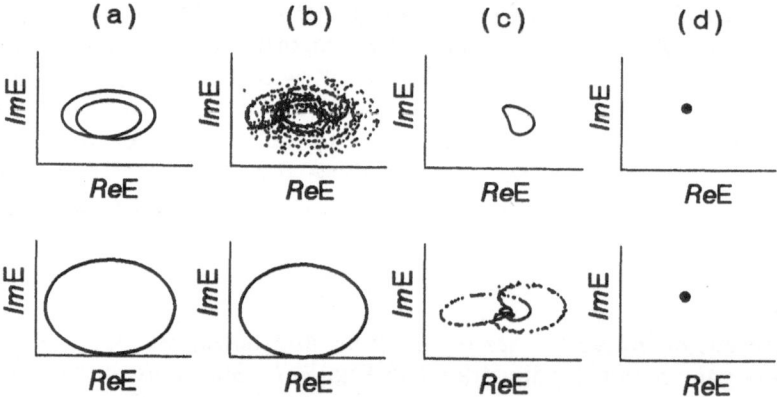

Fig. 7-9. Laser field phase diagram and spectrum for increasing injected intensity I_{inj} = (a) $10^{-5} \times$, (b) $10^{-3} \times$, (c) $10^{-2} \times I_{frl}$, where I_{frl} is the laser intensity in the absence of the injected field. The gain media are 10 nm $In_{.65}Ga_{.35}As$-InP (top figures) and 10 nm $In_{.56}Ga_{.44}As$-InP (bottom figures), giving a threshold gain of 8000/cm and α = 4 and 0.7, respectively.

jected intensity, we plot the laser field vector and its spectrum. The results are obtained by solving Eqs. (29), (60), and (61) numerically for a bulk GaAs gain medium. With a small injected intensity, the laser field oscillates essentially at its free-running laser frequency. In the reference frame of the injected signal, the vector representing the laser field rotates so that its tip traces an ellipse (Fig. 7-8a). For a sufficiently intense injected field, the laser is locked to the injected field frequency. In the injected field frame of reference, the laser field vector is stationary and its tip is represented by the dot in Fig. 7-8d. Comparing the spectra of Figs. 7-8a and 7-8d, we see that the laser frequency is shifted to that of the injected field. For intermediate values of injected intensity, the laser exhibits interesting dynamical behavior. Figure 7-8b shows a bifurcation, where the laser field vector repeatedly traces out the two different trajectories shown. The signature of bifurcation is the appearance of a frequency component that bisects the two existing frequencies, i.e., those of the free-running laser and the injected field. For a stronger injected field the laser becomes chaotic as shown in Fig. 7-8c. If one looks closely at the spectrum, one sees a continuum of frequencies.

The laser dynamics described in the previous paragraph depends on the gain medium. In Fig. 7-9, we repeat the computations for a $10nm$ $In_{.65}Ga_{.35}As$ quantum well and a 10 nm $In_{.56}Ga_{.44}As$ quantum well. The barrier material is InP in both cases. Comparison of the figures show both quantitative and qualitative differences in how the two gain media respond to the injected field. In these examples, the density fluctuations are sufficiently small to make the linear-density gain model valid. This allows us to use the linewidth enhancement factor α to characterize the gain media, thus making it easier to generalized our results. We found a greater dependence on the injected field and a larger instability region for gain media with large α. On the other hand for sufficiently small α, the instability region shrinks to a point where it becomes impossible to make the system bifurcate or go chaotic. The injected intensity required for a stable lock is, however, independent of α.

7-5. Coupled Resonators

So far we have worked with laser fields that are linear superpositions of eigenmodes of a conventional optical resonator, i.e., a two-mirror resonator or a ring resonator. Semiclassical laser theory is, in fact, more general in that the modal decomposition may be in terms of the eigenmodes of more complicated or unconventional resonator geometries. We illustrate this with an example of a semiconductor laser operating with two coupled optical resonators [see Rose *et al.* (1992)].

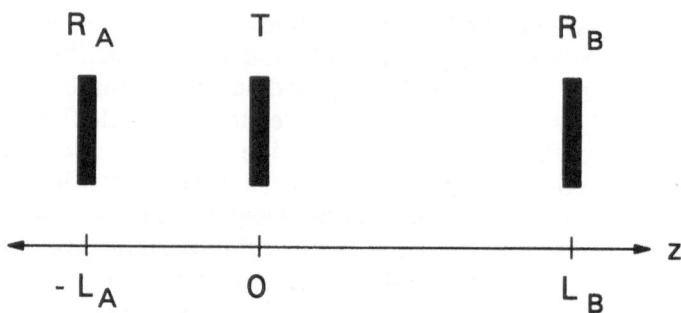

Fig. 7-10. Two coupled optical resonators. The mirrors are located at $z = -L_A$, L_B, and 0, respectively.

Consider the resonator configuration shown in Fig. 7-10, where three mirrors located at $z = -L_A$, L_B, and 0, respectively, define two optical resonators. The shared mirror with transmission T controls the amount of coupling between resonators. Figure 7-10 describes several different semiconductor laser experiments of current interest. For example, if only the left resonator has gain and the shared mirror has relatively high reflectivity, e.g., it is uncoated, we have a laser that is coupled to an external resonator (Fig. 7-11a). Such a configuration is of interest because of the possibility of the external resonator to control the operating frequency and linewidth of the laser, e.g., as in the cleaved-coupled-cavity (C^3) laser (more precisely, we may have to treat the C^3 laser as three coupled resonators). If the shared mirror is antireflection coated (Fig. 7-11b), we have an external resonator semiconductor laser. Treating the external resonator laser as coupled resonators allows us to study the effects of imperfect antireflection coating. Note that the resonators are strongly coupled because of the antireflection coating and therefore the usual perturbative treatment of coupled oscillators is invalid. If the mirror at $z = L_B$ has low reflectivity (Fig. 7-11c, we have a laser subject to feedback from an external mirror, which might model the end of an optical fiber. This configuration is widely used for studying laser instabilities. Finally if both resonators have gain (Fig. 7-11d), we have a coupled laser array, which when extended to multiple coupled lasers has high power applications.

To treat the coupled resonator problem for arbitrary coupling strength, we model the middle mirror as a dielectric "bump" at $z = 0$ with the dielectric function

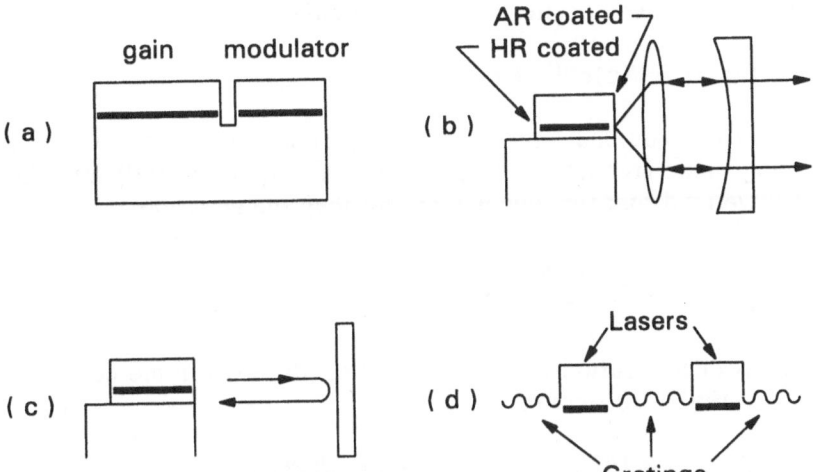

Fig. 7-11.   Different experimental configurations that may be described by coupled resonators. (*a*) Laser with coupled external resonator, (*b*) external resonator laser, (*c*) laser subjected to feedback from external mirror, and (*d*) longitudinally coupled laser array.

$$\epsilon(z) = \epsilon_0 \left[1 + \frac{\eta}{K} \delta(z) \right] , \qquad (88)$$

where the mirror coupling parameter $\eta = 2[(1-T)/T]^{1/2}$, and K is the light wave number. The background refractive index and its spatial variations, like in the case of a semiconductor connected to an external resonator, may be accounted for by making the replacement $L_{A/B} \rightarrow L_{A/B,eff} \equiv n_{A/B} L_{A/B}$ (with the real value of $n_{A/B} \neq 1$).

Our theory is based on a composite-cavity mode expansion of the electric field. By composite-cavity modes we mean the eigenmodes of the combined system of the two coupled cavities when no active material is present. The electric field in the passive coupled-cavity obeys

$$\frac{\partial^2}{\partial z^2} E(z,t) = \frac{\epsilon(z)}{c^2} \frac{\partial^2}{\partial t^2} E(z,t) . \qquad (89)$$

If the field is in a composite-cavity eigenmode, then

$$E(z,t) = U(z)e^{-i\Omega t} + \text{c.c.},$$ (90)

where $U(z)$ is a coupled-cavity mode eigenfunction and $\Omega = cK$ is the corresponding mode frequency. Substituting the expansion (90) into the wave Eq. (89), we find that the eigenfunctions obey the equation

$$\frac{d^2}{dz^2} U(z) = -\frac{\epsilon(z)}{c^2} \Omega^2 U(z) .$$ (91)

From Maxwell's equations, the boundary conditions for the coupled cavity problem of Fig. 7-10 require

$$U(-L_A) = U(L_B) = 0 ; \quad U(0^-) = U(0^+) ,$$ (92)

where 0^- and 0^+ denote the positions infinitesimally before and after the coupling mirror, respectively. Integrating Eq. (91) with respect to z, we find

$$\left.\frac{d}{dz} U(z)\right|_{z=0^+} - \left.\frac{d}{dz} U(z)\right|_{z=0^-} = -\eta K U(0) .$$ (93)

The conditions (92) are fulfilled by a discrete set of eigenfunctions

$$
\begin{aligned}
U_n(z) &= A_n \sin[K_n n_c(z + L_A)] & -L_A \le z \le 0 \\
&= B_n \sin[K_n(z - L_B)] & 0 \le z \le L_B, \\
&= 0 & \text{elsewhere}
\end{aligned}
$$ (94)

with

$$\frac{A_n}{B_n} = -\frac{\sin(K_n L_B)}{\sin(K_n n_c L_A)} \equiv \xi_n ,$$ (95)

where according to Eq. (93) the allowed K_n values satisfy

$$\sin(K_n n_c L_A)\cos(K_n L_B) + n_c \cos(K_n n_c L_A)\sin(K_n L_B) =$$

$$\eta \sin(K_n n_c L_A)\sin(K_n L_B) \ . \tag{96}$$

At this point, the remarks following Eq. (88) regarding the phenomenological treatment of an antireflection coated laser surface become obvious. The replacements discussed there amount to rewriting Eq. (96) as

$$\sin(K_n L_{A,eff})\cos(K_n L_B) + \cos(K_n L_{A,eff})\sin(K_n L_B) =$$

$$\eta \sin(K_n L_{A,eff})\sin(K_n L_B) \ . \tag{97}$$

The cavity eigenfrequencies corresponding to the K_n values determined from Eq. (96) are defined by $\Omega_n = cK_n$. Using Eq. (91) we can derive the exact orthogonality relation

$$\int_{-L_A}^{L_B} dz \ \epsilon(z) \ U_n(z) U_m(z) = \mathcal{N} \ \delta_{nm} \ , \tag{98}$$

where $\mathcal{N}$ is an arbitrary normalization constant, which we set equal to unity. To prove Eq. (98), write Eq. (91) for $U_n(z)$ and for $U_m(z)$, multiply the U_n equation by U_m and vice versa and subtract. This gives

$$U_m(z) \frac{d^2 U_n(z)}{dz^2} - U_n(z) \frac{d^2 U_m(z)}{dz^2} =$$

$$-\epsilon(z) [K_n^2 - K_m^2] U_m(z) U_n(z) \ .$$

Integrating both sides from $-L_A$ to L_B, we have

$$[K_m^2 - K_n^2] \int_{-L_A}^{L_B} dz \, \epsilon(z) U_m(z) U_n(z) =$$

$$\int_{-L_A}^{L_B} dz \frac{d}{dz} \left[U_m(z) \frac{d U_n(z)}{dz} - U_n(z) \frac{d U_m(z)}{dz} \right] = 0 \ ,$$

where we use the fact that $U_n(z)$ vanishes at the mirrors. Hence if $m \neq n$, the integral in Eq. (98) must vanish. This is a general result, independent of the explicit form of the dielectric constant $\epsilon(z)$. To achieve the normalization $\mathcal{N} = 1$, we use $\epsilon(z)$ of Eq. (88) to write the $m = n$ version of Eq. (98) as

$$1 = A_n^2 n_c{}^2 \int_{-L_A}^{0} dz \sin^2[K_n n_c (z+L_A)] + B_n^2 \int_{0}^{L_B} dz \sin^2[K_n(z-L_B)]$$

$$+ \frac{\eta}{K_n} A_n^2 \sin^2(K_n n_c L_A)$$

$$= \frac{L_A}{2} A_n^2 n_c{}^2 \left[1 - \frac{\sin(2K_n n_c L_A)}{2K_n n_c L_A} \right] + \frac{L_B}{2} B_n^2 \left[1 - \frac{\sin(2K_n L_B)}{2K_n L_B} \right]$$

$$+ \frac{\eta}{K_n} A_n^2 \sin^2(K_n n_c L_A) . \tag{99}$$

Eliminating B_n^2 using Eq. (95), we find

$$A_n^{-2} = \frac{L_A n_c{}^2}{2} \left[1 - \frac{\sin(2K_n n_c L_A)}{2K_n n_c L_A} \right] +$$

$$\frac{L_B}{2\xi_n{}^2} \left[1 - \frac{\sin(2K_n L_B)}{2K_n L_B} \right] + \frac{\eta}{K_n} \sin^2(K_n n_c L_A) \tag{100}$$

and

$$B_n^{-2} = \frac{L_A n_c{}^2 \xi_n{}^2}{2} \left[1 - \frac{\sin(2K_n n_c L_A)}{2K_n n_c L_A} \right] +$$

$$\frac{L_B}{2} \left[1 - \frac{\sin(2K_n L_B)}{2K_n L_B} \right] + \frac{\eta}{K_n} \sin^2(K_n L_B) . \tag{101}$$

Figure 7-12 shows a typical pair of eigenfunctions.

We now place an active medium into the left-hand cavity. This introduces the polarization $P(z,t)$ into the wave equation according to

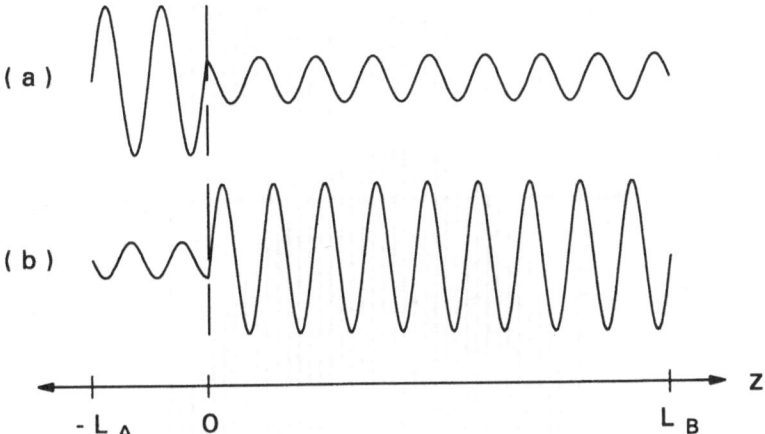

Fig. 7-12. Eigenfunctions for two amplitude ratios;   *a*) large amplitude ratio so that the amplitude in the laser cavity is large, *b*) small amplitude ratio so that the amplitude outside the laser cavity is large. The parameters are $n_c = 3.5$, $T = 0.5$, $L_A = 50\ \mu m$, $L_B = 0.4\ cm$.

$$\left[-\frac{\partial^2}{\partial z^2} + \frac{\epsilon(z)}{c^2}\frac{\partial^2}{\partial t^2} + \frac{\Gamma}{c^2}\frac{\partial}{\partial t} \right] E(z,t) = -\frac{4\pi}{c^2}\frac{\partial^2}{\partial t^2} P(z,t) , \qquad (102)$$

where Γ describes the outcoupling loss, which is modelled to be delocalized over the entire cavity. We expand the electric field in the eigenmodes of the cavity as

$$E(z,t) = \frac{1}{2}\sum_n \mathscr{E}_n(t) U_n(z) e^{-i\nu_n t} + \text{c.c.} . \qquad (103)$$

A semiconductor laser in an external high-quality cavity is analyzed using a composite-resonator eigenmode description of two coupled resonators. Laser threshold and lasing frequency are studied for various cavity lengths and coupling mirror transmissions controlling the amount of light feedback into the laser. The results are valid for all values of coupling. Pronounced mode hopping is predicted which leads to dramatic laser frequency shifts for small variations of the external cavity length.

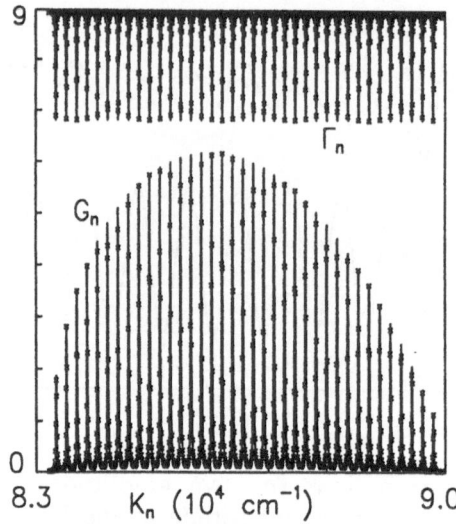

Fig. 7-13. Modal gain G_n and modal loss Γ_n showing the allowed modes (crosses) for a length of L_A = 50 μm and L_B = 0.4cm for n_c = 3.5 and T = 0.2. Here we have a carrier density of N = 1.82 × 10¹⁸ cm^{-3} and a loss Γ = 9 10⁹ s^{-1} chosen for clarity of presentation such that the modal loss always exceeds the gain. Under lasing conditions the modal gain and modal loss curves would intersect at the wavenumber of the lasing mode.

Figure 7-13 shows typical gain and loss profiles, here for a carrier density N = 1.82×10¹⁸ cm^{-3}, electron and hole effective masses m_e = .0665 m_0, m_h = .52 m_0, m_0 is the bare electron mass, and the dipole matrix element d_{cv}/e = 3 × 10⁻⁸ cm. The various peaks represent the longitudinal laser modes for a cavity length L_A. Hence the peaks are separated by the mode spacing of the laser cavity while the allowed K_n values (composite-cavity modes) are separated on average by the mode spacing of the entire coupled cavity length, L = L_A + L_B. Figure 7-14 shows the two central peaks of Fig. 7-13 and illustrates the position of the composite-cavity modes between the two peaks. The crosses and circles on the gain curve represent the composite-cavity modes for two values of the external cavity length L_B, which differ by .01 μm. Typically the length ratio $n_c L_A / L_B$ determines the number of composite-cavity modes between each peak. As the external cavity length L_B is varied on the order of a fraction of a wavelength, denoted as δ, the positions of the composite-cavity modes change continuously in a counter clockwise direction. Most of the compo-

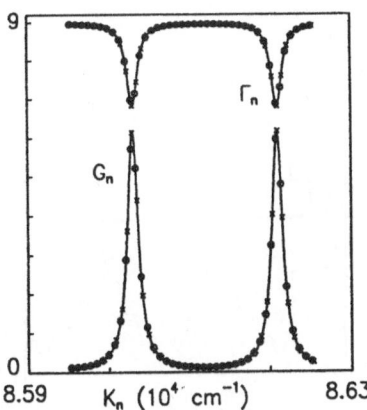

Fig. 7-14. Modal gain G_n and modal loss Γ_n showing the two central peaks of Fig. 7-13. In addition to the crosses from Fig. 7-13, the circles indicate the allowed modes at a cavity length L_b + .01 μm. All parameters are the same as in Fig. 7-13.

site-cavity modes reside in the troughs and are seen with increasing L_B to move quickly up the steep sides and slowly on the peaks and troughs. Circles on top of crosses indicate little change. In the limit when the transmission of the coupling mirror goes to zero, a single supermode resides in the vicinity of each peak and for changing L_B occasionally drops down quickly as another mode rises to replace it. In this sense the crosses and circles in Fig. 7-14 depict a mode hopping instability mechanism. Because the modal loss is essentially a straight line, the mode with the highest gain lases in our single-mode theory. Figures 7-13 and 7-14 show that the energy of the mode with the highest gain switches as the external cavity length changes slightly. As a result, the laser switches to the next mode with more gain. In the limit that the coupling transmission goes to unity, the modal gain function approaches the usual semiconductor laser gain curve.

In order to study the variations of the laser threshold frequency, we vary the external cavity length over a range of one micron for each value of the transmission. After each length step, we adjust the carrier density to satisfy the threshold condition. Figures 7-15 and 7-16 show the wavenumber of the lasing mode (solid line) versus the change δ in the external cavity length L_B for n_c = 3.5 (Fig. 7-15) and for an antireflection coated laser outcoupling mirror (Fig. 7-16). Figures 7-16a through 7-16c correspond to different values of the transmission of the coupling mirror. For

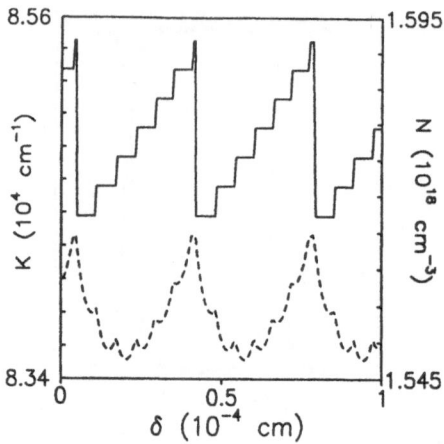

Fig. 7-15. Wavenumber of the lasing mode (solid) and threshold density (dashed) as a function of the external cavity length variation (δ) for $T = 10^{-3}$, $n_c = 3.5$, $L_A = 50\ \mu m$, and $L_B = 0.4\ cm$. The total length scale is one micron, $\delta(\mu m)$.

small values of the transmission (Fig. 7-16a), the mode may switch over one cavity mode for a short length and then switch back. As the transmission increases, the energy of the lasing mode switches over two cavity modes and then three, and so on. Figure 7-16b shows the energy curve for $T = 10^{-3}$. The energy appears to follow a staircase before switching back to the original laser mode. For this particular set of parameters, the maximum number of steps is five (Fig. 7-16c). As the length ratio increases that number is seen to increase. Because the threshold density increases with increasing transmission, the entire curve moves upward. The dashed lines in Fig. 7-16 show the threshold density curves. The density curve approaches a straight line (Fig. 7-16a) as the transmission goes to zero. Associated with each change in the slope of the density curve is the jump of the laser mode. The density curves acquire more structure as the transmission increases, which is consistent with the energy curves. The overall variation of the density is very small, and hence the related index changes turn out to be negligible.

   In conclusion, we see that the existence of a frequency instability against resonator length fluctuations in the semiconductor laser coupled cavity. In our model the instability sets in at very small transmission values ($T \simeq 10^{-5}$) as a jump between adjacent cavity modes and progresses at higher transmission values to jumping sequentially among several cavity modes. The instability mechanism can be understood from the structure of the

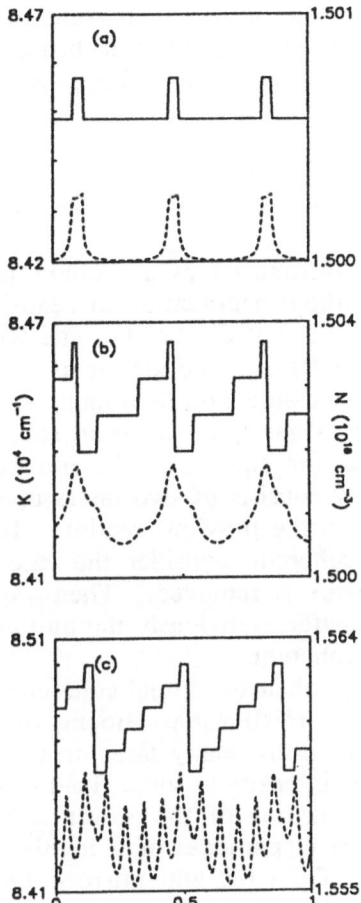

Fig. 7-16. Wavenumber of the lasing mode (solid) and threshold density (dashed) as a function of the external cavity length variation (δ) for $n_c = 1.0$, $L_A = 370$ μm, $L_B = 0.4$ cm, and a) $T = 10^{-5}$, b) $T = 10^{-3}$, c) $T = .1$. The total length scale is one micron, $\delta(\mu m)$.

modal gain. Related mode hopping instabilities in coupled ring cavity configurations have been investigated by Ikeda and Mizuno (1984) using delayed feedback equations. These authors speculate that the sensitivity of the frequency jumping may contribute to additional noise induced by the reflections off the external mirror and could thus be related to the experimentally observed feedback instability. In fact, we can imagine the situation in which the length fluctuations in an experimental situation could be quite prevalent, such that mode hopping would result. Each jump corresponds to several GHz, which is in the range of the observed linewidth broadening. Depending on the measurement technique this phenomena might be averaged over to give an effective line broadening and might thus be connected with the observation of "coherence collapse" in feedback experiments.

7-6. Laser Arrays

Laser arrays were introduced as a method to scale laser systems to higher output power without degradation in beam coherence. The concept was first used in gas lasers, but did not become widely known until it was applied to semiconductor lasers. The idea is to design a laser with as much power and coherence as possible and then make an array of such lasers in a way such that they act as one large coherent source. Ideally the far field peak intensity scales as the square of the number of lasers. A simple example of a laser array consists of two lasers operating with the coupled resonators as discussed in the previous section. To see that the laser outputs can be mutually coherent, consider the case of 100% coupling, i.e., when the coupling mirror is removed. Then we have two gain regions operating with one resonator. Obviously the outputs from both ends of the resonator are mutually coherent.

While high coupling enhances mutual coherence, it does not necessarily lead to the proper phase relationships among the different lasers of the array. For example, there are many laser array applications that require the optimization of laser intensity at some fixed position in the far field, or in practical terms, optimization of the power in a bucket. Figure 7-17a shows two representative eigenmodes of a highly coupled resonator array. Assuming some leakage from the end mirrors, the outputs are either in-phase (as in mode 1) or π out-of-phase (as in mode 2). Combination of the outputs in the far field gives the intensity distributions shown in Fig. 7-17b No manipulation of the optical path lengths can result in constructive interference at one far-field position for these two types of modes, so that half the array power is lost. Some form of mode discrimination is then necessary to prevent all in-phase or all out-of-phase modes from lasing simultaneously. One source of mode discrimination is the coupling due to mirror absorption or scattering losses, which have greater effect on the in-phase modes because of their greater amplitude at the coupling mirror. Hence, the problem is one of balancing the factors giving mutual coherence against those giving mode discrimination, and that usually involves choosing the optimal amount of coupling.

Figure 7-18 shows two different laser array coupling schemes. End-to-end coupling, which is an extension of the coupled-resonator case of Sec. 7-5, is used for lasers with external resonators, e.g., gas and chemical lasers. Semiconductor laser arrays are more typically coupled side-by-side via evanescent waves. The amount of evanescent wave coupling depends

(a)   (b)

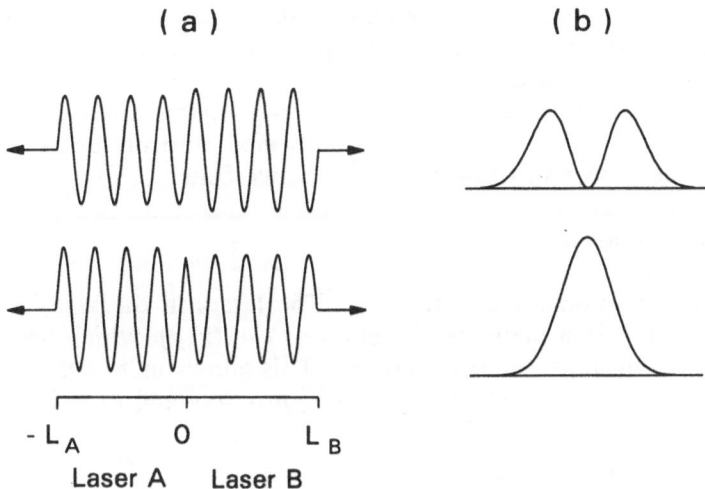

$-L_A$ O L_B

Laser A   Laser B

Fig. 7-17. (a) Two eigenmodes of a highly coupled laser array. (b) Far field distributions from the two eigenmodes. The optical paths travelled by the laser outputs are assumed equal. Note that only the in-phase mode contributes to the on-axis intensity.

(a)   (b)

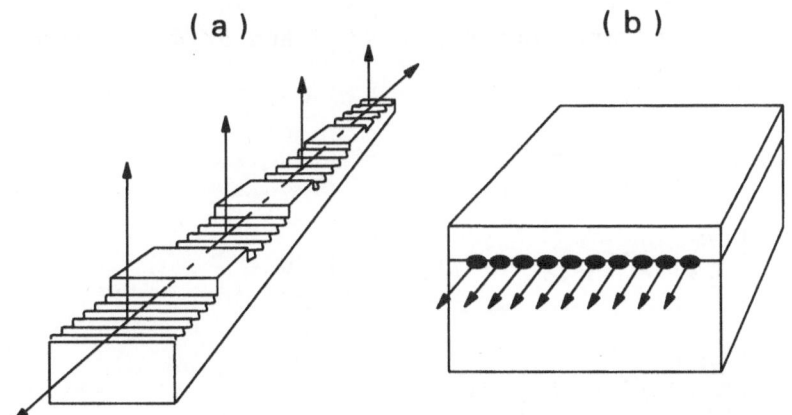

Fig. 7-18. Two general coupling schemes for laser arrays : (a) end-to-end and (b) side-by-side coupling.

on the laser separation and in the case of index guided arrays, on the index step separating the lasers. Maximum or 100% coupling gives a broad-area laser, which operates multimode.

This section shows how the composite-resonator analysis presented in the previous section can be applied to describe semiconductor laser arrays. Repeating the steps from Eqs. (89) to (91), we find

$$[\nabla^2 + \mu\epsilon(\mathbf{r})\Omega^2]u(\mathbf{r}) = 0 \qquad (104)$$

for the passive resonator eigenmodes. The lasers in the array are defined by the variations in a dielectric function $\epsilon(\mathbf{r}) = \epsilon(x,y)$, which has very weak y dependence in most heterostructures. This allows us to use what is commonly referred to as the *effective index approximation*, to write

$$u(\mathbf{r}) = X(x,y) \, Y(y) \, Z(z) \,, \qquad (105)$$

where X is a weak function of y. Substituting Eq. (105) into Eq. (104), we find

$$\frac{1}{X}\frac{\partial^2 X}{\partial x^2} + \frac{1}{Y}\frac{\partial^2 Y}{\partial y^2} + \frac{1}{Z}\frac{\partial^2 Z}{\partial z^2} + \mu\epsilon(x,y)\Omega^2 = 0 \,, \qquad (106)$$

where we ignore terms containing $\partial^2 X/\partial y^2$ and $\partial X/\partial y$. Equation (106) may be separated into

$$\frac{1}{Z}\frac{d^2}{dz^2} Z(z) = -\beta^2 \,, \qquad (107)$$

$$\left[\frac{d^2}{dx^2} + \mu\Omega^2\epsilon(x,y) - \eta^2(y)\right] X(x,y) = 0 \,, \qquad (108)$$

$$\left[\frac{d^2}{dy^2} + \eta^2(y) - \beta^2\right] Y(y) = 0 \,, \qquad (109)$$

where η and β are the separation constants and y is treated as a parameter in Eq. (108). The solution of Eq. (107) is straightforward and gives

$$Z_m(z) = \sin(k_{mz}z) \,, \qquad (110)$$

where the resonator extends from $z = 0$ to $z = L$. As discussed earlier, the boundary conditions $Z_m(0) = Z_m(L) = 0$ require $\beta \equiv k_{mz} = m\pi/L$, where m is an integer. Equation (108) is solved for different values of y. For

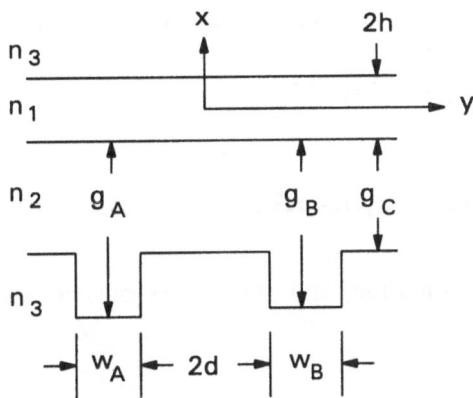

Fig. 7-19. Heterostructure arrangement for a two laser channel substrate planar (CSP) array.

example, for $g = g_A$, g_B, or g_C, the solution is

$$X = E\, e^{p_3(x + h + g)} \qquad\qquad \text{for } x < -h - g ,$$

$$= C\, \cos[p_2(x + h + g)] + D\, \sin[p_2(x + h + g)] \quad \text{for } -h - g < x < -h ,$$

$$= A\, \cos[p_1(x + h)] + B\, \sin[p_1(x + h)] \qquad \text{for } -h < x < h ,$$

$$= F\, e^{-p_3(x - h)} \qquad\qquad\qquad \text{for } x > h , \qquad (111)$$

where we use the boundary conditions $X(\pm\infty, y) = 0$, so that only exponentially decaying solutions are acceptable at the outer regions,

$$p_3{}^2 + k_0{}^2 n_3{}^2 = \eta^2 ,$$

$$- p_2{}^2 + k_0{}^2 n_2{}^2 = \eta^2 ,$$

$$- p_1{}^2 + k_0{}^2 n_1{}^2 = \eta^2 , \qquad (112)$$

and $k_0 c = \Omega$. Equation (112) may be used to eliminate η, giving

$$P_1 = \sqrt{k_0^2(n_2^2 - n_3^2) - p_3^2} \tag{113}$$

$$P_2 = \sqrt{k_0^2(n_1^2 - n_3^2) - p_3^2} . \tag{114}$$

Continuity of the functions and their first derivatives at the boundaries give

$$\left[\cos(p_2 g) + \frac{p_3}{p_2}\sin(p_2 g)\right][p_3\cos(2p_1 h) - p_1\sin(2p_1 h)]$$

$$= \left[\frac{p_2}{p_1}\sin(p_2 g) - \frac{p_3}{p_1}\cos(p_2 g)\right][p_1\cos(2p_1 h) + p_3\sin(2p_1 h)] , \tag{115}$$

which together with Eqs. (113) and (114) can be used to solve for p_1, p_2, and p_3. This is has to be done numerically. The boundary conditions also give the eigenfunction amplitudes

$$A = E\left[\cos(p_2 g) + \frac{p_3}{p_2}\sin(p_2 g)\right], \tag{116a}$$

$$B = E\left[\frac{p_3}{p_1}\cos(p_2 g) - \frac{p_2}{p_1}\sin(p_2 g)\right], \tag{116b}$$

$$C = E , \tag{116c}$$

$$D = \frac{p_3}{p_2}E \tag{116d}$$

and

$$F = E\left\{\left[\cos(p_2 g) + \frac{p_3}{p_2}\sin(p_2 g)\right]\cos(2p_1 h)\right.$$

$$\left. + \left[\frac{p_3}{p_1}\cos(p_2 g) - \frac{p_2}{p_1}\sin(p_2 g)\right]\sin(2p_1 h)\right\}. \tag{116e}$$

Figure 7-20a shows the transverse modes inside and between the channels. Note that there is negligible y dependence in the transverse eigenfunctions. It is useful to define an effective index

$$n_{eff}(y) = \frac{\eta(y)}{k_0}. \tag{117}$$

Figure 7-20b shows the y dependence of the effective index for our example.

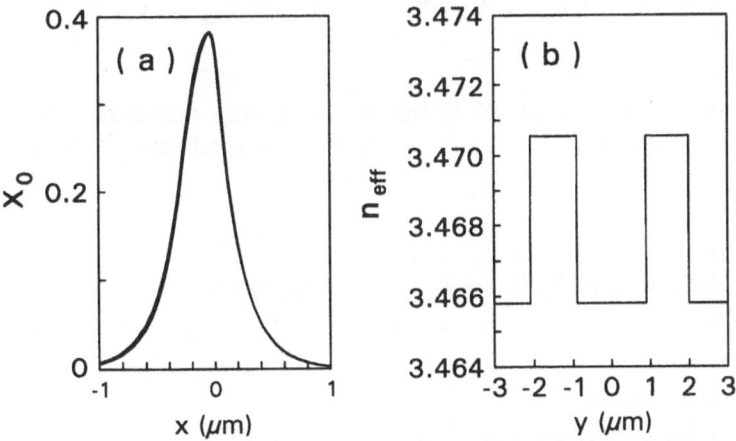

Fig. 7-20. (a) Transverse modes and (b) effective indices for heterostructure shown in Fig. 7-19. For the transverse modes, the solid curve is for the modes at $y = \pm (d + w)/2$ (inside the channels) and the dashed curve is for the mode $y = 0$ (between channels). We use $n_1 = 3.6$, $n_2 = 3.5$, $n_3 = 3.4$, $g_A = g_B = 0.26$ μm, $g_C = 0.2$ μm, and $h = 0.05$ μm.

The small differences in n_{eff} can have appreciable influence on the lateral modes. In terms of the effective index, Eq. (109) may be rewritten in the form

$$\frac{d^2Y}{dy^2} + [n_{eff}^2 k_0^2 - k_{mz}^2]Y = 0 .$$

(118)

The solution of Eq. (118) for Y is similar to that for the transverse modes, i.e., the solutions at the different sections for constant n_{eff} are either cosines, sines, or exponentials, and the boundary conditions determine the mode eigenvalues and relative amplitudes. We show some examples of the lateral eigenmodes later.

The derivation for the orthogonality relation is as described in the previous section. For the eigenmodes

$$u_{nlm}(\mathbf{r}) = X_n(x,y)Y_l(y)Z_m(z) ,$$

(119)

where n, l, and m are the mode indices, we get

$$\int d^3r \; \epsilon(\mathbf{r}) \; u_{nlm}(\mathbf{r}) \; u_{n'l'm'}(\mathbf{r}) = \mathcal{N}\delta_{n,n'}\delta_{l,l'}\delta_{m,m'} ,$$

(120)

where $\epsilon(y) = n^2(\mathbf{r})\epsilon_0$, and $\mathcal{N}$ is the normalization constant. Similarly, the orthogonality relations for the transverse and lateral fields may be derived, giving

$$\int dx \; n(x,y) \; X_n(x,y) \; X_{n'}(x,y) = n_{eff}(y) \; h ,$$

(121)

$$\int dy \; n_{eff}(y) \; Y_l(y) \; Y_l(y) = d + w ,$$

(122)

where we choose $\mathcal{N} = \epsilon_0(d + w)hL/2$.

For evanescent wave coupled arrays, the lateral modes Y_l are the ones of interest. They may be used to describe the different behaviors of an array. For the two-channel array shown in Fig. 7-19, there are two lateral modes for each longitudinal mode number k_{mz} (the transverse structure is designed so that there is only one transverse mode). They differ in fre-

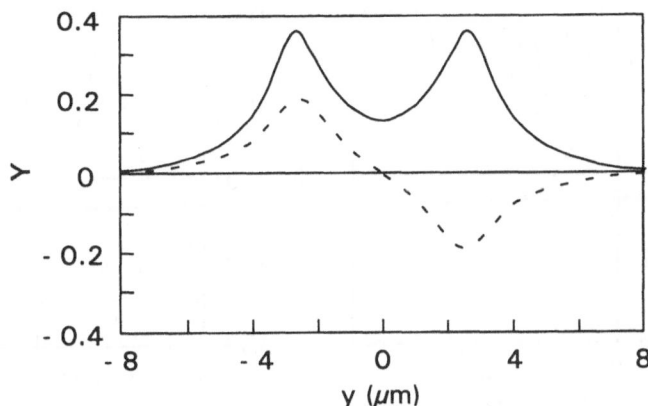

Fig. 7-21. In-phase (solid curve) and π-out-of-phase (dashed curve) mode for a two channel array. We use $w_A = w_B = 1 \ \mu m$, $d = 2 \ \mu m$ and the effective indices 3.47 for the channels and 3.465 elsewhere.

quency and lateral spatial dependence. The fields in the two channels are in phase for one array mode and π radians out of phase for the other (Fig. 7-21). One usually labels the array modes by integers starting with $n = 1$ for the lowest order mode. A phased locked array operates with only one of these modes. With the $n = 1$ mode, one gets a single-lobe far field intensity distribution and with the $n = 2$ mode, one gets a double lobe far field intensity distribution (Fig. 7-17b). For identical channels, which is the case assumed here, these modes are either symmetric or antisymmetric about $y = 0$.

When the channels are different, the array modes are asymmetrical. The differences between channels may be in their dimensions or refractive indices. For example, the channel widths may be different due to masking or etching inaccuracies during fabrication. Figure 7-22 shows that the array modes have different amplitudes in the different channels. The figure also shows that the difference in amplitudes depends on the coupling, i.e. the separation or index step between channels. In general, the smaller the coupling, the more sensitive the amplitudes are to channel differences. For weakly coupled channels, the frequency of mode 1 is approximately that of one of the isolated channel, while the frequency of mode 2 approximately equals that of the other. An unlocked array has a field that is a superposition of these two modes. This may be seen by

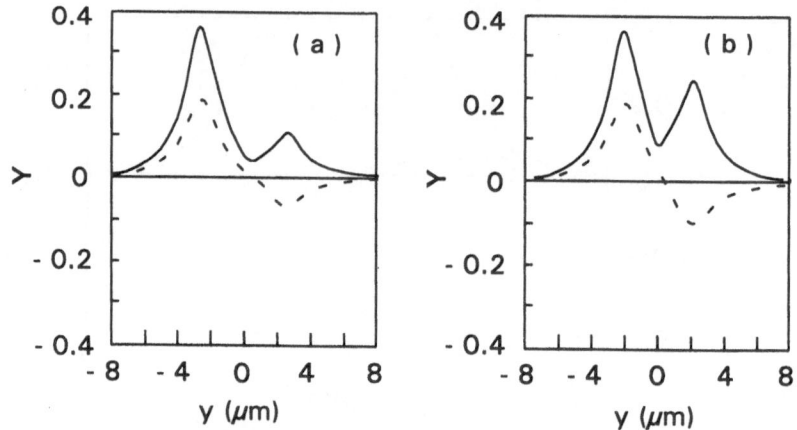

Fig. 7-22. Array eigenmodes for dissimilar channels for (a) weak and (b) strong coupling. Note that the difference in channel amplitudes increases with decreasing coupling.

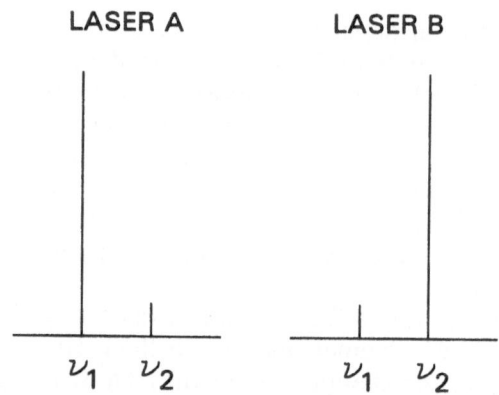

Fig. 7-23. The superposition of the array eigenmodes in Fig. 7-22 (a) gives an unlocked array.

looking at the spectra of channels A and B. Figure 7-23 shows that the frequency of channel A is essentially ν_2 and the frequency of channel B is essentially ν_1, i.e., they have different frequencies. The figure also shows that each channel has the leakage field from the neighboring channel.

In the language of array modes, the question of whether the array is phase locked translates to whether one has single or multi array mode operation. The answer requires the inclusion of the gain medium into the theory. Since the passive array modes form an orthonormal basis, like the standing or running-wave eigenmodes of the single laser, the procedure is identical to that used in Sec. 7-2. It is straightforward to show that the gain medium polarization is given by Eq. (22) with U_m^* being an array eigenmode. A general multimode theory that treats all these effects is complicated and is still under development. We discuss a simple low-intensity, quasiequilibrium approach in Chap. 8. At present, if we assume that, because of the possibility of considerable spatial differences in the array modes, spatial hole burning dominates, then we can compute the lockbands shown in the following figures.

Regardless of the gain model used, the array shows the following general characteristics. For an array that is uniformly pumped by broad area electrodes, and for which the losses vary little in the lateral dimension, Fig.

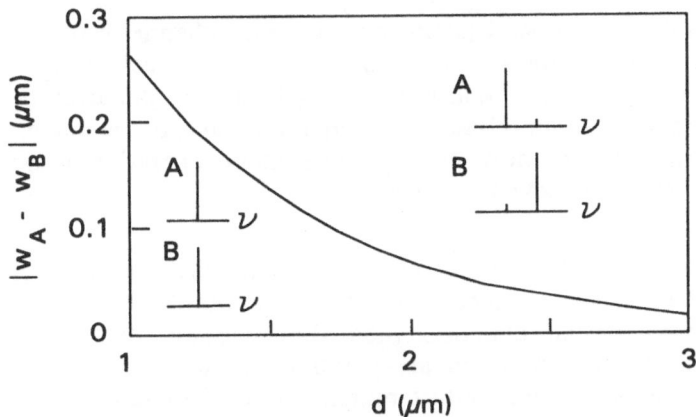

Fig. 7-24. Lockband for uniformly pumped array. The y-axis is the channel width difference $|w_A - w_B|$ and the x-axis is d, the separation between channels.

7-24 shows that there are two regions of operation. For some coupling range, there is a range of channel width differences where the array operates with one array mode, i.e., is frequency locked. The frequency locked region has bistable solutions: the in-phase ($n = 1$) mode and the π out-of-phase ($n = 2$) mode. Whichever solution one gets depends on which mode first reaches threshold. Outside of the locked region, both modes are above threshold.

When the pump or the losses are not spatially uniform, then a third region of operation emerges. An example is the case where we have more absorption or less gain in the regions between the channels than anywhere else. Figure 7-25 shows that there is now a third region of operation,

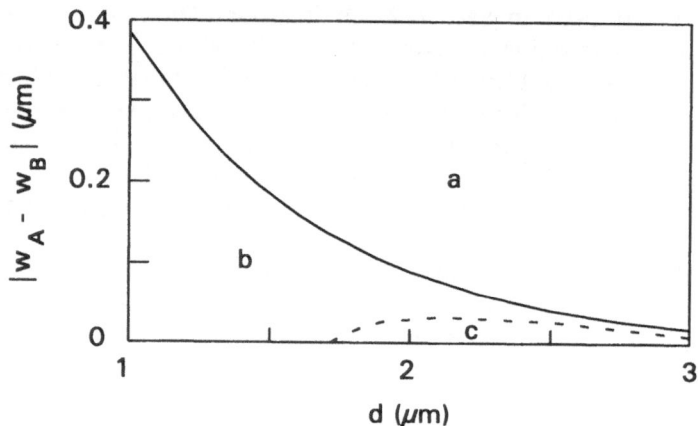

Fig. 7-25. Lockband (solid curve) versus channel separation for an array where the excitation between the channels is lower than in all other regions. Within the locked portion, the array operates only in the out-of-phase mode in region b. In region c, either only in-phase or only out-of-phase mode operation is possible. The array is unlocked in region a.

region b, where only the out-of-phase mode lases. This region exists because the excitation geometry favors the out-of-phase mode which is less concentrated between channels than the in-phase mode. Above this region, the coupling between array modes is sufficiently small to allow two-mode operation. Below this region, the mode coupling is sufficiently large to negate the difference in the small signal gain between array modes. Since the net linear gain (the small-signal gain minus the losses) is greater in the channels than elsewhere, we expect that the amplified spontaneous emission into the out-of-phase mode to be greater. Therefore we expect the array to operate with the out-of-phase mode in regions b and c. The difference between these two regions arises when, for example, an injected signal is used to force the array into operating with the in-phase mode. Upon termination of the injected signal, an array operating in region b returns to lasing in the out-of-phase mode, while an array operating in region c remains lasing in the in-phase mode. Region b increases in size when we increase the difference in the excitation levels or the losses. It is possible to extend region b all the way to $\Delta w = 0$.

An array may be made to operate always in the in-phase mode by tailoring the injection current so that the region between waveguides is further above threshold than elsewhere. This excitation geometry favors the in-phase mode because it has a larger amplitude between waveguides than the out-of-phase mode. Figure 7-26 depicts an example of the con-

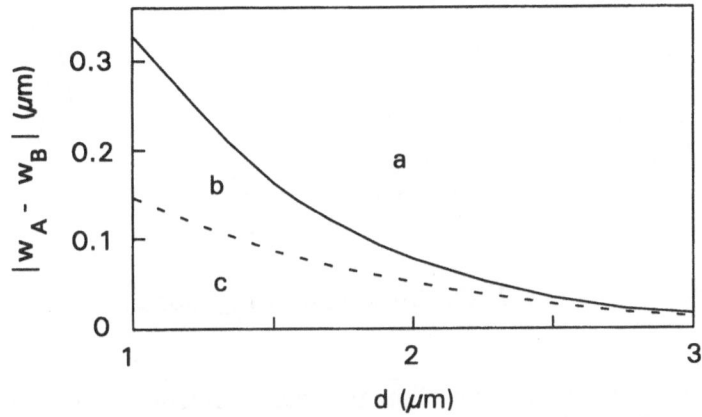

Fig. 7-26. Lockband (solid curve) versus channel separation for an active volume where the excitation between the channels is higher than in all other regions. Within the locked portion, the array operates with only the in-phase mod in region b. In region c, the array is bistable, and in region a, the array is unlocked.

ditions leading to only in-phase mode operation. Here in region b, we have array configurations that give in-phase mode operation. Note however that this kind of an array does not operate at optimum efficiency because the region between channels, which has high current and low laser intensity, operates without saturation.

Spatial pump inhomogeneities frequency lock the array modes. For example, consider the situation where only one channel is pumped. We show in the following discussion that this leads to a frequency locking term that may lock the two array modes to one another. If the modes lock in phase, only channel A lases, and if they lock with a π phase difference, only channel B lases (Fig. 7-27). To show the existence of a frequency locking term in the gain medium polarization, we first note that no such term exists for a symmetrically pumped two-channel array. This is no longer the case in larger arrays. In general, the first-order frequency locking terms vanish only when the active region is pumped uniformly over a

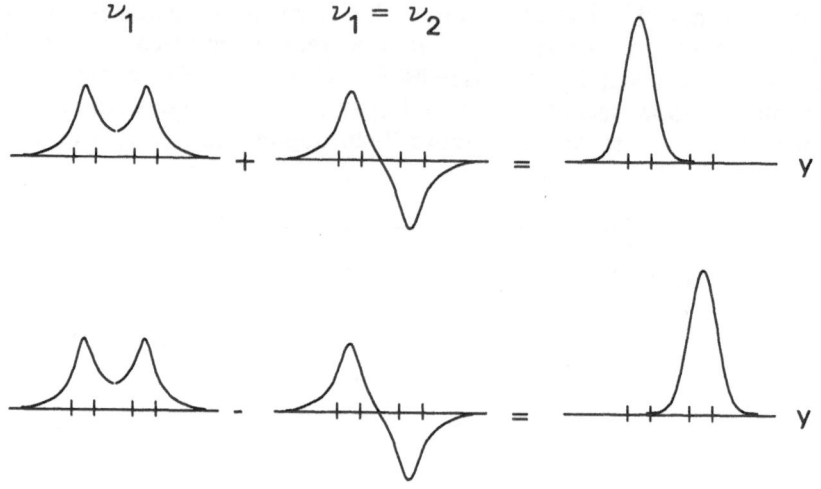

Fig. 7-27. Locking of array modes.

volume that is much broader than the lateral extent of the passive array eigenmodes. When that is not the case, we find that

$$- \frac{\nu}{2\epsilon_0} \frac{\mathcal{N}_n{}'}{\mathcal{N}} \mathcal{P}_n = (ig_n + \sigma_n)E_n + \sum_{m \neq n} B_{nm} E_m e^{i\psi_{nm}} , \qquad (123)$$

where $\mathcal{N}_n{}' = \int d^3r\, u_n{}^2(\mathbf{r})$ and ψ_{nm} is the phase difference between the fields in the two modes. The frequency locking coefficient is

$$B_{nm} = F_n \frac{1}{\mathcal{N}} \int d^3r\, n^2(x,y)\, u_n(\mathbf{r})u_m(\mathbf{r})$$

$$\times \sum_k \left[i + \frac{\omega_k - \nu}{\gamma} \right] \mathcal{L}(\omega_k - \nu)\, (n_{ek} + n_{hk} - 1). \qquad (124)$$

Since the passive array modes are orthogonal to one another, the locking coefficient becomes appreciable only when the array is pumped nonuniformly over a region where the passive array modes have noticeable amplitude.

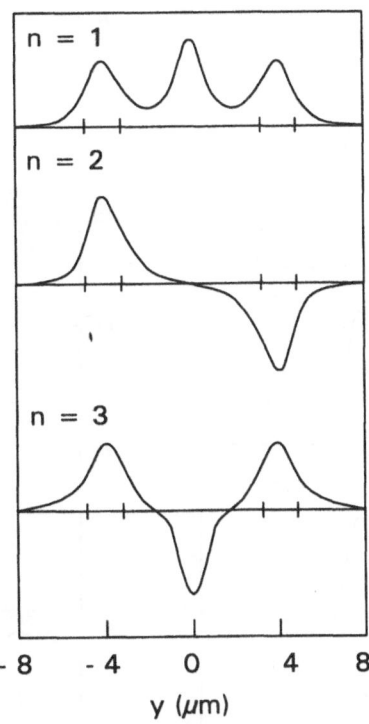

Fig. 7-28. Array modes of a three-channel array. The ticks indicate the boundaries of the waveguides and the channels are assumed to be identical.

The frequency locking of array modes describes gain guiding effects. To show this, let us consider the example of a three-channel array whose eigenmodes are shown in Fig. 7-28. If the electrodes are much wider than the lateral extent of the passive array eigenmodes, index guiding dominates. We find that there are two stable steady-state solutions. They are operation in an in-phase array field that consists predominantly of the $n = 1$ passive array eigenmode and operation with an out-of-phase array field that consists predominantly of the $n = 3$ passive array eigenmode. The near fields associated with these solutions are shown in Fig. 7-29a. In both cases we have a phase-locked array. The array does not operate in the $n = 2$ mode because the modal loaded gain for this mode is substantially smaller than the others. An array starting from spontaneous emission

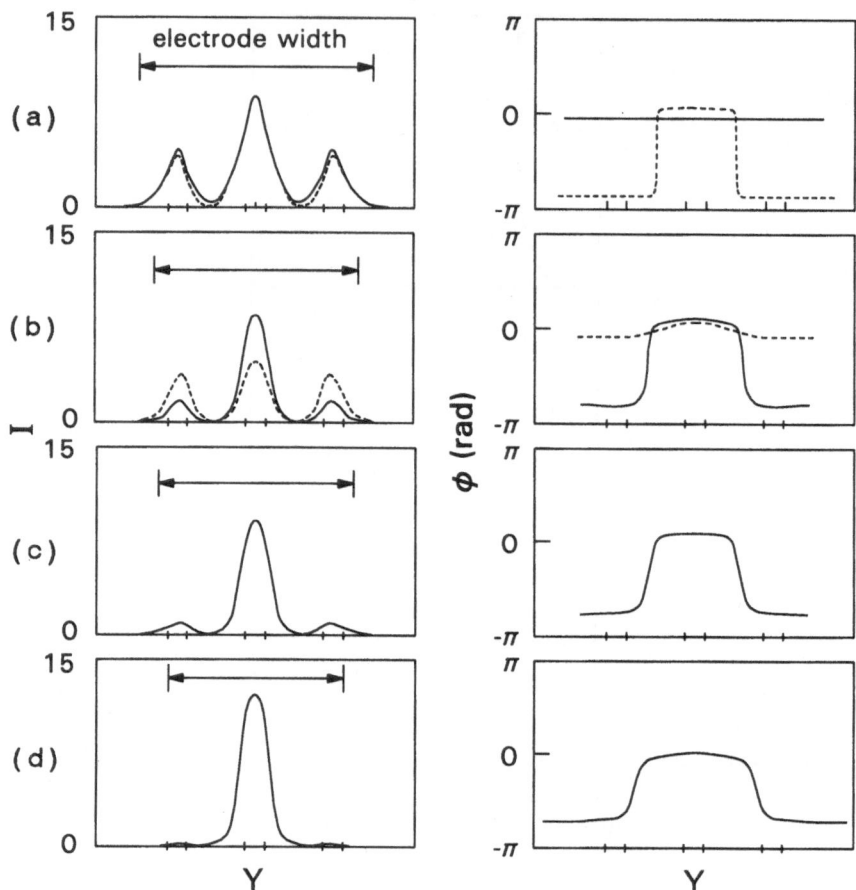

Fig. 7-29. Near field distributions of a three-channel array for decreasing electrode width. The electrode widths are (a) 12 μm, (b) 10.5 μm, (c) 10 μm, and (d) 9 μm. The distance between the other boundaries of the outer channels is 9 μm. The solid curves plot the field that evolves from spontaneous emission, and the dashed curves plot the other stable near-field solution. Note that the near field increasing deviates from the passive array modes.

operates with the in-phase array mode because its mode gain is slightly larger than that of the $n = 3$ mode. When we reduce the electrode width, the out-of-phase array field begins to peak at the central channel, while the in-phase array field begins to spread more to the outer channels. (The in-phase field is now a coherent superposition of array modes and the amplitudes in the three channels begin to equalize.) The shape changes lead to an increase in the gain of the out-of-phase array field and to a decrease in the gain of the in-phase field. At 10 μm electrode width, the out-of-phase array field exceeds that of the in-phase array field, so that the former becomes the lasing array field for the array starting from spontaneous emission (Fig. 7-29b). At this stage, operation with the in-phase array field becomes the second stable steady-state solution. The lasing array fields are now more of an admixture of the passive array eigenmodes. Further reduction in the electrode widths leads to the disappearance of the second stable steady-state solution. The array now operates in the out-of-phase array field, with most of its intensity in the central channel (Figs. 7-29c and 7-29d).

Hence for an array in which both index guiding and gain guiding are present, semiclassical laser theory accounts for the gain-guiding effects by the frequency locking of the passive array eigenmodes. When the refractive index varies spatially more than the gain, index guiding dominates. There is no degeneracy in the lasing frequencies of the array modes, and the near field of a phase-locked array is essentially that given by one of the passive array eigenmodes. When gain varies spatially more than the refractive index, gain guiding dominates. Gain-dependent dispersive effects give rise to frequency locking terms in the gain medium polarization that are sufficient to cause degeneracies among the array eigenmodes. The laser field, which is now given by a coherent superposition of the passive array eigenmodes, takes on a shape that is determined by the gain distribution. Of course, as is the case throughout this chapter, the description of the laser is based on the passive single-resonator or composite resonator-modes. Therefore, it eventually breaks down when gain guiding is considerably stronger than index guiding.

References

Haug, H. and S. W. Koch (1993), *Quantum Theory of the Optical and Electronic Properties of Semiconductors*, World Scientific Publ., Singapore, 2nd Edition.

Meystre, P. and M. Sargent III (1991), *Elements of Quantum Optics*, Springer-Verlag, Heidelberg, 2nd Edition.

Sargent, M., M. O. Scully, and W. E. Lamb, Jr. (1977), *Laser Physics*, Addison-Wesley Publishing Co., Reading MA.

For a discussion of composite resonator modes and laser instabilities see, e.g.,

Spencer, M. and W. Lamb, Jr. (1972), Phys. Rev. **A5**, 884.

Chow, W. W. (1986), IEEE J. Quantum Electron. QE-22, 1174.

Rose, M., M. Lindberg, W. W. Chow, S. W. Koch, and M. Sargent III (1992), Phys. Rev. **A46**, 603.

Yariv, A. (1985), *Optical Electronics* Holt, Rinehart and Winston, New York, 3rd Edition, Chap. 13.

Ikeda, K. and M. Mizuno (1984), Phys. Rev. Lett. **53**, 1340.

Chapter 8
MULTIMODE OPERATION

A major question in semiconductor laser physics is why and when a semiconductor laser oscillates with multiple modes. From the simplest point of view, one might think that the mode with the highest gain would oscillate alone by clamping the gain to the loss line and thereby causing all other modes to experience a net loss. However a number of mechanisms enter to change the gain that a mode experiences in the presence of other modes. These mechanisms include longitudinal and transverse spatial hole burning, spectral hole burning, and population pulsations.

In this chapter we start by considering the gain that incipient side-modes experience in the presence of a single oscillating mode. The simplest mode interactions fall into a subject known as multiwave mixing, which can be used for spectroscopy and phase conjugation in addition to laser mode interactions. Section 8-1 treats the interaction of two weak modes whose frequencies symmetrically straddle the frequency of the oscillating mode. The adjacent-mode beat frequency and the oscillating mode Rabi frequency are taken to be small compared to the carrier-carrier relaxation rates. These frequencies are furthermore assumed to be sufficiently small that carrier-phonon scattering is able to maintain the carrier distribution temperatures at the lattice temperature. With these conditions the dynamics of the interaction is completely specified by the response of the total carrier density to the multimode field. This allows us to express the mode gain and coupling coefficients as functions of the single-mode susceptibility. So expressed, the same formulas describe the gain and coupling coefficients of the homogeneously broadened atomic two-level medium with a short dipole decay time T_2; the differences between the two-level and semiconductor media are encapsulated in the respective single-mode susceptibilities.

Section 8-2 allows the beat frequency to be comparable to or greater than the intraband relaxation rates, so that the dynamics of the individual k-dependent carrier probabilities plays a role. This approach reveals the roles of spectral hole burning and inhomogeneously broadened population-pulsations and discusses whether the sidemodes can build up in the presence of the oscillating mode.

Since these models assume that the sidemodes do not saturate the response of the medium, they are only able to predict when a sidemode can build up; they cannot predict the resulting time development. For that, Sec. 8-4 offers a perturbation-theory method that describes multimode operation close enough to threshold that a third-order $(\chi^{(3)})$ expansion is valid. More generally, one can integrate coupled nonlinear differential equations numerically, as outline in Sec. 8-7. Sections 8-4 and 8-5 discusses one- and two-mode operation according to the third-order theory, revealing how spatial hole burning can modify the saturation that modes inflict upon themselves and others. Section 8-6 shows how in three-mode operation, mode locking can develop from internally generated combination tones.

8-1. Multiwave Mixing

Some of the basic characteristics of modal interactions in lasers are revealed by the multiwave mixing of weak (nonsaturating) modes and one or two relatively intense modes in a nonlinear medium. In this section, we study this mixing in a quasiequilibrium semiconductor medium. We consider a multimode electric field of the form

$$E(\mathbf{r},t) = \tfrac{1}{2} \mathcal{E}(\mathbf{r},t)e^{-i\nu t} + \text{c.c.}, \tag{1}$$

where the complex amplitude $\mathcal{E}(\mathbf{r},t)$ is a function of position $\mathbf{r}$ and is assumed to vary little not only in an optical period (the slowly-varying envelope approximation), but also in a dipole decay time T_2 (rate equation approximation), and ν is the central carrier frequency of the wave. We are primarily interested in treating the sidemode problem in semiconductor lasers, but for the sake of generality, we consider the case depicted in Fig. 8-1 for three waves with possibly different directions. This multiwave configuration described by the field envelope

$$\mathcal{E}(\mathbf{r},t) = \mathcal{E}_1(\mathbf{r})e^{i(\mathbf{K}_1\cdot\mathbf{r}+\Delta t)} + \mathcal{E}_2(\mathbf{r})U_2(\mathbf{r}) + \mathcal{E}_3(\mathbf{r})e^{i(\mathbf{K}_3\cdot\mathbf{r}-\Delta t)}, \tag{2}$$

where $U_2(\mathbf{r}) = \sin(\mathbf{K}_2\cdot\mathbf{r})$ for the four-wave mixing typically used in phase conjugation, $U_2(\mathbf{r}) = \exp(i\mathbf{K}_2\cdot\mathbf{r})$ for three-wave mixing, the pump wave amplitude $\mathcal{E}_2(\mathbf{r})$ can be arbitrarily intense, while the signal $\mathcal{E}_1(\mathbf{r})$ and conjugate $\mathcal{E}_3(\mathbf{r})$ amplitudes are assumed to be weak, i.e., are treated linearly, and Δ is the beat frequency between the pump and signal waves. The field (1) induces a polarization of the form

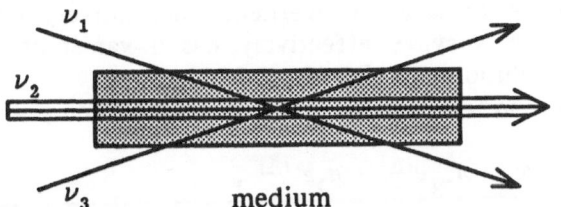

Fig. 8-1. Interaction of a strong wave and two weak (nonsaturating) waves with a nonlinear medium.

$$P(\mathbf{r},t) = \tfrac{1}{2}\mathscr{P}(\mathbf{r},t)e^{-i\nu t} + \text{c.c.}, \tag{3}$$

where for the multiwave mixing problem, we write the complex polarization $\mathscr{P}(\mathbf{r},t)$ envelope as

$$\mathscr{P}(\mathbf{r},t) = \mathscr{P}_1(\mathbf{r})e^{i(\mathbf{K}_1\cdot\mathbf{r}+\Delta t)} + \mathscr{P}_2(\mathbf{r})U_2(\mathbf{r}) + \mathscr{P}_3(\mathbf{r})e^{i(\mathbf{K}_3\cdot\mathbf{r}-\Delta t)} . \tag{4}$$

The components $\mathscr{P}_m(\mathbf{r})$ act as sources for the field amplitudes according to the slowly-varying Maxwell equations like

$$\frac{d\mathscr{E}_1}{dz} = \frac{iK}{2\epsilon}\mathscr{P}_1 , \tag{5}$$

where ϵ is the background permittivity of the medium, $K = \nu/c$, and z is the direction of propagation of the signal wave.

It is convenient to write the slowly-varying envelope $\mathscr{P}(\mathbf{r},t)$ of the polarization of the medium in terms of a slowly-varying complex susceptibility

$$\mathscr{P}(\mathbf{r},t) = \epsilon\chi(N(\mathbf{r},t))\mathscr{E}(\mathbf{r},t) , \tag{6}$$

where in the rate equation approximation $\chi(N(\mathbf{r},t))$ depends on $\mathbf{r}$ and t only through $N(\mathbf{r},t)$ [or in a homogeneously broadened two-level medium through the population difference $D(\mathbf{r},t)$]. At first the time dependence in this equation looks suspicious, since in the time domain a convolution is needed to account for nonzero dipole memory. However as far as χ is concerned, Eq. (6) *is in the frequency domain*; the slow time variations of

$\mathscr{E}$ and N are dc as far as χ is concerned. Alternatively, on the time scales that $\mathscr{E}(\mathbf{r},t)$ and N vary, χ effectively has a vanishing memory, which reduces the convolution to a simple product.

We expand N in terms of its first-order population pulsations as

$$N(\mathbf{r},t) = N_0 + n_{-1}e^{i\Delta t} + n_1 e^{-i\Delta t} , \tag{7}$$

where N_0 is the population difference for a constant field envelope ($\mathscr{E}_1 = \mathscr{E}_3 = 0$), n_{-1} is the population-pulsation contribution at the beat frequency Δ, and $n_1 = n_{-1}^*$. The population pulsations enter because the nonlinear carrier density attempts to respond to the superposition of the modes, which pulsates at the beat frequency $\Delta = \nu_2 - \nu_1$. Since we calculate the response only as a linear function of $\mathscr{E}_1$ and $\mathscr{E}_3$, the pulsations occur only at $\pm\Delta$; pulsations at $\pm 2\Delta$, for example, would require terms at least proportional to $\mathscr{E}_1^2$ or $\mathscr{E}_3^2$. The pulsations act as modulators (or like Raman "shifters"), putting sidebands onto the medium's response to the ν_2 mode. One of these sidebands falls precisely at ν_1, yielding a contribution to the sidemode gain coefficient. The other sideband influences the polarization of the sidemode placed symmetrically on the other side of the strong mode, namely at the frequency $\nu_2 + \nu_2 - \nu_1$.

The population-pulsation contributions $n_{\pm 1}$ are small since we assume that $\mathscr{E}_1$ and $\mathscr{E}_3$ are small and the $n_{\pm 1}$ are linearly proportional to these amplitudes. Hence we can expand the susceptibility $\chi(N)$ in the first-order Taylor series

$$\chi(N(\mathbf{r},t)) \simeq \chi(N_0) + \frac{\partial\chi(N)}{\partial N}\bigg|_{N_0} [n_{-1}e^{i\Delta t} + n_1 e^{-i\Delta t}] . \tag{8}$$

This expansion enables us to solve the intricate many-body multiwave mixing problem for sufficiently slowly-varying field envelopes. Specifically, once the single-mode $\chi(N_0)$ is known either analytically or numerically, we can calculate $\chi(N(\mathbf{r},t))$ including the signal and conjugate-wave components simply by calculating the population-pulsation components.

While this approach allows χ to be an arbitrary function of N, we have to specify the equation of motion for N itself, since it determines the dynamics of the medium. Substituting Eq. (1) into Eq. (7.27), we get

$$\dot{N} = \Lambda - \Gamma(N) + \frac{|\mathscr{E}(\mathbf{r},t)|^2}{2\hbar V}\sum_{\mathbf{k}} \text{Im}(\mu_{\mathbf{k}}^* \chi_{\mathbf{k}}) . \tag{9}$$

Further using Eqs. (2.116), (7.19), and (1), we find $\mathscr{P}(\mathbf{r},t)$ of Eq. (3) is given by

$$\mathscr{P}(\mathbf{r},t) = \mathscr{E}(\mathbf{r},t) \frac{1}{V} \sum_{\mathbf{k}} \mu_{\mathbf{k}}^{*} \chi_{\mathbf{k}}(t) , \tag{10}$$

which with Eq. (6) gives the susceptibility

$$\chi(N) = \frac{1}{\epsilon V} \sum_{\mathbf{k}} \mu_{\mathbf{k}}^{*} \chi_{\mathbf{k}}(t) . \tag{11}$$

Substituting this into Eq. (9), we have

$$\boxed{\dot{N} = \Lambda - \Gamma(N) + \frac{\epsilon}{2\hbar} |\mathscr{E}(\mathbf{r},t)|^2 \chi''} , \tag{12}$$

which is the same as Eq. (3.80) derived using free-carrier theory, but here we see it is valid for the full quasiequilibrium many-body theory. Substituting Eqs. (2), (7), and (8) into Eq. (10) and keeping only terms proportional to $e^{i\Delta t}$, we have

$$\left[i\Delta + \Gamma_1' - \mathscr{N}\nu \frac{\partial \chi''}{\partial N} \right] n_{-1} = \frac{\epsilon \chi''(N_0)}{\hbar} [\perp_1 \perp_2^{*} + \perp_2 \perp_3^{*}] ,$$

where for typographical simplicity, we write $\perp_1 = \mathscr{E}_1 \exp(i\mathbf{K}_1 \cdot \mathbf{r})$, $\perp_2 = \mathscr{E}_2 U_2(\mathbf{r})$, and $\perp_3 = \mathscr{E}_3 \exp(i\mathbf{K}_3 \cdot \mathbf{r})$. Solving for the population-pulsation component n_{-1}, we have

$$n_{-1} = \frac{\epsilon \chi''(N_0)}{\hbar} \frac{\perp_1 \perp_2^{*} + \perp_2 \perp_3^{*}}{i\Delta + \Gamma_1'} , \tag{13}$$

where the power-broadened relaxation rate Γ_1' is given by Eq. (7.45) already met in the theory of relaxation oscillations. For different relaxation and pumping schemes, this constant is changed appropriately.

Substituting Eq. (13) into Eq. (8), the result into Eq. (6), and projecting onto the signal propagation function $\exp(i\mathbf{K}_1 \cdot \mathbf{r})$, we have the signal polarization component

$$\mathscr{P}_1(z) = \epsilon\chi(N_0)\mathscr{E}_1(z) + \frac{\epsilon^2\chi''(N_0)}{\hbar}\frac{\partial\chi(N_0)}{\partial N_0}\frac{\mathscr{E}_1|\mathscr{E}_2|^2 + \mathscr{E}_2^2\mathscr{E}_3^* e^{2i\Delta Kz}}{i\Delta + \Gamma_1'}, \quad (14)$$

where the phase mismatch $2\Delta Kz$ is given by $(2\mathbf{K}_2 - \mathbf{K}_1 - \mathbf{K}_3)\cdot\mathbf{r}$. This corresponds to the three-wave mixing case of Fig. 8-1; the four-wave case with counter propagating pump waves has the smaller phase mismatch $2\Delta Kz = (\nu_3 - \nu_1)nz/c$, but requires an average over the pump-induced spatial holes. Inserting the polarization (14) into the slowly-varying Maxwell equation (5), we have the signal-wave propagation equation

$$\boxed{\frac{d\mathscr{E}_1}{dz} = \alpha_1\mathscr{E}_1 + \vartheta_1\mathscr{E}_3^* e^{2i\Delta Kz}}, \quad (15)$$

Interchanging the subscripts $_1$ and $_3$ (which implies replacing Δ by $-\Delta$), we have the conjugate-wave propagation equation

$$\frac{d\mathscr{E}_3^*}{dz} = \alpha_3^*\mathscr{E}_3^* + \vartheta_3^*\mathscr{E}_1 e^{-2i\Delta Kz}, \quad (16)$$

In this equations, the signal complex gain coefficient α_1 is given by

$$\boxed{\alpha_1 = \tfrac{1}{2} iK\chi(N_0) + \bar{\vartheta}_1}, \quad (17)$$

the conjugate-wave coupling coefficient ϑ_1 is given by

$$\boxed{\vartheta_1 = \frac{iK\epsilon\chi''(N_0)}{2\hbar}\frac{\partial\chi(N_0)}{\partial N_0}\frac{\mathscr{E}_2^2}{i\Delta + \Gamma_1'}}, \quad (18)$$

$\bar{\vartheta}_1$ is given by ϑ_1 with $\mathscr{E}_2^2$ replaced by $|\mathscr{E}_2|^2$, α_3 and ϑ_3 are given by α_1 and ϑ_1, respectively, with Δ replaced by $-\Delta$. Equations (15) through (18) and most of their derivation also apply to the homogeneously broadened two-level medium with N_0 replaced by the population difference D and with a different value for Γ_1'.

In terms of the linewidth enhancement factor

$$\alpha = \frac{\partial\chi'(N)/\partial N}{\partial\chi''(N)/\partial N},$$

the coupling coefficient ϑ_1 is given by

$$\vartheta_1 = \frac{K\epsilon\chi''(N_0)}{2\hbar} \frac{\partial\chi''(N_0)}{\partial N_0} \mathcal{E}_2^2 \frac{-1 + i\alpha}{i\Delta + \Gamma_1'} . \tag{19}$$

The reverse is also possible and interesting, namely that according to Eq. (18), the linewidth enhancement factor is given by the ratio

$$\alpha = \frac{Im\{\vartheta_1\}}{Re\{\vartheta_1\}}\bigg|_{\Delta=0}, \tag{20}$$

which is the ratio of the imaginary to the real parts of the mode coupling coefficient ϑ_1 provided $\mathcal{E}_2$ is real. Alternative interpretations are: 1) the ratio of the degenerate induced probe index and gain gratings, 2) the ratio of the real and imaginary parts of the degenerate population-pulsation contributions, and 3) the ratio (3.90) of the real to imaginary parts of the single-mode $\chi^{(3)}$ of Eq. (3.86).

For beat frequencies Δ small compared to intraband and dipole relaxation rates and a pump-wave intensity small enough to avoid spectral hole burning, Eqs. (17) and (18) give the nondegenerate semiclassical multiwave mixing coefficients for an *arbitrary* semiconductor susceptibility, which can therefore include all the standard many-body contributions.

Asymmetric Sidemode Gain

We illustrate the gain formula (17) in Fig. 8-2. As $|\Delta|$ increases, the population pulsation term ϑ_1 goes to zero due to falloff of the population-pulsation complex Lorentzian $1/(\Gamma_1' + i\Delta)$. This causes α_1 to approach the background single-mode value. However for smaller $|\Delta|$, we see that the sidemode gain is greater below the pump wave frequency than above, and is, in fact, bigger than the background single-mode gain. This behavior results from the real part of ϑ_1 and can lead to build up of an appropriately tuned sidemode.

We can use our formulas to explain why the real part of ϑ_1 is larger for $\Delta > 0$ (probe mode tuned below pump mode) than it is for $\Delta < 0$. Specifically, the contribution to gain goes like

$$Re\{\vartheta_1\} \propto Re\left\{\frac{i\chi''}{\Gamma_1' + i\Delta}\frac{\partial\chi(N_0)}{\partial N_0}\right\} \propto -\Delta\frac{\partial\chi'(N_0)}{\partial N_0} - \Gamma_1'\frac{\partial\chi''(N_0)}{\partial N_0} . \tag{21}$$

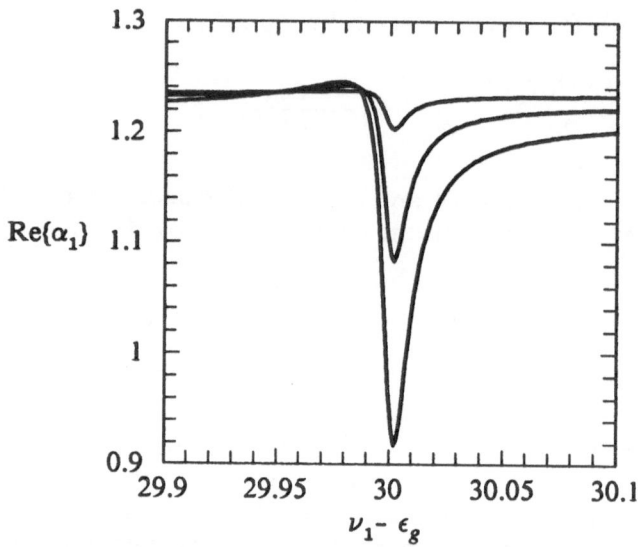

Fig. 8-2. Real part of probe absorption coefficient versus probe detuning above the band gap for a pump-wave detuning $\nu_2 = \varepsilon_g + 30 \ meV$, Rabi flopping frequencies $\hbar\mathcal{R}_0 = .2, .45,$ and $.7 \ meV$ (in order of increasing modulation), a temperature $T = 300$, $\hbar\gamma = 4 \ meV$, $\hbar\gamma_{nr} = .005 \ meV$, $\hbar B = 7{\times}10^{-4} \ meV$, $N = 3{\times}10^{18} \ cm^{-3}$, and the carrier masses $m_e = 1.167m$, $m_h = 6.669m$.

This yields an enhancement for $\Delta > 0$, i.e., for probe tuned below pump mode provided $\partial\chi'(N_0)/\partial N_0$ is negative at the pump mode frequency. This is, in fact, the case, since for GaAs although g_h is a relatively insensitive function of energy, g_e peaks above the electron chemical potential (see Fig. 7-7). This is a little above the gain region, so that the integrated weighted value of the complex lineshape function $\mathcal{D}$ has a negative imaginary part, which gives $\partial\chi'(N)/\partial N$ a negative value. As Eq. (7.48) shows, g_e is $\partial f_e/\partial N$, so that the origin of the asymmetry is the rapid change of the electron Fermi-Dirac distribution as a function of the carrier density that takes place at the electron quasiequilibrium chemical potential. Note that the two-level gain coefficient can display asymmetries favoring probe tuning on either side of the pump wave, since the pump wave can be tuned above or below the atomic resonance.

The extrema of Eq. (17) are determined by those of $Re\{\vartheta_1\}$, which is given by

$$Re\{\vartheta_1\} \propto \frac{\alpha\Delta - \Gamma_1'}{\Delta^2 + \Gamma_1'^2} \; .$$

Equating the derivative of this with respect to Δ to zero, we find the quadratic equation

$$\Delta^2 - 2\Gamma_1'\Delta/\alpha - \Gamma_1'^2 = 0 \; ,$$

which has the roots

$$\Delta = \frac{\Gamma_1'}{\alpha}[1 \pm \sqrt{\alpha^2 + 1}] \; . \tag{22}$$

To a good approximation, this gives a linear dependence on pump-field intensity $|\mathcal{E}_2|^2$ due to Γ_1'. Some variation from linearity exists unless N is held constant as the intensity is varied.

Solution of the Coupled-Mode Equations

To solve the coupled-mode Eqs. (15) and (16), we transform out the exponentials using the variables

$$\mathcal{E}_1'(z) = \mathcal{E}_1(z)e^{-i\Delta Kz} \tag{23}$$
$$\mathcal{E}_3'^*(z) = \mathcal{E}_3^*(z)e^{i\Delta Kz} \tag{24}$$

This gives the propagation equations

$$\frac{d\mathcal{E}_1'}{dz} = \alpha_1'\mathcal{E}_1' + \vartheta_1\mathcal{E}_3'^* \; , \tag{25}$$

$$\frac{d\mathcal{E}_3'^*}{dz} = \alpha_3'^*\mathcal{E}_3'^* + \vartheta_3^*\mathcal{E}_1' \; , \tag{26}$$

where

$$\alpha_m' = \alpha_m + i\Delta K \; . \tag{27}$$

Taking the Laplace transforms of these equations, we find the matrix equation

$$
\begin{bmatrix} s-\alpha_1' & -\vartheta_1 \\ -\vartheta_3^* & s-\alpha_3'^* \end{bmatrix} \begin{bmatrix} \mathcal{E}_1'(s) \\ \mathcal{E}_3^{*}(s) \end{bmatrix} = \begin{bmatrix} \mathcal{E}_1(0) \\ \mathcal{E}_3^{*}(0) \end{bmatrix},
\tag{28}
$$

where s is the Laplace transform variable. Using the method of determinants, we have

$$
\begin{aligned}
\mathcal{E}_1'(s) &= \frac{\mathcal{E}_1(0)(s-\alpha_3'^*) + \vartheta_1\mathcal{E}_3^{*}(0)}{s^2 - (\alpha_1'+\alpha_3'^*)s + \alpha_1'\alpha_3'^* - \vartheta_1\vartheta_3^*} \\
&= \frac{\mathcal{E}_1(0)(s-\alpha_3'^*) + \vartheta_1\mathcal{E}_3^{*}(0)}{(s-s_+)(s-s_-)} = \frac{\mathcal{E}_1(0)}{s-s_+} + \frac{\mathcal{E}_1(0)(s_--\alpha_3'^*) + \vartheta_1\mathcal{E}_3^{*}(0)}{(s-s_+)(s-s_-)} \\
&= \frac{\mathcal{E}_1(0)}{s-s_+} + \frac{\mathcal{E}_1(0)(s_--\alpha_3'^*) + \vartheta_1\mathcal{E}_3^{*}(0)}{s_+-s_-}\left[\frac{1}{s-s_+} - \frac{1}{s-s_-}\right],
\end{aligned}
\tag{29}
$$

where

$$
s_\pm = \tfrac{1}{2}(\alpha_1' + \alpha_3'^*) \pm \sqrt{\alpha^2 + \vartheta_1\vartheta_3^*} \equiv a \pm w
\tag{30}
$$

$$
\alpha = \tfrac{1}{2}(\alpha_1' - \alpha_3'^*).
\tag{31}
$$

For the α_m' given by Eq. (27), $a = \tfrac{1}{2}(\alpha_1 + \alpha_3^*)$ and $\alpha = \tfrac{1}{2}(\alpha_1 - \alpha_3^*) + iK\Delta$. Taking the inverse Laplace transform of Eq. (29), we find

$$
\begin{aligned}
\mathcal{E}_1'(z) &= e^{-az}[\mathcal{E}_1(0)e^{wz} + (-\mathcal{E}_1(0)(\alpha + w) + \vartheta_1\mathcal{E}_3^{*}(0))\sinh wz/w] \\
&= e^{-az}[\mathcal{E}_1(0)\cosh wz + (-\alpha\mathcal{E}_1(0) + \vartheta_1\mathcal{E}_3^{*}(0))\sinh wz/w].
\end{aligned}
\tag{32}
$$

We can derive $\mathcal{E}_3^{*}(z)$ similarly or simply note that it is given by Eq. (32) by interchange of 1 and 3 and taking the complex conjugate (this leaves a and w unchanged and changes the sign of α). We find

$$
\mathcal{E}_3^{*}(z) = e^{-az}[\mathcal{E}_3^{*}(0)\cosh wz + (\alpha\mathcal{E}_3^{*}(0) + \vartheta_3^*\mathcal{E}_1(0))\sinh wz/w],
\tag{33}
$$

The amplitudes $\mathcal{E}_1(z)$ and $\mathcal{E}_3^{*}(z)$ are found using Eqs. (23) and (24).

In summary, for short-T_2 two-level media and quasiequilibrium semiconductor media, this section derives multiwave coefficients that inherit all of the generality of the single-mode models. The semiconductor models can include the Coulomb enhancement and bandgap renormalization predicted by the quasiequilibrium many-body theory of Chaps. 4 - 6. This surprising generality is based on the fact that the quasiequilibrium semi-conductor susceptibility depends explicitly on the total carrier density and only implicitly through this density on the slowly-varying field envelope. Hence in the quasiequilibrium limit, the dynamics of the problem are completely specified by the evolution of the total carrier density, which is relatively simple to treat even for the quasiequilibrium many-body theories. We have used the general multiwave coefficients to interpret the linewidth enhancement factor as the ratio of the real and imaginary parts of the degenerate coupling coefficient, and have explained the observed asymmetry in the probe gain spectra in terms of the rapid change of the electron Fermi-Dirac distribution as a function of the carrier density that takes place above the gain region.

Multiwave mixing is interesting in its own right as a method of nonlinear spectroscopy to study properties of matter, a way to achieve phase conjugation, and, with quantization of the sidemodes, a way to generate squeezed states. The main application of multiwave mixing in this book is to help answer the question as to when the sidemodes can build up.

For this, we choose $\mathcal{E}_m = \mathcal{E}_m(t)$ rather than $\mathcal{E}_m(z)$, assuming thereby that the amplitude is described by a mode with a uniform value along the z axis. In general $\mathcal{E}_m$ is a function of both z and t, but this generality complicates the analysis substantially and is not needed for a sufficiently high-Q cavity. The same equations of motion (15) and (20) apply, but with d/dz replaced by $(n/c)d/dt$ and with the addition of cavity-loss terms $-(\nu/2Q_m)\mathcal{E}_m$. The sidemodes fail to grow, i.e., mode 2 oscillates stably alone, provided the eigenvalues (28) have negative real parts. Note that the situation is more complicated than looking at the real parts of gain coefficients α_1 and α_3 alone, since the sidemodes work together in bringing about an instability.

Since we have explicit values for these eigenvalues for the running-wave case (unidirectional ring), we are in a position to analyze the stability in the quasiequilibrium limit. As Fig. 3-18 reveals, a sidemode tuned below the main mode can have greater gain than the main mode, and therefore could in principle grow. We postpone further analysis until the end of the next section, in which we consider modes spaced sufficiently widely apart that the quasiequilibrium approximation is inadequate.

8-2. Short-Cavity Sidemode Interactions

For long semiconductor lasers obtainable with external cavities, the modal beat frequencies can be sufficiently small that the quasiequilibrium approximation is valid and the theory of Sec. 8-1 can be used to describe weak sidemode interactions. For these interactions, the total carrier density pulsates at the modal beat frequencies, thereby forcing the quasiequilibrium Fermi-Dirac distributions to effectively "breathe" in and out, driven by the field variations through the total carrier density and therefore through the carrier chemical potentials.

For short semiconductor lasers, the modes are spaced correspondingly farther apart, so that the intermode beat frequencies can approach the intraband carrier relaxation rates and the dipole relaxation rate. The total carrier density N has a bandwidth given essentially by Γ_1', which is on the order of 200 psec or more. Hence N cannot follow these relatively large beat frequencies. Nevertheless, the carrier probabilities $n_{\alpha k}$ can in principle follow such variations. In addition, we can relax the requirement that the strong-mode Rabi frequency be small compared to the fast relaxation rate constants.

Even before reaching the departure from quasiequilibrium made possible by the finite carrier-carrier scattering rates, we may find deviations from the simple $\chi(N)$ theory due to "Pauli blocking" of the injection current in the gain region. By its very nature the gain region is characterized by an abundance of occupied states, which are not available to incoming carriers. Hence as Eq. (3.20) shows, instead of evenly replacing carriers that recombine by stimulated emission, the pump term $\Lambda_{\alpha k}$ injects carriers preferentially above the gain region. Carrier-carrier scattering drives the resulting carrier distributions towards Fermi-Dirac distributions, but with temperatures higher than the lattice temperature. In turn, carrier-phonon scattering attempts to equilibrate the carrier and lattice temperatures, but typically requires times on the order of many picoseconds to do so. Hence if the mode beat frequencies or Rabi frequencies approach the carrier-phonon scattering rates, χ no longer depends on N alone; it depends on induced temperature differences between the lattice and carrier distributions as well. It remains to be seen as to whether a theory similar to that in this section can help in treating this effect, but there is little doubt that such a temperature relaxation process will change both short-cavity mode-competition phenomena and strong single-mode saturation in significant ways.

To treat many-body effects with such large beat frequencies, we need to carry out extensive numerical calculations. To gain some analytic insight, we pursue a less ambitious course that replaces the full Boltzmann equation of Chap. 4 with the exponential relaxation approximation of Eq. (3.22). Nevertheless, the analysis is more complicated than that in Sec. 8-1, so the reader might want to skip the derivation on first reading.

We choose the individual intraband electron and hole decay rates γ_{ek} and γ_{hk} to be those predicted by the many-body Boltzmann analysis of Chap. 4, although in Eq. (3.22) specifically a constant value must be used to preserve the zero sum of that equation. We find that since the dipole decay constant

$$\gamma_{\mathbf{k}} \equiv \frac{1}{T_2(\mathbf{k})} = \tfrac{1}{2}[\gamma_{e\mathbf{k}} + \gamma_{h\mathbf{k}}] \,, \tag{34}$$

has essentially the same size as the level decay constants $\gamma_{\alpha\mathbf{k}}$, spectral hole burning and population pulsations each have roughly equal influence on the mode coupling. In the limit of beat frequencies large compared to the interband relaxation times, we find that the coefficients have forms similar to those for inhomogeneously broadened two-level media subjected to an arbitrarily intense wave. The population pulsations sharpen the spectral holes burned by the strong mode and double their depth. This suppresses modes close to the oscillating mode, while encouraging the oscillation of modes further away.

We consider a semiconductor medium subjected to an arbitrarily intense mode and one or two nonsaturating sidemodes. We assume that the saturating mode intensity is constant throughout the interaction region and ignore transverse variations. As in Sec. 8-1, we label the sidemodes by the indices 1 and 3 and the saturator mode by 2. We consider a multimode plane-wave electric field of the form

$$E(z,t) = \frac{1}{2} \sum_m \mathcal{E}_m(t)\, e^{i(K_m z - \nu_m t)} + \text{c.c.}, \tag{35}$$

where the mode amplitudes $\mathcal{E}_m(t)$ are in general complex and K_m are the wave propagation vectors. For simplicity we take mode functions appropriate for a unidirectional ring laser. The mode index equals 1, 2, or 3. The field (35) induces the complex polarization

$$P(z,t) = \frac{1}{2} \sum_m \mathcal{P}_m(t)\, e^{i(K_m z - \nu_m t)} + \text{c.c.}, \tag{36}$$

where $\mathcal{P}_m(t)$ is a complex polarization coefficient that yields index and absorption/gain characteristics for the sidemode and saturator waves. The polarization $P(z,t)$ in general has other components, but we are interested only in those given by Eq. (36). In particular, strong wave interactions induce components not only at the frequencies ν_1, ν_2, and ν_3 but at $\nu_1 \pm j(\nu_2 - \nu_1)$ as well, where j is an integer. In this linear sidemode approximation, other pairs of modes placed symmetrically about ν_2 do not interact with the ν_1 ν_3 pair, and hence the present theory describes the onset of general multimode operation.

The problem reduces to determining the sidemode polarization $\mathcal{P}_1(t)$, which drives the sidemode amplitude $\mathcal{E}_n$ according to the equation of motion

$$\dot{\mathcal{E}}_m = \frac{i\nu_m}{2\epsilon} \mathcal{P}_m - \left[\frac{\nu}{2Q_m} - i(\Omega_m - \nu_m)\right]\mathcal{E}_m , \tag{37}$$

where ϵ is the permittivity of host medium, Q_m is the cavity Q for the mth mode, and Ω_m is the passive cavity frequency of the mth mode.

We use Eqs. (3.14) and (3.16) for the microscopic polarization $\dot{p}_{\mathbf{k}}$. We use Eq. (3.15) with (3.22) for $\dot{n}_\alpha$, that is,

$$\dot{n}_{\alpha\mathbf{k}} = \lambda_\alpha(\mathbf{k}) - \gamma_{nr}n_{\alpha\mathbf{k}} - B_{\mathbf{k}}n_{e\mathbf{k}}n_{h\mathbf{k}} - \gamma_\alpha(\mathbf{k}_2)[n_{\alpha\mathbf{k}} - f_{\alpha\mathbf{k}}]$$
$$+ \frac{i}{\hbar}(\mu_{\mathbf{k}}p_{\mathbf{k}}^* - \mu_{\mathbf{k}}^* p_{\mathbf{k}})E(z,t) . \tag{38}$$

Section 8-1 shows how to calculate the response of a medium to beat frequencies on the order of the interband relaxation rates or smaller. In this section, we suppose that the beat frequencies are large compared to the interband relaxation rates. As such, we suppose that the total carrier density N and hence the quasiequilibrium Fermi-Dirac distributions cannot follow the intermode beat frequencies and can therefore be approximated by their steady-state values in the presence of the pump field $\mathcal{E}_2$ alone. To find the response for such large beat frequencies, we Fourier analyze the polarization component $p_{\mathbf{k}}$, the number probabilities $n_{\alpha\mathbf{k}}$, and the probability difference $D_{\mathbf{k}} = n_{e\mathbf{k}} + n_{h\mathbf{k}} - 1$. Specifically we write

$$p_{\mathbf{k}} = \exp[i(K_1 z - \nu_1 t)] \sum_m p_{\mathbf{k},m+1} \exp\{im[(K_2 - K_1)z - \Delta t]\} , \tag{39}$$

$$n_{\alpha\mathbf{k}} = \sum_j n_{\alpha\mathbf{k}j} \exp\{ij[(K_2 - K_1)z - \Delta t]\} , \quad \alpha = e, h. \tag{40}$$

$$D_{\mathbf{k}}(z,t) \equiv n_{e\mathbf{k}}(z,t) + n_{h\mathbf{k}}(z,t) - 1$$

$$= \sum_j d_{\mathbf{k}j}(z,t) \exp\{ij[(K_2 - K_1)z - \Delta t]\} , \tag{41}$$

A calculation that allows for the beat frequencies to be as small as the interband relaxation rates, Fourier analyzes the total carrier density N and the quasiequilbrium Fermi-Dirac functions $f_{\alpha\mathbf{k}}$ as well.

We substitute the expansions (39) through (41) into the equations of motion (3.14) and (38) and identify coefficients of common exponential frequency factors. In the approximation that $\mathcal{E}_1$ and $\mathcal{E}_3$ do not saturate, only $p_{\mathbf{k}1}$, $p_{\mathbf{k}2}$, and $p_{\mathbf{k}3}$ occur in the polarization expansion (39), and only $j = 0, \pm 1$ appear in the probability expansions (40) and (41). Calculating the coefficient of $\exp(iK_z z - i\nu_2 t)$ for the saturating mode by neglecting the nonsaturating sidemode fields, we find

$$-i\nu_2 p_{\mathbf{k}2} = -(i\omega + \gamma)p_{\mathbf{k}2} - i(\mu_{\mathbf{k}}\mathcal{E}_2/2\hbar)d_{\mathbf{k}0}$$

which gives

$$p_{\mathbf{k}2} = -i(\mu_{\mathbf{k}}/2\hbar)\mathcal{E}_2 \mathcal{D}_{\mathbf{k}2} d_{\mathbf{k}0} , \tag{42}$$

where $\mathcal{D}_{\mathbf{k}2}$ is the $m = 2$ case of the complex Lorentzian

$$\mathcal{D}_{m\mathbf{k}} = \frac{1}{\gamma + i(\omega_{\mathbf{k}} - \nu_m)} . \tag{43}$$

We calculate the dc probability Fourier component $n_{e\mathbf{k}0}$ and $n_{h\mathbf{k}0}$ saturated by the saturator wave $\mathcal{E}_2$ alone. To simplify the equations of motion for $n_{\alpha\mathbf{k}}$, we note that the γ_α term dominates the other decay processes by several orders of magnitude, causing $n_{\alpha\mathbf{k}}$ to be nearly equal to $f_{\alpha\mathbf{k}}$. Hence we approximate Eq. (38) by

$$\dot{n}_{\alpha\mathbf{k}} = \lambda_{\alpha\mathbf{k}} + \gamma_\alpha f_{\alpha\mathbf{k}} - \gamma_\alpha' n_{\alpha\mathbf{k}} + \frac{i}{\hbar}(\mu_{\mathbf{k}} p_{\mathbf{k}}^* - \mu_{\mathbf{k}}^* p_{\mathbf{k}})E(z,t) , \tag{44}$$

where the decay constant γ_α' is nearly equal to γ_α and is given by

$$\gamma_\alpha' \simeq \gamma_\alpha + \gamma_{nr} , \alpha = e, h . \tag{45}$$

Here we have neglected the small $B_{\mathbf{k}} n_{e\mathbf{k}} n_{h\mathbf{k}}$ radiative recombination term, although with some small N dependence, its effects can be lumped into γ_α'. With neglect of these small contributions, Eqs. (44) and (45) have the same form as the inhomogeneously-broadened two-level system and can be solved in similar ways [see Meystre and Sargent (1991), Chap. 8]. The major difference is the closure relation (3.17), which determines the carrier number density N and introduces important carrier-density pulsation contributions. Since the γ_e and γ_h contributions cancel out in the N equation, the γ_{nr} and $B_{\mathbf{k}}$ cannot be neglected for this equation, and in fact they play crucial roles.

Substituting Eqs. (35), (39), (40) and (42) into (44) and making the rotating-wave approximation, we have for the $j = 0$ term

$$0 = \lambda_{\alpha k} + \gamma_{\alpha k} f_{\alpha k 0} - \gamma_{\alpha k}' n_{\alpha k 0} + [i(\mu_k \mathcal{E}_2/2\hbar)p_{k2}^* + \text{c.c.}],$$

which gives

$$n_{\alpha k 0} = \frac{\lambda_{\alpha k} + \gamma_\alpha f_{\alpha k 0}}{\gamma_\alpha'} - \frac{|\mu_k \mathcal{E}_2/\hbar|^2}{2\gamma\gamma_\alpha'} \mathcal{L}_2 d_{k0}. \tag{46}$$

Here $\mathcal{L}_{2k}$ is the saturator dimensionless Lorentzian

$$\mathcal{L}_{2k} = \frac{\gamma^2}{\gamma^2 + (\omega_k - \nu_2)^2}. \tag{47}$$

The $\mathcal{E}_1$ and $\mathcal{E}_3$ contributions are ignored, since we assume the sidemodes do not saturate. Using Eq. (46) for $\alpha = e$ and h, we find

$$d_{k0} = \frac{\lambda_{ek} + \gamma_e f_{ek0}}{\gamma_e'} + \frac{\lambda_{hk} + \gamma_h f_{hk0}}{\gamma_h'} - I_2 \mathcal{L}_{2k} d_{k0} - 1$$

$$= \frac{\dfrac{\lambda_{ek} + \gamma_e f_{ek0}}{\gamma_e'} + \dfrac{\lambda_{hk} + \gamma_h f_{hk0}}{\gamma_h'} - 1}{1 + I_2 \mathcal{L}_{2k}} \simeq \frac{f_{ek0} + f_{hk0} - 1}{1 + I_2 \mathcal{L}_{2k}}, \tag{48}$$

where the dimensionless saturator intensity

$$I_2 = \mathcal{R}_0^2 T_{1f} T_2, \tag{49}$$

$\mathcal{R}_0$ is the Rabi frequency $|\mu_k \mathcal{E}_2/\hbar|$, T_2 is the dipole decay time ($\equiv \gamma^{-1}$) given by Eq. (34), and T_{1f} is the population-difference decay time

$$T_{1f} = \frac{1}{2}\left[\frac{1}{\gamma_e'} + \frac{1}{\gamma_h'}\right], \tag{50}$$

The intensity (49) is usually less than unity, since the γ_α are larger than the Rabi frequency for typical laser operation. The semiconductor would breakdown for average internal cw field intensities I greater than about 6 MW/cm². Writing I as

$$I = \tfrac{1}{2}c\epsilon_0 E^2 = \tfrac{1}{2}c\epsilon_0 |\hbar \mathscr{R}_0/\mu_k|^2 \; , \tag{51}$$

we find the Rabi frequency

$$\mathscr{R}_0 = |\mu_k/\hbar| \sqrt{2I/c\epsilon_0} \; . \tag{52}$$

We express our frequencies in *meV*. For this, note that $\hbar_{m\,e\,V} = 1000\hbar/e$ and $\mu_k = er_{cv}$, where e is the magnitude of the electron charge in coulombs. This gives

$$\hbar_{m\,e\,V}\,\mathscr{R}_0 = 2.75 \times 10^5 \times r_{cv}\sqrt{I} \; . \tag{53}$$

For $r_{cv} = 3 \times 10^{-8}$ *cm* and $I = 6$ MW/cm², this gives $\hbar_{m\,e\,V}\,\mathscr{R}_0 \simeq 2$ *meV*. We see shortly that for laser excitations, the γ_α are several times larger than this. On the other hand, acoustic-phonon-carrier scattering rates are substantantially smaller than this, so that the carrier distributions are likely to have different temperatures from the lattice.

The coefficient of $\exp(iK_1 z - i\nu_1 t)$ for the sidemode wave ($m = 0$ term in Eq. (39)) includes an extra term $\mathscr{E}_2 d_{k,-1}$

$$-i\nu_1 p_{k1} = -(i\omega + \gamma)p_{k1} - i(\mu_k/2\hbar)[\mathscr{E}_1 d_{k0} + \mathscr{E}_2 d_{k,-1}] \; ,$$

giving

$$p_{k1} = -i(\mu_k/2\hbar)\mathscr{D}_{1k}[\mathscr{E}_1 d_{k0} + \mathscr{E}_2 d_{k,-1}] \; , \tag{54}$$

The $\mathscr{E}_2 d_{k,-1}$ term gives the scattering of $\mathscr{E}_2$ into the $\mathscr{E}_1$ mode by the probability-pulsation component $d_{k,-1}$. The polarization component p_{k0} remains zero when only d_{k0} and $d_{k,\pm 1}$ are nonzero, since it is proportional to $\mathscr{E}_1 d_{k,-1}$, which involves at least two $\mathscr{E}_1$'s.

Similarly the component p_{k3} has the nonzero value

$$p_{k3} = -i(\mu_k/2\hbar)\mathscr{D}_{3k}[\mathscr{E}_3 d_{k0} + \mathscr{E}_2 d_{k1}] \; , \tag{55}$$

while $p_{k,j>3}$ vanishes since $d_{k,j>1}$ would be involved.

Proceeding with the probability-pulsation terms $n_{\alpha k,-1}$ and $d_{k,-1}$, we substitute the expansions (35) and (38) through (40) into Eq. (44) (ignoring the small time dependences introduced by the $\delta\varepsilon_g$ terms in the $\mathscr{D}_m$'s) and identify coefficients of $e^{i\Delta t}$ to find

$$i\Delta n_{\alpha k,-1} = \gamma_\alpha f_{\alpha k,-1} - \gamma_\alpha' n_{\alpha k,-1}$$
$$+ i(\mu_k/2\hbar)[\mathcal{E}_1 p_{k2}^* + \mathcal{E}_2 p_{k3}^* - \mathcal{E}_2^* p_{k1} - \mathcal{E}_3^* p_{k2}].$$

For $|\Delta|$ large compared to the interband relaxation rates, the total carrier density sees the average field given by $\mathcal{E}_2$ alone and hence the N_{-1} contributions can be neglected. In view of Eq. (45), this means that $f_{\alpha k,-1}$ is negligible. Solving for $d_{k,-1} = n_{ek,-1} + n_{hk,-1}$ in this limit, we have

$$d_{k,-1} = i(\mu_k/\hbar)T_{1f}\mathcal{F}(\Delta)[\mathcal{E}_1 p_{k2}^* + \mathcal{E}_2 p_{k3}^* - \mathcal{E}_2^* p_{k1} - \mathcal{E}_3^* p_{k2}], \quad (56)$$

where the high-frequency dimensionless population pulsation factor $\mathcal{F}(\Delta)$ is defined by

$$\mathcal{F}(\Delta) = \frac{1}{2T_{1f}}\left[\frac{1}{\gamma_e' + i\Delta} + \frac{1}{\gamma_h' + i\Delta}\right]. \quad (57)$$

Here $\mathcal{F}(\Delta)$ is normalized such that $\mathcal{F}(0) = 1$. Substituting the polarizations (42), (54), and (55) into Eq. (56) and using the fact that $d_{k1} = d_{k,-1}^*$, we have

$$d_{k,-1} = -\frac{d_{k0}|\mu_k|^2 \frac{T_{1f}}{2\hbar^2}\mathcal{F}(\Delta)[(\mathcal{D}_1 + \mathcal{D}_2^*)\mathcal{E}_1\mathcal{E}_2^* + (\mathcal{D}_2 + \mathcal{D}_3^*)\mathcal{E}_2\mathcal{E}_3^*]}{1 + \tfrac{1}{2}I_2\mathcal{F}(\Delta)\gamma(\mathcal{D}_1 + \mathcal{D}_3^*)}. \quad (58)$$

Our calculation is self-consistent, since only d_{k0} and $d_{k,\pm1}$ can obtain nonzero values from p_{k1}, p_{k2}, p_{k3}, and vice versa.

Coupled–Mode Equations

Substituting Eq. (58) into (54), setting $\mathcal{P}_1 = 2V^{-1}\Sigma_k \mu_k^* p_{k1}$, and using Eq. (48), we find the coupled sidemode equation of motion

$$\dot{\mathcal{E}}_1 = [\alpha_1 - \nu/2Q_1 - i(\Omega_1-\nu_1)]\mathcal{E}_1 + \vartheta_1\mathcal{E}_3^*, \quad (59)$$

Similarly the coupled sidemode amplitude $\mathcal{E}_3$ obeys the equation

$$\dot{\mathcal{E}}_3^* = [\alpha_3^* - \nu/2Q_3 + i(\Omega_3-\nu_3)]^*\mathcal{E}_3^* + \vartheta_3^*\mathcal{E}_1, \quad (60)$$

where the coefficients α_3 and ϑ_3 are given by α_1 and ϑ_1, respectively, by interchanging the subscripts $_1$ and $_3$ (note this implies replacing Δ by $-\Delta$).

These equations are very similar to those (15) and (16) of Sec. 8-1, but describe time-varying amplitudes that are assumed to be uniform throughout a laser cavity, while Eqs. (15) and (16) apply to steady-state propagation.

For the large-Δ limit, i.e., $|\Delta|$ comparable to γ_α and γ, we use $d_{\mathbf{k},-1}$ given by Eq. (58), which gives

$$\alpha_1 = \frac{\nu|\mu_\mathbf{k}|^2}{2\hbar\epsilon V}\sum_\mathbf{k}\mathcal{D}_1 d_{\mathbf{k}0}\left[1 - \frac{\frac{1}{2}I_2\mathcal{F}(\Delta)\gamma(\mathcal{D}_1 + \mathcal{D}_2^*)}{1 + \frac{1}{2}I_2\mathcal{F}(\Delta)\gamma(\mathcal{D}_1 + \mathcal{D}_3^*)}\right] \qquad (61)$$

$$\vartheta_1 = -\frac{\nu_1|\mu_\mathbf{k}|^2}{2\hbar\epsilon V}\left(\frac{\mu_\mathbf{k}\mathcal{E}_2}{\hbar}\right)^2\sum_\mathbf{k}\frac{\frac{1}{2}\mathcal{D}_1 d_{\mathbf{k}0}T_{1f}\mathcal{F}(\Delta)(\mathcal{D}_2 + \mathcal{D}_3^*)}{1 + \frac{1}{2}I_2\mathcal{F}(\Delta)\gamma(\mathcal{D}_1 + \mathcal{D}_3^*)} . \qquad (62)$$

Except for the appearance of $\Sigma_\mathbf{k} d_{\mathbf{k}0}$ instead of $\int d\omega\, \mathcal{W}(\omega)/(1+I_2\mathcal{L}_2)$, these formulas have the same form as those found for inhomogeneously broadened two-level systems [see Chaps. 8 and 9 of Meystre and Sargent (1991)], and can be evaluated in the inhomogeneously-broadened limit using contour integration. Since this limit reveals some of the features of the large-Δ case with substantially less computational effort, we give those formulas and illustrate them numerically as well as the numerical evaluation of Eqs. (61) and (62) in Sec. 8-4. Specifically, from Meystre and Sargent Eqs. (8.28), (5.52), and (8.48), we have

$$\alpha_1 = \alpha_{inc} + \alpha_{coh} , \qquad (63)$$

where the incoherent and coherent contributions

$$\alpha_{inc} = \alpha_0'(\nu_1)\left[1 - \frac{\gamma^2 I_2}{\gamma'(\upsilon + \gamma')}\right] \qquad (64)$$

$$\alpha_{coh} = -\frac{1}{2}\alpha_0'(\nu_2)\gamma I_2\mathcal{F}\frac{\gamma + \upsilon}{\gamma'^2 - \beta^2}\left[\frac{(\gamma'+\gamma)(\gamma'-\upsilon)}{\gamma'(\gamma'+\upsilon)} - \frac{(\beta+\gamma)(\beta-\upsilon)}{\beta(\beta+\upsilon)}\right] , \qquad (65)$$

the complex frequency $\upsilon = \gamma + i\Delta$, and the complex frequency factor

$$\beta^2 = \upsilon(\upsilon + \gamma I_2\mathcal{F}) . \qquad (66)$$

Similarly the coupling coefficient ϑ_1 can be evaluated as

$$\vartheta_1 = -\frac{\alpha_0'\gamma^2 I_2 \mathscr{F}(\Delta)(\gamma + \upsilon)}{2\gamma'\beta(\gamma' + \beta)}.$$ (67)

We investigated the sidemode gain evaluating α_1 and ϑ_1 both numerically and using the inhomogeneously broadened limit and found that population pulsations play just as important a role as spectral hole burning for mode spacings comparable to the intraband relaxation rates. For the carrier-carrier relaxation rates of Binder *et al.*, we predict that sidemode gain is smaller than the main mode gain, leading to single-mode operation. However for somewhat smaller intraband relaxation rates, sidemode gain was readily found that exceeds the single-mode gain, which would encourage multimode operation to occur. Furthermore the effects due to carrier temperature variations due to beat and Rabi frequencies larger than the acoustic-phonon-carrier scattering rates could well modify the mode competition and allow multimode operation to occur.

For example, Fig. 8-3 plots α_1, α_{inc}, and α_2 (dot-dashed line) versus probe detuning $\hbar\nu_1 - \varepsilon_g$ for the decay constants $\hbar\gamma_e = \hbar\gamma_h = 3\ meV$. We see that for these values and models, sidemodes could oscillate quite easily. Comparing α_{inc} with α_1, we see that the effects of population pulsations are just as big as the power-broadened spectral hole burned by the strong mode. They cause modes within γ_α of the strong mode to be suppressed, while modes further away have increased gain. In the linear sidemode approximation used here, population pulsations redistribute the gain, and do not change the value obtained from integrating over ν_1. The α_2 curves show that the strong mode itself does not perceive the hole it burns.

At the onset of multimode laser operation, population pulsations induced by a weak sidemode at frequency ν_1 and a strong mode (at frequency ν_2) scatter energy between those modes and as well as between the strong mode and the weak mode (at frequency ν_3) symmetrically placed on the other side of the strong mode. This second scattering couples the weak modes as described by the coupled-mode equations (59) and (60). Figure 8-4 plots the real parts of α_{inc} (dashed), α_2 (dot-dashed), $\alpha_1 - \vartheta_1$, α_1, and $\alpha_1 + \vartheta_1$ (solid in order of increasing dip depth) versus probe detuning $\nu_1 - \varepsilon_g$ in meV for the parameters in Fig. 8-3. The $\alpha_1 \pm \vartheta_1$ values are the approximate eigenvalues for dual sidemode buildup described by Eqs. (59) and (60). In particular the $\alpha_1 + \vartheta_1$ value shows increased sidemode suppression in the immediate vicinity of ν_2, and increased dual sidemode gain outside of this region.

The onset of multimode operation in semiconductor laser diodes is determined by a number of factors, such as transverse field variations, saturation by the fluorescence of modes below threshold, and temperature variations, in addition to the mechanisms dealt with in this section. What we see here is that according to a simple carrier-carrier relaxation model, pop-

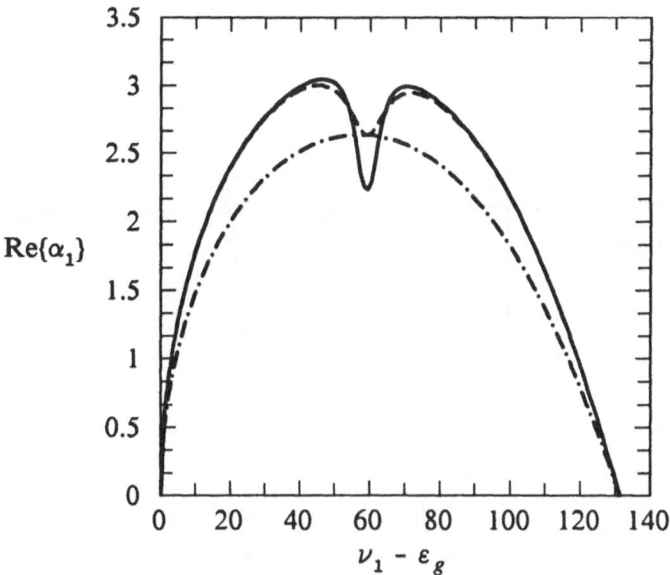

Fig. 8-3. Inhomogeneously broadened limit probe absorption α_1 (solid line) and incoherent part α_{inc} (dashed line), and strong-mode gain α_2 (dot-dashed line) versus probe detuning $\hbar\nu_1 - \varepsilon_g$. The total carrier density $N = 4 \times 10^{18}$ cm^{-3}, the carrier masses m_e and $m_h = 1.28$ m and 4.53 m, respectively, the temperature $T = 300$ K, pump Rabi energy 2 meV (which corresponds to 5.9 MW/cm^3), pump detuning $\delta_2 = -59$ meV, and $\hbar\gamma_e = \hbar\gamma_h = 3$ meV. [From Sargent, Koch, and Chow (1992)].

ulation pulsations play as important a role as spectral hole burning in the sidemode gain and coupling coefficients for intermode beat frequencies comparable to the intraband carrier-carrier scattering rates. We have also neglected the saturation of the total carrier density by the spontaneous emission of modes below threshold. As such this simple model cannot give a precise prediction of the onset of sidemode oscillation in a semiconductor laser, but it is computationally simple and predicts some of the main features of mode coupling in semiconductor lasers without having to fit macroscopic parameters such as the linear gain and the linewidth enhancement factor.

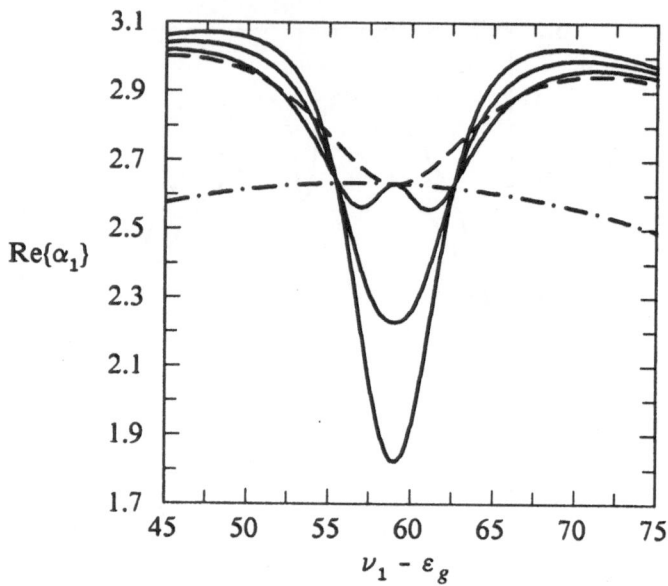

Fig. 8-4. Real parts of α_{coh} and ϑ_1 (in meV) versus probe detuning $\nu_1 - \varepsilon_g$ in meV for parameters in Fig. 8-3. The curves are similar and nearly symmetric about the pump frequency ν_2. [From Sargent, Koch, and Chow (1992)].

8-3. Third-Order Multimode Equations

The sidemode models of Secs. 8-1 and 8-2 can predict a sidemode laser instability, but since they are linear theories, they cannot predict more than an initial transient behavior, nor can they predict the relative sizes of mode amplitudes in steady-state operation. In general one must integrate the coupled multimode equations of motion including that for the total carrier density (see Sec. 8-7). Nevertheless, we can gain some analytic insight by carrying out a calculation up to third order in the laser field mode amplitudes, i.e., one valid up to any product of three $\mathcal{E}_m$'s. The analysis applies to the operation of both bidirectional and unidirectional ring lasers, as well as two-mirror lasers.

The method is very similar to that used in Sec. 8-1, but expands the total carrier density N and the susceptibility $\chi(N)$ about their zero-field values, N_0 and $\chi(N_0)$, respectively, rather than about those saturated by the strong mode 2. This N_0 can be generated by an injection current, optical pumping above the interaction region, or a combination of both. The first nonlinear term is called $\chi^{(3)}$, which, as we see shortly, is a strong function of N_0, field frequency, and temperature. As is Sec. 8-1, the derivation is valid for the quasiequilibrium many-body theory and for all carrier densities and temperatures. The main approximations are that the field intensity must be small enough to be treated by a third-order theory and the total field bandwidth must be small compared to the intraband decay rate constants.

We suppose that the field (1) propagates along the z axis, so we write

$$E(z,t) = \tfrac{1}{2}\mathcal{E}(z,t)e^{-i\nu t} + \text{c.c.,} \tag{68}$$

where the complex amplitude $\mathcal{E}(z,t)$ is a function of position z and is assumed to vary little not only in an optical period (the slowly-varying envelope approximation), but also in a dipole decay time T_2 (rate equation approximation). We write $\mathcal{E}(z,t)$ as the multimode expansion

$$\mathcal{E}(z,t) = \sum_m \mathcal{E}_m(t)U_m(z)e^{i(\nu-\nu_m)t} , \tag{69}$$

where we assume that the mode amplitudes $\mathcal{E}_n(t)$ vary little in the decay time of the total carrier density $N(t)$. This assumption implies a relatively high-Q cavity; the single-mode theory of Sec. 7-3 treats the time response of the field mode and the carrier density on an equal footing. The field (68) induces a polarization of the form

$$P(z,t) = \tfrac{1}{2}\mathcal{P}(z,t,\nu)e^{-i\nu t} + \text{c.c.,} \tag{70}$$

where $\mathcal{P}(z,t)$ is a complex polarization that varies little in an optical period. Corresponding to the multimode expansion (69), we write

$$\mathcal{P}(z,t,\nu) = \sum_m \mathcal{P}_m(t)U_m(z)e^{i(\nu-\nu_m)t-i\phi_m} , \tag{71}$$

where ϕ_m is defined by $\mathcal{E}_m(t) = |\mathcal{E}_m(t)|\exp[-i\phi_m(t)]$. It is convenient to write the slowly-varying envelope $\mathcal{P}(z,t,\nu)$ of the polarization of the medium in terms of a slowly-varying complex susceptibility

$$\mathcal{P}(z,t,\nu) = \epsilon\chi(N(z,t),\nu)\mathcal{E}(z,t,\nu) , \tag{72}$$

where in the rate equation approximation $\chi(N(z,t),\nu)$ depends on z and t

only through $N(z,t)$ [or in a homogeneously broadened two-level medium through the population difference $D(z,t)$].

We write the total carrier density N as

$$N(t) = N^{(0)} + N^{(2)}(t)$$ (73)

where $N^{(n)} \propto |\mathcal{E}|^n$, and expand the susceptibility $\chi(N)$ in the first-order Taylor series

$$\chi(N(t)) \simeq \chi(N_0) + \left.\frac{\partial \chi(N)}{\partial N}\right|_{N_0} N^{(2)}(t) .$$ (74)

Here for typographical simplicity, we write $N_0 \equiv N^{(0)}$ and suppress the ν argument in χ and $\mathcal{P}$, although they continue to depend on ν. To find the $N^{(2)}$ resulting from weak field saturation, we expand the total carrier-density equation of motion to second order in $|\mathcal{E}(z,t)|$. The equation of motion for N is Eq. (12) with $\mathcal{E}(\mathbf{r},t)$ replaced by $\mathcal{E}(z,t)$, that is

$$\dot{N} = \Lambda - \Gamma(N) + \frac{\epsilon}{2\hbar}\chi''(N)|\mathcal{E}(z,t)|^2 ,$$ (75)

where Λ represents the optical pumping or carrier injection, $\Gamma(N)$ is a decay function including both radiative and nonradiative decay, and ϵ is the host susceptibility. To zeroth order there is no time variation, which gives

$$\Lambda = \Gamma(N_0) .$$ (76)

To second order in $|\mathcal{E}(z,t)|$, we have

$$\frac{d}{dt}N^{(2)}(t) = - N^{(2)}(t)\Gamma_1 + \frac{\epsilon}{2\hbar}\chi''(N_0)|\mathcal{E}(z,t)|^2 ,$$ (77)

where the decay rate coefficient

$$\Gamma_1 = \left.\frac{\partial \Gamma(N)}{\partial N}\right|_{N=N_0} .$$ (78)

We multiply Eq. (77) by the integrating factor $\exp(\Gamma_1 t)$ and integrate formally to find

$$N^{(2)}(t) = \frac{\epsilon}{2\hbar}\chi''(N_0)\int_{-\infty}^{t} dt' e^{-i\Gamma_1(t-t')}|\mathcal{E}(z,t')|^2 . \tag{79}$$

Substituting the multimode expansion (69) and assuming that the mode amplitudes $\mathcal{E}_n(t)$ vary little in the decay time $1/\Gamma_1$, we evaluate the time integral to find

$$N^{(2)}(t) = \frac{\epsilon}{2\hbar\Gamma_1}\chi''(N_0)\sum_{\rho}\sum_{\sigma}\mathcal{E}_\rho^*(t)\mathcal{E}_\sigma(t)U_\sigma(z)U_\rho^*(z)\mathcal{F}(\nu_\rho-\nu_\sigma)e^{i(\nu_\rho-\nu_\sigma)t} \tag{80}$$

Here we have used the traditional Lamb mode indices ρ and σ, and the complex population-pulsation Lorentzian $\mathcal{F}(\nu_\rho-\nu_\sigma)$ is given by

$$\mathcal{F}(\nu_\rho-\nu_\sigma) = \frac{\Gamma_1}{\Gamma_1 + i(\nu_\rho - \nu_\sigma)} . \tag{81}$$

Equation (80) contains the mode spatial and temporal interferences contained in the total intensity expression $|\mathcal{E}(z,t)|^2$. For example, the term $U_\sigma(z)U_\rho^*(z)$ has a spatial variation for either standing or running waves if $\sigma \neq \rho$. For $\sigma = \rho$ and a standing wave, we have $\sin^2 K_\sigma z$, which implies a sinusoidal variation in $N(z,t)$, unless the carriers can diffuse sufficiently rapidly in the interband relaxation time that they experience an average field intensity. Should this be the case, one might replace the $\sin^2 K_\sigma z$ by $\frac{1}{2}$. The complex population-pulsation Lorentzian $\mathcal{F}(\nu_\rho-\nu_\sigma)$ expresses the attempt of the medium to follow the beat frequency between modes σ and ρ. Of course, for $\sigma = \rho$, this factor reduces to unity.

Substituting Eq. (80) into Eq. (74) and the result into Eq. (72), we find the complex polarization envelope

$$\mathcal{P}(z,t) = \epsilon\chi(N(z,t))\mathcal{E}(z,t)$$

$$= \epsilon\chi(N_0)\mathcal{E}(z,t) + \epsilon\chi^{(3)}(N_0)\mathcal{E}(z,t)$$

$$\times \sum_{\rho}\sum_{\sigma}\mathcal{E}_\rho^*(t)\mathcal{E}_\sigma(t)U_\sigma(z)U_\rho^*(z)\mathcal{F}(\nu_\rho-\nu_\sigma)e^{i(\nu_\rho-\nu_\sigma)t} , \tag{82}$$

where $\chi^{(3)}(N_0)$ is the same contribution as found in the single-mode gain-saturation discussion of Sec. 3-5, namely

$$\chi^{(3)}(N_0) = \frac{\epsilon \chi''(N_0)}{2\hbar\Gamma_1} \frac{\partial\chi(N)}{\partial N}\bigg|_{N=N_0} . \tag{3.86}$$

Projecting the polarization (82) onto the mth mode, we have from Eq. (71)

$$\mathscr{P}_m(t) = \frac{1}{\mathscr{M}}\int_0^L dz\, U_m^*(z)\mathscr{P}(z,t)e^{i(\nu_m-\nu)t+i\phi_m} = \mathscr{P}_m^{(1)}(t) + \mathscr{P}_m^{(3)}(t) , \tag{83}$$

where $\mathscr{M}$ is the mode normalization factor

$$\mathscr{M} = \int_0^L dz\,|U_m(z)|^2 = \begin{cases} \tfrac{1}{2}L & \text{, two-mirror cavity} \\ L & \text{, ring cavity} \end{cases}, \tag{84}$$

which is independent of the mode index m. Note that for ring lasers, K_m is positive for running waves propagating along the z direction and negative for propagation in the opposite direction. The first-order contribution $\mathscr{P}_m^{(1)}(t)$ is given by

$$\mathscr{P}_m^{(1)}(t) = \frac{\epsilon}{\mathscr{M}}\sum_\mu E_\mu(t)\exp[i\Psi_{m\mu}(t)]\int_0^L dz\, U_m^*(z)U_\mu(t)\chi(N_0) , \tag{85}$$

where the relative phase angle

$$\Psi_{m\mu} = (\nu_m - \nu_\mu)t + \phi_m(t) - \phi_\mu(t) . \tag{86}$$

The third-order contribution is given by

$$\mathscr{P}_m^{(3)}(t) = \frac{\epsilon}{\mathscr{M}}\sum_\mu\sum_\rho\sum_\sigma E_\mu(t)E_\rho(t)E_\sigma(t)\, e^{i\Psi_{m\mu\rho\sigma}(t)}\,\mathscr{F}(\nu_\rho-\nu_\sigma)$$

$$\times \int_0^L dz\, U_m^*(z)U_\mu(z)U_\rho^*(z)U_\sigma(z)\chi^{(3)}(N_0) , \tag{87}$$

where the relative phase angle

$$\psi_{m\mu\rho\sigma}(t) = \phi_m(t) - \phi_\mu(t) + \phi_\rho(t) - \phi_\sigma(t) + (\nu_m - \nu_\mu + \nu_\rho - \nu_\sigma)t . \quad (88)$$

To evaluate the space integral in Eq. (85), we suppose that $\chi(N_0)$ is uniform throughout a region defined by the spatial distribution function $S(z)$. For external-mirror lasers, $S(z)$ vanishes outside the medium, and the placement of the medium can affect the mode coupling between standing waves. $S(z)$ can also model media thin compared to an optical wavelength, such as a stack of quantum wells along the laser z axis. For standard semiconductor laser diodes, $S(z) = 1$ throughout the cavity. With uniformity of $\chi(N_0)$, Eq. (85) becomes

$$\mathscr{P}_m^{(1)}(t) = \epsilon\chi(N_0) \sum_\mu E_\mu(t)S_{m\mu}e^{i\Psi_{mn}} , \quad (89)$$

where the mode factor

$$S_{m\mu} = \frac{1}{\mathscr{M}} \int_0^L dz\, U_m^*(z)U_\mu(z)S(z) . \quad (90)$$

Similarly evaluating the space integral in Eq. (87), we have

$$\mathscr{P}_m^{(3)}(t) = \epsilon\chi^{(3)}(N_0) \sum_{\mu,\rho,\sigma} E_\mu(t)E_\rho(t)E_\sigma(t)e^{i\Psi_{m\mu\rho\sigma}(t)} \mathscr{F}(\nu_\rho-\nu_\sigma)S_{m\mu\rho\sigma} , \quad (91)$$

where the spatial mode function

$$S_{m\mu\rho\sigma} = \frac{1}{\mathscr{M}} \int_0^L dz\, U_m^*(z)U_\mu(z)U_\rho^*(z)U_\sigma(z)S(z) . \quad (92)$$

Substituting Eqs. (83), (89), and (91) into the self-consistency equations (7.10) and (7.11), we find the laser multimode amplitude- and frequency-determining equations

$$\frac{dE_m}{dt} = -\frac{\nu}{2Q_m}E_m + \sum_{\mu} \text{Re}\{\alpha_{m\mu}e^{i\Psi_{m\mu}}\}E_{\mu}$$

$$- \sum_{\mu,\rho,\sigma} \text{Im}\{\vartheta_{m\mu\rho\sigma}e^{i\psi_{m\mu\rho\sigma}}\}E_{\mu}E_{\rho}E_{\sigma} , \qquad (93)$$

$$\nu_m + \frac{d\phi_m}{dt} = \Omega_m - \sum_{\mu} \text{Im}\{\alpha_{m\mu}e^{i\Psi_{m\mu}}\}E_{\mu}/E_m$$

$$- \sum_{\mu,\rho,\sigma} \text{Re}\{\vartheta_{m\mu\rho\sigma}e^{i\psi_{m\mu\rho\sigma}}\}E_{\mu}E_{\rho}E_{\sigma}/E_m , \qquad (94)$$

where the complex linear-gain coefficient

$$\alpha_{m\mu} = \tfrac{1}{2}i\nu\chi(N_0)S_{m\mu} , \qquad (95)$$

and the complex third-order coefficients

$$\vartheta_{m\mu\rho\sigma} = \tfrac{1}{2}\nu\chi^{(3)}(N_0)\mathscr{F}(\nu_{\rho}-\nu_{\sigma})S_{m\mu\rho\sigma} . \qquad (96)$$

where $\chi^{(3)}(N_0)$ is given by (3.86). By writing these coefficients in terms of $\chi^{(3)}(N_0)$, and hence by Eq. (3.84) in terms of $\chi(N_0)$, we can describe both the many-body quasiequilibrium semiconductor medium as well as the short-T_2 homogeneously broadened two-level medium. In fact, one can easily show that the third-order coefficients in Table 9-1 of Sargent, Scully, and Lamb (1977) reduce to the $m = \mu - \rho + \sigma$ coefficients of Eq. (96) in the limit that $\gamma \ll \gamma_a$ and γ_b, the decay constants for the levels.

Note that for running waves, $i\vartheta_{1232}E_2{}^2K/\nu$ equals the multiwave-mixing coupling coefficient ϑ_1, except that for ϑ_1, N_0 includes saturation by the pump wave E_2. Similarly $-[\alpha_{11} + i\vartheta_{1212}E_2{}^2]K/\nu$ corresponds to the multiwave-mixing absorption coefficient α_1. We see that in both the quasiequilibrium semiconductor and the short-T_2 homogeneously broadened two-level medium, the complex saturation coefficients $\vartheta_{m\mu\rho\sigma}$ differ from one another only because of the population-pulsation ($\mathscr{F}$) and spatial ($S_{m\mu\rho\sigma}$) factors. As we see in Sec. 8-5, this allows the mode coupling to be described in a fairly universal way.

In general, the spatial factor $S_{m\mu}$ has the values

$$S_{m\mu} = \begin{cases} \dfrac{1}{L}\displaystyle\int_0^L dz\,\exp[i(K_\mu - K_m)z]S(z), & \text{ring cavity} \\[3em] \dfrac{2}{L}\displaystyle\int_0^L dz\,\sin(K_\mu z)\sin(K_m z)S(z), & \text{two-mirror cavity} \end{cases} \tag{97}$$

Similarly $S_{m\mu\rho\sigma}$ given by Eq. (92) has the running-wave value

$$S_{m\mu\rho\sigma} = \frac{1}{L}\int_0^L dz\,\exp[-i(K_m - K_\mu + K_\rho - K_\sigma)z]S(z) . \tag{98}$$

For standing waves, we use trigonometric identities to show

$$\begin{aligned} S_{m\mu\rho\sigma} &= \frac{1}{4\mathcal{M}}\int_0^L dz\,S(z)\{\cos[(K_m - K_\mu)z] - \cos[(K_m + K_\mu)z]\} \\ &\quad \times\{\cos[(K_\rho - K_\sigma)z] - \eta\cos[(K_\rho + K_\sigma)z]\} \\ &= \frac{1}{8\mathcal{M}}\int_0^L dz\,S(z)\{\cos[(K_m - K_\mu + K_\rho - K_\sigma)z] + \cos[(K_m - K_\mu - K_\rho + K_\sigma)z] \\ &\quad + \eta\cos[(K_m + K_\mu - K_\rho - K_\sigma)z] + \text{rapidly varying terms}\} , \end{aligned} \tag{99}$$

where we include the spatial-hole-burning efficiency factor η to allow for possible washing out of the spatial holes due to diffusion neglected in our analysis. $\eta = 1$ gives full spatial hole burning according to the third-order theory, while $\eta = 0$ corresponds to complete washout of the holes. There are five rapidly varying terms in (99) that vary on the order of an optical wavelength and can be neglected provided $S(z)$ varies little in a wavelength. Such an approximation is *not* valid for some periodically layered structures such as can be fabricated with molecular-beam epitaxy techniques. For such structures, the $S_{m\mu\rho\sigma}$ should include all eight contributions.

The values simplify if $S(z) = 1$ throughout the cavity. Then for either running wave functions $U_m = \exp(iK_m z)$ or standing-wave functions $U_m(z) = \sin K_m z$, $S_{m\mu} = \delta_{m\mu}$. Similarly for $S(z) = 1$ throughout the cavity, the third-order spatial function (99) reduces to

$$S_{m\mu\rho\sigma} = \begin{cases} \delta_{m,\mu-\rho+\sigma}, \text{ ring cavity} \\ \tfrac{1}{4}[\delta_{m,\mu-\rho+\sigma} + \delta_{m,\mu+\rho-\sigma} + \eta\delta_{m,-\mu+\rho+\sigma}] \text{ , 2-mirror cavity .} \end{cases} \tag{100}$$

Alternatively if the modes have sufficiently large Q's that $\nu/2Q_m \ll |\nu_m - \nu_\mu|$, $\Psi_{m\mu}(t)$ for $m \neq \mu$ varies too fast for the amplitude and frequency-determining equations to follow, so that only Eq. (100) contributes. For the third-order contributions, $\Psi_{m\mu\rho\sigma}(t)$ varies too fast unless $m = \mu-\rho+\sigma$, which restricts the summations in Eqs. (93) and (94) in the same way as the condition $S(z) = 1$ throughout the cavity does for the running-wave case.

The multimode equations (93) and (94) have essentially the same form and the multimode Lamb equations [see Sargent, Scully, and Lamb (1977)], and can be used to study one, two, three and general multimode laser operation provided the mode intensities are small enough to be treated using a third-order theory. The use of the coefficient formulas (95) and (96) further restricts the range of multimode frequencies to be much less than the dipole decay rate γ constant and the amplitudes $\mathcal{E}_n(t)$ to vary little in the carrier-density decay time $1/\Gamma_1$. In the following sections, we consider one, two, and three-mode operation as described by Eqs. (93) and (94) with the coefficients (95) and (96).

8-4. Single-Mode Operation

At first glance, it seems as though we are wasting our time to consider a single-mode third-order theory, since Sec. 7-2 treats single-mode operation for a field intensity limited only by the requirement that it not burn spectral holes (else the quasiequilibrium approximation becomes inaccurate). However the simple theory of Sec. 7-2 is effectively based on a local gain approximation that assumes that the field and gain are uniform throughout the interaction region. The third-order theory of Sec. 8-3 is more general in that it treats longitudinal spatial variations caused by the mode factors $U_m(z)$ and by the medium spatial distribution function $S(z)$.

In particular, for $S(z) = 1$ in the standing-wave case, the spatial factor $S_{mmmm} = (2+\eta)/4$, which ranges from 3/4 for full spatial hole burning to $\tfrac{1}{2}$ for complete washout of the spatial holes. The additional saturation given by $S_{mmmm} = 3/4$ is due to destructive scattering of the running waves that comprise the standing wave off the grating that their fringe pattern induces in the nonlinear population difference, $f_{e\mathbf{k}} + f_{h\mathbf{k}} - 1$. Experience shows that improvements to the model of the medium can often be treated relatively easily with a third-order perturbation theory. With the understanding gained, one can then embark on the possibly more difficult task of treating more intense field intensities. We also need the single-mode results to understand the multimode results of Secs. 8-5 and 8-6.

Our approach in this and the following sections parallels that given in Sargent, Scully, and Lamb (1977), here with emphasis on an active medium consisting of a quasiequilibrium semiconductor. For single-mode operation the amplitude-determining equation (93) gives

$$\frac{dE_m}{dt} = E_m \left[g_m - \frac{\nu}{2Q_m} - \beta_m I_m \right] , \tag{101}$$

where the modal gain coefficient

$$g_m = \text{Re}\{\alpha_{mm}\} = -\tfrac{1}{2}\nu\chi''(N_0) , \tag{102}$$

the self-saturation coefficient β_m

$$\beta_m = \text{Im}\{\vartheta_{mmmm}\} \frac{\gamma\Gamma_1\hbar^2}{|\mu|^2} = \frac{\nu\epsilon\hbar\gamma}{4|\mu_k|^2}\chi''(N_0) \frac{\partial\chi''(N_0)}{\partial N_0} S_{mmmm}$$

$$= -g_m \frac{\epsilon\hbar\gamma}{2|\mu_k|^2} \frac{\partial\chi''(N_0)}{\partial N_0} S_{mmmm} , \tag{103}$$

the dimensionless intensity

$$I_m = \frac{|\mu_k E_m/\hbar|^2}{\gamma\Gamma_1} , \tag{104}$$

and k has the value at the laser operating frequency. Note that this intensity is given in units of the "slow" interband saturation intensity

$$I_{ssat} = c\epsilon_0 |\hbar/\mu_\mathbf{k}|^2 \gamma\Gamma_1 , \tag{105}$$

which is orders of magnitude smaller than the "fast" saturation intensity

$$I_{sat} = c\epsilon_0 |\hbar/\mu_\mathbf{k}|^2 \gamma/T_{1f} , \tag{106}$$

which is relevant for spectral hole burning and is the unit for the dimensionless intensity (49). The present third-order perturbation theory assumes $I_m \ll 1$, which certainly implies that the intensity (104) is much smaller than (49).

In steady state, Eq. (101) predicts the mode intensity

$$I_m = \frac{g_m - \nu/2Q_m}{\beta_m} = \frac{\nu}{2Q_m \beta_m}(\mathcal{N}_m - 1) \,, \tag{107}$$

where the relative excitation

$$\mathcal{N}_m = \frac{g_m}{\nu/2Q_m} \,. \tag{108}$$

Substituting $\mathcal{N}_m$ into Eq. (103), we find

$$\beta_m = -\mathcal{N}_m \frac{\nu \epsilon \hbar \gamma}{4|\mu|^2 Q_m} \frac{\partial \chi''(N_0)}{\partial N_0} S_{mmmm} \,. \tag{109}$$

Equation (7.83) shows that $\partial \chi''/\partial N < 0$, so β_m is positive. Combining Eqs. (109) and (107), we have

$$I_m = \frac{2|\mu|^2/\epsilon \hbar \gamma}{|\partial \chi''(N_0)/\partial N_0| S_{mmmm}} \frac{\mathcal{N}_m - 1}{\mathcal{N}_m} \,. \tag{110}$$

Here we see one of the problems with third-order theory, namely that it saturates so strongly that it approaches a constant value with increasing $\mathcal{N}$, instead of continuing to increase as does Eq. (7.35). At least it doesn't blow up, which is what fifth-order perturbation theory does for large $\mathcal{N}$!

Figure 8-5 illustrates the intensity Eq. (110) as a function of tuning above the band gap with the corresponding gain curves for reference. Such tuning curves are commonplace with many kinds of lasers, but are virtually unknown for laser diodes, which typically have many modes under the gain curve and would mode hop long before we could finish recording a tuning curve. Nevertheless, with very short lasers on the order of one to several wavelengths long, e.g., VCSELs (vertical cavity serface emitting lasers), it's straightforward to achieve a single resonance under the entire gain curve. For such lasers, Fig. 8-5 reveals how sensitive the near threshold output is to tuning relative to the gain maximum. One can detect some asymmetry in the tuning curves, featuring a more gradual slope on the higher detuning side. This is due to the fact that the $|\partial \chi''/\partial N|$, and hence the third-order saturation, increases throughout the gain region (see Fig. 7-5).

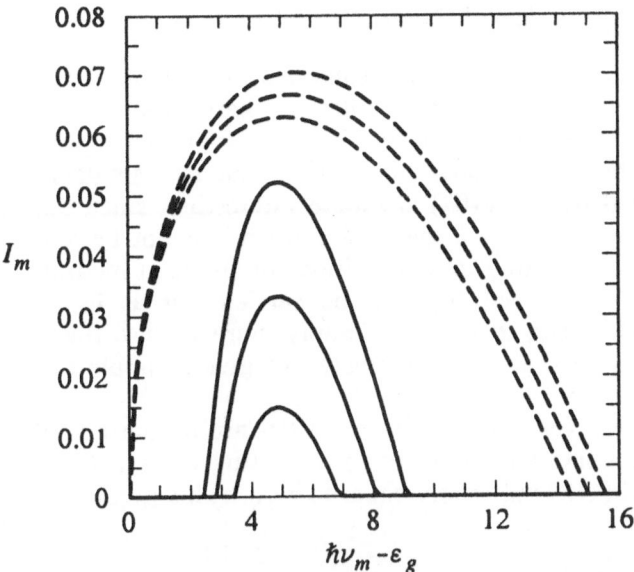

Fig. 8-5. Third-order intensity (solid lines) I_m given by Eq. (110) and corresponding gain (dashed lines in arbitrary units) versus tuning with respect to the band gap. The δ-function lineshape approximation free-carrier model was used. The relative excitations are $\mathcal{N}$ = 1.0495, 1.1115, 1.1746 corresponding to N = 1.3×10^{18}, 1.31×10^{18}, and 1.32×10^{18} cm^{-3}, respectively, (in order of increasing peaks), T = 300 K, m_e = 1.167m, m_h 6.669m, and $\nu/2Q_m$ = .06.

8-5. Two-Mode Operation

We consider two-mode operation as described by the amplitude-determining equation (93). We suppose that the mode frequencies are far enough apart that the linear coupling terms are negligible, and we multiply through by $2E_m\mu^2/\hbar^2\gamma\Gamma_1$ to find the intensity equations of motion

$$\dot{I}_1 = 2I_1[g_1 - \nu/2Q_1 - \beta_1 I_1 - \theta_{12}I_2] \,, \tag{111}$$

$$\dot{I}_2 = 2I_2[g_2 - \nu/2Q_2 - \beta_2 I_2 - \theta_{21}I_1] \,, \tag{112}$$

where the cross-saturation coefficients θ_{mn} are given by

$$\theta_{mn} = \text{Im}\{\vartheta_{mmnn} + \vartheta_{mnnm}\}\gamma\Gamma_1\hbar^2/|\mu_k|^2 . \tag{113}$$

In general, four steady-state solutions ($dI_m/dt = 0$) are possible, since either mode may or may not oscillate. The solutions are physical only if they are positive or zero. The case for which neither mode oscillates is trivial: both modes are below threshold. The two single-mode cases with solutions given by Eq. (107) are more interesting, since both modes might oscillate in the absence of the other, but might not be able to oscillate in the other's presence due to competition for the gain medium. The analysis of the single-mode stability can be carried out as in Secs. 8-1, which allows a single mode to be considerably more intense than the third-order approximation permits. Alternatively, we give a simple third-order stability analysis below.

The solution for which both intensities are nonzero results from solving the pair of coupled equations given by setting the bracketed expressions in Eq. (111) and (112) separately equal to zero. This gives

$$I_m = \frac{\alpha'_m/\beta_m}{1 - C} , \tag{114}$$

where the effective net-gain coefficients

$$\alpha'_m = g_m - \frac{\nu}{2Q_m} - \theta_{mn}\frac{g_n - \nu/2Q_n}{\beta_n} , \tag{115}$$

where $n \neq m$ and the *coupling parameter C* is defined by

$$C = \frac{\theta_{12}\theta_{21}}{\beta_1\beta_2} . \tag{116}$$

Here α'_1 is called an *effective net gain coefficient* since it is the linear net gain for I_1 *minus* the saturation induced by I_2 oscillating alone.

Before considering the specific values given by Eqs. (113) and (96), we summarize the linear stability analysis of these equations in a fashion similar to that for the single-mode cases of Sec. 7-3. Specifically we substitute $I_m(t) = I_m^{(s)} + \epsilon_m(t)$ into Eq. (111) to find

$$\dot{\epsilon}_1 = 2\epsilon_1[g_1 - \nu/2Q_1 - \beta_1 I_1^{(s)} - \theta_{12}I_2^{(s)}] - 2I_1^{(s)}(\beta_1\epsilon_1 + \theta_{12}\epsilon_2) , \tag{117}$$

with a similar equation for ϵ_2. Consider first the stability of the $I_1^{(s)} = 0$ solution in the presence of I_2 oscillating alone with the value $(g_2-\nu/2Q_2)/\beta_2$. In other words, we ask, can I_1 build up in the presence of

I_2? For this case, Eq. (117) reduces to $\dot{\epsilon}_1 = 2\alpha_1' \epsilon_1$. Hence if I_1's effective net gain is positive, I_1 builds up. It may do so and then either suppress I_2 or coexist with it.

To find out which of these combination actually take place, we consider the stability of the two-mode solution of Eq. (114). In this case the term inside the [...] in Eq. (117) vanishes, reducing Eq. (117) and its companion for ϵ_2 to the matrix equation

$$\frac{d}{dt} \begin{bmatrix} \epsilon_1 \\ \epsilon_2 \end{bmatrix} = \Theta \begin{bmatrix} \epsilon_1 \\ \epsilon_2 \end{bmatrix},$$

(118)

where the stability matrix Θ is defined by

$$\Theta = -\frac{2}{1-C} \begin{bmatrix} \alpha_1' & \theta_{12}\alpha_1'/\beta_1 \\ \theta_{21}\alpha_2'/\beta_2 & \alpha_2' \end{bmatrix}.$$

(119)

For two-mode stability, Θ has to have eigenvalues with negative real parts. Setting $\det(\Theta - \lambda I) = 0$, we find the eigenvalues

$$\lambda_{1,2} = -\frac{\alpha_1' + \alpha_2'}{1 - C} \pm \sqrt{\frac{(\alpha_1' + \alpha_2')^2}{(1 - C)^2} - \frac{4\alpha_1'\alpha_2'}{1 - C}}.$$

(120)

Equation (114) contains two possibilities for physical (positive) intensities: 1) both effective $\alpha's > 0$ and the coupling parameter $C < 1$, and 2) both effective $\alpha's < 0$ and $C > 1$. The eigenvalues (120) for the former both have negative real parts, since the radicand is smaller than the lead term. Hence two-mode operation is possible for $\alpha_m' > 0$ and $C < 1$. For $\alpha_m' < 0$ and $C > 1$, one eigenvalue has a positive real part, implying that two-mode operation is unstable. However both single-mode solutions are stable, that is, we have *bistability*. Which mode oscillates depends on the initial conditions.

Section 8-4 discusses g_m and β_m for the quasiequilibrium semiconductor. In both the effective net gain of Eq. (115) and the coupling parameter (116), the coupling ratio θ_{mn}/β_n appears. We evaluate this important ratio for the unidirectional case and for standing-wave cases with and without spatial hole burning. Using Eqs. (96), (103), and (113), we have

$$\frac{\theta_{mn}}{\beta_n} = \frac{\text{Im}\{\vartheta_{mmnn} + \vartheta_{mnnm}\}}{\text{Im}\{\vartheta_{nnnn}\}}$$

which gives

$$\frac{\theta_{mn}}{\beta_n} = \frac{\text{Im}\{[S_{mmnn} + S_{mnnm}\mathscr{F}(\nu_n-\nu_m)]\partial\chi/\partial N\}}{S_{nnnn}\partial\chi''/\partial N}$$

$$= \frac{S_{mmnn} + S_{mnnm}\mathscr{L}_N(\nu_n-\nu_m)[1 - (\nu_n-\nu_m)\alpha/\Gamma_1]}{S_{nnnn}} , \qquad (121)$$

where α is the linewidth enhancement factor (3.89) and $\mathscr{L}_N$ is the dimensionless Lorentzian function

$$\mathscr{L}_N(\nu_n-\nu_m) = \frac{\Gamma_1{}^2}{\Gamma_1{}^2 + (\nu_n-\nu_m)^2} . \qquad (122)$$

Unidirectional Ring Laser Operation

For unidirectional operation, $S_{m\mu\rho\sigma} = 1$, which gives the ratio

$$\frac{\theta_{mn}}{\beta_n} = 1 + \mathscr{L}_N(\nu_n-\nu_m)[1 - (\nu_n-\nu_m)\alpha]) , \qquad (123)$$

and the running-wave coupling parameter

$$C_{RW} = [1 + \mathscr{L}_N(\nu_2-\nu_1)]^2 - [\mathscr{L}_N(\nu_2-\nu_1)(\nu_2-\nu_1)\alpha/\Gamma_1]^2 , \qquad (124)$$

This is never less than unity unless α is unexpectedly large ($> \sqrt{5}$). If $C_{RW} < 1$ and both effective $\alpha's$ are positive, two-mode operation is stable.

Two-Mirror Standing-Wave Laser Operation

For two-mirror standing wave operation, Eq. (99) gives

$$S_{mmnn} = \frac{1}{8\mathscr{M}}\int_0^L dz\,\{2 + \eta\cos[2(K_m-K_n)z]\}S(z) = \tfrac{1}{2}\bar{S} + \tfrac{1}{4}\eta S_{2(m-n)} , \qquad (125)$$

where

$$\bar{S} = \frac{1}{L}\int_0^L dz\, S(z) , \qquad (126)$$

$$S_{2(m-n)} = \frac{1}{L} \int_0^L dz \ \cos[2(K_m - K_n)z]S(z) \ . \tag{127}$$

Similarly S_{mnnm} is given by

$$S_{mnnm} = \tfrac{1}{4}(1+\eta)\bar{S} + \tfrac{1}{4}S_{2(m-n)} \ . \tag{128}$$

With appropriate choice of $S(z)$, $S_{2(m-n)}$ can vary all the way from $\bar{S}$ to $-\bar{S}$. The dimensionless factors $\bar{S}$ and $S_{2(m-n)}$ correspond to the Lamb-theory $\bar{N}$ and $N_{2(m-n)}$ density factors, respectively, in the usual Lamb theory [see Sargent, Scully, and Lamb (1977)]. Substituting Eqs. (126) and (127) into Eq. (121), we have

$$\frac{\theta_{mn}}{\beta_n} = \frac{1}{2+\eta} \left\{ 2 + \eta \frac{S_{2(m-n)}}{\bar{S}} \right.$$
$$\left. + \left[1 + \eta + \frac{S_{2(m-n)}}{\bar{S}} \right] \mathscr{L}_N(\nu_n - \nu_m) \left[1 - \frac{(\nu_n - \nu_m)\alpha}{\Gamma_1} \right] \right\}. \tag{129}$$

Note that if the mode beat frequencies are expressed in units of the decay constant Γ_1, the only medium-dependent parameter in this coupling expression is the linewidth enhancement factor α of Eq. (3.89).

For full spatial hole burning ($\eta = 1$), this gives the coupling parameter

$$C = \frac{(2 + S_{2(m-n)}/\bar{S})^2}{9} C_{RW} \ , \tag{130}$$

which can easily be less than unity by filling the cavity completely, for which $S_{2(m-n)} = 0$, or even more easily by filling the middle of the cavity only, for which $S_{2(m-n)}/\bar{S} \rightarrow -1$. This shows that spatial hole burning is one mechanism that lends stability to multimode operation.

However if the spatial holes are completely washed out ($\eta = 0$),

$$\frac{\theta_{mn}}{\beta_n} = 1 + \frac{1}{2} \left[1 + \frac{S_{2(m-n)}}{\bar{S}} \right] \mathscr{L}_N(\nu_n - \nu_m) \left[1 - \frac{(\nu_n - \nu_m)\alpha}{\Gamma_1} \right] . \tag{131}$$

As for the running-wave case, this is not likely to lead to $C < 1$, i.e., single-mode operation is more likely. This is illustrated by the coupling

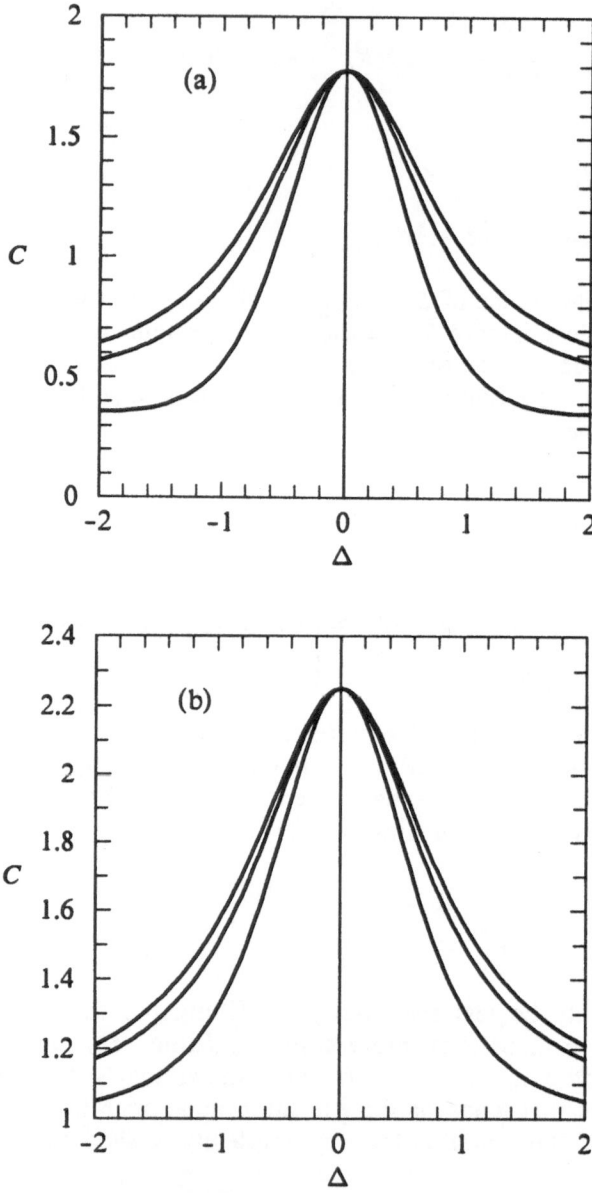

Fig. 8-6. (*a*) Coupling parameter versus intermode beat frequency $\Delta = \nu_m - \nu_n$ for full spatial hole burning ($\eta = 1$), a filled cavity ($S_2 = 0$), and linewidth enhancement factors of $\alpha = 0, 1, 2$ (in order of decreasing coupling). (*b*) same with $\eta = 0$ (spatial holes washed out)

parameter in Fig. 8-6, where we see that with full spatial-hole burning (η = 1), $C < 1$, allowing multimode operation. On the other hand for $\eta = 1$, $C > 1$ indicating bistable operation. Hence if spatial holes are washed out, one would expect a laser satisfying the assumptions of this model to exhibit hysteresis: instead of mode hopping as one finds with short diode lasers, one should find that the oscillating mode continues to suppress others around it as the mode frequency is tuned.

Two-mode Bidirectional Ring-Laser Operation

As a final case, consider the bidirectional ring laser. This is described by the running-wave case with $K_1 = K_+$, for the positive traveling running wave, and $K_2 = -K_-$, for the negative traveling wave. Equation (98) gives $S_{mmnn} = 1$ and $S_{mnnm} = \eta$, the grating visibility. This gives the same result as the unidirectional ring, except for the appearance of η, that is, instead of Eq. (124), we have the bidirectional coupling parameter

$$C_{BD} = [1 + \eta \mathscr{L}_N(\nu_2 - \nu_1)]^2 - [\eta \mathscr{L}_N(\nu_2 - \nu_1)(\nu_2 - \nu_1)\alpha/\Gamma_1]^2 . \tag{132}$$

Here we see that if spatial hole burning is washed out ($\eta = 0$), the population pulsations are washed out too, and C_{BD} reduces to 1. This typically results in single-mode operation, although not in bistable operation. In contrast, a bidirectional He-Ne ring laser containing equal amounts of Ne[20] and Ne[22] and tuned midway between the two isotopic line centers has a C_{BD} substantially less than unity. This readily allows both counterpropagating waves to oscillate, which is what you want for a ring laser gyro, whose output beat frequency is proportional to the rotation rate. The analysis here implies that even if spatial-hole burning is washed out, single-mode operation is still the most probable situation in a semiconductor ring laser, thereby ruling out the very existence of a beat frequency. If spatial hole burning plays a role, bistable operation should be observed. One should bear in mind that this is a plane-wave theory and transverse variations can have important consequences, and short-cavity mode competition can be affected by temperature variations induced by Pauli blocking of injection pumping.

8-6. Three-Mode Operation and Mode Locking

In three-mode operation, in addition to the β_m self-saturation and θ_{mn} cross-saturation terms, there are terms like $\vartheta_{1232} E_2^2 E_3$, which are similar in character to the $\vartheta_1 \mathscr{E}_3^*$ term in the coupled-mode equation (15). Physically, modes 2 and 3 induce a population pulsation that scatters mode 2 with appropriate frequency shifts into mode 1. These terms were originally called *combination tones* by Lamb (1964). In the laser they act like injected signals (see Sec. 7-4) that attempt to convince a mode to oscillate at the combination-tone frequency. In particular, if the ϑ_{1232} term convinces mode 1 to oscillate at its frequency $(\nu_2 - \nu_3 + \nu_2)$, we have the relation

$$\nu_1 = \nu_2 - (\nu_3 - \nu_2) , \tag{133}$$

which would cause the beat frequency between modes 2 and 1 to equal that between modes 3 and 2. Such a locking of beat frequencies is called mode locking, and can happen spontaneously in lasers, totally independently of external influences, such as a modulator. Such a mode locked state yields a periodic output, which, depending on the amplitudes and phases, can give pulses, FM operation, or a host of periodic waveforms in between.

In semiconductor lasers, the situation is not so simple, because there can be appreciable differences between the adjacent modal beat frequencies due to anomalous dispersion and to mode pulling. To see the effects of the dispersion, we calculate the adjacent-mode beat frequency in a laser including the effects of the dispersive term $\lambda dn/d\lambda$. In a two-mirror cavity of length L, $\lambda_m = 2L/m$, so $\nu_m = c/n_m \lambda_m = mc/2L n_m$ and $\partial \lambda_m /\partial m = -\lambda_m /m$. Using a first-order Taylor series, we have

$$\Delta = \nu_{m+1} - \nu_m = \frac{\partial \nu_m}{\partial m} = \frac{c}{2L} \frac{\partial}{\partial m} \frac{m}{n_m} = \frac{c}{2L n_m}\left[1 - \frac{m}{n_m}\frac{\partial n_m}{\partial m}\right]$$

$$= \frac{c}{2L n_m}\left[1 - \frac{m}{n_m}\frac{\partial n_m}{\partial \lambda_m}\frac{\partial \lambda_m}{\partial m}\right] = \frac{c}{2L n_m}\left[1 + \frac{\lambda_m}{n_m}\frac{\partial n_m}{\partial \lambda_m}\right] . \tag{134}$$

To illustrate this value, we consider a cavity of length $L = \frac{1}{4}\times10^{-3}$ m, an index of refraction $n = 3.5$, and $\lambda dn/d\lambda = -1.7$. This gives

$$\Delta = \frac{3\times10^8}{\frac{1}{4}\times10^{-3}\times3.5}\left[1 - \frac{1.7}{3.5}\right] = 171.43\times10^9\times.5143 = 88.2 \text{ GHz} ,$$

which is about half what it would be in the absence of the dispersion term.

The change in the adjacent-mode beat frequency from one mode pair to the next is given by

$$\frac{\partial^2 \nu_m}{\partial m^2} = \frac{c}{2L\,n_m^2}\frac{\partial n_m}{\partial \lambda_m}\frac{\lambda_m}{m}\left[1 + \frac{\lambda_m}{n_m}\frac{\partial n_m}{\partial \lambda_m}\right]$$

$$-\frac{c}{2L\,n_m^2}\frac{\lambda_m}{m}\left[\frac{\partial n_m}{\partial \lambda_m} - \frac{\lambda_m}{n_m}\left(\frac{\partial n_m}{\partial \lambda_m}\right)^2 + \lambda_m\frac{\partial^2 n_m}{\partial \lambda_m^2}\right]$$

$$= \frac{c}{L\,n_m^2}\frac{\lambda_m^2}{m}\left[\frac{1}{n_m}\left(\frac{\partial n_m}{\partial \lambda_m}\right)^2 - \frac{1}{2}\frac{\partial^2 n_m}{\partial \lambda_m^2}\right]. \tag{135}$$

To estimate the size of the beat-frequency variation, we ignore the term proportional to $\partial^2 n_m / \partial \lambda_m^2$ and use the same parameters as above to find

$$\frac{c}{L\,m\,n_m^3}\left(\frac{\lambda_m \partial n_m}{\partial \lambda_m}\right)^2 = \frac{3\times10^8\times1.7^2}{\tfrac{1}{4}\times10^{-3}\times250\times3.5^3} \simeq .324 \text{ GHz.}$$

The third-order equations of motion for three-mode operation as predicted by Eqs. (93) and (94) are

$$\dot{E}_1 = E_1\left[g_1 - \frac{\nu}{2Q_1} - \sum_{m=1}^{3}\theta_{1m}I_m\right] - \text{Im}\{\vartheta_{1232}e^{-i\Psi}\}E_2^2 E_3 \tag{136}$$

$$\dot{E}_2 = E_2\left[g_2 - \frac{\nu}{2Q_2} - \sum_{m=1}^{3}\theta_{2m}I_m\right] - \text{Im}\{(\vartheta_{2123} + \vartheta_{2321})e^{i\Psi}\}E_1 E_2 E_3 \tag{137}$$

$$\dot{E}_3 = E_3\left[g_3 - \frac{\nu}{2Q_3} - \sum_{m=1}^{3}\theta_{3m}I_m\right] - \text{Im}\{\vartheta_{3212}e^{-i\Psi}\}E_2^2 E_1 \tag{138}$$

$$\dot{\Psi} = a + b_s\sin\Psi + b_c\cos\Psi = a + b\sin(\Psi - \Psi_0), \tag{139}$$

where the relative phase angle $\Psi(t)$ is given by

$$\Psi(t) = \Psi_{2123} = (2\nu_2 - \nu_1 - \nu_3)t + 2\phi_2 - \phi_1 - \phi_3, \tag{140}$$

and the coefficients are

$$a = 2\text{Im}\{\alpha_{22} - \alpha_{11} - \alpha_{33}\} - \sum_{m=1}^{3} [2\tau_{2m} - \tau_{1m} - \tau_{3m}]I_m$$

$$\tau_{nm} = \text{Re}\{\vartheta_{mmnn} + \vartheta_{mnnm}\}\gamma\Gamma_1\hbar^2/|\mu|^2$$

$$b_s = \text{Im}\left\{2E_1E_3(\vartheta_{2123} + \vartheta_{2321}) + \left[\frac{\vartheta_{1232}E_3}{E_1} + \frac{\vartheta_{3212}E_1}{E_3}\right]E_2^{\,2}\right\}$$

$$b_c = \text{Re}\left\{-2E_1E_3(\vartheta_{2123} + \vartheta_{2321}) + \left[\frac{\vartheta_{1232}E_3}{E_1} + \frac{\vartheta_{3212}E_1}{E_3}\right]E_2^{\,2}\right\}$$

$$b = \sqrt{b_s^2 + b_c^2}$$

$$\Psi_0 = -\tan^{-1}(b_c/b_s) \ .$$

We see that Eqs. (136) through (139) are similar to the mode-locking equations of Sec. 7-4, but here the "injected signals" are combination tones induced by mode beating in the nonlinear medium. If one can use the decoupled approximation, most of the analysis is already done in Sec. 7-4. Suffice it to say here that if $|a| \leq |b|$ in Eq. (139) that mode locking occurs, that is, the three-mode output is periodic in the cavity round-trip time.

8-7. Higher-Order Operation

As noted in Sec. 8-3, we can solve the coupled multimode equations of motion including that for the total carrier density. While this reduces to a numerical approach, it allows the field amplitudes to be larger than those permitted in third-order, and it does not adiabatically eliminate the carrier density, which can vary on the same time scale as the field amplitudes. Ultimately one can integrate the coupled Maxwell semiconductor Bloch equations. In this section, we consider a much simpler approach that is valid so long as the polarization can be adiabatically eliminated, that is, so long as the quasiequilibrium approximation is valid.

Substituting Eqs. (72) and (69) into Eq. (83), we find

$$\mathcal{P}_m(t) = \frac{1}{\mathcal{M}} \int_0^L dz\, U_m^*(z)\,\mathcal{P}(z,t)\,e^{i(\nu_m - \nu)t + i\phi_m}$$

$$= \frac{\epsilon}{\mathcal{M}} \sum_\mu E_\mu(t)\,e^{i\Psi_{m\mu}} \int_0^L dz\, U_m^*(z)\,U_\mu(z)\,\chi(N)$$

$$= \epsilon \sum_\mu E_\mu(t)\,e^{i\Psi_{m\mu}}\,\chi_{m\mu}(N) \,, \tag{141}$$

where the susceptibility component $\chi_{m\mu}(N)$ is defined by

$$\chi_{m\mu}(N) = \frac{1}{\mathcal{M}} \int_0^L dz\, U_m^*(z)\,U_\mu(z)\,\chi(N) \,. \tag{142}$$

Substituting the field envelope (69) into (75), we find the carrier density equation of motion

$$\dot{N} = \Lambda - \Gamma(N) + \frac{\epsilon}{2\hbar}\chi''(N) \sum_{\rho,\sigma} E_\rho(t)E_\sigma(t)\,U_\rho^*(z)\,U_\sigma(z)\,e^{i\Psi_{\rho\sigma}} \,. \tag{143}$$

Substituting this into the self-consistency equations (7.10) and (7.11), we have

$$\dot{E}_m = -\frac{\nu}{2Q_m}E_m - \frac{\nu}{2} \sum_\mu E_\mu(t)\,\mathrm{Im}\{e^{i\Psi_{m\mu}}\chi_{m\mu}(N)\} \tag{144}$$

$$\nu_m + \dot{\phi}_m = \Omega_m - \frac{\nu}{2} \sum_\mu E_\mu(t)\,\mathrm{Re}\{e^{i\Psi_{m\mu}}\chi_{m\mu}(N)\}/E_m \,. \tag{145}$$

The coupled set of equations (143) through (145) can be integrated numerically using the desired single-mode χ, e.g., for a linear-density susceptibility, a free-carrier susceptibility, or for a many-body susceptibility. The analysis is simplified significantly if χ has a uniform spatial distribution.

References

Binder, R., D. Scott, A.E. Paul, M. Lindberg, K. Henneberger, and S.W. Koch (1992), Phys. Rev. **B45**, 1107.

Bogatev, A. P., P. G. Eliseev, and B. N. Sverdlov (1975), IEEE J. Quant. Electron. **QE-11**, 510.

Lamb, W. E., Jr. (1964), Phys. Rev. **134**, A1429.

Meystre, *P*, and M. Sargent III (1991), Springer-Verlag, Heidelberg, Second Edition.

Sargent, M. III, S. W. Koch, and W. W. Chow (1992), J. Opt. Soc. Am **B9**, 1288.

Sargent, M. III (1993), Phys. Rev. **A48**, 717.

Sargent, M. III, M. O. Scully, and W. E. Lamb, Jr. (1974), *Laser Physics*, Addison-Wesley Publishing Co., Reading, MA.

Chapter 9
QUANTUM THEORY OF THE LASER

For many purposes, the electromagnetic field can be treated classically, while the semiconductor medium is treated quantum mechanically. However two general characteristics of semiconductor laser operation require at least some level of field quantization. These are spontaneous emission and the laser linewidth. This chapter introduces the necessary additional formalism to treat these problems and shows some simple ways of calculating them.

Section 9-1 summarizes the quantization of a single-mode of the electromagnetic field in terms of boson annihilation and creation operators. The photon number states are defined and used together with the two-band semiconductor states to form a basis for analyzing semiconductor-field interactions. Section 9-2 derives the Weisskopf-Wigner coefficient that describes spontaneous emission into the modes of free space. Section 9-3 develops the quantum Langevin formalism needed to treat these problems. Section 9-4 applies this formalism to the field-semiconductor operator equations of motion to derive the diffusion coefficients. Section 9-5 derives the relative intensity noise (RIN) and frequency-noise spectra, and the laser linewidth. The linewidth is larger than that for lasers containing two-level media, because the index of refraction is sensitive to fluctuations in the carrier density primarily due to spontaneous emission. The relaxation oscillations driven by spontaneous emission also lead to sidebands in the laser spectrum.

9-1. Single-Mode Field Quantization

To quantize the electromagnetic field, we consider a cavity of volume V, closed by perfectly reflecting mirrors. A classical monochromatic, single-mode electromagnetic field polarized in the $\hat{x}$-direction has the form

$$\mathbf{E}(z,t) = \hat{x}q(t)[2\Omega^2/\epsilon_0 V]^{1/2}\sin Kz , \tag{1}$$

Here Ω is the single-mode field oscillation frequency, K is the wave number Ω/c, and $q(t)$ is a measure of the field amplitude.

The electromagnetic field satisfies Maxwell's equations in a vacuum, namely Eqs. (2.59) through (2.62). Substituting Eq. (1) into Eq. (2.61), we find the corresponding magnetic field

$$\mathbf{B}(z,t) = \frac{\hat{y}}{c^2 K} \dot{q}(t) \left[\frac{2\Omega^2}{\epsilon_0 V} \right]^{1/2} \cos Kz .$$
(2)

The classical electromagnetic energy density is given by

$$\mathscr{U} = \frac{1}{2} [\epsilon_0 E^2 + B^2/\mu_0] ,$$
(3)

with the corresponding Hamiltonian

$$\mathscr{H} = \int_V dV [\epsilon_0 E^2 + B^2/\mu_0] ,$$
(4)

where dV is a volume element and E and B are the magnitudes of $\mathbf{E}$ and $\mathbf{B}$, respectively. Inserting Eqs. (1) and (2) into Eq. (4), we find

$$\mathscr{H} = \tfrac{1}{2}(\Omega^2 q^2 + p^2),$$
(5)

which is formally identical with the Hamiltonian for a simple harmonic oscillator with unit mass. We can therefore immediately quantize a single mode of the electromagnetic field by using the theory of the quantization of the simple harmonic oscillator [see, for example, Sec. 3-4 of Meystre and Sargent (1991)].

It is convenient to express the results in terms of the boson annihilation and creation operators a and $a^\dagger$ defined by

$$a = 1/\sqrt{2\hbar\Omega} \ (\Omega q + ip)$$
(6)

$$a^\dagger = 1/\sqrt{2\hbar\Omega} \ (\Omega q - ip) .$$
(7)

The single-mode electromagnetic field Hamiltonian (5) becomes

$$\boxed{\mathscr{H} = \hbar\Omega(a^\dagger a + \tfrac{1}{2})} .$$
(8)

The corresponding eigenstates $|n\rangle$ of the field satisfy

$$\mathscr{H}|n\rangle = \hbar\Omega(n + \tfrac{1}{2})|n\rangle , \quad n = 0, 1, 2,...,$$
(9)

where n may be loosely interpreted as the "number of photons" in the state $|n\rangle$. The corresponding state vector is a linear superposition of these energy eigenstates

$$|\psi\rangle = \Sigma_n \, c_n \, |n\rangle \, , \tag{10}$$

and the quantized-field density operator is defined by Eq. (2.97) using Eq. (10).

Solving Eqs. (7) and (8) for q and substituting the result into Eq. (1), we find the electric field operator

$$\boxed{E(z,t) = \mathcal{E}_\Omega (a + a^\dagger) \sin Kz} \, , \tag{11}$$

where the "electric field per photon"

$$\mathcal{E}_\Omega \equiv [\hbar\Omega/\epsilon_0 V]^{1/2} \, . \tag{12}$$

The names annihilation and creation operators come from the effect that a and $a^\dagger$ have on the energy eigenstates $|n\rangle$ according to

$$a|n\rangle = \sqrt{n}\,|n-1\rangle \tag{13}$$

$$a^\dagger|n\rangle = \sqrt{n+1}\,|n+1\rangle \, . \tag{14}$$

Here Eq. (13) shows that a annihilates a photon, i.e., goes from a state $|n\rangle$ of n photons to the state $|n-1\rangle$ of $n-1$ photons. Similarly $a^\dagger$ creates a photon. Somehow the originators of this notation got the † on the wrong operator!

A fully quantal density matrix that includes the field photon-number states $|n\rangle$ is given by the outer product of these states with the four semiconductor states of Sec. 3-1 $|n_e \, n_h \, \mathbf{k}\rangle$. This gives the basis $|n_e \, n_h \, \mathbf{k}\rangle|n\rangle \equiv |n_e \, n_h \, \mathbf{k} \, n\rangle$, i.e., $|3n\rangle$, $|2n\rangle$, $|1n\rangle$, $|0n\rangle$. As in Sec. 3-1, we can calculate the equation of motion for matrix elements of the semiconductor-field density operator in this basis. In the following section, we use this approach to calculate the equation of motion for the photon-number probability.

With the quantized electric field (11), the single-mode field-semiconductor interaction energy in Eq. (3.2) becomes in the rotating-wave approximation

$$\mathcal{V} = \hbar \sum_{\mathbf{k}} (g_{\mathbf{k}}^* a^\dagger b_{-\mathbf{k}} a_{\mathbf{k}} + g_{\mathbf{k}} a_{\mathbf{k}}^\dagger b_{-\mathbf{k}}^\dagger a) \, , \tag{15}$$

where $g_{\mathbf{k}} = -\mu_{\mathbf{k}}\mathcal{E}_\Omega/\hbar$. This interaction energy has nonvanishing matrix elements between the states $|3n\rangle$ and $|0n+1\rangle$, for all n, but all other matrix elements vanish, such as between $|2n\rangle$, $|1n\rangle$, and other states. The interaction energy $\mathcal{V}$ is chosen to conserve energy. Specifically $a^\dagger b_{-\mathbf{k}} a_{\mathbf{k}}$ annihilates an electron-hole pair and creates a photon, that is, it transfers one photon of energy from the semiconductor to the field. Similarly $a_{\mathbf{k}}^\dagger b_{-\mathbf{k}}^\dagger a$ creates an electron-hole pair by annihilating a photon. $a_{\mathbf{k}}^\dagger b_{-\mathbf{k}}^\dagger a^\dagger$ and its adjoint are not included in Eq. (15), since they do not conserve energy. By examining the time development of these operators in the Heisenberg picture, one can show that neglecting these latter two combinations is the rotating-wave approximation in disguise.

In the absence of the interaction energy $\mathcal{V}$, the annihilation operator a for the field obeys the Heisenberg equation of motion

$$\dot{a}(t) = \frac{i}{\hbar}[\mathcal{H}_s, a] = -i\Omega a(t) , \tag{16}$$

which has the integral

$$a(t) = a(0)e^{-i\Omega t} . \tag{17}$$

The corresponding equations for the creation operator $a^\dagger$ are just the adjoints of Eqs. (16) and (17), so that $a^\dagger(t) = a^\dagger(0)e^{i\Omega t}$. Similar equations describe the semiconductor spin-flip operators $b_{-\mathbf{k}} a_{\mathbf{k}} \equiv |0\rangle\langle 3|$ and $a_{\mathbf{k}}^\dagger b_{-\mathbf{k}}^\dagger \equiv |3\rangle\langle 0|$. Hence we see that the interaction-energy term $a^\dagger(t)b_{-\mathbf{k}} a_{\mathbf{k}}$ has the time dependence

$$a^\dagger(t)b_{-\mathbf{k}}(t)a_{\mathbf{k}}(t) \propto e^{i(\Omega-\omega_{\mathbf{k}})t} . \tag{18}$$

In contrast, a term like $a^\dagger(t)a_{\mathbf{k}}^\dagger(t)b_{-\mathbf{k}}^\dagger(t) \propto e^{i(\Omega+\omega_{\mathbf{k}})t}$, which relative to Eq. (18), averages itself to zero in the rotating-wave approximation. Hence the interaction energy (15) is written in the rotating-wave approximation. The neglected terms $a^\dagger(t)b_{-\mathbf{k}}(t)a_{\mathbf{k}}(t)$ and $a(t)a_{\mathbf{k}}^\dagger(t)b_{-\mathbf{k}}^\dagger(t)$ do not conserve energy, since instead of exchanging energy they imply creating quanta in both the field and the semiconductor or annihilating quanta in both. Nevertheless over sufficiently short times, the uncertainty principle allows such energy nonconservation subject to greatly reduced probability.

9-2. Spontaneous Emission

For a quantized electromagnetic field, the semiconductor polarization operator $b_{-\mathbf{k}} a_{\mathbf{k}}$ of Eq. (3.9) is replaced by $a^\dagger b_{-\mathbf{k}} a_{\mathbf{k}}$, which annihilates an electron-hole pair and creates a photon. In calculating the equation of motion for $a^\dagger b_{-\mathbf{k}} a_{\mathbf{k}}$, we encounter the commutator

$$[a^\dagger b_{-\mathbf{k}} a_{\mathbf{k}}, \, a_{\mathbf{k}}^\dagger b_{-\mathbf{k}}^\dagger a] = a^\dagger a(1 - b_{-\mathbf{k}}^\dagger b_{-\mathbf{k}})(1 - a_{\mathbf{k}}^\dagger a_{\mathbf{k}}) - a a^\dagger a_{\mathbf{k}}^\dagger a_{\mathbf{k}} b_{-\mathbf{k}}^\dagger b_{-\mathbf{k}}$$

$$= a^\dagger a(1 - b_{-\mathbf{k}}^\dagger b_{-\mathbf{k}} - a_{\mathbf{k}}^\dagger a_{\mathbf{k}}) - a_{\mathbf{k}}^\dagger a_{\mathbf{k}} b_{-\mathbf{k}}^\dagger b_{-\mathbf{k}} \,. \quad (19)$$

This is like Eq. (3.8), except that the field number operator $a^\dagger a$ appears explicitly *and* there is the extra contribution $a_{\mathbf{k}}^\dagger a_{\mathbf{k}} b_{-\mathbf{k}}^\dagger b_{-\mathbf{k}}$, which occurs because $a a^\dagger = a^\dagger a + 1$. This extra contribution leads to spontaneous emission. The quantized field version of Eq. (3.9) is given in the rotating-wave approximation by

$$\frac{d}{dt} a^\dagger b_{-\mathbf{k}} a_{\mathbf{k}} = -i(\omega_{\mathbf{k}} - \Omega) a^\dagger b_{-\mathbf{k}} a_{\mathbf{k}}$$

$$+ i g_{\mathbf{k}}[a^\dagger a(1 - b_{-\mathbf{k}}^\dagger b_{-\mathbf{k}} - a_{\mathbf{k}}^\dagger a_{\mathbf{k}}) - a_{\mathbf{k}}^\dagger a_{\mathbf{k}} b_{-\mathbf{k}}^\dagger b_{-\mathbf{k}}] \,, \quad (20)$$

Unlike Eq. (3.9), this equation is in an interaction picture since the frequency difference $\omega_{\mathbf{k}} - \Omega$ appears instead of the optical frequency $\omega_{\mathbf{k}}$ alone. Taking the expectation value of Eq. (20) and including an average over carrier-carrier scattering, we have

$$\frac{d}{dt} \langle a^\dagger b_{-\mathbf{k}} a_{\mathbf{k}} \rangle = -[\gamma_k + i(\omega_{\mathbf{k}} - \Omega)] \langle a^\dagger b_{-\mathbf{k}} a_{\mathbf{k}} \rangle$$

$$+ i g_{\mathbf{k}}[\langle a^\dagger a(1 - b_{-\mathbf{k}}^\dagger b_{-\mathbf{k}} - a_{\mathbf{k}}^\dagger a_{\mathbf{k}}) \rangle - \langle a_{\mathbf{k}}^\dagger a_{\mathbf{k}} b_{-\mathbf{k}}^\dagger b_{-\mathbf{k}} \rangle] \,. \quad (21)$$

For processes like spontaneous emission that are slow compared to the carrier-carrier scattering rates, this equation can be solved in "steady state" by setting the time derivative to zero and solving, that is

$$\langle a^\dagger b_{-\mathbf{k}} a_{\mathbf{k}} \rangle = \frac{i g_{\mathbf{k}} \langle a^\dagger a(1 - b_{-\mathbf{k}}^\dagger b_{-\mathbf{k}} - a_{\mathbf{k}}^\dagger a_{\mathbf{k}}) - a_{\mathbf{k}}^\dagger a_{\mathbf{k}} b_{-\mathbf{k}}^\dagger b_{-\mathbf{k}} \rangle}{\gamma_k + i(\omega_{\mathbf{k}} - \Omega)} \,. \quad (22)$$

To see how the excited-state operator $a_{\mathbf{k}}^\dagger b_{-\mathbf{k}}^\dagger b_{-\mathbf{k}} a_{\mathbf{k}} = a_{\mathbf{k}}^\dagger a_{\mathbf{k}} b_{-\mathbf{k}}^\dagger b_{-\mathbf{k}}$ decays, we write its equation of motion as

$$\frac{d}{dt} a_{\mathbf{k}}^{\dagger} b_{-\mathbf{k}}^{\dagger} b_{-\mathbf{k}} a_{\mathbf{k}} = - i[g_{\mathbf{k}}^{*} a^{\dagger} b_{-\mathbf{k}} a_{\mathbf{k}} a_{\mathbf{k}}^{\dagger} b_{-\mathbf{k}}^{\dagger} b_{-\mathbf{k}} a_{\mathbf{k}} - g_{\mathbf{k}} a_{\mathbf{k}}^{\dagger} b_{-\mathbf{k}}^{\dagger} b_{-\mathbf{k}} a_{\mathbf{k}} a_{\mathbf{k}}^{\dagger} b_{-\mathbf{k}}^{\dagger} a]$$

$$= - i[g_{\mathbf{k}}^{*} a^{\dagger} b_{-\mathbf{k}} a_{\mathbf{k}} - g_{\mathbf{k}} a_{\mathbf{k}}^{\dagger} b_{-\mathbf{k}}^{\dagger} a] . \qquad (23)$$

Taking the expectation value of Eq. (22) for a state with *no* photons (the vacuum state) and including a formal collision term, we have

$$\frac{d}{dt} \langle a_{\mathbf{k}}^{\dagger} b_{-\mathbf{k}}^{\dagger} b_{-\mathbf{k}} a_{\mathbf{k}} \rangle_{vac} = \frac{d}{dt} \langle a_{\mathbf{k}}^{\dagger} b_{-\mathbf{k}}^{\dagger} b_{-\mathbf{k}} a_{\mathbf{k}} \rangle \bigg|_{col} - [i g_{\mathbf{k}}^{*} \langle a^{\dagger} b_{-\mathbf{k}} a_{\mathbf{k}} \rangle_{vac} + \text{adj.}] . (24)$$

In the vacuum state, Eq. (22) gives

$$\langle a^{\dagger} b_{-\mathbf{k}} a_{\mathbf{k}} \rangle_{vac} = - \frac{i \langle g_{\mathbf{k}} a_{\mathbf{k}}^{\dagger} a_{\mathbf{k}} b_{-\mathbf{k}}^{\dagger} b_{-\mathbf{k}} \rangle_{vac}}{\gamma_k + i(\omega_{\mathbf{k}} - \Omega)} . \qquad (25)$$

Substituting this and its adjoint into Eq. (24) and moving the $a_{\mathbf{k}}$ to the left two operators, we find

$$\frac{d}{dt} \langle a_{\mathbf{k}}^{\dagger} a_{\mathbf{k}} b_{-\mathbf{k}}^{\dagger} b_{-\mathbf{k}} \rangle_{vac} = \frac{d}{dt} \bigg|_{col} \langle a_{\mathbf{k}}^{\dagger} a_{\mathbf{k}} b_{-\mathbf{k}}^{\dagger} b_{-\mathbf{k}} \rangle_{vac} - \frac{2\gamma_k |g_{\mathbf{k}}|^2 \langle a_{\mathbf{k}}^{\dagger} a_{\mathbf{k}} b_{-\mathbf{k}}^{\dagger} b_{-\mathbf{k}} \rangle_{vac}}{\gamma_k^{\,2} + (\omega_{\mathbf{k}} - \Omega)^2} .$$

$$(26)$$

Hence we see that in the presence of a single mode of the radiation field, the excited state decays even in the absence of photons, i.e., spontaneous emission occurs.

If the semiconductor is in free space, there are modes in all 4π steradians and with all frequencies. To calculate the total spontaneous emission "induced" by all these vacuum modes, we need to sum Eq. (26) over all modes. The easiest way to do this is to convert the sum into an integral by introducing the density of states for the electromagnetic field. This density of states is derived in a fashion similar to that for the carriers in Sec. 2-3 [see Eqs. (16) through (19)], but the relationship between energy and wave number is quite different, namely $K = \Omega n/c$. More specifically, consider a large box of dimensions L_x, L_y, and L_z. The x-axis modes in this box are characterized by wave numbers K_x such that

$$K_x = \frac{2\pi n}{L_x} , \tag{27}$$

where n is an integer ranging from $-\infty$ to ∞. For L_x sufficiently large, we can assume an essentially continuous range of values for K_x. Then we can replace the summation over K_x by an integral

$$\sum_x^K \rightarrow \int_{-\infty}^{\infty} dK_x \, \frac{dn}{dK_x} , \tag{28}$$

where

$$\frac{dn}{dK_x} = \frac{L_x}{2\pi}$$

is the number of states within the interval K_x and $K_x + dK_x$. Using a similar argument for the other two components of $\mathbf{k}$ gives

$$\sum_{\mathbf{K}} f(\mathbf{K}) = \frac{V}{(2\pi)^3} \int_{-\infty}^{\infty} dK_x \int_{-\infty}^{\infty} dK_y \int_{-\infty}^{\infty} dK_z f(\mathbf{K})$$

$$= \frac{Vn^3}{(2\pi c)^3} \int_0^{\infty} d\Omega \Omega^2 \int_0^{\pi} d\theta \sin\theta \int_0^{2\pi} d\phi f(\Omega,\theta,\phi) , \tag{29}$$

where the volume $V = L_x L_y L_z$, $f(\mathbf{K})$ is a function we wish to sum over, and in the second line we switch to spherical coordinates and use $K = \Omega n/c$. For spontaneous emission into all the modes of free space, we need to sum Eq. (26) over $\mathbf{K}$, that is, we need to evaluate

$$\sum_{\mathbf{K}} \frac{2\gamma_k |g_{\mathbf{k}}|^2}{\gamma_k^2 + (\omega_{\mathbf{k}} - \Omega)^2} = \frac{Vn^3}{(2\pi c)^3} \int_0^{\infty} d\Omega \Omega^2 \int_0^{\pi} d\theta \sin\theta \int_0^{2\pi} d\phi \frac{2\gamma_k |g_{\mathbf{k}}(\Omega,\theta)|^2}{\gamma_k^2 + (\omega_{\mathbf{k}} - \Omega)^2} . \tag{30}$$

First note that for a plane-wave mode propagating in the $\mathbf{K}$ direction, two transverse field polarizations are possible. Denoting these by the unit vectors $\mathbf{e}_\sigma$, we have

$$|g_{\mathbf{k}}(\Omega,\theta)|^2 = \hbar^{-2} \sum_{\sigma=1}^{2} |\langle 1\ 1|e\mathbf{r}\cdot\mathbf{e}_\sigma|0\ 0\rangle\mathcal{E}_\Omega|^2$$

$$= |\mathcal{E}_\Omega \mu_k/\hbar|^2 \sin^2\theta \left|\cos^2\phi + \sin^2\phi\right| = |\mathcal{E}_\Omega \mu_k \sin\theta/\hbar|^2 \ , \quad (31)$$

which is independent of the azimuthal coordinate ϕ. For running waves, the square of the "electric field per photon" is given by

$$\mathcal{E}_\Omega{}^2 = \hbar\Omega/2\epsilon_0 V \ . \tag{32}$$

In substituting Eq. (31) into Eq. (30), we encounter the integral

$$\int_0^\pi d\theta \sin^3\theta = \int_{-1}^{1} d(\cos\theta) (1 - \cos^2\theta) = \frac{4}{3} \ . \tag{33}$$

Substituting Eqs. (31) and (32) into (30) and using (33), we find

$$\sum_{\mathbf{K}} \frac{2\gamma_k |g_{\mathbf{k}}|^2}{\gamma_k{}^2 + (\omega_{\mathbf{k}} - \Omega)^2} = \frac{n^3}{6\pi^2\hbar c^3\epsilon_0} \int_0^\infty d\Omega\,\Omega^3 |\mu_k|^2 \frac{2\gamma_k}{\gamma_k{}^2 + (\omega_{\mathbf{k}} - \Omega)^2}$$

$$\simeq \frac{\omega_{\mathbf{k}}^3 |\mu_k|^2 n^3}{3\pi\hbar c^3\epsilon_0} \ , \tag{34}$$

where we suppose that $\Omega^3 |\mu_k|^2$ varies sufficiently little in the range γ_k that they can be evaluated at the peak $\omega_{\mathbf{k}}$ of the Lorentzian lineshape function $\gamma_k/(\gamma_k{}^2 + (\omega_{\mathbf{k}} - \Omega)^2)$ and taken outside the integral.

In our derivation, we have assumed that carrier-carrier collisions are independent of the electromagnetic vacuum, which is an excellent approximation. However we have also assumed that all plane-wave modes of space are possible. This assumption is undoubtedly poor in some semiconductor structures, particularly those that are constructed to suppress field propagation in various directions. By suppressing the field modes, the spontaneous emission coefficient (34) is correspondingly reduced in size. This is called *supressed spontaneous emission.*

Inserting Eq. (34) into Eq. (26), we have

$$\frac{d}{dt}\langle a_{\mathbf{k}}^{\dagger}a_{\mathbf{k}}b_{-\mathbf{k}}^{\dagger}b_{-\mathbf{k}}\rangle_{vac} = \frac{d}{dt}\bigg|_{col}\langle a_{\mathbf{k}}^{\dagger}a_{\mathbf{k}}b_{-\mathbf{k}}^{\dagger}b_{-\mathbf{k}}\rangle_{vac} - B_k\langle a_{\mathbf{k}}^{\dagger}a_{\mathbf{k}}b_{-\mathbf{k}}^{\dagger}b_{-\mathbf{k}}\rangle_{vac} \ , \quad (35)$$

where

$$B_k = \frac{\omega_{\mathbf{k}}^3|\mu_k|^2n^3}{3\pi\hbar c^3\epsilon_0} = \frac{1}{4\pi\epsilon_0}\frac{4\omega_{\mathbf{k}}^3|\mu_k|^2n^3}{3\hbar c^3} \qquad (36)$$

is the Weisskopf-Wigner spontaneous emission coefficient for a semiconductor medium with index of refraction n. In cgs units, the factor $1/4\pi\epsilon_0$ is missing. The electron probability function $n_{e\mathbf{k}} \equiv \langle a_{\mathbf{k}}^{\dagger}a_{\mathbf{k}}\rangle$ is given by

$$n_{e\mathbf{k}} = \langle a_{\mathbf{k}}^{\dagger}a_{\mathbf{k}}b_{-\mathbf{k}}^{\dagger}b_{-\mathbf{k}}\rangle + \langle a_{\mathbf{k}}^{\dagger}a_{\mathbf{k}}(1 - b_{-\mathbf{k}}^{\dagger}b_{-\mathbf{k}})\rangle \ .$$

The expectation value $\langle a_{\mathbf{k}}^{\dagger}a_{\mathbf{k}}(1 - b_{-\mathbf{k}}^{\dagger}b_{-\mathbf{k}})\rangle$ doesn't couple to the radiation field and hence the equation of motion for $n_{e\mathbf{k}}$ and similarly that for $n_{h\mathbf{k}}$ include the spontaneous emission contribution in Eq. (35) as assumed phenomenologically in Eq. (3.15).

9-3. Quantum Langevin Equations

In this section, we develop a quantum Langevin formalism to treat noise problems in semiconductor lasers. The formalism can also be used to describe noise in other problems, such as resonance fluorescence and the generation of squeezed states. The basic idea is that we have a system of interest, such as a single-mode laser field, that is coupled to many other variables whose evolution is interesting only insofar as it affects the system of interest. For example, the laser mode is coupled to the many modes of free space via transmission through the cavity mirrors. In principle, one can describe all the variables, those for the system and the others, by writing an infinite coupled set of Heisenberg equations of motion. The Langevin method simplifies this infinite set of equations subject to the assumption that the "other" variables act as a reservoir that is unperturbed by the system variables. This reservoir is assumed to have many degrees of

8g

freedom and a sufficiently wide bandwidth that it responds much faster than the system variables can respond. As such, we can adiabatically eliminate the reservoir variables from the coupled set of Heisenberg equations of motion, thereby obtaining modified equations of motion for the system variables. These modified equations are called Langevin equations and consist of the sum of a drift operator and a noise operator. This reduction is the Heisenberg-picture counterpart to the reduced density operator approach in the Schrödinger picture. Both methods dramatically reduce the number of variables one has to deal with and in the process introduce noise. In the reduced density operator approach, the noise shows up as density operator mixtures, instead of the original pure case.

To render the quantum Langevin method more concrete, we first treat the problem of a quantized simple harmonic oscillator coupled to reservoir of such oscillators. Among other applications, this problem models the coupling of a laser mode to the modes of free space. Starting with the coupled Heisenberg equations of motion, we adiabatically eliminate the reservoir using a Weisskopf-Wigner approximation. Due to the simplicity of the simple harmonic oscillator, we obtain an explicit formula for the noise operators. We show how integrating the correlation function of the noise operators gives the damping constant of the system oscillator. This is an example of the fluctuation-dissipation theorem. We then discuss a general coupled set of Langevin equations and derive the generalized Einstein relation that enables us to calculate the diffusion coefficients in terms of the drift operators.

The simple harmonic oscillator coupled to a reservoir of harmonic oscillators is described by the total Hamiltonian

$$\mathcal{H} = \mathcal{H}_s + \mathcal{H}_r + \mathcal{V},$$ (37)

where the unperturbed Hamiltonian of the system is given by

$$\mathcal{H}_s = \hbar\Omega \, a^\dagger a,$$ (38)

the unperturbed Hamiltonian of the reservoir is given by

$$\mathcal{H}_r = \sum_j \hbar\omega_j \, b_j^\dagger b_j,$$ (39)

and the system-reservoir interaction energy is given by

$$\mathcal{V} = \hbar \sum_j (g_j a^\dagger b_j + g_j{}^* b_j{}^\dagger a) \, . \tag{40}$$

As such, the exchange of energy between system and reservoir is assumed to consist of the simultaneous creation of a quantum of excitation of the system with annihilation of a quantum in the jth mode of the reservoir, or vice versa. Note that for simplicity we drop the $\frac{1}{2}\hbar\Omega$ and $\frac{1}{2}\hbar\omega_j$ that appear in the usual simple-harmonic oscillator Hamiltonians, since these terms just lead to a shift in the energy zero, which gives a trivial overall phase factor. As for the interaction energy (15), we see that Eq. (40) is written in the rotating-wave approximation.

Using the total Hamiltonian of Eq. (37), we find the equations of motion

$$\dot{a}(t) = -i\Omega a(t) - i \sum_j g_j b_j(t) \, , \tag{41}$$

$$\dot{b}_j(t) = -i\omega_j b_j(t) - ig_j{}^* a(t) \, . \tag{42}$$

Integrating Eq. (42) formally, we find

$$b_j(t) = b_j(0)e^{-i\omega_j t} - ig_j{}^* \int_0^t dt' a(t')e^{-i\omega_j(t-t')}$$

$$\equiv b_{free}(t) + b_{radiated}(t) \, . \tag{43}$$

The b_{free} term on the RHS of Eq. (43) is the unperturbed solution of Eq. (42), and describes the free evolution of $b_j(t)$ in the absence of interaction with the system. The $b_{radiated}$ term gives the modification of this free evolution due to the coupling with the system, and shows that $a(t)$ is a source of $b_j(t)$. Inserting Eq. (43) into Eq. (41), we find

$$\dot{a}(t) = -i\Omega a(t) - i \sum_j g_j b_j(0)e^{-i\omega_j t} - \sum_j |g_j|^2 \int_0^t dt' a(t')e^{-i\omega_j(t-t')} \, . \tag{44}$$

Here the first summation gives fluctuations and the second gives the radiation reaction.

To separate the simple, rapid free evolution of $a(t)$ at the frequency Ω [see Eq. (42)] from the fast evolution due to the large bandwidth of the reservoir, we introduce the slowly varying operator

$$A(t) = a(t)e^{i\Omega t} , \quad \text{with} \quad [A(t), A^{\dagger}(t)] = 1 . \tag{45}$$

This transformation amounts to going into a Heisenberg interaction picture, where $A(t)$ contains the time dependence due to the interaction energy (40). From Eq. (44), the evolution of $A(t)$ is given by

$$\dot{A}(t) = - \sum_j |g_j|^2 \int_0^t dt' A(t') e^{-i(\omega_j - \Omega)(t - t')} + F(t) , \tag{46}$$

where $F(t)$ is the *noise operator*

$$F(t) = -i \sum_j g_j b_j(0) e^{i(\Omega - \omega_j)t} . \tag{47}$$

Note that this operator varies rapidly in time due to the presence of all the reservoir frequencies.

An integral similar to that on the RHS of Eq. (46) occurs in the Weisskopf–Wigner theory of spontaneous emission. To evaluate the integral, we replace the sum over j by an integral over ω with a density of states $\mathcal{D}(\omega)$ and suppose that $A(t)$ varies little in the time given by the inverse of the reservoir bandwidth. This allows us to extend the limit of integration to infinity. Using the representation

$$\lim_{t \to \infty} \int_0^t dt' e^{-i(\Omega - \omega)(t - t')} = \pi\delta(\Omega - \omega) - \mathcal{P}\left[\frac{i}{\Omega - \omega}\right] \tag{48}$$

of the delta-function, we obtain

$$\boxed{\dot{A}(t) = - \frac{\gamma}{2} A(t) + F(t)} , \tag{49}$$

where the decay constant γ is given by

$$\gamma = 2\pi \mathscr{D}(\Omega) |g(\Omega)|^2 \ . \tag{50}$$

Here we neglect the small shift due to the principal part in Eq. (49), which is the equivalent of the Lamb shift for our simple harmonic oscillator. Equation (49) is a quantum Langevin equation for the annihilation operator $A(t)$. It consists of the drift term $-\frac{1}{2}\gamma A(t)$ plus the noise operator $F(t)$.

We suppose that the reservoir is in thermodynamic equilibrium, that is, that it is described by the diagonal density operator

$$\rho = \frac{\exp(-\beta \mathscr{H}_r)}{\text{tr}\{\exp(-\beta \mathscr{H}_r)\}} = \frac{1}{Z} \prod_j e^{-\beta \hbar \omega_j b_j^\dagger b_j} \ , \tag{51}$$

where $Z = \text{tr}\{\exp(-\beta \mathscr{H}_r)\}$. Since ρ is diagonal, $\langle F(t') \rangle_r$ vanishes (has zero mean) and

$$\langle \dot{A} \rangle = -\frac{\gamma}{2} \langle A \rangle \ . \tag{52}$$

On the other hand, calculating the correlation function $\langle F^\dagger(t)F(t') \rangle_r$, we have [using Eq. (47)]

$$\langle F^\dagger(t)F(t') \rangle_r = \sum_{j,k} g_j^* \langle b_j^\dagger(0)b_k(0) \rangle_r e^{-i(\Omega - \omega_j)t} e^{i(\Omega - \omega_k)t'} \ . \tag{53}$$

Using ρ of Eq. (48), we have

$$\langle b_j^\dagger(0)b_k(0) \rangle_r = \bar{n}_j \delta_{jk} \ , \tag{54}$$

where the average number of thermal quanta in mode j is given by

$$\bar{n}_j = \text{tr}\{\rho b_j^\dagger b_j\} = \frac{1}{e^{\beta \hbar \omega_j} - 1} \ . \tag{55}$$

Substituting Eq. (54) into Eq. (53), we have the correlation function

$$\langle F^\dagger(t)F(t') \rangle_r = \sum_j |g_j|^2 \bar{n}_j e^{-i(\Omega - \omega_j)(t - t')} \ .$$

The form of this quantity is very similar to that of Eq. (46), except that there is no integral over time and $\bar{n}_j$ appears instead of $A(t')$. Accordingly converting the Σ_j to an integral over ω and noting that $\bar{n}(\omega)$ is a slowly varying function of ω, we use Eq. (48) to find

$$\langle F^\dagger(t)F(t')\rangle_r = \gamma\bar{n}(\Omega)\delta(t-t')$$

$$= 2\langle D_{A^\dagger A}\rangle_r\delta(t-t') , \tag{56}$$

where γ is given by Eq. (50). Further noting that $[b_j, b_j^\dagger] = 1$, we have

$$\langle F(t)F^\dagger(t')\rangle_r = \gamma(\bar{n}(\Omega)+1)\delta(t-t')$$

$$= 2\langle D_{AA^\dagger}\rangle_r\delta(t-t') , \tag{57}$$

and since $\langle b_j^\dagger(0)b_k^\dagger(0)\rangle_r = \langle b_j(0)_k(0)\rangle_r = 0$ [unless the reservoir is *squeezed*: see Sec. 16-4 of Meystre and Sargent (1991)],

$$\langle F^\dagger(t)F^\dagger(t')\rangle_r = \langle F(t)F(t')\rangle_r = 0 . \tag{58}$$

These correlation functions are characteristic of Markoffian random processes, which means that they represent fluctuations in a reservoir with an essentially zero memory. Here, of course, as everywhere that a Dirac delta function is used to model the time dependence of a physical process, the process has a nonzero response time, but that time is so short compared to other times that to all intents and purposes, you can set it equal to zero.

The δ-function weight factors, $\langle D_{A^\dagger A}\rangle_r$ and $\langle D_{AA^\dagger}\rangle_r$, are called *diffusion coefficients* and as Eqs. (56) and (57) show are intimately related to the drift coefficient γ. Alternatively, we can integrate Eq. (56) to find the simple fluctuation dissipation theorem

$$\gamma = \frac{1}{\bar{n}(\Omega)}\int_{-\infty}^{\infty} dt'\, \langle F^\dagger(t)F(t')\rangle_r . \tag{59}$$

Our main interest in correlation functions like Eqs. (56) and (57) is because they occur in the semiconductor-laser noise spectra and the laser linewidth. Before proceeding to calculate such quantities, we note that the noise operators are also needed to preserve the unity value of the commutator $[A(t), A^\dagger(t)]$. Specifically, if we drop the $F(t)$ in Eq. (49), we would have

$$\frac{d}{dt}[A(t),\, A^\dagger(t)] = -\, \gamma[A(t),\, A^\dagger(t)] \,, \tag{60}$$

which predicts that the commutator decays to zero in horrible violation of the laws of quantum mechanics! The noise operator $F(t)$ saves the unity value of the commutator and tells us how the number operator $A^\dagger A$ evolves in time. Since the expectation value of this quantity can model the laser mode intensity when appropriately coupled to the semiconductor medium, we consider it explicitly. The method is an example of the approach used in the derivation of the more general fluctuation-dissipation theorem that follows.

The system number operator obeys the equation of motion

$$\begin{aligned}
\frac{d}{dt}(A^\dagger(t)A(t)) &= \dot{A}^\dagger(t)A(t) + A^\dagger(t)\dot{A}(t) \\
&= -\, \gamma A^\dagger(t)A(t) + F^\dagger(t)A(t) + A^\dagger(t)F(t) \,.
\end{aligned} \tag{61}$$

To evaluate the correlation $F^\dagger(t)A(t)$, we write $A(t)$ as

$$A(t) = A(t-\Delta t) + \int_{t-\Delta t}^{t} dt'\, \dot{A}(t') \,, \tag{62}$$

where Δt is an interval much smaller than $1/\gamma$, but much larger than the correlation time of the reservoir. As such, the correlation $\langle A(t-\Delta t)F(t)\rangle_r = 0$, since $A(t-\Delta t)$ doesn't know about the whims of the future noise operator $F(t)$. Substituting the Langevin Eq. (49) into Eq. (62), we have the expectation value

$$\langle F^\dagger(t)A(t)\rangle_r = \langle F^\dagger(t)A(t-\Delta t)\rangle_r + \int_{t-\Delta t}^{t} dt'\, \langle F^\dagger(t)[-\tfrac{1}{2}\gamma A(t') + F(t')]\rangle_r$$

$$= \int_{t-\Delta t}^{t} dt'\, \langle F^\dagger(t)F(t')\rangle_r = \gamma\bar{n}(\Omega) \,, \tag{63}$$

where we use Eq. (56) and drop the $\tfrac{1}{2}\gamma\langle F^\dagger(t)A(t')\rangle_r$, since it is nonzero only over a set of measure zero, i.e., at $t' = t$. $\langle A^\dagger(t)F(t)\rangle_r$ is also given by Eq. (63), since that equation is real. Substituting these values into Eq. (61), we have

$$\frac{d}{dt}(A^\dagger(t)A(t)) = -\gamma A^\dagger(t)A(t) + \gamma\bar{n}(\Omega) + F_{A^\dagger A}(t) , \tag{64}$$

where the intensity noise operator $F_{A^\dagger A}(t)$ has zero mean. Similarly

$$\frac{d}{dt}(A(t)A^\dagger(t)) = -\gamma A(t)A^\dagger(t) + \gamma(\bar{n}(\Omega)+1) + F_{AA^\dagger}(t) , \tag{65}$$

which implies that

$$\frac{d}{dt}[A(t), A^\dagger(t)] = -\gamma [A(t), A^\dagger(t)] + \gamma + F_{AA^\dagger}(t) - F_{A^\dagger A}(t) .$$

The expectation value of this equation

$$\frac{d}{dt}\langle[A(t), A^\dagger(t)]\rangle = \gamma - \gamma \langle[A(t), A^\dagger(t)]\rangle . \tag{66}$$

Therefore if $[A(t), A^\dagger(t)] = 1$ for any time t, we have

$$\frac{d}{dt}\langle[A(t), A^\dagger(t)]\rangle = 0$$

for all time and $\langle[A(t), A^\dagger(t)]\rangle$ is conserved.

Even more generally, we note that Eq. (66) holds without the reservoir average. Using Eq. (61) and its companion for $AA^\dagger$, we have

$$\frac{d}{dt}[A(t), A^\dagger(t)) = -\gamma[A(t), A^\dagger(t)] + [F(t), A^\dagger(t)] + [A(t), F^\dagger(t)] . \tag{67}$$

Note from Eq. (47) that

$$F(t)F^\dagger(t') = \sum_{j,k} g_j g_k^* b_j(0)b_k^\dagger(0)e^{i(\Omega-\omega_j)t} e^{-i(\Omega-\omega_k)t'}$$

$$= \sum_{j,k} g_j g_k^* b_k^\dagger(0)b_j(0)e^{i(\Omega-\omega_j)t} e^{-i(\Omega-\omega_k)t'} + \sum_j |g_j|^2 e^{i(\Omega-\omega_j)(t-t')}$$

$$= F^\dagger(t')F(t) + \gamma\delta(t-t') ,$$

i.e., $[F(t), F^\dagger(t')] = \gamma\delta(t - t')$. With Eqs. (62) and (49), we have

$$[F(t), A^\dagger(t)] = \int_{t-\Delta t}^{t} dt'[F(t), F^\dagger(t')] = \frac{\gamma}{2} \, ,$$

since $[F(t), A^\dagger(t-\Delta t)] = 0$. Similarly $[A(t), F^\dagger(t)] = \frac{1}{2}\gamma$, which gives Eq. (66) without reservoir averages.

Generalized Langevin Equations and Einstein Relations

We can write these results for an arbitrary set of system operators $A_\mu(t)$ coupled to an arbitrary set of Markoffian reservoirs. This generalization is needed for the semiconductor laser, since in addition to the field-cavity problem describable with the Langevin formalism given up to here, we need to couple the field to the semiconductor operators, which leads to additional noise sources such as spontaneous emission and pump fluctuations.

We write the Langevin equations of motion for the system operators as

$$\dot{A}_\mu = D_\mu(t) + F_\mu(t) \, , \tag{68}$$

where D_μ is the drift term and F_μ is the noise operator, which we suppose has the correlation functions

$$\langle F_\mu(t')F_\nu(t'')\rangle = 2\langle D_{\mu\nu}\rangle\delta(t'-t'') \, , \tag{69}$$

where $\langle D_{\mu\nu}\rangle$ is a diffusion coefficient. The problem is to determine $\langle D_{\mu\nu}\rangle$ given the equation of motion (68). Unlike the simple harmonic oscillator coupled to a reservoir of such oscillators, we do *not* in general know what $F_\mu(t)$ is. For the purpose of calculating noise spectra, however, we don't need to know what $F_\mu(t)$ is; all we need to know is the expectation value of the second moments, that is, the $\langle D_{\mu\nu}\rangle$.

To calculate the $\langle D_{\mu\nu}\rangle$, we note the identity

$$A_\mu(t) = A_\mu(t-\Delta t) + \int_{t-\Delta t}^{t} dt' \, \dot{A}_\mu(t') \, , \tag{70}$$

which gives the system-operator, noise-operator correlation function

$$\langle A_\mu(t)F_\nu(t)\rangle = \langle A_\mu(t-\Delta t)F_\nu(t)\rangle + \int_{t-\Delta t}^t dt' \langle (D_\mu(t') + F_\mu(t'))F_\nu(t)\rangle . \quad (71)$$

Because the operator $A_\mu(t')$ at time t' cannot be affected by a fluctuation at a later time t, the first term on the RHS of Eq. (71) is zero. Similarly the correlation $\langle D_\mu(t')F_\nu(t)\rangle$ is zero except at the point $t'=t$, but the integration interval is zero (it's a set of measure zero). All that remains is

$$\langle A_\mu(t)F_\nu(t)\rangle = \int_{t-\Delta t}^t dt' \langle F_\mu(t')F_\nu(t)\rangle . \quad (72)$$

Substituting Eq. (69) into (72) we have

$$\langle A_\mu(t)F_\nu(t)\rangle = \langle D_{\mu\nu}\rangle , \quad (73)$$

where the 2 in Eq. (69) is cancelled by the $\frac{1}{2}$ encountered in integrating only half way through the δ function. Similarly we find

$$\langle F_\mu(t)A_\nu(t)\rangle = \langle D_{\mu\nu}\rangle . \quad (74)$$

Thus we see that the system operator $A_\mu(t)$ may be correlated with the noise operator $F_\nu(t)$ at the time t, although $\langle A_\mu(t-\Delta t)F_\nu(t)\rangle$ vanishes. Note that many of the $\langle D_{\mu\nu}\rangle$ also vanish, since fluctuations in different reservoirs are typically uncorrelated, and others vanish as in Eq. (58).

We use Eqs. (73) and (74) to determine the equation of motion for the average $\langle A_\mu A_\nu\rangle$. From Eq. (68), we have

$$\frac{d}{dt}\langle A_\mu A_\nu\rangle = \langle \dot{A}_\mu A_\nu\rangle + \langle A_\mu \dot{A}_\nu\rangle$$
$$= \langle D_\mu A_\nu\rangle + \langle F_\mu A_\nu\rangle + \langle A_\mu D_\nu\rangle + \langle A_\mu F_\nu\rangle . \quad (75)$$

Substituting Eqs. (73) and (74) and rearranging, we find

$$\boxed{2\langle D_{\mu\nu}\rangle = \frac{d}{dt}\langle A_\mu A_\nu\rangle - \langle A_\mu D_\nu\rangle - \langle D_\mu A_\nu\rangle} . \quad (76)$$

This equation is called the *generalized Einstein relation*. It shows that the diffusion coefficients $\langle D_{\mu\nu}\rangle$ are related to the drift coefficients D_μ and D_ν and thus comprises a quantum fluctuation-dissipation theorem. This equation makes it possible to calculate the diffusion coefficients from the drift coefficients, provided one can *independently* calculate the equation of motion for $\langle A_\mu A_\nu \rangle$. If you use the first line of Eq. (75) to calculate this equation of motion, obviously you just get an identity. In the next section, we use this method to calculate the diffusion coefficients needed for the semiconductor laser linewidth and noise spectra.

In our discussion of intensity noise spectra, we derive the drift terms and diffusion coefficients for the field intensity and phase from those for the annihilation and creation operators. To do this, we note that the equation of motion for an arbitrary function $M(A)$ of the system operators $A = (A_1, A_2, ...)$ can be written using the chain rule as

$$\frac{dM(A)}{dt} = \frac{\partial M}{\partial t} + \sum_\nu \frac{\partial M}{\partial A_\nu} \frac{dA_\nu}{dt} = \frac{\partial M}{\partial t} + \sum_\nu \frac{\partial M}{\partial A_\nu}[D_\nu + F_\nu] . \qquad (77)$$

This value and others that follow may not preserve correct oprator ordering; we use these formulas in a partially classical way, in which the quantum character is preserved in the diffusion coefficients. In general the noise operator defined by

$$\sum_\nu \frac{\partial M}{\partial A_\nu} F_\nu \qquad (78)$$

has a nonzero mean. We can distill this mean out of Eq. (78) and include it in the formula for the new drift term. To this end, we expand $\partial M(A(t))/\partial A_\nu$ in a first-order Taylor series

$$\frac{\partial M(A(t))}{\partial A_\nu} \simeq \frac{\partial M(A(t-\Delta t))}{\partial A_\nu} + \sum_\mu [A_\mu(t) - A_\mu(t-\Delta t)] \frac{\partial^2 M}{\partial A_\mu \partial A_\nu} .$$

Inserting this into Eq. (78) and taking the reservoir average, we have

$$\sum_\nu \langle \frac{\partial M}{\partial A_\nu} F_\nu \rangle = \sum_{\mu,\nu} \langle A_\mu(t) F_\nu(t) \frac{\partial^2 M}{\partial A_\mu \partial A_\nu} \rangle = \sum_{\mu,\nu} \langle D_{\mu\nu} \frac{\partial^2 M}{\partial A_\mu \partial A_\nu} \rangle , \qquad (79)$$

where as for Eq. (71), we drop correlations between system quantities at the time $t - \Delta t$ and the future noise operator $F_\nu(t)$. Substituting Eq. (79) into Eq. (77), we have the Langevin equation

$$\frac{dM(\mathbf{A})}{dt} = \frac{\partial M}{\partial t} + \sum_\nu \frac{\partial M}{\partial A_\nu} D_\nu + \sum_{\mu,\nu} \langle D_{\mu\nu} \frac{\partial^2 M}{\partial A_\mu \partial A_\nu} \rangle + F_M , \qquad (80)$$

where F_M is given by Eq. (78) minus Eq. (79), so that $\langle F_M \rangle = 0$.

In particular, the operator $A'_\rho(\mathbf{A})$ belonging to the set of transformed system operators $\mathbf{A}' = (A_1', A_2', ...)$ obeys the Langevin equation

$$\frac{dA'_\rho}{dt} = \frac{\partial A'_\rho}{\partial t} + \sum_\mu \frac{\partial A'_\rho}{\partial A_\mu} D_\mu + \sum_{\mu,\nu} \langle D_{\mu\nu} \frac{\partial^2 A'_\rho}{\partial A_\mu \partial A_\nu} \rangle + F'_\rho$$

$$= D'_\rho + F'_\rho . \qquad (81)$$

To get the transformed diffusion coefficients, we use Eq. (78) with $M = A'_\sigma$ along with the expansion

$$A'_\rho(\mathbf{A}(t)) \simeq A'_\rho(\mathbf{A}(t-\Delta t)) + \sum_\mu [A_\mu(t) - A_\mu(t-\Delta t)] \frac{\partial A'_\rho}{\partial A_\mu}$$

to evaluate the quantity $\langle A'_\rho(\mathbf{A}(t)) F'_\sigma(t) \rangle$. We find

$$\langle A'_\rho(\mathbf{A}(t)) F'_\sigma(t) \rangle = \sum_{\mu,\nu} \langle A_\mu(t) F_\nu(t) \frac{\partial A'_\rho}{\partial A_\mu} \frac{\partial A'_\sigma}{\partial A_\nu} \rangle ,$$

where correlations between system variables at time $t - \Delta t$ are $F_\nu(t)$ dropped. This gives the transformed diffusion coefficients $\langle D'_{\rho\sigma} \rangle$ in terms of the original set $\langle D_{\mu\nu} \rangle$ as

$$\langle D'_{\rho\sigma} \rangle = \sum_{\mu,\nu} \langle D_{\mu\nu} \frac{\partial A'_\rho}{\partial A_\mu} \frac{\partial A'_\sigma}{\partial A_\nu} \rangle . \qquad (82)$$

9-4. Semiconductor Langevin Equations

With the fully quantized interaction energy (15), the free-carrier semi-conductor Langevin equations are similar to the semiclassical equations of Sec. 3-1, but include quantum noise operators and the field annhilation (a) and creation ($a^\dagger$) operators. We first find the equation of motion and diffusion coefficients for the dipole operator $b_{-\mathbf{k}} a_{\mathbf{k}}$. We use these quantities to adiabatically eliminate $b_{-\mathbf{k}} a_{\mathbf{k}}$ from the field and carrier-density equations of motion.

Including decay, the dipole operator equation of motion is

$$\frac{d}{dt} b_{-\mathbf{k}} a_{\mathbf{k}} = - (\gamma + i\omega_{\mathbf{k}}) b_{-\mathbf{k}} a_{\mathbf{k}} - g_{\mathbf{k}} a (a_{\mathbf{k}}^\dagger a_{\mathbf{k}} + b_{-\mathbf{k}}^\dagger b_{-\mathbf{k}} - 1) + f_{\sigma_{\mathbf{k}}} , \quad (83)$$

where f_σ is a rapidly varying k-dependent noise operator. As in Eq. (45), it is convenient to work with slowly varying operators. Choosing a frame rotating at the field frequency ν, we define the slowly varying annihilation operator $A(t) = a(t) e^{i\nu t}$ and the slowly varying "spin-flip" operator

$$\sigma_{\mathbf{k}} = b_{-\mathbf{k}} a_{\mathbf{k}} e^{i\nu t} . \quad (84)$$

This operator makes a transition from the excited state $|3\rangle \equiv |1_e 1_h \mathbf{k}\rangle$ to the unexcited state $|0\rangle \equiv |0_e 0_h \mathbf{k}\rangle$. In terms of $\sigma_{\mathbf{k}}$ and A, Eq. (83) becomes the quantum Langevin equation

$$\dot{\sigma}_{\mathbf{k}} = - (\gamma + i\omega_{\mathbf{k}} - i\nu)\sigma_{\mathbf{k}} + ig_{\mathbf{k}} A (a_{\mathbf{k}}^\dagger a_{\mathbf{k}} + b_{-\mathbf{k}}^\dagger b_{-\mathbf{k}} - 1) + F_{\sigma_{\mathbf{k}}} , \quad (85)$$

where the k-dependent noise operator $F_{\sigma_{\mathbf{k}}} = f_{\sigma_{\mathbf{k}}} e^{i\nu t}$. Moving into this rotating frame is using a Heisenberg interaction picture.

To find the dipole-operator diffusion coefficient $\langle D_{\sigma_{\mathbf{k}}^\dagger \sigma_{\mathbf{k}}} \rangle$, we use the Einstein relation (76) and encounter the combination in Eq. (3.6), namely

$$\sigma_{\mathbf{k}}^\dagger \sigma_{\mathbf{k}} = a_{\mathbf{k}}^\dagger b_{-\mathbf{k}}^\dagger b_{-\mathbf{k}} a_{\mathbf{k}} = a_{\mathbf{k}}^\dagger a_{\mathbf{k}} b_{-\mathbf{k}}^\dagger b_{-\mathbf{k}} \equiv \sigma_{33} . \quad (86)$$

This operator is the projection operator for the excited state $|3\rangle$. Substituting this value along with Eq. (85) and its adjoint into Eq. (76), we find the diffusion coefficient

$$2\langle D_{\sigma_{\mathbf{k}}^\dagger \sigma_{\mathbf{k}}} \rangle = (\gamma + i\omega_{\mathbf{k}} - i\nu)\sigma_{33} + (\gamma - i\omega_{\mathbf{k}} + i\nu)\sigma_{33} + \dot{\sigma}_{33} \simeq 2\gamma \sigma_{33} , \quad (87)$$

where in quasiequilibrium we neglect $\dot{\sigma}_{33}$ compared to $2\gamma \sigma_{33}$. Note that the electric-dipole terms drop out immediately because combinations like $a_k a_k$ are null operators. Similarly

$$2\langle D_{\sigma_k \sigma_k}{}^\dagger\rangle = (\gamma + i\omega_k - i\nu)\sigma_{00} + (\gamma - i\omega_k + i\nu)\sigma_{00} + \dot{\sigma}_{00} \simeq 2\gamma\sigma_{00}, \quad (88)$$

where

$$\sigma_{00} = (1 - a_k^\dagger a_k)(1 - b_{-k}^\dagger b_{-k}) \qquad (89)$$

is the projection operator for the unexcited state $|0\rangle$.

In quasiequilbrium, we use the steady-state solution to Eq. (85), namely

$$\sigma_k = \frac{ig_k A(a_k^\dagger a_k + b_{-k}^\dagger b_{-k} - 1) + F_{\sigma_k}}{\gamma + i\omega_k - i\nu} = \frac{ig_k A[\sigma_{33} - \sigma_{00}] + F_{\sigma_k}}{\gamma + i\omega_k - i\nu}. \quad (90)$$

At first glance, the inclusion of F_{σ_k} in this expression seems suspicious, since F_{σ_k} contains fluctuations that are fast compared to the time scales over which σ_k itself varies. As such it would appear that $\dot{\sigma}_k$ cannot be ignored as we have done in Eq. (90). However we wish to use this equation in the relatively slowly varying field and carrier density equations of motion. A formal integral Eq. (85) yields a noise operator with a correlation function that decays exponentially with the dipole lifetime T_2. Since this is very short compared to the times over which the field and carrier density vary appreciably, we can convert this correlation function into a Dirac δ-function. The resulting diffusion coefficients are the same as those found by using F_{σ_k} as given in Eq. (90). Physically, using F_{σ_k} in Eq. (90) gives σ_k a whiter (broader bandwidth) noise spectrum than it really has. But the slowly varying field and carrier density operators are not sensitive to the difference.

The field operator equation of motion consists of Eq. (49) plus the contribution due to the interaction energy (15) and the change from the frame rotating at the passive-cavity frequency Ω to that rotating at the laser oscillation frequency ν. We find

$$\dot{A}(t) = -\,[\nu/2Q + i(\Omega - \nu)]A(t) - i\sum_k g_k^* \sigma_k + F(t), \qquad (91)$$

where we use ν/Q for the field damping constant in place of the γ in Eq.

(49). Using Eq. (90) to adiabatically eliminate $\sigma_\mathbf{k}$, we have

$$\dot{A}(t) = - [\nu/2Q + i(\Omega - \nu)]A(t) + \sum_\mathbf{k} |g_\mathbf{k}|^2 \mathscr{D}_\mathbf{k}[\sigma_{33} - \sigma_{00}]A + F_A(t) , \quad (92)$$

where the complex Lorentzian $\mathscr{D}_\mathbf{k}$ is given by Eq. (24) and the new noise operator $F_A(t)$ is given by

$$F_A(t) = F(t) - i \sum_\mathbf{k} g_\mathbf{k}^* \mathscr{D}_\mathbf{k} F_{\sigma_\mathbf{k}}(t) . \quad (93)$$

The projection operators σ_{33} and σ_{00} in Eq. (92) also contribute noise to the field, but for the sake of simplicity, we consider these contributions using a classical noise equation for the total carrier density fluctuations. The correlation function for $F_A(t)$ is given by

$$\langle F_A{}^\dagger(t)F_A(t')\rangle = \langle F^\dagger(t)F(t')\rangle + \sum_\mathbf{k} \sum_{\mathbf{k}'} g_\mathbf{k} g_{\mathbf{k}'}^* \mathscr{D}_\mathbf{k}^* \mathscr{D}_{\mathbf{k}'} \langle F_{\sigma_\mathbf{k}}^\dagger(t)F_{\sigma_{\mathbf{k}'}}(t)\rangle$$

$$= \langle F^\dagger(t)F(t')\rangle + \frac{1}{\gamma^2} \sum_\mathbf{k} |g_\mathbf{k}|^2 \mathscr{L}_\mathbf{k} \langle F_{\sigma_\mathbf{k}}^\dagger(t)F_{\sigma_\mathbf{k}}(t)\rangle$$

$$= \left[\frac{\nu}{Q}\bar{n}(\nu) + \frac{1}{\gamma} \sum_\mathbf{k} |g_\mathbf{k}|^2 \mathscr{L}_\mathbf{k} \langle \sigma_{33}\rangle \right]\delta(t-t') , \quad (94)$$

where we assume that the cavity-damping and carrier-scattering reservoirs are uncorrelated, i.e., $\langle F^\dagger(t)F_{\sigma_\mathbf{k}}\rangle = 0$ and furthermore that noise operators for different $\mathbf{k}$ are uncorrelated. Equation (94) gives the diffusion coefficient

$$2\langle D_{A\dagger A}\rangle = \frac{\nu}{Q}\bar{n}(\nu) + R_{sp} , \quad (95)$$

where the spontaneous emission rate R_{sp} into the laser mode is given by

$$R_{sp} = \frac{2}{\gamma} \sum_{\mathbf{k}} |g_{\mathbf{k}}|^2 \mathscr{L}_{\mathbf{k}} \langle \sigma_{33} \rangle .$$ (96)

Similarly,

$$2\langle D_{A\,A\dagger} \rangle = \frac{\nu}{Q}[\bar{n}(\nu) + 1] + R_{abs} ,$$ (97)

where the absorption rate R_{abs} is given by

$$R_{abs} = \frac{2}{\gamma} \sum_{\mathbf{k}} |g_{\mathbf{k}}|^2 \mathscr{L}_{\mathbf{k}} \langle \sigma_{00} \rangle .$$ (98)

The semiclassical intensity gain G is given by $R_{sp} - R_{abs}$.

We can also use Eq. (92) as in Eq. (64) to find the equation of motion for the photon number $\langle A\dagger A \rangle$. We find

$$\frac{d}{dt}\langle A^\dagger(t)A(t) \rangle = -\left[\frac{\nu}{Q} - (R_{sp} - R_{abs}) \right]\langle A^\dagger(t)A(t) \rangle + \frac{\nu}{Q}\bar{n}(\Omega) + R_{sp} .$$ (99)

In steady state, this gives the relation

$$\frac{\nu}{Q} - (R_{sp} - R_{abs}) = \frac{\frac{\nu}{Q}\bar{n}(\Omega) + R_{sp}}{\langle A^\dagger(t)A(t) \rangle} ,$$ (100)

which shows that the loss exceeds the gain by a small amount almost completely due to spontaneous emission (since at optical frequencies $\bar{n}(\Omega) \simeq 0$).

If we're only interested in fluctuations that vary little in the relaxation oscillation time of the semiconductor laser, we have all that we need to proceed. However to study fluctuations on the order of the relaxation oscillation frequency, we need to calculate the diffusion coefficients for the total carrier number as well. To this end, we consider the Langevin equation for the electron number operator $a_{\mathbf{k}}^\dagger a_{\mathbf{k}}$, which is essentially Eq. (3.10) plus damping, pumping, and noise terms

$$\frac{d}{dt}a^{\dagger}_{\mathbf{k}}a_{\mathbf{k}} = \Lambda_{e\mathbf{k}}(1 - a^{\dagger}_{\mathbf{k}}a_{\mathbf{k}}) - B_{\mathbf{k}}a^{\dagger}_{\mathbf{k}}a_{\mathbf{k}}b^{\dagger}_{-\mathbf{k}}b_{-\mathbf{k}} - \gamma_{nr}a^{\dagger}_{\mathbf{k}}a_{\mathbf{k}} - \gamma_e(a^{\dagger}_{\mathbf{k}}a_{\mathbf{k}} - (a^{\dagger}_{\mathbf{k}}a_{\mathbf{k}})_{eq})$$

$$+ ig^*_{\mathbf{k}}a^{\dagger}b_{-\mathbf{k}}a_{\mathbf{k}} - ig_{\mathbf{k}}a^{\dagger}_{\mathbf{k}}b^{\dagger}_{-\mathbf{k}}a + F_{e\mathbf{k}} . \tag{101}$$

Here we approximate the carrier-carrier relaxation by the relaxation-time approximation of Sec. 3-1 and $(a^{\dagger}_{\mathbf{k}}a_{\mathbf{k}})_{eq}$ is the quasiequilibrium value of the number operator $a^{\dagger}_{\mathbf{k}}a_{\mathbf{k}}$. Noting that $a^{\dagger}_{\mathbf{k}}a_{\mathbf{k}}a^{\dagger}_{\mathbf{k}}a_{\mathbf{k}} = a^{\dagger}_{\mathbf{k}}a_{\mathbf{k}}$ and using the generalized Einstein relation (76), we have the diffusion coefficient

$$2\langle D_{ee} \rangle = \frac{d}{dt}\langle a^{\dagger}_{\mathbf{k}}a_{\mathbf{k}}\rangle + 2B_{\mathbf{k}}\langle a^{\dagger}_{\mathbf{k}}a_{\mathbf{k}}b^{\dagger}_{-\mathbf{k}}b_{-\mathbf{k}}\rangle + 2\gamma_{nr}\langle a^{\dagger}_{\mathbf{k}}a_{\mathbf{k}}\rangle$$

$$+ 2\gamma_e\langle a^{\dagger}_{\mathbf{k}}a_{\mathbf{k}}\rangle[1 - (a^{\dagger}_{\mathbf{k}}a_{\mathbf{k}})_{eq}] + ig^*_{\mathbf{k}}a^{\dagger}\langle b_{-\mathbf{k}}a_{\mathbf{k}}\rangle - ig_{\mathbf{k}}\langle a^{\dagger}_{\mathbf{k}}b^{\dagger}_{-\mathbf{k}}\rangle a$$

$$= \langle \Lambda_{\alpha\mathbf{k}}(1 - a^{\dagger}_{\mathbf{k}}a_{\mathbf{k}})\rangle + B_{\mathbf{k}}\langle a^{\dagger}_{\mathbf{k}}a_{\mathbf{k}}b^{\dagger}_{-\mathbf{k}}b_{-\mathbf{k}}\rangle + \gamma_{nr}\langle a^{\dagger}_{\mathbf{k}}a_{\mathbf{k}}\rangle$$

$$+ \gamma_e\langle a^{\dagger}_{\mathbf{k}}a_{\mathbf{k}}[1 - (a^{\dagger}_{\mathbf{k}}a_{\mathbf{k}})_{eq}]\rangle + \gamma_e\langle(a^{\dagger}_{\mathbf{k}}a_{\mathbf{k}})_{eq}(1 - a^{\dagger}_{\mathbf{k}}a_{\mathbf{k}})\rangle . \tag{102}$$

This equation is the *sum* of the pumping and decay rates. This additive character of the rates increasing and decreasing a probability operator is typical of such diffusion coefficients.

The total carrier-density operator is given by the sum of Eq. (101) over k divided by the volume. Substituting Eq. (90) to adiabatically eliminate the dipole operators, we have

$$\dot{N} = \Lambda - \frac{1}{V}\sum_{\mathbf{k}} B_{\mathbf{k}}a^{\dagger}_{\mathbf{k}}a_{\mathbf{k}}b^{\dagger}_{-\mathbf{k}}b_{-\mathbf{k}} - \gamma_{nr}N$$

$$- \frac{2A^{\dagger}A}{\gamma V}\sum_{\mathbf{k}} |g_{\mathbf{k}}|^2\mathscr{L}_{\mathbf{k}}[a^{\dagger}_{\mathbf{k}}a_{\mathbf{k}} + b^{\dagger}_{-\mathbf{k}}b_{-\mathbf{k}} - 1] + F_N , \tag{103}$$

where the carrier-density noise operator $F_N(t)$ is given by

$$F_N(t) = \frac{1}{V}\sum_{\mathbf{k}}\left[ig^*_{\mathbf{k}}A^{\dagger}\mathscr{D}_{\mathbf{k}}F_{\sigma_{\mathbf{k}}} - ig_{\mathbf{k}}A\mathscr{D}^*_{\mathbf{k}}F^{\dagger}_{\sigma_{\mathbf{k}}} + F_{e\mathbf{k}}\right] . \tag{104}$$

Note that the expectation value of Eq. (103) doesn't feature the $\frac{1}{2}$ that appears in the carrier-density equation of motion (7.31). This is due to the use of Eq. (15) for $\mathcal{V}$. For consistency with the current quantized-field calculations, Sec. 9-6 doesn't include this $\frac{1}{2}$ factor.

To calculate noise spectra, we use the semiclassical transformation

$$A = \frac{E}{\mathcal{E}}e^{-i\phi}, \; A^\dagger = \frac{E}{\mathcal{E}}e^{i\phi} , \tag{105}$$

where $\mathcal{E}$ is the electric field per photon (12). Noting that $e^{-2i\phi} = A/A^\dagger$, we have the inverse relations

$$E = \mathcal{E}\sqrt{A^\dagger A} ,$$

$$\phi = \tfrac{1}{2}i\log(A/A^\dagger) = \tfrac{1}{2}i[\log A - \log A^\dagger] , \tag{106}$$

Basically this approximation keeps enough of the quantum mechanical correlations to get reasonable spectra in appropriate limits. It would be nicer, but substantially more difficult, to stick with the annihilation operator approach. However unless extremely careful, we would miss the important amplitude phase couplings that lead to the linewidth enhancement factor, and, in agreement with the Schawlow-Townes theory, we would overestimate the linewidth by a factor of two when operating way above the laser threshold. The key is to include just the right mix of amplitude and phase fluctuations, and unfortunately, this requires a fair amount of pure intuition (hand waving?). Remember: intuition is necessary, but not sufficient! The diffusion coefficients used in the next section are derived in App. E.

9-5. Power Spectra and Laser Linewidth

We determine various semiconductor laser noise spectra by using the classical field decomposition of Eq. (106) in the quantized-field equations (92) and (103) and using appropriate diffusion coefficients determined by the fully quantal treatment of Sec. 9-5. We linearize these equations of motion to find equations for the fluctuations

$$\frac{d}{dt}\delta E(t) = - \frac{\nu}{2}E_s \frac{\partial \chi_s''}{\partial N}\delta N(t) + F_E(t) = \tfrac{1}{2}E_s G_N \delta N(t) + F_E(t) \tag{107}$$

$$\frac{d}{dt}\delta N(t) = \frac{4\epsilon E_s}{\hbar\nu}\nu\chi_s''\delta E(t) - \Gamma_1'\delta N(t) + F_N(t)$$

$$= -\frac{4\epsilon E_s G}{\hbar\nu}\delta E(t) - \Gamma_1'\delta N(t) + F_N(t). \qquad (108)$$

$$\frac{d}{dt}\delta\phi(t) = -\frac{\nu}{2}\frac{\partial\chi_s'}{\partial N}\delta N(t) + F_\phi(t) = \tfrac{1}{2}\alpha G_N\delta N(t) + F_\phi(t), \qquad (109)$$

where $G = -\nu\chi_s'' = R_{sp} - R_{abs}$, $G_N = -\nu\partial\chi_m''/\partial N$, and $\Gamma_1' = \Gamma_1 + \tfrac{1}{2}\epsilon I G_N/\hbar\nu$ $= \Gamma_1 + \tfrac{1}{2}I G_N/V\mathcal{E}^2$. We Fourier transform Eqs. (107) and (108), to find the matrix equation

$$\begin{bmatrix} i\omega & -\tfrac{1}{2}G_N E \\ \epsilon GE/\hbar\nu & i\omega+\Gamma_1' \end{bmatrix} \begin{bmatrix} \delta\tilde{E} \\ \delta\tilde{N} \end{bmatrix} = \begin{bmatrix} \tilde{F}_E \\ \tilde{F}_N \end{bmatrix},$$

where for typographical simplicity, we drop the subscript s from here on. This matrix equation has the solutions

$$\delta\tilde{E}(\omega) = \frac{(i\omega+\Gamma_1')\tilde{F}_E + \tfrac{1}{2}G_N E\tilde{F}_N}{i\omega(i\omega + \Gamma_1') + 2\epsilon E^2 GG_N/\hbar\nu} = \frac{(i\omega+\Gamma_1')\tilde{F}_E + \tfrac{1}{2}G_N E\tilde{F}_N}{\Omega_R^2 - \omega^2 + i\omega\Gamma_1'} \qquad (110)$$

$$\delta\tilde{N}(\omega) = \frac{i\omega\tilde{F}_N - (4\epsilon GE/\hbar\nu)\tilde{F}_E}{\Omega_R^2 - \omega^2 + i\omega\Gamma_1'} \qquad (111)$$

$$\tilde{\phi}(\omega) = \frac{\tilde{F}_\phi}{i\omega} + \frac{\alpha G_N}{2i\omega}\delta\tilde{N}(\omega) = \frac{\tilde{F}_\phi}{i\omega} + \alpha\frac{\tfrac{1}{2}G_N\tilde{F}_N - (\Omega_R^2/i\omega E)\tilde{F}_E}{\Omega_R^2 - \omega^2 + i\omega\Gamma_1'}, \qquad (112)$$

where in this notation, the approximate (neglecting the Γ_1' shift) relaxation oscillation frequency $\Omega_R^2 = 2\epsilon E^2 GG_N/\hbar\nu$.

Consider first the relative intensity noise (RIN), which is defined by

$$\text{RIN} = \frac{\mathcal{S}_I(\omega)}{I^2}, \qquad (113)$$

where in the Wiener-Khintchine stationary noise approximation, the spectrum $\mathcal{S}_I(\omega)$ is given by

$$\mathcal{S}_I(\omega) = \int_{-\infty}^{\infty} d\tau \langle \delta I(\tau)\delta I(0)\rangle \ e^{-i\omega\tau} = \langle |\delta\tilde{I}(\omega)|^2 \rangle \,. \tag{114}$$

The RIN can be measured using a wide bandwidth photodiode and a spectrum analyzer. $\delta\tilde{I}(\omega)$ can be determined in a fashion similar to the way we find $\delta\tilde{E}(\omega)$; just some factors of $\frac{1}{2}$ and E are moved around. The equations of motion for the fluctuations $\delta I(t)$ and $\delta N(t)$ are given by

$$\frac{d}{dt}\delta I(t) = I G_N \delta N(t) + F_I(t) \tag{115}$$

$$\frac{d}{dt}\delta N(t) = -\frac{2\epsilon G}{\hbar\nu}\delta I(t) - \Gamma_1'\delta N(t) + F_N(t) \,, \tag{116}$$

with the solutions

$$\delta\tilde{I}(\omega) = \frac{(i\omega+\Gamma_1')\tilde{F}_I + G_N I \tilde{F}_N}{\Omega_R^2 - \omega^2 + i\omega\Gamma_1'} \tag{117}$$

$$\delta\tilde{N}(\omega) = \frac{i\omega\tilde{F}_N - (2\epsilon G/\hbar\nu)\tilde{F}_I}{\Omega_R^2 - \omega^2 + i\omega\Gamma_1'} \,. \tag{118}$$

Substituting Eq. (117) into Eqs. (114) and (113), we have

$$\text{RIN} = \frac{(\omega^2+\Gamma_1'^2)2\langle D_{II}\rangle + I^2 G_N^2 2\langle D_{NN}\rangle + 2\Gamma_1' G_N I 2\langle D_{NI}\rangle}{I^2[(\Omega_R^2 - \omega^2)^2 + \omega^2\Gamma_1'^2]} \,. \tag{119}$$

This formula is illustrated in Fig. 9-1. Note the peak near the approximate relaxation-oscillation frequency Ω_R.

Frequency noise is important in communication applications. We can calculate its spectrum by the formula

$$\mathcal{S}_{\dot{\phi}}(\omega) = \langle |\omega\tilde{\phi}(\omega)|^2 \rangle \tag{120}$$

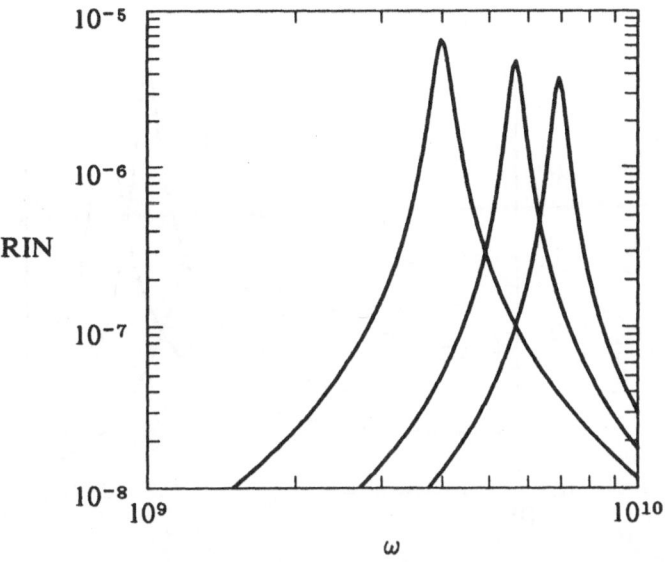

Fig. 9-1. Relative Intensity Noise (RIN) *vs* Fourier-spectrum frequency ω for intensities I of 1, 2, and 3 mW according to Eq. (119) in combination with a free-carrier model in the δ-function lineshape limit. All frequencies are in sec^{-1}. Various basic parameters are given by $T = 300$ K, $m_e = 1.167m$, $m_h = 6.669m$, nonradiative decay rate γ_{nr} 3×10^9 s^{-1}, carrier density $N = 2\times10^{18}$ cm^{-3}, cavity volume $V = .1\times4\times250$ $\mu m^3 = 10^{-10}$ cm^3, confinement factor $\Gamma = 0.15$, and $\nu/2Q = 1.42\times10^{11}$ s^{-1}. Various computed parameters are photon number $n_0 = 34273$ (for 3 mW), carrier number $= 2\times10^8$, intensity gain $G = \nu/Q = 2.84\times10^{11}$ s^{-1}, change of G with respect to carrier number $G_N = 3694$ s^{-1}, relaxation oscillation frequency $\Omega_R = 6\times10^9$ s^{-1} (for $I = 3$ mW), and $\delta N(t)$ decay rate $\Gamma_1' = 4.27\times10^8$ s^{-1} (for $I = 3$ mW). The $\langle D_{ee} \rangle$ contribution to $\langle D_{NN} \rangle$ was ignored.

and use Eq. (112). This gives

$$\mathcal{S}_{\dot{\phi}}(\omega) = 2\langle D_{\phi\phi} \rangle + \alpha^2 \frac{\frac{1}{2}G_N^2\omega^2\langle D_{NN} \rangle + (2\Omega_R^4/I)\langle D_{EE} \rangle}{(\Omega_R^2 - \omega^2)^2 + \omega^2\Gamma_1'^2}. \tag{121}$$

This formula is illustrated in Fig. 9-2.

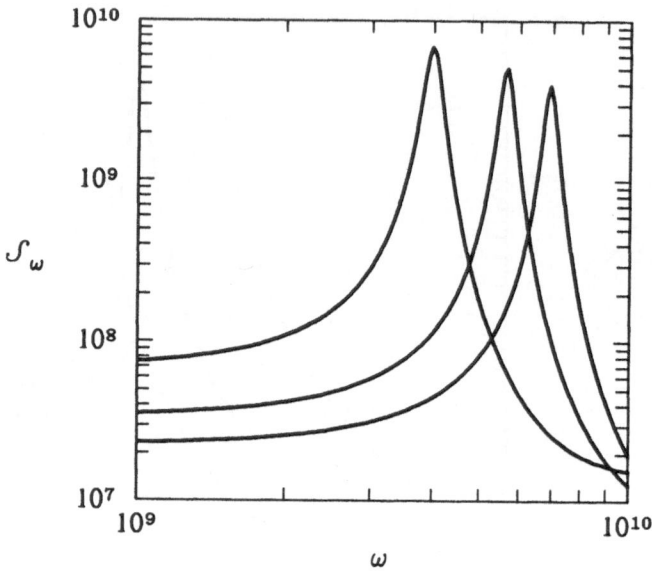

Fig. 9-2.   Frequency-noise spectrum $\mathcal{S}_\omega$ vs Fourier-spectrum frequency ω for intensities I of 1, 2, and 3 mW according to Eq. (121) with a free-carrier model in the δ-function lineshape limit. The parameters have the same values as in Fig. 9-1 and the linewidth-enhancement factor α has the value 2, which is roughly correct for these parameters.

Laser Linewidth

In the Langevin approach the laser linewidth is determined by the width of the spectrum

$$I(\omega) = \int_{-\infty}^{\infty} dt\, e^{-i\omega t} \langle A^\dagger(t) A(0) \rangle . \tag{122}$$

Far enough above threshold, the fluctuations in the amplitude of $A(t)$ are constrained due to the steady-state saturated-gain-equals-loss condition. Hence we approximate Eq. (122) by

$$I(\omega) = I\int_{-\infty}^{\infty} dt e^{-i\omega t} \langle e^{i(\phi(t) - \phi(0))}\rangle = I\int_{-\infty}^{\infty} dt\, e^{-i\omega t}\, \langle e^{i\Delta\phi(t)}\rangle\ ,$$

which gives

$$I(\omega)= I\int_{-\infty}^{\infty} dt e^{-i\omega t} \exp\{-\tfrac{1}{2}\langle[\Delta\phi(t)]^2\rangle\}\ ,\qquad (123)$$

where I is the steady-state intensity value and the second line is derived for Gaussian noise processes in Eqs. (E.13) through (E.15).

To calculate $\langle[\Delta\phi(t)]^2\rangle$, we note that

$$\Delta\phi(t) \equiv \phi(t) - \phi(0) = \frac{1}{\sqrt{2\pi}}\int_{-\infty}^{\infty} d\omega \tilde{\phi}(\omega)[e^{i\omega t} - 1]\ ,\qquad (124)$$

and for real Langevin noise sources

$$\langle\tilde{F}_\mu(\omega)\tilde{F}_\nu(\omega')^*\rangle = \frac{1}{2\pi}\int_{-\infty}^{\infty} dt \int_{-\infty}^{\infty} dt' e^{-i\omega t}\, e^{i\omega' t'}\, \langle F_\mu(t)F_\nu(t')\rangle$$

$$= \frac{2\langle D_{\mu\nu}\rangle}{2\pi}\int_{-\infty}^{\infty} dt e^{-i(\omega-\omega')t} = 2\langle D_{\mu\nu}\rangle\delta(\omega - \omega')\ .\qquad (125)$$

Using Eqs. (111), (112), and (125), we have

$$\langle\tilde{\phi}(\omega)\tilde{\phi}(\omega')^*\rangle = \langle|\tilde{\phi}(\omega)|^2\rangle\delta(\omega - \omega')\ ,$$

where $\omega^2\langle|\tilde{\phi}(\omega)|^2\rangle$ is given by Eq. (121). Equation (124) is real, so we can square it with the second version containing the complex conjugate of the integrand. This gives

$$\langle[\Delta\phi(t)]^2\rangle = \frac{1}{2\pi}\int_{-\infty}^{\infty} d\omega \int_{-\infty}^{\infty} d\omega'\, \langle\tilde{\phi}(\omega)\tilde{\phi}(\omega')^*\rangle(e^{i\omega t} - 1)(e^{-i\omega' t} - 1)$$

or

$$\langle[\Delta\phi(t)]^2\rangle = \frac{1}{\pi}\mathrm{Re}\left\{\int_{-\infty}^{\infty} d\omega \langle|\tilde{\phi}(\omega)|^2\rangle(1 - e^{i\omega t})\right\}$$

$$= \frac{1}{\pi}\mathrm{Re}\left\{\int_{-\infty}^{\infty} d\omega \left[\frac{2\langle D_{\phi\phi}\rangle}{\omega^2} + \frac{\alpha^2\Omega_R^4 2\langle D_{EE}\rangle}{E^2\omega^2|\Delta|^2} + \frac{\alpha^2 G_N^2\langle D_{NN}\rangle}{2|\Delta|^2}\right]\right.$$

$$\left. \times [1 - e^{i\omega t}]\right\}, \tag{126}$$

where $\Delta = \omega^2 - \Omega_R^2 - i\omega\Gamma_1'$.

Equation (126) can be evaluated by countour integration as demonstrated in Eqs. (E.16) through (E.18). Substituting Eq. (E.18) into (126), we have the linewidth formula

$$I(\omega) = I\exp\left[-\frac{\alpha^2 R_{sp}\cos3\delta}{8\Gamma I\cos\delta}\right]\int_{-\infty}^{\infty} dt e^{-i\omega t}$$

$$\times \exp\left\{-\frac{R_{sp}}{4I}\left[(1 + \alpha^2)|t| - \frac{\alpha^2 e^{-\Gamma|t|}\cos(\Omega|t|-3\delta)}{2\Gamma\cos\delta}\right]\right\}. \tag{127}$$

If the relaxation-oscillation contribution is neglected, this is the Fourier transform of an exponentially decaying function. This gives a Lorentzian power spectrum with the full width at half maximum of

$$\Delta\nu = \frac{R_{sp}}{2I}(1 + \alpha^2) = \tfrac{1}{2}(1 + \alpha^2)\Delta\nu_{ST}, \tag{128}$$

where $\Delta\nu_{ST}$ is the Schawlow-Townes linewidth. The $\tfrac{1}{2}$ comes from the stabilization of amplitude fluctuations above threshold, and the α^2 comes from phase fluctuations resulting from saturated index fluctuations caused by amplitude fluctuations, in turn, caused by spontaneous emission. The spectrum given by Eq. (127) is illustrated in Fig. 9-3. Note the relaxation sidebands at $\omega = \pm\Omega$ and, for the smallest Γ, at $\omega = \pm2\Omega$.

More generally, the amplitude fluctuations do enter the linewidth calculation (122) and Hemstead and Lax (1967) have shown using a nonlinear

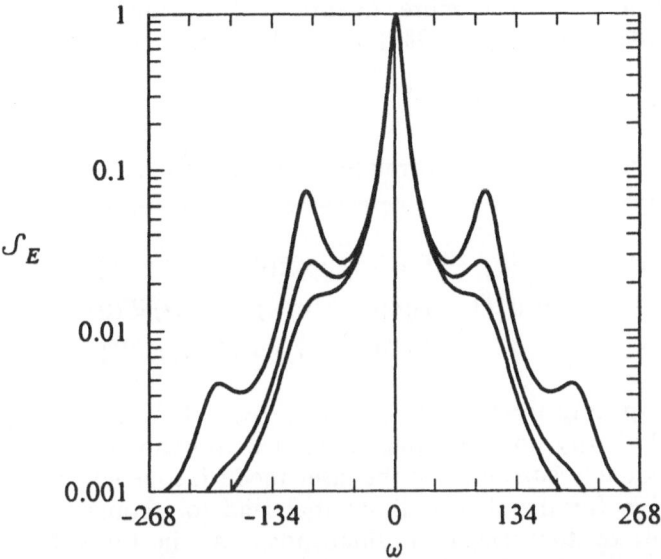

Fig. 9-3. Laser linewidth spectrum obtained by evaluating Eq. (127) with a 256-point FFT for Γ = 8, 16, 24. Other parameters used are $\Omega = (\Omega_r^2 - \Gamma^2)^{1/2}$ = 100, R_{sp} = 1, and α = 2. The time axis runs from -1.5 to 1.5, so that the frequency axis runs from $-256\pi/3$ to $256\pi/3 \simeq 268$. These values are illustrative only and have not been chosen to correspond to a particular laser configuration.

rotating van der Pol analysis that the linewidth is reduced by a multiplicative factor of 2 as the pump is varied from ten times below threshold to ten times above. This is a fairly slowly varying change, so the constant intensity assumption of Eq. (123) is likely to underestimate the linewidth by as much as a factor of 1.5 near threshold.

Amplitude-phase correlations also affect the relative height of the relaxation-oscillation sidebands, with the sideband below the laser frequency somewhat larger than its counterpart above the laser frequency. We see such an asymmetry in the sidemode gain spectrum of Sec. 8-1. There the complex population pulsation resonance factor $\mathcal{F}(\Delta)$ is multiplied by the complex gain derivative $\partial\chi/\partial N$, which peaks above the gain region where the Fermi-Dirac distribution functions change most rapidly. This mixes in an index contribution that favors the sidemode gain tuned below the pump-wave frequency. In this frame of mind, one is tempted to attribute the larger size of the lower relaxation-oscillation sideband to increased gain.

To get a somewhat more quantitative feeling for the asymmetry, Vahala, Harder, and Yariv (1983) wrote the laser field amplitude as

$$E(t) = [E_s + \delta E(t)]e^{-i\phi(t)}$$

in terms of which the autocorrelation $\langle E^*(t)E(0)\rangle$ needed for Eq. (122) is given by

$$
\begin{aligned}
\langle E^*(t)E(0)\rangle &= \langle [E_s + \delta E(t)][E_s + \delta E(0)]e^{i[\phi(t)-\phi(0)]}\rangle \\
&\simeq E_s^2\{1 - \tfrac{1}{2}\langle[\phi(t)-\phi(0)]^2\rangle\} + \langle\delta E(t)\delta E(0)\rangle \\
&\quad + iE_s[\langle\delta E(0)\phi(t)\rangle - \langle\delta E(t)\phi(0)\rangle] \ .
\end{aligned}
\tag{129}
$$

Here we keep only terms up to second order. This "small angle" approximation is obviously poor for long times, which contribute to the frequency range immediately surrounding the laser mode frequency. But it is a better approximation for the shorter times that lead to frequency components in the vicinities of the relaxation sidebands. As in Eq. (122), we Fourier transform Eq. (129) to find the intensity spectrum

$$I(\omega) = E_s^2\{\delta(\omega) + \tfrac{1}{2}\langle|\widetilde{\phi}(\omega)|^2\rangle + \langle|\delta\widetilde{E}(\omega)|^2\rangle + [i\langle\delta\widetilde{E}(\omega)\widetilde{\phi}(\omega)\rangle + \text{c.c.}]\} \ . \tag{130}$$

Using Eqs. (110) and (112) and the appropriate diffusion coefficients, one can evaluate $i\langle\delta\widetilde{E}(\omega)\widetilde{\phi}(\omega)\rangle$ + c.c., which usually gives an asymmetry that favors the lower relaxation sideband. Ultimately one would like to evaluate Eq. (122) directly, rather than invoking various semiclassical approximations. The linewidth problem is important and has attracted a great deal of attention over the years, but to our knowledge noone has succeeded it really doing it right.

The linewidth is influenced by other factors as well, such as low mirror reflectivities, transverse variations particularly in gain guided systems, and optical feedback. In particular even in the plane-wave approximation, the spontaneous emission rate, and hence the linewidth as well, should be multiplied by the factor

$$K = \left[\frac{\left(\sqrt{R_1} + \sqrt{R_2}\right)\left(1 - \sqrt{R_1 R_2}\right)}{\sqrt{R_1 R_2}\ln(R_1 R_2)}\right]^2 , \tag{131}$$

where R_1 and R_2 are the mirror reflectivities. As shown in Fig. 9-4, this factor becomes important for reflectivities less than 20%.

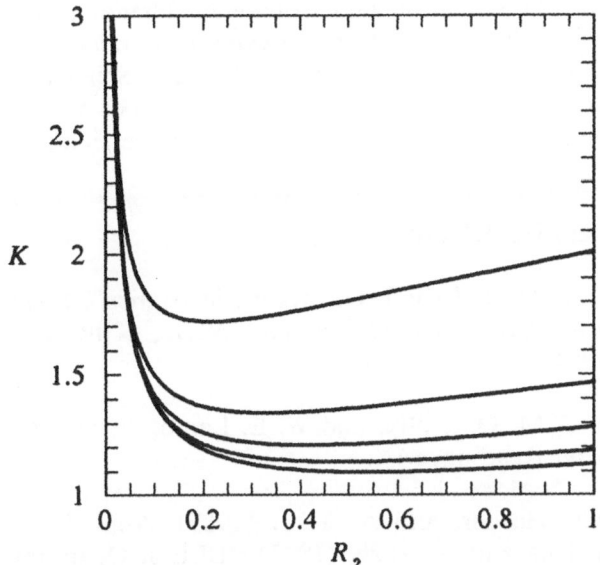

Fig. 9-4. K factor of Eq. (131) versus the right mirror reflectivity R_2 for the values of R_1 = .05, .1, .2, .3, and .8, in order of decreasing K values. For $R_1 = R_2 = 1$, $K = 1$.

References

Goldberg, P., P. W. Milonni, and B. Sundaram (1991), Phys. Rev. A**44**, 1969 and 4556.

Haug, H. (1967), Z. Phys. **200**, 57, **206**, 163.

Hemstead, R. D. and M. Lax (1967), Phys. Rev. **161**, 350.

Henry, C. H. (1983), IEEE J. Quant. Electron. QE-19, 1391. See also his review article of (1990) entitled, "Line Broadening of Semiconductor Lasers," in *Coherence, amplification, and quantum effects in semiconductor lasers*, Ed. by Y. Yamamoto. John Wiley & Sons, New York.

Lax, M. (1968), "Fluctuation and Coherence Phenomena in Classical and Quantum Physics", *pp.* 269-478 in *Brandeis University Summer Institute in Theoretical Physics, 1966*, Gordon and Breach, Science Publishers, New York. This reference contains an overview of Lax's work on classical and quantum noise as well as references to his journal articles on the subject.

Meystre, P. and M. Sargent III (1991), *Elements of Quantum Optics*, Springer-Verlag, Heidelberg.

Petermann, K. (1979), IEEE J. Quantum Electronics **QE-15**, 566. See also his book, *Laser Diode Modulation and Noise*, Kluwer Academic Pub., Dordrecht, (1991).

Sargent III, M., M. O. Scully, and W. E. Lamb, Jr. (1977), *Laser Physics*, Addison-Wesley, Reading, MA.

Vahala, K., Ch. Harder, and A. Yariv (1983), Appl. Phys. Lett. **42**, 211. See also K. Vahala and A. Yariv (1983), IEEE J. Quantum Electron. **QE-19**, 1096 and 1102.

Chapter 10
PROPAGATION EFFECTS

The laser theory discussed in Chaps. 7 through 9 is based on the uniform amplitude approximation, which assumes that the laser field can be written as a superposition of the passive resonator modes. In general this approximation is valid only for mirror reflectivities close to unity, which are typically not used in most commercial edge-emitting lasers or in laser amplifiers, where the facets are likely to be uncoated or antireflection coated. For these devices, propagation effects may be important, which necessitates a different method of analysis. A conceptually straightforward approach, that was originally developed by W. W. Rigrod (1963), involves writing the laser field as the sum of forward and backward travelling waves, whose amplitudes may vary considerably along the laser axis. This approach is widely used in gas and solid state lasers and has been proven to be accurate in many cases. When the homogeneously broadened gain formula $g_0/(1 + I/I_s)$ applies, analytical solutions may exist. For the more complicated semiconductor gain, the Rigrod equations have to be solved numerically.

This chapter analyzes semiconductor lasers and amplifiers for which propagation effects are important. Section 10-1 treats the one-dimensional problem where transverse and lateral field variations are neglected. The problem of amplified spontaneous emission in an amplifier is treated here as an example. Section 10-2 introduces the more general problem where lateral and transverse field variations are also taken into account. These variations are driven by diffraction and filamentation. The role of the semiconductor gain medium in causing filamentation of the laser field is discussed. Section 10-3 describes two methods for treating diffraction effects during field propagation: one is based on Fourier transformation and the second is based on a finite difference solution. Section 10-4 deals with the problem of filamentation in an amplifier. The role of carrier diffusion is also discussed in this section. Finally, Sec. 10-5 describes the unstable resonator semiconductor laser. The lateral, longitudinal and temporal laser field variations are taken into account and the effects of the gain medium on laser stability is discussed. All examples in this chapter are computed using the free-carrier theory unless stated otherwise.

10-1. Longitudinal Field Dependence

The Rigrod analysis treats the laser field inside the resonator as the sum of a forward and a backward travelling wave. Each travelling wave is written as in Eq. (2.67). The typical Rigrod analysis ignores spatial grating effects as well as the time variations in the slowly varying field amplitude, and we do so here. However for pulse propagation problems, the analysis can straightforwardly be extended to treat a slowly spatially and temporally varying field amplitude. Equation (2.72) gives the equation of motion for the amplitude of both travelling waves. Furthermore, the Rigrod analysis works with intensities, which prevents it from handling nonlinear effects dependent on amplitude interference, such as spatial hole burning and various aspects of distributed feedback lasers. Rewriting the equation in terms of the forward and backward travelling-wave intensities, we have

$$\frac{d}{dz} I_{L1}(z) = [\Gamma G(z) - \alpha_{abs}] I_{L1}(z) \tag{1}$$

$$- \frac{d}{dz} I_{L2}(z) = [\Gamma G(z) - \alpha_{abs}] I_{L2}(z) , \tag{2}$$

where z is along the laser axis and the subscripts 1 and 2 denote the forward and backward propagation directions, respectively. We have generalized Eq. (2.72) to include α_{abs}, which accounts for the internal optical losses, and the confinement factor Γ, which accounts for the mismatch between the active region and laser-field cross sections. We introduce the intensity gain $G(z) = 2g(z)$, where $g(z)$ is the amplitude gain defined in Eq. (2.75). For the pulse problem, the total z-derivative becomes $d/dz = \partial/\partial z \pm (n/c)\partial/\partial t$, where n is the medium refractive index. The field intensities, carrier density, and local gain are the transversely and laterally averaged quantities within the gain layer, which we write as

$$I_{Ln}(z) = \frac{1}{hw} \int_{-\infty}^{\infty} dy \int_0^h dx \, I_{Ln}(x,y,z) , \tag{3}$$

where h is the active region thickness and w is an effective lateral beam width. The total beam power is

$$P_{Ln}(z) = \frac{hw}{\Gamma} I_{Ln}(z) \tag{4}$$

Solutions of Eqs. (1) and (2) must satisfy the boundary conditions at the mirrors

$$I_{L1}(0) = R_2 I_{L2}(0) \tag{5}$$

$$I_{L2}(L) = R_2 I_{L1}(L) , \tag{6}$$

where L is the resonator length.

The Rigrod analysis is a very useful tool. To demonstrate this, we use it to study the effects of amplified spontaneous emission in semiconductor amplifiers. In this problem, the circulating fields inside the amplifier are the amplified input laser field and the amplified spontaneous emission

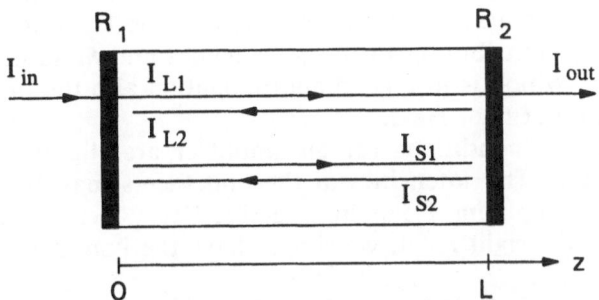

Fig. 10-1. Laser amplifier with facet amplitude reflectivities R_1 and R_2. I_{L1} and I_{L2} are the intensities of the amplified input field. I_{S1} and I_{S2} are the intensities of the ASE fields.

(ASE) field (see Fig. 10-1). For a single-pass amplifier, the amplified input laser field is comprised of a forward and a backward travelling wave, with the former being much larger than the latter. Equations (1) and (2) describe the amplification of these travelling waves. Even for the very best antireflection coating, the amplifier facets have non-zero reflectivities, which give rise to a weak optical resonator. The build-up of spontaneous emission intensity in the resonator modes by the amplifier gain medium is described by the equations

$$\frac{d}{dz} I_{S1} = [\Gamma \, G\,(z) - \alpha_{abs}] \, I_{S1}(z) + \Gamma S(z) \tag{7}$$

$$- \frac{d}{dz} I_{S2} = [\Gamma\, G(z) - \alpha_{abs}]\, I_{S2}(z) + \Gamma S(z) \;, \tag{8}$$

where I_{Sn} is the sum of all the amplifier modal intensities originating from spontaneous emission. An important difference between the amplified input field and the ASE field is that the former is correlated with the input laser field while the latter is not. The spontaneous emission power into the ASE modes per volume of gain medium is

$$S(z) = f\hbar\nu BN^2 \;, \tag{9}$$

where BN^2 is the radiative recombination rate (see Eq. 3.26). In the above equation, B is an empirical bimolecular radiative recombination rate coefficient, N is the saturated carrier density, f is the fraction of spontaneous emission power contributing to the ASE modes and $\hbar\nu$ is the average ASE photon energy, which is to a good approximation also the photon energy of the amplified input laser field.

The boundary conditions for an amplifier are slightly different from those of a laser. The intensities at the end facets may be determined by examining the circulating fields in an active Fabry-Perot etalon. At one of the facets, e.g. the right facet, we simply have the boundary conditions

$$I_{S2}(L) = R_2 I_{S1}(L)$$
$$I_{L2}(L) = R_2 I_{L1}(L) \;. \tag{10}$$

Then at the left facet we have to consider the resonance condition due to the non-zero facet reflectivities. The incident field is actually the sum of fields arriving at the left facet after completing one to several round trips. If they are exactly resonant with the amplifier, i.e. their phases are sub-multiples of twice the amplifier optical path length, their phases are multiples of 2π from one another and we have coherent addition of the fields. Otherwise, the phase differences are distributed between 0 and 2π. In general, the incident field at the left facet is given by the sum

$$\sqrt{R_2 I_1(0)}\; e^{GL - i\phi}\; \frac{1}{M} \sum_{m=0}^{M-1} (R_1 R_2)^{m/2}\, e^{m(G_{av}L - i\phi)}$$

which gives the intensity

$$I_1(0)R_2 e^{2G_{av}L} \left[\frac{1}{M^2} \frac{1 + \Lambda^M - 2\Lambda^{M/2}\cos(M\phi)}{1 + \Lambda - 2\Lambda^{1/2}\cos(\phi)} \right],$$ (11)

where $I_1(0)$ is $I_{L1}(0)$ or $I_{S1}(0)$, $\Lambda = R_1 R_2 \exp(2G_{av}L)$, G_{av} is the average saturated gain coefficient, ϕ is the roundtrip phase shift, and $M \simeq -1/ln(\Lambda)$ is the number of roundtrips made during a photon lifetime. At steady state, and for a relatively high-gain amplifier, $\Lambda \leq 1$. For a resonant field, $\cos\phi = 1$ and the quantity inside the square brackets in (11) equals unity. The expression given by (11) is then the incident field intensity calculated using Eqs. (1), (2), (7) and (8) in one round trip after steady state is reached, i.e. $I_{L2}(0)$ or $I_{S2}(0)$. We note that the dominant ASE fields are those which are resonant with the amplifier, so that

$$I_{S1}(0) = R_1 I_{S2}(0) .$$ (12)

Similarly, if the input laser field is resonant with the amplifier

$$I_{L1}(0) = (1 - R_1) I_{in} + R_1 I_{L2}(0) ,$$ (13)

where I_{in} is the input laser intensity and field interference effects have been ignored (mean field approximation). On the other hand, if it is outside the amplifier lockband, $\cos\phi < 1$ and the circulating field at the input laser frequency becomes negligible. By lockband, we mean the width of the amplifier resonance, which depends on the empty resonator and the inversion in the gain medium. For this case, the left facet boundary condition is

$$I_{L1}(0) = (1 - R_1) I_{in} .$$ (14)

Equations (13) and (14) represent the best and worst cases in terms of frequency matching between the input laser field and the amplifier. With them one can estimate the sensitivity of amplifier performance to frequency detuning without having to keep track of the phases of the radiation fields.

The amplifier analysis involves numerically solving the intensity Eqs. (1), (2), (7) and (8) together with the carrier-density Eq. (3.54). The intensity and carrier equations are coupled by the gain, which may be given by one of gain expressions in Chaps. 3 through 6. An important goal of this study is to determine how the amplifier gain depends on excitation. Figure 10-2 plots the amplifier gain versus excitation for different input laser intensities. The amplifier gain is

$$\mathcal{G} = \frac{P_o}{P_i} = (1 - R_2)\,\frac{I_{L1}(L)}{I_{in}}\,, \tag{15}$$

where P_i is the power of the input laser field and P_o is the portion of the

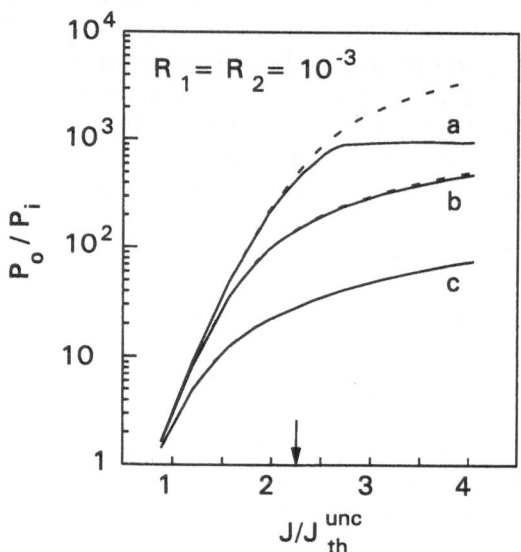

Fig. 10-2. Amplifier gain versus excitation (current divided by the threshold current for the uncoated free-running device) for input laser intensities of a) 0.001, b) 0.01 and c) 0.1 MW/cm^2. For the solid curves, the input laser field is outside the amplifier lockband, and for the dashed curves, the input laser frequency coincides with one of the amplifier resonances. The arrow indicates the free-running coated device threshold. We assume an undoped bulk GaAs active layer, $R_1 = R_2 = 10^{-3}$, $\Gamma = 0.3$ and $L = 500$ μm. [From Chow and Craig (1990).]

amplifier output that is correlated with the input laser field. To facilitate comparisons with experiments, we reference the excitation to that needed to reach oscillation threshold of an uncoated free-running device, i.e. $R_1 = R_2 = 0.32$ and $I_{in} = 0$. The oscillation threshold may be computed by the formula [see Eq. (1.7)]

$$\Gamma G_{th}L = \alpha_{abs}L - \tfrac{1}{2}\,ln(R_1 R_2)\,, \tag{16}$$

where G_{th} is the threshold gain for the ASE field. The solid curves are obtained using Eq. (14), which assumes that the input laser frequency is outside the amplifier lockband, while the dashed curves are obtained using (13), which assumes that the input laser frequency is tuned exactly to an amplifier resonance. When the dashed and solid curves coincide, resonator effects in the amplifier are negligible, which considerably relaxes the frequency matching requirement between the input laser field and the amplifier. The arrow in the figure indicates the oscillation threshold for the coated free-running device. Above this threshold, resonator effects begin to appear and their extent depends on the input laser intensity. For instance, for $I_{in} > 0.01$ MW/cm, we find negligible difference between the dashed and solid curves. Note that when operating along the dashed curve to the right of the arrow, the device is essentially an injection locked laser.

For a constant I_{in}, the amplifier gain initially increases with excitation. With an unlocked amplifier, it saturates abruptly beyond some value of excitation, as shown by the solid curve (a) in Fig. 10-2. This sudden transition in slope, which occurs above the free-running device threshold, marks the ASE oscillation threshold of the coated device (R_1, $R_2 \neq 0.32$ and $I_{in} \neq 0$). While the ASE oscillation threshold increases with increasing I_{in}, the value of the maximum amplifier gain is independent of $_{in}$. One can understand the reason for this behavior by noting first that the amplifier output power is proportional to

$$P_0 = (1 - R_1)(1 - R_2) P_i \exp \left\{ \int_0^L dz \, [\Gamma G(z) - \alpha_{abs}] \right\}. \qquad (17)$$

Here, the integral increases with excitation until the ASE oscillation threshold is reached. At that point, the amplifier becomes an oscillator, and the ASE intensity adjusts itself to clamp the gain at the threshold level, Eq. (7.36). Therefore the maximum average gain is given by the ASE oscillation condition

$$R_1 R_2 \exp \left\{ 2 \int_0^L dz \, [\Gamma G_{max}(z) - \alpha_{abs}] \right\} = 1, \qquad (18)$$

which is a generalization of Eq. (1.6). From Eqs. (15), (17) and (18), the maximum amplifier gain for an off-resonant input laser field is

$$\mathcal{G}_{max} = \frac{(1 - R_1)(1 - R_2)}{\sqrt{R_1 R_2}}, \tag{19}$$

which depends only on the antireflection coating reflectivities. For $R_1 = R_2 = 10^{-3}$, Eq. (19) gives $\mathcal{G}_{max} = 1000/cm$, which agrees with Fig. 10-2. For an amplifier that is locked to the input laser field, this maximum gain limit is absent, as is revealed by the dashed curve (a) in Fig. 10-2.

Another quantity of interest for an amplifier is the excitation level before the ASE power becomes significant. Figure 10-3 shows

$$\frac{P_o}{P_{ase}} = \frac{I_{L1}(L)}{I_{S1}(L)}, \tag{20}$$

which is a measure of the signal to noise ratio (S/N), versus excitation for different input intensities. For general applications that do not necessarily involve square-law detection, the noise P_{ase} consists of the portion of the amplifier output that is uncorrelated with the input laser field. At low excitation, P_o/P_{ase} increases with excitation because of the increase in amplifier gain. Then it decreases because of the increase in ASE power. The maxima in P_o/P_{ase} occur at excitation levels that are noticeably below the ASE oscillation threshold because of the large value for the spontaneous emission in a semiconductor gain medium. Gain saturation leads to a decrease in ASE with increasing I_{in}. The results show that a $P_o/P_{ase} = 20$ dB requires $I_{in} > 20\ kW/cm^2$. To achieve this with a 0.1 mm × 100 mm active region cross section and a 0.3 fill factor requires approximately 7 mW of input laser power. With a 30% coupling efficiency, this would require a 25 mW input power. Unlike the amplifier gain and efficiency, P_o/P_{ase} is sensitive to the fraction f of spontaneous emission power contributing to the ASE. This is especially true below the ASE oscillation threshold, where P_o/P_{ase} reaches its maximum. So far we have considered a gain-guided amplifier and let $f = 10^{-4}$. The value of f, which is not precisely known, is possibly larger, especially with broad-area amplifiers where the ASE field is likely to be multimode. We find that a factor of ten decrease in f leads to approximately a 10 dB increase in the P_o/P_{ase} maximum. After the onset of ASE oscillation, P_o/P_{ase} becomes independent of f because the output intensity of a laser oscillator operating above threshold is independent of the level of startup noise. With an index-guided amplifier, f may be smaller by an order of magnitude, in which case $I_{in} = 2\ kW/cm^2$ is sufficient to achieve $P_o/P_{ase} = 20\ dB$. However for this smaller input intensity, P_o/P_{ase} is a sensitive function of the exci-

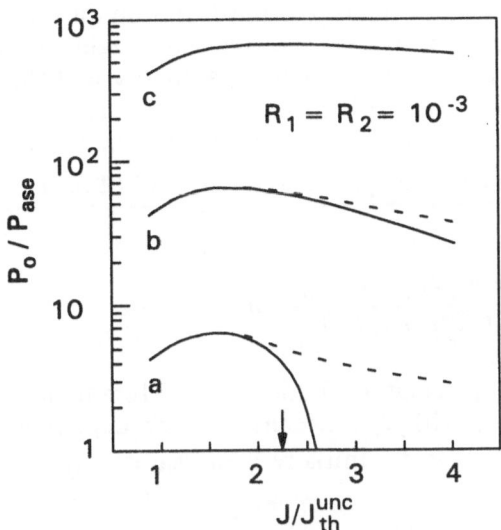

Fig. 10-3. P_o/P_{ase} versus excitation for input laser intensities of a) 0.001, b) 0.01, and c) 0.1 MW/cm^2. For the solid curves, the input laser field is outside the amplifier lockband, and for the dashed curves, the input laser frequency coincides with one of the amplifier resonances. All parameters are similar to those in Fig. 10-2. [From Chow and Craig (1990).]

tation level and of the difference between the input laser field and amplifier resonance frequencies.

To estimate the amplifier efficiency, we assume that the input electrical power is used to create an inversion and overcome losses due to the electrical resistance R_e in the bulk semiconductor material and electrical contacts. We define the amplifier efficiency as

$$\eta = \frac{P_o - P_i}{P_e} , \tag{21}$$

where P_e is the electrical power into the amplifier. The input electrical power is

$$P_e = I_e V_b + I_e^2 R_e , \tag{22}$$

where $I_e = ehwL(BN_0^2 + \gamma_{nr}N_0)$ is the injection current, N_0 is the unsaturated carrier density, w is an effective gain region width, and V_b is the bias voltage estimated from the bandgap energy and the quasi-Fermi level energies of the electron and hole distributions. The losses due to electrical resistance in the bulk material and electrical connections are lumped into the Omhic resistance term $I_e^2 R$. From (21) and (22), the efficiency is

$$\eta = \frac{(1 - R_2)I_{L1}(L) - I_{in}}{\Gamma I_e(V_b + I_e R_e)} \, wh \, . \tag{23}$$

Figure 10-4 plots η versus excitation for different input intensities. The efficiency increases with I_{in}, because carrier depletion is more complete with a more intense field. Initially η increases with excitation because as

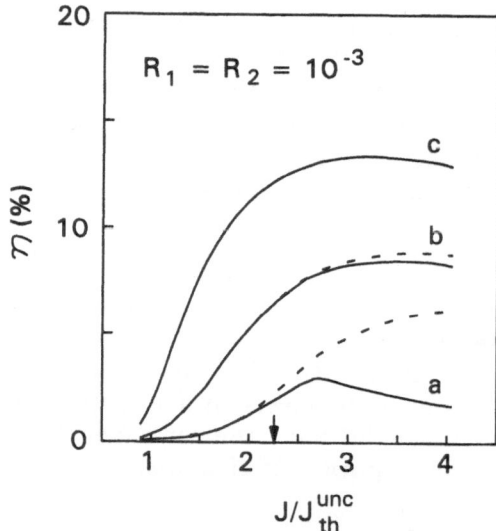

Fig. 10-4. Efficiency versus excitation for input laser intensities of a) 0.001, b) 0.01, and c) 0.1 MW/cm^2. For the solid curves, the input laser field is outside the amplifier lockband, and for the dashed curves, the input laser frequency coincides with one of the amplifier resonances. We use $R_e = 0.5\Omega$ for a $100mm$ wide and $500mm$ long active region. All parameters are similar to those in Fig. 10-2. [From Chow and Craig (1990).]

the input electrical power grows, the fraction of input electrical power that contributes to the amplifier output increases relative to the fraction that is spent in achieving transparency in the gain medium. However, the amplifier efficiency eventually decreases with excitation because a larger fraction of the pump power is lost to ASE and to heat via the series resistance R_e. With increasing I_{in}, η becomes less sensitive to resonator effects owing to gain saturation. Also a larger I_{in} allows an amplifier to operate at a higher excitation without being dominated by ASE. Therefore for large input laser intensity, it is especially important to minimize R_e. For the cases considered, the maximum efficiency is primarily limited by the fraction of input electrical power spent on achieving transparency in the semiconductor active layer. This leads to the belief that semiconductor laser amplifier performance may be improved with quantum-well structures.

Figure 10-5 shows how the amplifier behavior depends on facet reflectivities. With $R_1 = R_2 = 10^{-4}$, resonator effects are negligible since both

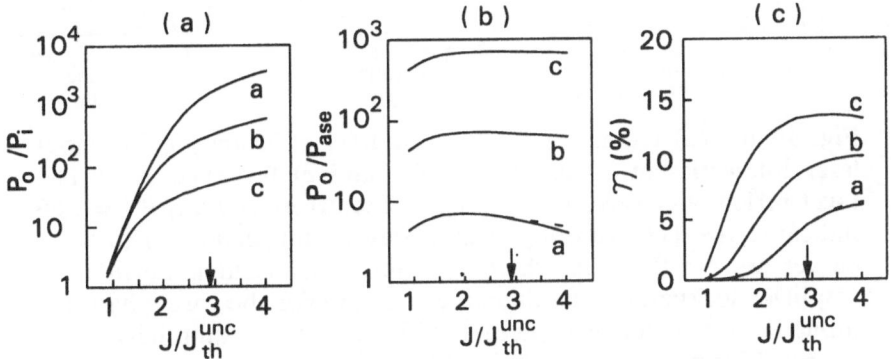

Fig. 10-5. (a) Gain, (b) P_o/P_{ase}, and (c) efficiency for input laser intensities of a) 0.001, b) 0.01, and c) 0.1 MW/cm^2. The facet reflectivity is $R_1 = R_2 = 10^{-4}$. The arrow at $J/J_{th}{}^{unc} = 2.92$ indicates the free-running coated device threshold. There is basically no difference between operating within (dashed curves) and outside (solid curves) the amplifier lockband. [From Chow and Craig (1990).]

boundary conditions (13) and (14) give essentially the same results. Resonator effects appear only after the current densities become too high to be of practical interest. Also, the maximum amplifier gain given by Eq. (19) is only reached for values of I_{in} that are too small and excitation densities that are too high to be encountered in practice. The maximum values for

P_o/P_{ase} remain approximately the same (Fig. 10-5b). However the decrease in P_o/P_{ase} with excitation is noticeably more gradual, especially for small I_{in}. Figure 10-5c indicates that a small input laser intensity benefits most from a reduction in facet reflectivity.

Figure 10-6 shows that ASE and resonator effects become more important with high facet reflectivity. When the input laser field is outside the amplifier lockband, the gain again reaches a maximum that is given by

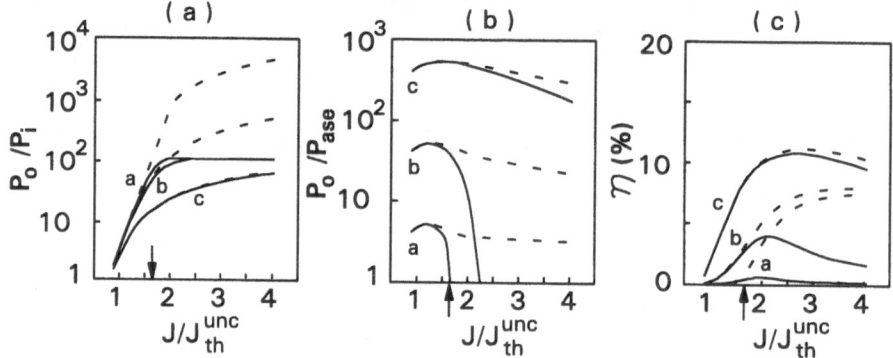

Fig. 10-6. (a) Gain, (b) P_o/P_{ase}, and (c) efficiency η for input laser intensities of a) 0.001, b) 0.01, and c) 0.1 MW/cm^2. The facet reflectivity is $R_1 = R_2 = 10^{-2}$. The arrow at $J/J_{th}^{unc} = 1.66$ indicates the free-running coated device threshold. The solid curves are for the case where the input laser field is outside the amplifier lockband. The dashed curves are for the case where the input laser frequency coincides with one of the amplifier resonances. [From Chow and Craig (1990).]

Eq. (19). The maximum gain is now only 20 dB, which is considerably smaller than that for $R_1 = R_2 = 10^{-4}$. ASE oscillation occurs at lower excitation levels, which leads to a more abrupt decrease in P_o/P_{ase} with injected current for $I_{in} < 0.1$ MW/cm^2. The efficiency is also reduced for $I_{in} < 0.1$ MW/cm^2. The dashed curves in Fig. 10-6 indicate that amplifier performance may be improved considerably by tuning the input laser frequency to within the amplifier lockband, i.e., by injection locking. While injection locking can compensate for the increased facet reflectivity, we should emphasize that the active frequency control requirements on the input laser field and amplifier are nontrivial, especially in phased arrays. Furthermore, an injection locked laser with 10^{-2} facet reflectivity performs no better than an amplifier with 10^{-4} facet reflectivity, which is insensitive to the input field frequency.

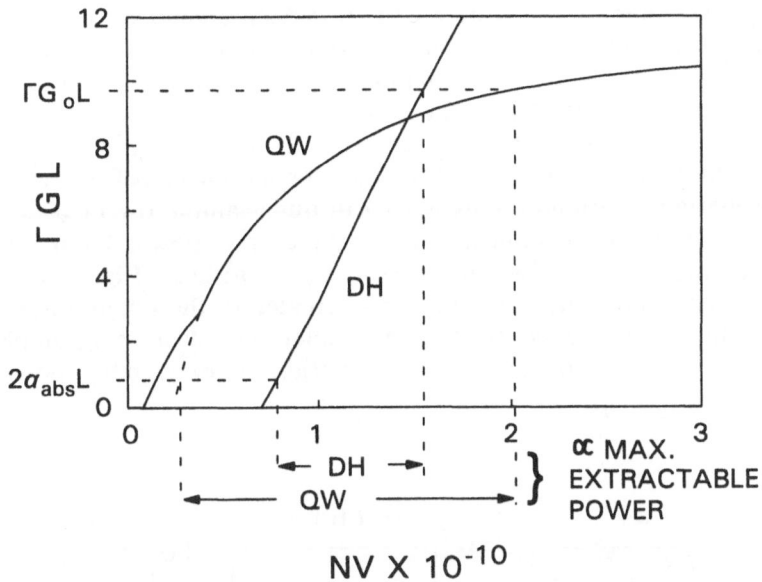

Fig. 10-7. ΓGL versus NV for bulk semiconductor and $10nm$ GaAs quantum-well gain media.

When comparing Figs. 10-2 through 10-6, we find that for $I_{in} = 0.1$ MW/cm^2, the gain medium is sufficiently saturated in one pass that amplifier performance is insensitive to facet reflectivity and detuning between the input field and the amplifier. However, an input laser intensity this high reduces the amplifier gain and places a practical constraint on the scalability of the devices.

Semiconductor amplifier performance depends on the nature of the gain medium. Chapters 5 to 6 discuss the different local gain behavior that can be obtained by introducing quantum confinement and strain. A basic difference between quantum-well and bulk semiconductor gain media can be seen by comparing Figs. 5-3 and 5-8. The bulk gain increases linearly with carrier density, while the quantum-well gain rolls over at high carrier density. This gain roll over is due to quantum confinement and it allows an amplifier to operate at a high excitation level without exceeding the oscillation threshold. It may result in low small-signal gain and high extractable power, which are desirable characteristics for power amplifiers. To see this, we plot in Fig. 10-7, ΓGL versus NV, where V is the volume of the active region. For a given $\Gamma G_0 L$, which should be below the ASE oscillation threshold, we can obtain from the figure the corresponding values for $N_0 V$. To a good approximation, $N_0 V$ is proportional to the input electrical power. Note that because of the smaller and decreasing slope of the quantum-well curve, a quantum-well amplifier may be operated at a higher excitation than a bulk semiconductor amplifier before reaching the ASE oscillation threshold. This is true in our example for $\Gamma G_0 L > 9$. With sufficient input optical intensity, one may extract power from the amplifiers until the net amplifier gain $\Gamma G - \alpha_{abs}$ vanishes. Figure 10-7 shows that the maximum extractable power is greater in the quantum-well amplifier. Another consequence of the smaller and decreasing slope of the quantum-well curve is the greater variation in extractable power for a given change in $\Gamma G_0 L$.

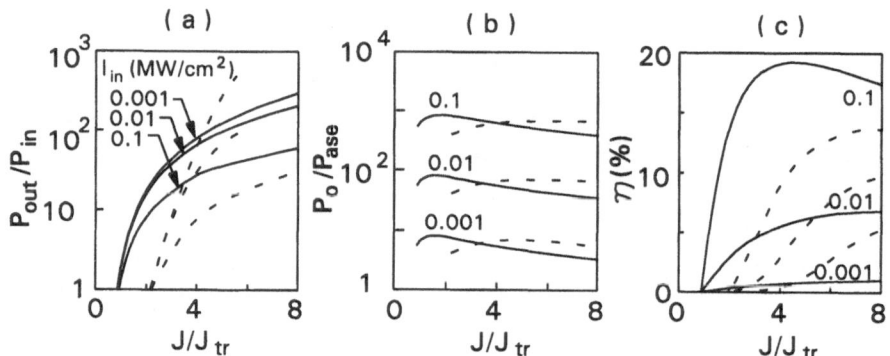

Fig. 10-8. Amplifier (a) gain, (b) S/N, and (c) efficiency for a $10 nm$ single quantum-well GaAs (solid curves) and a bulk GaAs (dashed curves) amplifier. For both amplifiers, $R_1 = R_2 = 10^{-4}$, $L = 500$ μm and $f = 10^{-4}$. The confinement factors are $\Gamma = 0.3$ and 0.03 for the bulk and quantum-well amplifiers, respectively. There is essentially no difference between operating within and outside the lockbands because of the low facet reflectivities. [From Chow and Craig (1991).]

Unfortunately the argument fails to take into account the effects of spontaneous emission. Carbon dioxide and atomic iodine are examples of good amplifier media because the optically extractable power increases faster with excitation than the spontaneous emission. The question is whether this is also the case for quantum-well semiconductor amplifiers. When we performed a Rigrod analysis with a quantum-well amplifier with a facet reflectivity of 10^{-2}, which is sufficiently high for ASE effects to become important, we find no significant difference between quantum-well and bulk semiconductor amplifier performance, except that the quantum-well operates more efficiently. The higher efficiency is not due to gain roll over but is due to the lower carrier density needed to reach transparency. The increase in ASE with excitation negates any advantage gain roll over may have in increasing the amplifier extractable power. Repeating the Rigrod analysis for the facet reflectivities $R_1 = R_2 = 10^{-4}$, we find several differences in quantum-well and bulk amplifier performance. Most important is that the overall quantum-well amplifier performance falls below that of the bulk amplifier at low input laser intensity and high excitation (compare solid and dashed curves in Figs. 10-8). This is because the low facet reflectivity and gain rollover lead to a low amplified input laser intensity inside the quantum-well amplifier, resulting in an appreciable fraction of the gain medium being unsaturated. In the bulk amplifier, the high local gain (because of the absence of roll over) compensates for the low facet reflectivity so that the amplified input laser field remains relatively intense.

Fig. 10-9. A double-pass amplifier. Typically the left mirror is antireflection coated and the right mirror is coated for maximum reflection. [From Chow and Craig (1991).]

The detrimental effects of the quantum-well gain roll over may be mitigated by increasing the circulating laser intensity with the double pass amplifier configuration shown in Fig. 10-9. A double-pass amplifier also has the practical advantage of requiring optical access to only one facet, which simplifies considerably the mounting of the amplifier to the heat sink. Figure 10-10 shows significant improvement in the quantum-well amplifier gain and efficiency, especially for small input laser intensities. Figure 10-10b shows that while the maximum values for P_0/P_{ase} remain essentially the same as those for $R_1 = R_2 = 10^{-4}$, and there is a slower decrease in P_0/P_{ase} with excitation. To obtain the more gradual degradation of P_0/P_{ase} with excitation for an input laser intensity of less than 0.01 MW/cm^2, the double pass amplifier has to be locked to the input laser field. For higher input laser intensities, the double-pass amplifier output becomes insensitive to the input laser field frequency. While further increasing R_1 improves power extraction, it also results in increased frequency selectivity.

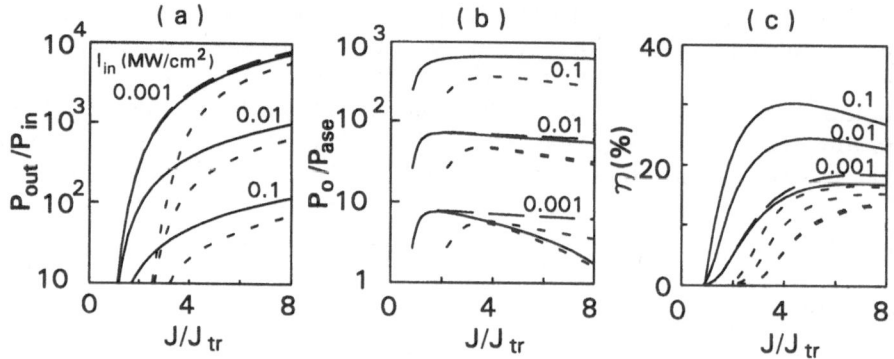

Fig. 10-10. (a) Gain, (b) P_o/P_{ase}, and (c) efficiency for a quantum-well double-pass amplifier with $R_1 = 10^{-4}$ and $R_2 = 0.99$. The solid curves are for an unlocked quantum-well amplifier and the long dashed curves are for a locked quantum-well amplifier and the corresponding curves for an equivalent bulk GaAs amplifier are denoted by short dashes.

In summary, we have shown in this section is that the Rigrod analysis provides a straightforward method for treating practical problems. For example, it shows the importance of ASE in semiconductor amplifiers. According to the Rigrod analysis an amplifier should operate at as high an excitation as possible so as to maximize the fraction of pump power that contributes to the amplification of the input laser field relative to that spent in overcoming internal losses and in achieving transparency in the gain medium. In addition to thermal considerations, the maximum pump power is limited by the increase in ASE, which leads to quenching of the amplifier gain, efficiency, and coherence. To effectively compete with ASE for the inversion, a sufficiently intense input laser field is needed. A large input intensity results in an amplifier output that is less dominated by ASE, and less sensitive to excitation and input laser frequency. ASE effects may also be mitigated by decreasing the amplifier facet reflectivities or, in some cases, tuning the input laser frequency to resonance with the amplifier.

When coupled to a gain theory described in the earlier chapters, the Rigrod analysis is able to demonstrate that gain roll over in a quantum-well gain medium does not result in better amplifier performance at high pump power because the rate of increase in spontaneous emission power feeding the ASE modes with carrier density is greater than or equal to the corresponding rate of increase in extractable optical power. Gain roll over, however, does make it more important to have an intense amplified input laser field to ensure saturation of the gain medium. This may be accomplished with either a high input laser intensity or a double-pass amplifier. While the multiple-pass amplifiers or injection-locked lasers obtained with higher facet reflectivities deplete the gain more effectively, they also have more stringent frequency matching requirements between the amplifier and the input laser field. The net result is that when compared to bulk semiconductor amplifiers, quantum-well amplifiers are likely to have comparable gain and signal to noise ratio, and significantly higher efficiency. The reason for the better efficiency, which in itself is a good reason for using a quantum-well gain medium, is the smaller carrier density needed to reach transparency in a quantum-well gain medium. This smaller carrier density also allows quantum-well amplifiers to operate at lower injected currents, which should help in reducing thermal effects.

Finally, the study reveals that the desirable amplifier configuration is characterized by very high excitation and very low resonator Q. A high excitation gives a device high efficiency. While the amplifier may be operating close to its free-running oscillation threshold, we are able to mitigate the frequency matching requirement associated with injection-locked lasers by having very low facet reflectivities and by maintaining a sufficiently intense input laser field. The above amplifier characteristics makes propagation effects important. The Rigrod analysis takes care of the field variations along the propagation direction. However it ignores lateral and transverse mode effects. While these effects are relatively unimportant in the single-pass amplifier problem treated in this section, they are important for broad-area lasers and multiple-pass amplifiers, whose lateral field profiles tend to be characterized by considerable structure. The remaining sections of this chapter deal with this problem.

10-2. Lateral Field Distributions

Lateral mode instabilities arise because high output power and single mode operation have contradictory design requirements. For high power operation, a large optical mode cross section is needed to circumvent the material damage threshold of $\simeq 10 \ MW/cm^2$. On the other hand, one is not completely free to change the transverse (perpendicular to the plane of the p-n junction) mode dimension because it is determined by the heterostructure, whose primary purpose is carrier confinement. Furthermore typical heterostructures support only the lowest transverse mode. This leaves the lateral (in the plane of the p-n junction) mode dimension as the only transverse degree of freedom. The lateral mode dimension may be increased by having a wide channel with weak lateral waveguiding. The weak optical confinement in these lasers, which are called broad area lasers, usually leads to multilateral mode operation. We consider multilateral mode operation to be an instability because it gives rise to spectral broadening and high spatial frequencies in the lateral field distribution. Many applications are unaffected by this instability because the laser output can still be coupled to a multimode optical fiber with reasonable efficiency. However as we expand the range of applications to include, for example, free-space communications, then it is desirable, if not necessary, for a high power semiconductor laser to be able to operate in a single mode.

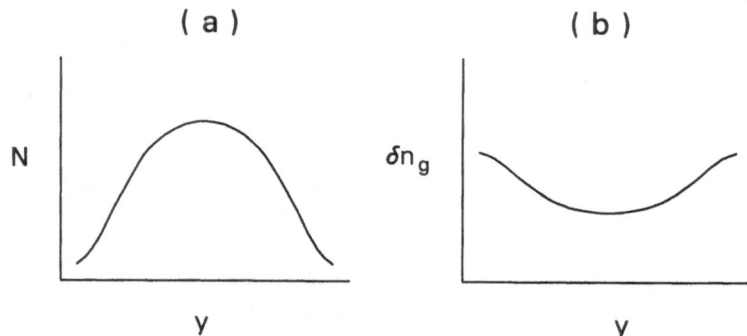

Fig. 10-11. Lateral (*a*) carrier and (*b*) refractive index distributions. These curves are for a laser operating close to threshold, where the effects of gain saturation are negligible.

There are several factors affecting lateral mode stability. In broad area lasers the gain medium plays a important role through filamentation. Filamentation results from self-focusing due to gain saturation. Its origin may

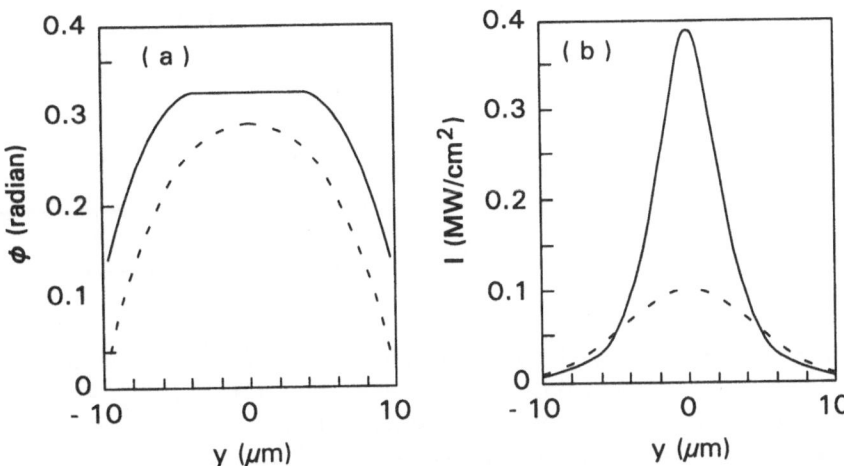

Fig. 10-12. Lateral (left, *a*) phase and (right, *b*) intensity distributions prior to (dashed curve) and after the onset (solid curve) of filamentation. The results are for a bulk GaAs laser with 10 μm wide electrodes. [From Chow and Depatie (1988).]

be understood by recalling from earlier chapters how the inhomogeneous character of the laser transition and the asymmetrical electron momentum distribution of the population inversion about the peak gain frequency lead to an appreciable carrier-induced refractive index δn_g that decreases with increasing carrier density. Near threshold, where the carrier density decreases with distance from the center of the laser beam (Fig. 10-11a), δn_g causes the refractive index in gain-guided lasers to increase with distance from the beam center (Fig. 10-11b). The resulting refractive index distribution acts as a diverging lens (antiguiding). While the net consequence of antiguiding and diffraction is a diverging wavefront (dashed curve in Fig. 10-12a), the laser still has a finite steady-state beam width because of gain guiding (dashed curve Fig. 10-12b). Further above threshold, the higher laser peak intensity creates a dip in the carrier density because of gain saturation (Fig. 10-13a). Figure 10-13b shows that due to δn_g, the net refractive index has a bump in the middle. Consequently a waveguide is created that focuses the laser beam, resulting in a higher peak intensity, which in turn creates a stronger waveguide. For semiconductor lasers, this self-focusing phenomenon is called *filamentation*. A steady-state filament size is reached when the self-focusing is balanced by diffraction and carrier diffusion. The filament is considerably narrower than the electrode width, and it has a flat wavefront, which is typical of index guided modes (compare solid and dashed curves in Fig. 10-12).

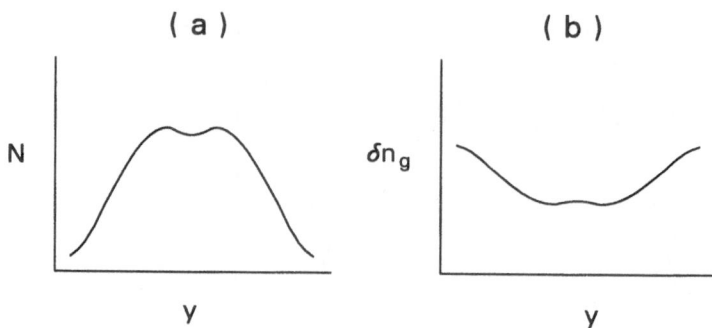

Fig. 10-13. Lateral phase (left) and intensity (right) distributions prior to (J/J_{th} = 1.05, dashed curve) and after (J/J_{th} = 1.07, solid curve) the onset of filamentation. These results are obtained using the free-carrier theory for a bulk GaAs laser with 10 μm wide electrodes.

The transition to a filamentary output with increasing excitation is depicted in Fig. 10-14, which plots the full width at half maximum $w_{1/2}$ of the lateral intensity profiles versus excitation for bulk GaAs lasers with different electrode widths, all chosen so that the lasers operate in a single lateral mode near threshold. The curves show a drastic reduction in the lateral field width after the onset of filamentation. Note that filamentation occurs very close to threshold. At high excitations the filament width reaches an asymptotic value that is independent of electrode width. The asymptotic filament size is governed by δn_g, carrier diffusion, and laser wavelength (via diffraction). At even higher excitations more than one filament appear in the lateral field distribution, which does not settle to steady-state.

An indication of the gain structure effects on filamentation is shown in Fig. 10-15, which plots $w_{1/2}$ versus excitation for 10nm GaAs quantum-well lasers with different electrode widths. Comparison with Fig. 10-14 shows that filamentation occurs further above threshold and that the asymptotic filament width is larger. While there is negligible narrowing in the intensity profile for the 10 μm electrode width, the wavefront flattens noticeably at high excitation, after the onset of filamentation.

The detrimental effects of filamentation are material damage due to increased peak intensity and increased spontaneous emission losses due to reduced overlap between laser beam and gain regions. The effects of the latter on laser performance are shown in Fig. 10-16, which plots the L-I

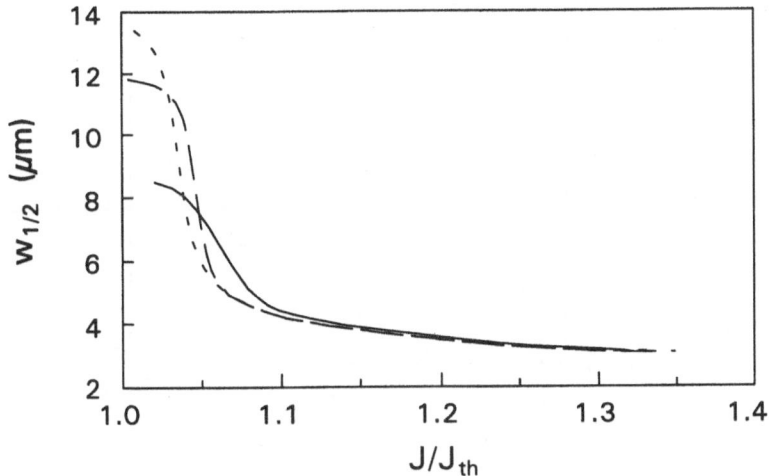

Fig. 10-14. Full width at half maximum of the lateral intensity profiles versus excitation for bulk GaAs lasers with electrode widths 10 μm (solid), 15 μm (long dashed), and 20 μm (short dashed). J_{th} is the threshold current density. The free-carrier theory was used for these calculations. The curves terminate at the threshold of the first-order lateral mode. [From Chow and Depatie (1988).]

curves for the bulk and quantum-well lasers. The kinks indicate the onset of filamentation. The degradation in laser performance is evident from the smaller slope of the L-I curve after the kink. No sharp kinks appear in the quantum-well lasers because of the more gradual transition to a filamentary output.

All of the above results were obtained using the free-carrier gain model for a two-band semiconductor. In the case of quantum-well lasers, also the important band mixing effects are neglected. Figure 10-17 shows the many-body effects on an L-I curve of a bulk GaAs laser. Note that the effects of filamentation is basically unchanged when many-body Coulomb effects are included. This is confirmed in Fig. 10-18, which plots the FWHM $w_{1/2}$ of the lateral intensity profile as a function of injected carrier density and in Fig. 10-19, which plots the lateral intensity and phase profiles at $N_0/N_{th} = 1.20$. The strong resemblance in field profiles computed using the free-carrier theory and the quasiequilibrium many-body Coulomb theory shows the cancellation between the modifications caused by bandgap renormalization and Coulomb enhancement. In the lit-

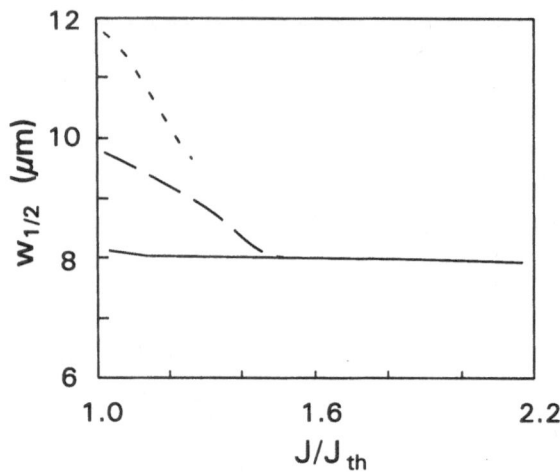

Fig. 10-15. Full width at half maximum of the lateral intensity profiles versus excitation for $10nm$ single quantum-well GaAs lasers with electrode widths of 10 μm (solid), 15 μm (long dashed), and 20 μm (short dashed). The curves terminate at the threshold for the first-order lateral mode. [From Chow and Depatie (1988).]

erature, the free-carrier theory has been used with an ad hoc inclusion of bandgap renormalization. Such an approach is inconsistent because both effects come from the same Coulomb interaction term in the Hamiltonian. Furthermore, as the long dashed lines in Figs. 10-17 to 10-19 show, this inconsistency leads to significantly different predictions, which have to be considered erroneous. A similar comparison is not available for the quantum-well laser. In terms of filamentation effects, we expect that the predictions of the free-carrier theory are qualitatively correct, i.e. that filamentation effects are weaker in quantum wells than in bulk semiconductor lasers and amplifiers. Also, since bandstructure details are ignored, the comparison of the different gain structures performed in this section involves only the differences between the 2- and 3-dimensional densities of states. The use of the many-body gain theory to investigate lateral mode effects in bulk and quantum-well lasers is described in Sec. 10-5 for unstable resonator semiconductor lasers.

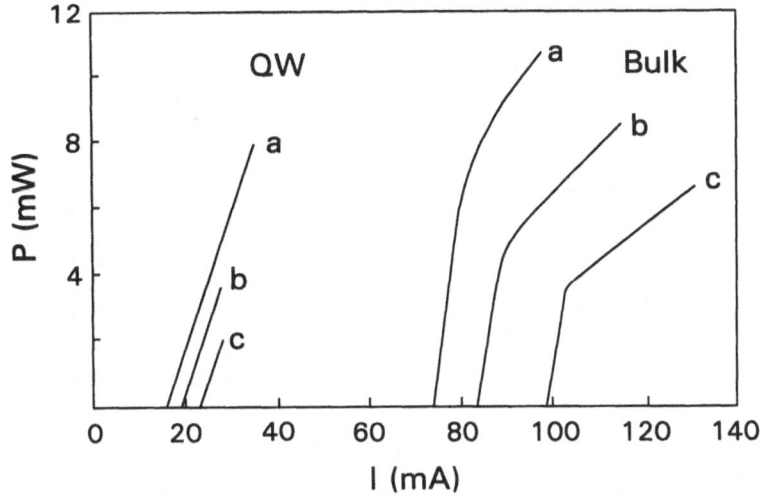

Fig. 10-16. *L-I* curves for the bulk and 10*nm* single-quantum-well GaAs lasers in Figs. 10-14 and 10-15. The electrode widths are (*a*) 10μm, (*b*) 15μm, and (*c*) 20μm.

10-3. Diffraction Effects

A physical optics laser model is necessary for investigating filamentation effects. This involves solving the wave equation (2.66), for the laser field. In this section, we present two commonly used approaches, the split-step method and the Crank-Nicolson method

Integral Method for Field Propagation

The most common and consequently the most developed method is the "split-step" method based on Fourier transformation. To describe this method, we start with Eq. (2.66). An important assumption is that diffraction and gain effects are separable, so that we can solve Eq. (2.66) to determine the effects of diffraction without the gain contribution, i.e., perform a propagation step using

$$-\nabla^2 \mathbf{E} + \left(\frac{n}{c}\right)^2 \frac{\partial^2 \mathbf{E}}{\partial t^2} = 0 , \tag{24}$$

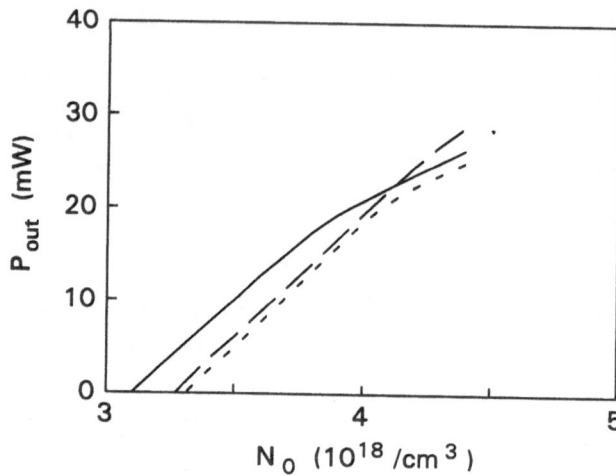

Fig. 10-17. Output power versus excitation for a 15 μm stripe width bulk GaAs laser. The solid curve is the many-body result, the long dashed curve is the free-carrier result with band-gap renormalization, and the short dashed curve is the free-carrier result. [From Chow *et al.* (1990).]

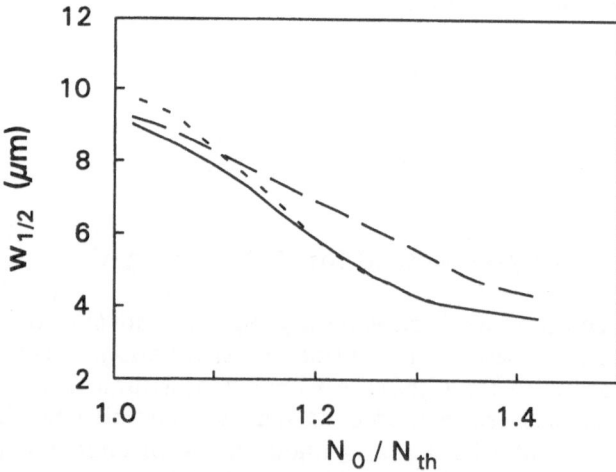

Fig. 10-18. Full width at half maximum of the lateral intensity profile versus excitation for a 15 μm stripe width bulk GaAs laser. The notation is the same as that in Fig. 10-17. [From Chow *et al.* (1990).]

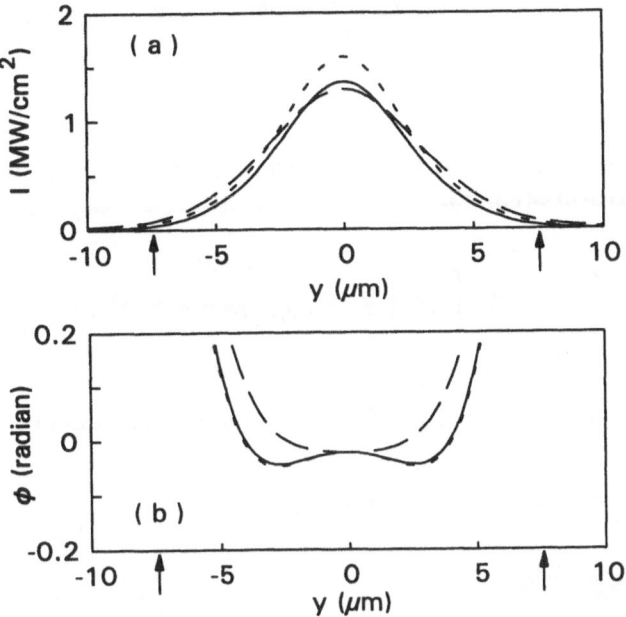

Fig. 10-19. Lateral (*a*) intensity and (*b*) phase distributions after the onset of filamentation for a 15 μm stripe width bulk GaAs laser. The arrows indicate the lateral extent of the electrodes. N_0/N_{th} = 1.2 and the curves are labeled similar to Fig. 10-17. [From Chow *et al.* (1990).]

and then repeat the same propagation step to determine the gain effects in the absence of diffraction. The resulting changes to the amplitudes are given by the sum of the changes incurred in both propagation steps. To propagate the field according to Eq. (24), we begin by writing the electric field as

$$E(x,y,z,t) = \frac{1}{2} \, \mathcal{E}(x,y,z,t) \, e^{-i\nu t} + \text{c.c.} \, , \tag{25}$$

where the field amplitude, $\mathcal{E}(x,y,z,t)$ contains the slowly varying time changes in the field and we assume a fixed field polarization. A time dependent field amplitude may be treated numerically by dividing $\mathcal{E}$ into several time slices and then propagating each slice separately. In most problems, we look at only one time slice or wavefront and we do that here. By substituting Eq. (25) into Eq. (24), we get the Helmholtz equation for each wavefront

$$\nabla^2 \mathcal{E} + K^2 \mathcal{E} = 0 ,$$

(26)

where $K = \nu/c$. To solve Eq. (26), we express $\mathcal{E}(x,y,z)$ in terms of its Fourier transform $u(\xi,\eta,z)$,

$$\mathcal{E}(x,y,z) = \frac{1}{2\pi} \int d\xi \int d\eta \, u(\xi,\eta,z) e^{-i(\xi x + \eta y)} .$$

(27)

Substituting Eq. (27) into Eq. (26), we encounter the equation

$$\frac{d^2 u}{dz^2} = - (K^2 - \xi^2 - \eta^2) u ,$$

(28)

which has the solution

$$u(\xi,\eta,z_2) = u(\xi,\eta,z_1) \exp\left[\pm iK \sqrt{1 - \frac{\xi^2 + \eta^2}{K^2}} (z_2 - z_1) \right] .$$

(29)

Expanding the square root, and taking only the forward propagating solution, we obtain from Eq. (27),

$$\mathcal{E}(x,y,z_2) = \frac{1}{2\pi} \int d\xi \int d\eta \, e^{i(z_2 - z_1)(K - \frac{\xi^2 + \eta^2}{2K})} e^{-i(\xi x + \eta y)} u(\xi,\eta,z_1) .$$

(30)

Using the inverse transform for $u(\xi,\eta,z_1)$ and after some algebra, we get

$$\mathcal{E}(x,y,z_2) = \frac{K}{i2\pi(z_2 - z_1)} \exp\left[iK(z_2 - z_1) + \frac{iK(x^2 + y^2)}{2(z_2 - z_1)} \right]$$

$$\times \int dx' dy' \, \mathcal{E}(x',y',z_1) \exp\left[-\frac{iKyy'}{z_2 - z_1} \right] \exp\left[-\frac{iKxx'}{z_2 - z_1} \right] .$$

(31)

An important result of Eq. (31) is that the integral is just the Fourier transform of the wavefront at position z_1. Since in practice $\mathcal{E}(x',y',z_1)$ is not an analytic function, this integral has to be evaluated numerically. Figure 10-20 shows the result of evaluating Eq. (31) for a top hat intensity distribution that has propagated a distance equivalent to 1 diffraction length.

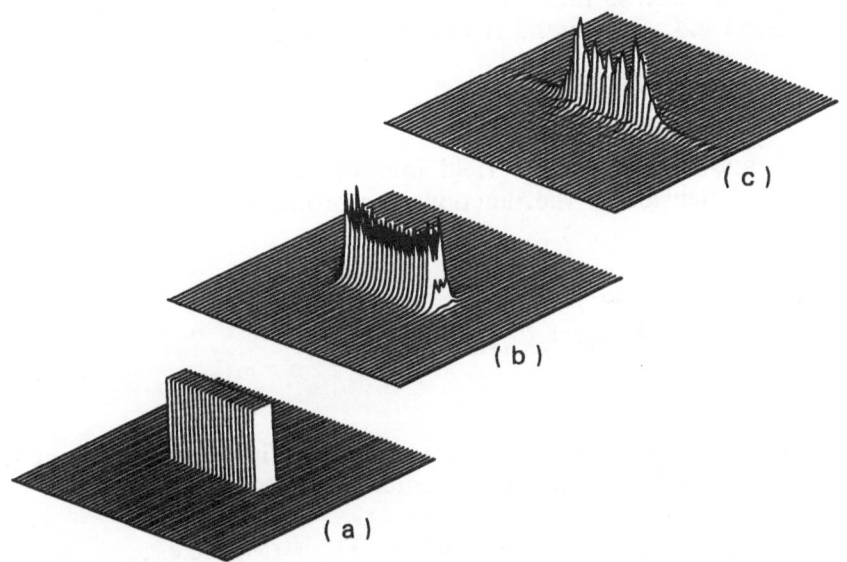

Fig. 10-20. Intensity distributions calculated from Eq. (31) at $z =$ (a) 0, (b) 0.1 μm and (c) 1 μm of a 1 $\mu m \times 4 \mu m$ rectangular beam profile. The wavelength is 0.81 μm.

Finite Difference Methods for Field Propagation

A weakness of the split-step Fourier transform method is that it cannot treat diffraction and gain effects simultaneously. We correct this by considering a finite difference approach known as the Crank-Nicolson method. Numerically, this approach works best when the x and y dimensions are separable, which is fortunate because to a good approximation, heterostructure edge emitters satisfy this condition. To see this we begin by noting that in most heterostructure lasers, the refractive index $n(\mathbf{r}) = n(x,y)$ is a weak function of y, so that the electric field may be written as

$$E(\mathbf{r},t) = \hat{x} \sqrt{\frac{1}{2\,\epsilon_0\,c\,n_{av}}}\; u(x,y)\; v\,(y,z)\; e^{i(Kz\,-\,\nu t)} + \text{c.c.}\;, \qquad (32)$$

where we assume field polarized in the $\hat{x}$ direction, $K = n_{av}K_0$, K_0 is the laser wavevector in vacuum, and ν is its frequency. The average refractive index n_{av} is chosen so that $\exp(iKz)$ accounts for most of the z dependence of the laser field. The lateral field dependence is given by

$$v\,(y,z) = \sqrt{I_g(y,z)}\; e^{-i\phi(y,z)}\;, \qquad (33)$$

where I_g and ϕ are the laser field intensity and phase averaged over the transverse dimension of the junction. According to Eqs. (32) and (33), the beam power is

$$P(z) = \frac{h}{\Gamma} \int_{-\infty}^{\infty} dy\; I_g(y,z)\;, \qquad (34)$$

where h is the thickness of the junction and Γ is the confinement factor. Similarly, we can write the active medium polarization as

$$\mathbf{P}(\mathbf{r},t) = \tfrac{1}{2}\hat{x}\,u_g(x,y)\mathscr{P}(y,z)e^{i[Kz\,-\,\nu t\,-\,\phi(y,z)]} + \text{c.c.}\;, \qquad (35)$$

where u_g is a weak function of y that vanishes outside the junction and $\mathscr{P}(y,z)$ is complex to account for the difference in phase between the medium polarization and the laser field.

Substituting Eqs. (32) and (35) into Eq. (2.66), we find

$$-\frac{1}{u}\frac{\partial^2 u}{\partial x^2} - \frac{1}{v}\left[\frac{\partial^2 v}{\partial y^2} + 2iK\frac{\partial v}{\partial z} + \frac{\mu_0\nu^2 u_g}{u}\mathscr{P}e^{i\phi}\right] + K^2 - (K_0 n)^2 = 0\;, \qquad (36)$$

where we have neglected all terms containing $\partial u/\partial y$ and have used the slowly varying amplitude and phase approximation. By treating y as a parameter in u, we can separate Eq. (36) into two equations, one giving the transverse field dependence for each y slab

$$-\frac{1}{u}\frac{d^2 u}{dx^2} + [K^2 - (K_0 n)^2] = \xi\;, \qquad (37)$$

and the other giving the lateral field dependence

$$\frac{1}{v}\left[\frac{\partial^2 v}{\partial y^2} + iK\,\frac{\partial v}{\partial z} + \mu_0\nu^2\Gamma\mathscr{P}e^{i\phi}\right] = -\xi \ , \tag{38}$$

where we use the property that the transverse field varies little as a function of y. In these equations, ξ is an integration constant. We can rewrite (37) as

$$-\frac{1}{u}\frac{d^2u}{dx^2} + K_0^2\,[n(x,y)^2 - n_{eff}(y)^2] = 0 \ , \tag{39}$$

in which the effective index n_{eff} is given by

$$K^2 + \xi = (K_0\,n_{eff})^2 \ . \tag{40}$$

Then

$$\frac{\partial v}{\partial z} = \frac{i}{2n_{av}K_0}\,\frac{\partial^2 v}{\partial y^2} + [\,\Gamma(g + iK_0\delta n_g) + iK_0\delta n_r - \alpha_{abs}]v \ , \tag{41}$$

where

$$\delta n_r(y) = n_{eff}(y) - n_{av} \tag{42}$$

is the fabricated real index distribution

$$\frac{\nu}{2\epsilon_0 cn}P = -ig + K_0\delta n_g \ , \tag{43}$$

g is the local amplitude gain, δn_g is the local carrier induced refractive index and α_{abs} accounts for all other optical losses.

We can write Eq. (41) as a finite difference equation and use it to propagate a forward and a backward travelling field inside a resonator until convergence is reached. A problem is that the stability of the numerical computation is very sensitive to how we evaluate the second derivative, $\partial^2 v/\partial y^2$. To expand on this, we rewrite Eq. (41) in the following more general form

$$\frac{\partial v}{\partial z} = D\,\frac{\partial^2 v}{\partial y^2} + F \ . \tag{44}$$

We begin with a guess of $v(y_j, z_n)$ for all y_j's at a given z_n. At first inspection, for the propagation from z_n to z_{n+1}, it would seem reasonable to compute $\partial^2 v/\partial y^2$ using the values of v at z_n, that is

$$\frac{\partial^2 v}{\partial y^2}(y_j, z_n) = \frac{v(y_{j+1}, z_n) - 2v(y_j, z_n) + v(y_{j-1}, z_n)}{\Delta y^2}. \tag{45}$$

However, doing so causes the numerical computation to become unstable after a few propagation steps. Fortunately, if we take the average of $\partial^2 v/\partial y^2$ computed at z_n and z_{n+1}, so that

$$\frac{\partial^2 v}{\partial y^2}(y_j, z_n) = [v(y_{j+1}, z_{n+1}) - 2v(y_j, z_{n+1}) + v(y_{j-1}, z_{n+1}) + v(y_{j+1}, z_n)$$

$$- 2v(y_j, z_n) + v(y_{j-1}, z_n)]/2\Delta y^2, \tag{46}$$

the results are more likely to remain stable. Of course, we do not yet know the values of v at z_{n+1}. So first we need to write (44) as

$$-\eta v(y_{j+1}, z_{n+1}) + (1 + 2\eta)v(y_j, z_{n+1}) - \eta v(y_{j-1}, z_{n+1}) = r(y_j z_n), \tag{47}$$

where

$$\eta = \frac{D\Delta z}{2\Delta y^2}, \tag{48}$$

$$r(y_j, z_n) = \eta v(y_{j+1}, z_n) + (1 - 2\eta)v(y_j, z_n) + \eta v(y_{j-1}, z_n) + \Delta z F(y_j, z_n) \tag{49}$$

$$F(y_j, z_n) = [g(y_j, z_n) + iK_0 \delta n_g(y_j, z_n)]v(y_j, z_n). \tag{50}$$

Then for the entire wavefront we have

$$\mathscr{D}\mathscr{V} = \mathscr{R}, \tag{51}$$

where

$$\mathscr{D} = \begin{bmatrix} 1+2\eta & -\eta & 0 & . & 0 & 0 \\ -\eta & 1+2\eta & -\eta & . & . & . \\ 0 & -\eta & 1+2\eta & . & . & . \\ . & 0 & -\eta & . & 0 & . \\ . & . & 0 & . & -\eta & 0 \\ . & . & . & . & 1+2\eta & -\eta \\ 0 & 0 & 0 & . & -\eta & 1+2\eta \end{bmatrix}, \tag{52}$$

$$\mathcal{V} = \begin{Bmatrix} v(y_2, z_{n+1}) \\ v(y_3, z_{n+1}) \\ v(y_4, z_{n+1}) \\ \cdot \\ \cdot \\ \cdot \\ v(y_{N-1}, z_{n+1}) \end{Bmatrix}, \tag{53}$$

$$\mathcal{R} = \begin{Bmatrix} r(y_2, z_n) \\ r(y_3, z_n) \\ r(y_4, z_n) \\ \cdot \\ \cdot \\ \cdot \\ r(y_{N-1}, z_n) \end{Bmatrix}. \tag{54}$$

If $1 \leq j \leq N$, $\mathcal{D}$ is an $(N-2) \times (N-2)$ matrix, while $\mathcal{V}$ and $\mathcal{R}$ are $(N-2) \times 1$ column matrices. The tridiagonal matrix Eq. (51) may be solved readily with a simple computer routine.

An advantage of the Crank-Nicolson method is that diffraction and gain effects are treated simultaneously. The price one pays is that the numerical procedure is simple only when we have a two dimensional $(y\text{-}z)$ problem, i.e., when the transverse and lateral dimensions are separable. On the other hand, the split-step Fourier transform method treats the three-dimensional problem readily. In practice, the separability of the transverse and lateral coordinates is a good assumption for almost all heterostructure lasers.

10-4. Filamentation in Amplifiers

We now apply the techniques of the previous section to treat two problems, one involving filamentation in amplifiers and the other involving single lateral mode stability in unstable resonators. Starting with the amplifier problem, we follow an approach that is similar to that used in Sec. 10-1. The amplifier field is separated into forward and backward travelling waves so that Eq. (32) becomes

$$E(\mathbf{r}, t) = E_+(\mathbf{r}, t) + E_-(\mathbf{r}, t) , \tag{55}$$

where

$$E_\pm(\mathbf{r}, t) = \sqrt{\frac{1}{2\epsilon_0 c n_{av}}} \, u(x, y) v_\pm(y, z) e^{i(\pm Kz - \nu t)} + \text{c.c.} \tag{56}$$

Repeating the steps from Eq. (36) to (41), we find

$$\pm \frac{\partial v_\pm}{\partial z} = \frac{i}{2K_0 n_{av}} \frac{\partial^2 v_\pm}{\partial y^2} + iK_0 \delta n_r(y) v_\pm + \Gamma[g(N) + iK_0 \delta n_g(N)] v_\pm - \alpha_{abs} v_\pm . \tag{57}$$

The starting field distribution is that of the input field to the amplifier.

The analysis begins by solving Eq. (39) for the transverse modes, $u(x, y)$ and the effective index $n_{eff}(y)$ at several lateral positions. If the amplifier is purely gain guided, then both the transverse mode and the effective index are independent of y. Using the effective index in Eq. (57), and starting with the intensity distribution of the input field to the amplifier, we solve Eq. (57) numerically by propagating the field inside the amplifier, with either the Fourier transform method or the finite difference method, until convergence is reached. Each propagation step requires the local gain and carrier induced refractive index. They are given by one of the gain formulas in Chaps. 3 through 6. The local values of the carrier density are required to compute the gain. We obtain the carrier density distribution by averaging out the transverse variations

$$N(y, z) = \frac{1}{h} \int_{-h/2}^{h/2} dx \, N(x, y, z) \tag{58}$$

and solving

$$\frac{\partial N}{\partial t} = D_f \nabla^2 N + \frac{J(y)\eta}{ed} - BN^2 - \gamma_{nr} N - \frac{g(N)}{\hbar\nu}(|v_+|^2 + |v_-|^2) , \tag{59}$$

where D_f is the carrier diffusion coefficient.

A rigorous treatment of carrier transport is computer time consuming and sometimes unnecessary. For some problems, simplifications are possible. For example, the unsaturated carrier density distribution at the junction may be estimated using a *current spreading model*. In such a model one assumes that the lateral spreading of carriers was due to diffusion in the active layer. The carrier density within and outside a stripe of width w

is given by

$$N_0(y) = \bar{N}_0 \left[1 - A\cosh(y/l_1) \right], \qquad (-\tfrac{1}{2}w < x < \tfrac{1}{2}w) \qquad (60)$$

$$N_0(y) = \bar{N}_0 B \exp\left[-\frac{y}{l_2} \right], \qquad (x < -\tfrac{1}{2}w \text{ or } x > \tfrac{1}{2}w) \qquad (61)$$

respectively, where $\bar{N}_0$ is the injected carrier density if there were no lateral diffusion and we ignore any z-dependence in the unsaturated carrier density. In Eqs. (60) and (61), l_1 and l_2 are the diffusion lengths inside and outside the stripe, respectively, and the coefficients A and B are given by

$$A = [\cosh(w/2l_1) + \zeta\sinh(w/2l_1)]^{-1} \qquad (62)$$

$$B = A\zeta\sinh(w/2l_1) \exp(w/2l_2), \qquad (63)$$

where

$$\zeta = \frac{D_1 l_2}{D_2 l_1} = \sqrt{\frac{D_1 \tau_2}{D_2 \tau_1}}. \qquad (64)$$

D_1 and D_2 are the diffusion constants for the regions inside and outside the stripe, and τ_1 and τ_2 are the corresponding carrier lifetimes. Figure 10-21 shows an example of a carrier distribution computed using Eqs. (60) and (61).

The unsaturated carrier distribution is modified by the laser field [last term in Eq. (59)] and by diffusion [first term on the left hand side of Eq. (59)]. The net result can be determined for a given laser field distribution by numerically solving Eq. (59). There is an alternate approach that gives an analytic expression for the saturated carrier distribution and provides better physical insight how the two mechanisms interact. This approach is based on the assumption that, especially when operating with high gains, the effect of diffusion may be treated as a perturbation. First, we rewrite Eq. (59) into the form

$$\gamma_s[N_0(y) - N(y,z)] + \zeta D\nabla^2 N(y,z) = \frac{2g(y,z)I_g(y,z)}{\hbar\nu}, \qquad (65)$$

where $I_g = (|v_+|^2 + |v_-|^2)/2$ is the laser field intensity averaged over the transverse dimension of the junction and ζ is an expansion parameter, which is set equal to one at the end of the calculation. For simplicity, we

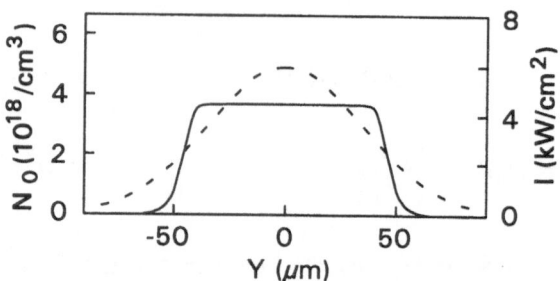

Fig. 10-21. Lateral distribution (solid line) of unsaturated carrier density for an electrode width of 100 μm and a carrier diffusion length of 3 μm. The dashed curve shows a typical input laser field distribution. The injected carrier concentration at the junction (without sideways diffusion) is $3.7 \times 10^{18}/cm^3$.

assume that all the carrier recombination terms may be lumped into an exponential decay term. The solution is a power series in ς,

$$N(y,z) = N_1(y,z) + \varsigma N_2(y,z) + \varsigma^2 N_3(y,z) + \dots . \tag{66}$$

Substituting Eq. (66) into Eq. (65) and equating terms proportional to ς, we find the first-order expressions

$$\gamma_s [N_0(y) - N_1(y,z)] = \frac{2g(y,z)I_g}{\hbar \nu} \tag{67}$$

$$N_2(y,z) = L_D^2 \frac{d^2 N_1}{dy^2} \left[1 + \frac{2I_g}{\gamma_s \hbar \nu} \frac{dg}{dN}(N_1) \right]^{-1} , \tag{68}$$

where $L_D = \sqrt{D/\gamma_s}$ is the carrier diffusion length. Note that within the junction, the laser field varies very little in the transverse direction, and that spatial hole burning in the longitudinal direction is washed out by carrier diffusion. We may interpret N_1 as the saturated carrier density distribution due to the interaction with the laser field, and N_2 as the modification to N_1 due to the smoothing effect of diffusion. The denominator in Eq. (68) indicates that the effects of diffusion are quenched by an intense laser field, which motivates the introduction of the effective diffusion length

$$L_{D,eff} = \frac{L_D}{\sqrt{1 + \dfrac{2I_g}{\gamma_s \hbar \nu}\dfrac{dg}{dN}}}, \tag{69}$$

which decreases with laser intensity.

Now we apply the equations derived in this section to study filamentation in a double-pass amplifier. The double-pass amplifier is introduced in our discussion on ASE effects in amplifiers, where we show that it improves extraction efficiency. On the other hand, a double-pass amplifier operates with strong counter propagating fields, which increases the likelihood of filamentation. Figure 10-22 shows plots of output power and efficiency versus input optical power for a bulk GaAs double-pass amplif-

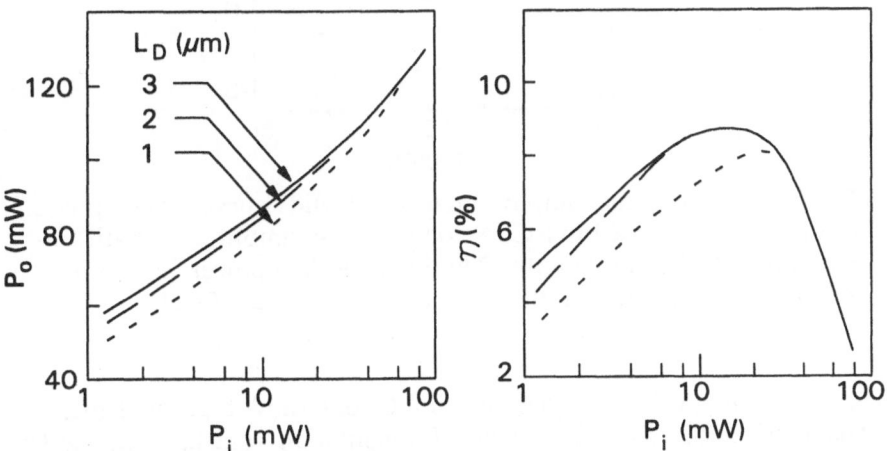

Fig. 10-22. (left) Output power and (right) efficiency versus input optical power for a double-pass bulk GaAs amplifier operating with $\Gamma G_0 L = 4$. L_D is the carrier diffusion length. The amplifier electrodes have a constant current density and a dimension of $100\ \mu m \times 500\ \mu m$.

ier. The sensitivity of the results to the carrier diffusion length, especially for small input laser intensity, indicates the presence of filamentation. This is confirmed in Fig. 10-23, which shows the existence of filaments for $L_D < 3\mu m$. We also found that for $L_D < 1mm$ the lateral beam profile is independent of the diffusion length. Here the filament width is determined by the effects of diffraction and the carrier induced refractive

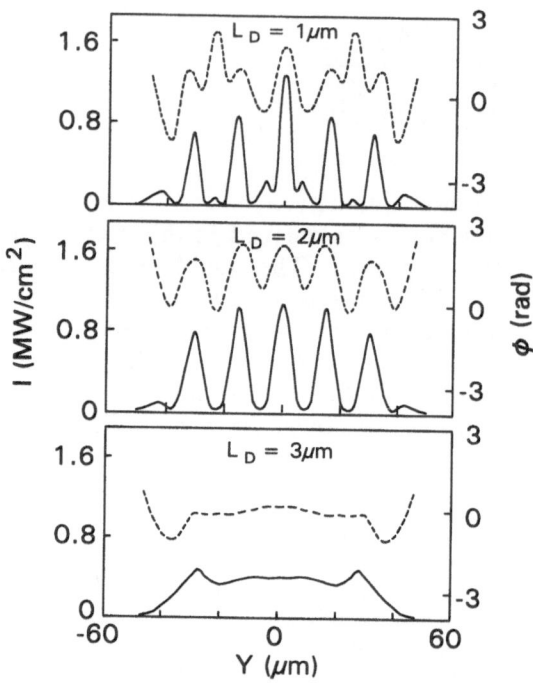

Fig. 10-23. Lateral output intensity (solid curve) and phase (dashed curve) of a bulk GaAs double-pass amplifier for different carrier diffusion lengths. The input optical power is 1.06 mW. All other parameters are the same as those in Fig. 10-22.

index. For 1 μm < L_D < 3 μm, the simulations show a gradual increase in the filament width. For L_D > 3 μm, filamentation vanishes and the lateral beam profile is again essentially independent of L_D.

Referring back to Fig. 10-22, we see that the diffusion length dependence vanishes at high input laser intensity. This suggests a P_i dependence in filamentation, which we confirm by looking at the amplifier peak intensity as a function of P_i. The behavior shown by the dashed curves in Fig. 10-24 may be explained as follows. When P_i is less than $4mW$, the output beam consists of several discrete filaments [Fig. 10-22b]. For P_i > 4 mW, the filament peaks decrease, while the intensity in the regions between filaments increases. When P_i exceeds 50 mW, the filaments vanish and the output beam profile becomes spatially smooth. The disappearance of fila-

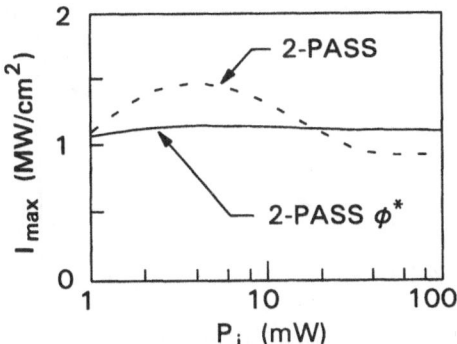

Fig. 10-24. Peak intensity versus input optical power for a bulk semiconductor double-pass amplifier operating with (solid curve) and without (dashed curve) a phase conjugator. We assume a phase conjugator reflectivity of 0.5, and a carrier diffusion length of 2 μm. $\Gamma G_0 L$ = 4 and 9 for the double-pass and phase conjugate double-pass amplifiers, respectively. The output power for the phase conjugate double-pass amplifier varies from 4x to 1.5x that of the double-pass amplifier. The peak intensity of the double-pass bulk semiconductor amplifier varies because of filamentation.

ments is linked to the decrease in the amplifier gain with increasing P_i. The amplifier output is affected less by the gain medium, i.e., by δn_g, and more by the input beam when the gain is low.

The elimination of filaments is a problem of practical interest. One option is to use quantum-well amplifiers. Figures 10-25 and 10-26 show that the output of a QW double-pass amplifier is relatively insensitive to L_D and also filament free. These results are obtained with the free-carrier gain model. However the small magnitude of δn_g in a QW gain medium, which is responsible for the lack of filamentation, is generally true regardless of many-body effects.

Another possible solution uses phase conjugation in conjunction with a double-pass amplifier. Figure 10-27 depicts a phase conjugate amplifier, where the laser beam makes two passes through the gain medium. The principle of a phase conjugate amplifier is that after the beam propagates once through the amplifier, a phase conjugate mirror reflects the beam back through the amplifier. If no amplification were involved, this would result in the cancellation of the phase aberrations introduced by the medium. With amplification, the phase aberrations are cancelled in part.

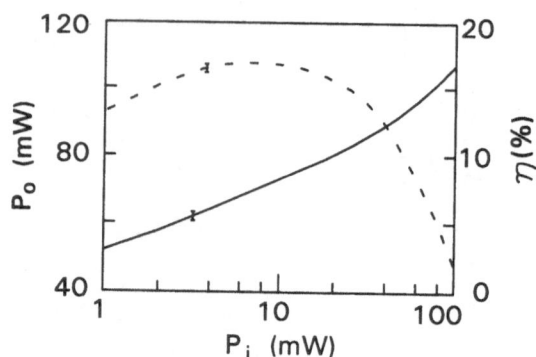

Fig. 10-25. Output power (solid curve) and efficiency (dashed curve) versus input optical power for a double-pass quantum-well GaAs amplifier operating with $\Gamma G_0 L = 4$. The error bars indicate the variation in the curves when the carrier diffusion length is varied between 0 and 4 μm.

To estimate the oscillation threshold, we note that in the absence of an incident laser beam, the reflectivity of a self-pumped phase conjugator is zero. We also assume that the semiconductor gain medium saturates homogeneously, and that for the range of input optical power considered, the amplified input field can effectively suppress the amplified spontaneous emission in the amplifier. Using the above assumptions, we calculate for $L = 500$ μm and $R_1 = R_2 = 10^{-4}$, an oscillation condition of $\Gamma G_0 L < 9.7$.

Figure 10-28 depicts a typical output beam profile of a double-pass bulk GaAs amplifier that would have a filamented output without phase conjugation. The absence of filamentation in a phase conjugate amplifier configuration leads to lower peak intensity in the output beam for the same output power, and to a more gradual change in the peak intensity with output power than is the case without a phase conjugator (solid curve Fig. 10-24). Also we find that because of phase conjugation, the absolute phase of the output beam always equals that of the input beam, which is important in some phased array schemes for scaling semiconductor lasers. Finally, the phase conjugator appears to be successful in preventing filamentation, regardless of the magnitude for the carrier diffusion length.

So far we have been dealing with double-pass amplifiers. We now show that filamentation becomes even stronger with regenerative amplifiers and injection-locked lasers. We do so by considering increased values of the mirror reflectivity R_1. Regenerative amplifiers and injection-locked lasers differ in that a regenerative amplifier operates below the oscillation

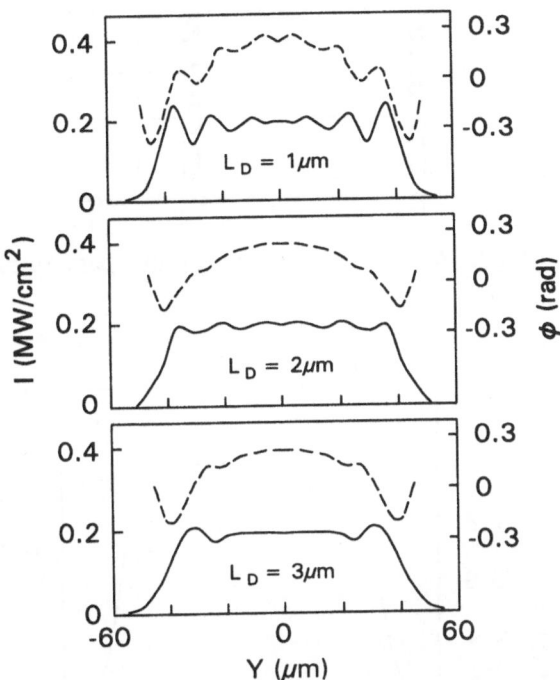

Fig. 10-26. Lateral output intensity (solid curve) and phase (dashed curve) of a double-pass QW amplifier for different carrier diffusion lengths. Note that the phase variations are approximately one tenth of that in the double-pass bulk GaAs amplifier. All parameters are the same as those in Fig. 10-25.

threshold, while an injection-locked laser operates above the threshold. Accordingly we choose the level of excitation so that for each configuration, the device operates as a regenerative amplifier when $R_1 < 10^{-4}$ and as an injection-locked laser when $R_1 > 10^{-4}$. For the latter, we use a resonant input power beam in order to maximize the effects of changing R_1.

Figure 10-29 plots the net amplifier gain versus R_1 for an input laser power of $1mW$. We found no difference in the output beams for $0 < R_1 < 10^{-4}$. For $R > 10^{-4}$ there is a slight increase in the output beam power with R_1. The increase in gain with R_1 is small in all cases because a $1\ mW$ input beam can extract almost all the available power in an amplifier in two passes.

There is little change in the output beam profiles, except for the injection-locked DH laser. Figure 10-30 shows a drastic increase in the output

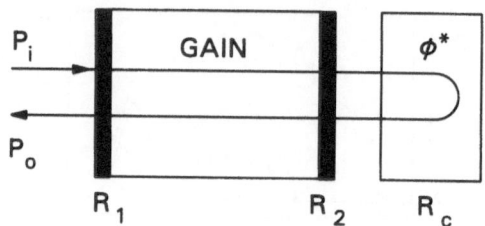

Fig. 10-27. A phase conjugate double-pass amplifier.

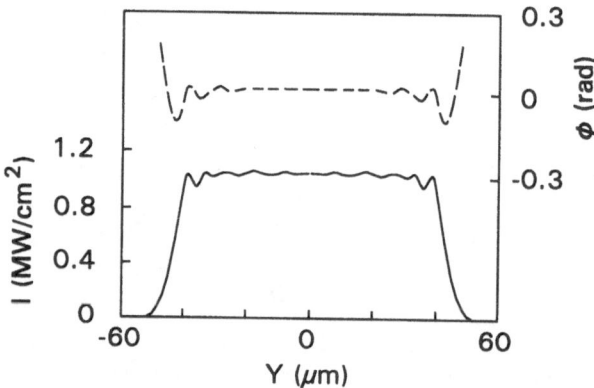

Fig. 10-28.   Lateral output intensity (solid curve) and phase
(dashed curve) of a phase conjugate double-pass bulk GaAs amplifier. These curves are essentially independent of the carrier diffusion length.   The input optical power is 1 mW and the phase
conjugator reflectivity is 0.5.

peak intensity of the injection-locked DH laser at R_1 = 0.004. The reason
for the rise in peak intensity is shown in Fig. 10-31, where we see a transition from a relatively smooth beam profile at R_1 = 10^{-4} to a highly filamented output beam profile at R_1 = 0.01.

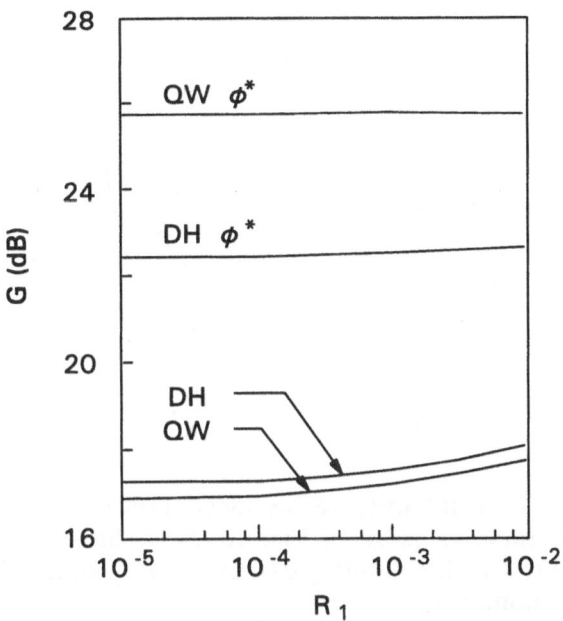

Fig. 10-29. Amplifier net gain versus facet reflectivity R_1 for the different types of amplifiers. The input optical power is 1 mW, and the carrier diffusion length is 3 μm. The amplifiers with phase conjugators are operated with $\Gamma G_0 L$ = 8.9 and the phase conjugator reflectivity of 0.5, while those without phase conjugators are operated with $\Gamma G_0 L$ = 4.0 and a high reflectivity coating of R_2 = 1.

10-5. Unstable Resonator Lateral Mode Stability

For a second demonstration of a laser wave optical analysis, we study the problem of semiconductor lasers operating with unstable resonators. Unstable resonator semiconductor lasers were introduced in 1985 to solve the filamentation problem in broad-area lasers. Reduction of filamentation was demonstrated and roughly five years later, high power (> 1 Watt) nearly diffraction-limited performance was achieved.

We apply our analytical tools to the unstable resonator semiconductor laser to identify and to understand the basic material and structural properties that govern its performance. Following Sec. 10-4, we write the laser field as

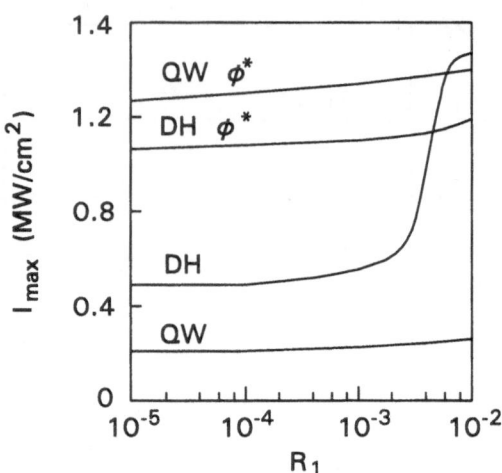

Fig. 10-30. Peak intensity versus facet reflectivity R_1. All parameters are same as those in Fig. 10-29. The sharp rise in the peak intensity for the double-pass bulk GaAs amplifier is because of filamentation.

$$E(\mathbf{r},t) = \sqrt{\frac{1}{2\epsilon_0 c n_{av}}}\,[v_+(y,z,t)e^{iKz} + v_-(y,z,t)e^{-iKz}]u(x,y)e^{-i\nu t} + \text{c.c.,} \qquad (70)$$

where we have assumed a TE polarization. Furthermore, we include the time dependence because we are interested in the stability of the laser. From the wave equation, we can derive the equations of motion for the forward and backward propagating lateral field distributions

$$\frac{1}{c}\frac{\partial v_\pm}{\partial t} \pm \frac{\partial v_\pm}{\partial z} = \frac{i}{2nK_0}\frac{\partial^2 v_\pm}{\partial y^2} + iK_0\delta n_r(y)v_\pm$$

$$+ \Gamma[g(N) + iK_0\delta n_g(N)]v_\pm - \alpha_{abs}v_\pm .$$

$$(71)$$

Transformation to the new coordinate system given by $z'= z$ and $t' = t \pm z/c$ changes Eq. (71) to Eq. (57). The rate equation for the carrier density remains unchanged. The time dependence is treated by dividing $v_\pm$ into several time slices in the new coordinate system, where each time slice propagates according to Eq. (57). Then the numerical solution of the field and medium equations follows the same procedures as outlined in the pre-

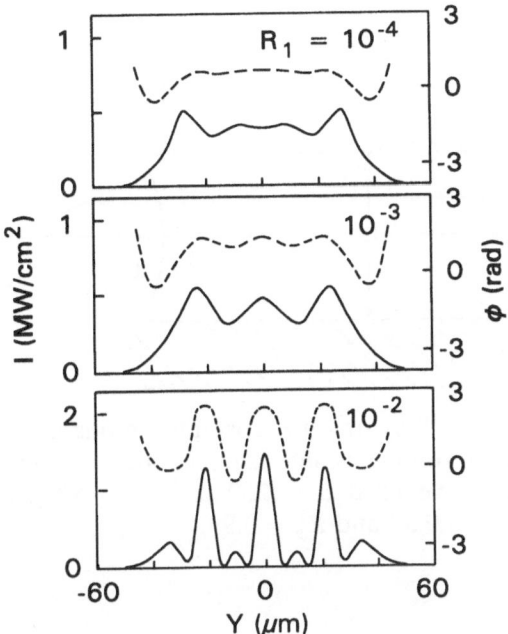

Fig. 10-31. Lateral output intensity (solid curves) and phase (dashed curves) for a bulk GaAs amplifier with facet reflectivity $R_1 = 10^{-4}$, 10^{-3}, and 10^{-2}. All other parameters are the same as those in Fig. 10-29.

vious two sections. The resonator boundary conditions are

$$v_+(x,0,t) = - \sqrt{R_1}\, v_-(x,0,t) \tag{72}$$

$$v_-(x,L,t) = - \sqrt{R_2}\, e^{-iKx^2/\rho}\, v_+(x,L,t) \,, \tag{73}$$

where R_1 and R_2 are the mirror reflection coefficients. In this study of unstable resonator lasers, we use the quasiequilibrium many-body theory of Chaps. 4 to 6 to compute the gain and carrier-induced refractive index.

To compare unstable resonator performance for different gain structures, we perform the calculations for bulk GaAs, a 5nm GaAs quantum-well, and a 5nm InGaAs strained quantum-well gain medium. Using the theory of Chap. 6, we compute the bandstructures for these materials (see Fig. 10-33). The strained quantum-well structure, which consists of a 5nm

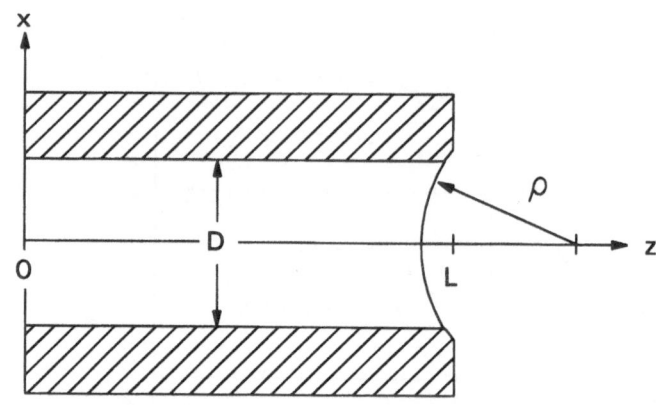

Fig. 10-32. A half-symmetry unstable resonator semiconductor laser. The gain volume is indicated by the unshaded region. In our simulations, we used L = 250 μm, D = 80 μm, and mirror reflectivities R_1 = 0.32 and R_2 = 0.99.

$In_{.1}Ga_{.9}As$ layer between GaAs barriers, was chosen because it gives the "ideal" laser bandstructure (two-band system with nearly equal electron and hole masses).

The lateral field distributions fall into two basic categories: single lateral mode or multilateral mode. Most high power lasers operate multi-lateral mode, and for this case, our stimulations give lateral field intensity distributions that never reach steady state and consist of sharp peaks or filaments. For a properly choosen unstable resonator configuration, single lateral mode operation is possible. In this case, the lateral intensity distributions remain constant from pass to pass and are relatively smooth.

One important question about the design of unstable resonators is, how critical is the resonator design? Figure 10-35 shows the spectra of lasers with different curve mirror radii of curvature. The gain medium is bulk GaAs. Each spectrum is obtained by taking the space-time Fourier transform of the forward propagating field at $z=L$, after a sufficiently large number of passes (approximately 4000). The plots with $\rho = \infty$ show the spectra of broad-area lasers with plane mirrors. According to the figure, significant spectral narrowing occurs when the curve mirror radius of curvature is reduced to 5 mm. Single mode operation is achieved with $\rho = 3.2mm \pm 0.1mm$, which corresponds to a magnification of $\simeq 1.7$. Further reduction in the mirror curvature causes higher order unstable resonator lateral modes to go above threshold.

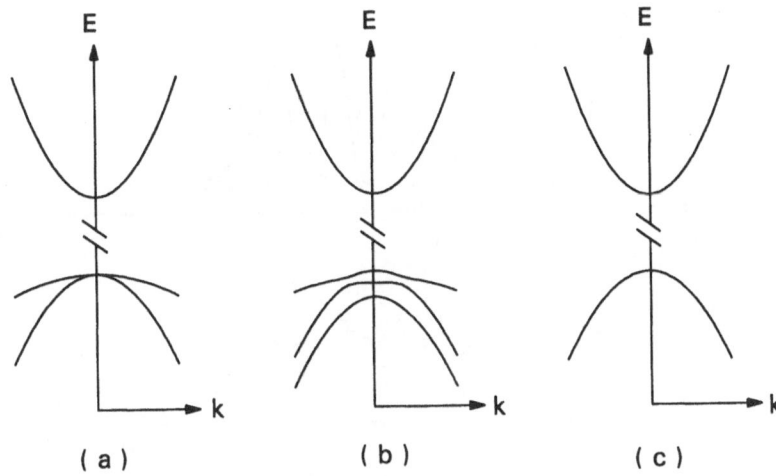

Fig. 10-33. The bandstructures of (*a*) bulk GaAs, (*b*) 5*nm* GaAs between $Ga_{.8}Al_{.2}As$ barriers and (*c*) 5*nm* $In_{.1}Ga_{.9}As$ between GaAs barriers. Each band is doubly degenerate.

Figure 10-36 depicts the results for the quantum-well unstable resonator. For this system the spectra appear cleaner than those for the bulk and more importantly, the resonator requirements for single lateral mode operation are less restrictive. The range of mirror curvature for single mode operation is increased to $2mm \leq \rho \leq 4mm$ (magnification 1.6 to 2.0).

Figure 10-37 shows that even better performance is possible with the strained quantum-well structure that we have choosen. The values of ρ for single lateral mode operation now ranges from 5*mm* to 30*mm* or a magnification of between 1.2 and 1.6. Also note that significant spectral narrowing is achieved with a value of ρ as big as 50*mm*. Of course in practical terms, this property is meaningless since fabricating a 5*cm* radius of curvature mirror with a diameter of 80μm is very difficult, if not impossible.

Another important piece of information is the excitation dependence of unstable resonator semiconductor lasers. Figure 10-38 shows the spectra of the bulk GaAs unstable resonator laser at two different excitation levels. The unstable resonator has a curve mirror with $\rho = 3.2mm$, which places it in the middle of the stable region. According to the figure, the laser eventually operates multimode. The threshold for multimode operation is $N/N_{th} = 2.3$. Similar simulations show that the thresholds for multilateral mode operation are significantly higher with the quantum-well and strained quantum-well gain media. We chose resonator configurations that are

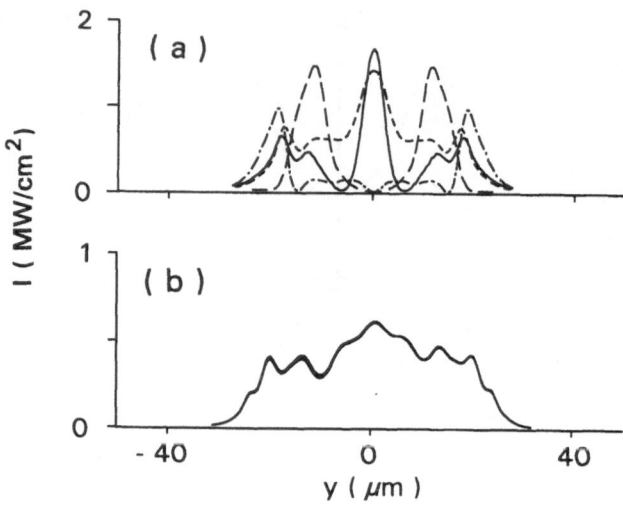

Fig. 10-34. Lateral field distributions showing (*a*) unstable and
(*b*) stable laser operation. Plotted are the field distributions for
four consecutive passes through the unstable resonator. The gain
medium for (*a*) is bulk GaAs and for (*b*) is GaAs quantum-well.
For both plots, the mirror curvature is $\rho = 2.5mm$ and the stripe
is $40\mu m$ wide and $250\mu m$ long.

in the middle of their stability regions. In both cases, the lasers remained
single lateral mode past $N/N_{th} = 10$. We did not extend our simulations
further to find the multilateral mode threshold because even at $N/N_{th} = 10$, effects neglected in our model, e.g., heating and pump inhomogeneities, become important.

In summary, we have shown that the gain medium can significantly
affect the lateral mode stability of an unstable resonator semiconductor
laser. In particular, quantum confinement or strain can result in single lateral mode operation over wider ranges of mirror curvature and excitation.
These results should be quite general. Bulk and quantum-well structures
differ notably in their bandstructures and in the nature of many-body
effects, which are factors that contribute to a smaller linewidth enhancement factor α in quantum-well structures. The threshold values of α for
the bulk, quantum-well, and strained quantum-well lasers operating within
the stable regions are 2.1, 1.1, and 0.8, respectively. However, one must be
careful about putting too much emphasis on α, which is based on a linearized gain model. The α parameter does not appear in our analysis because

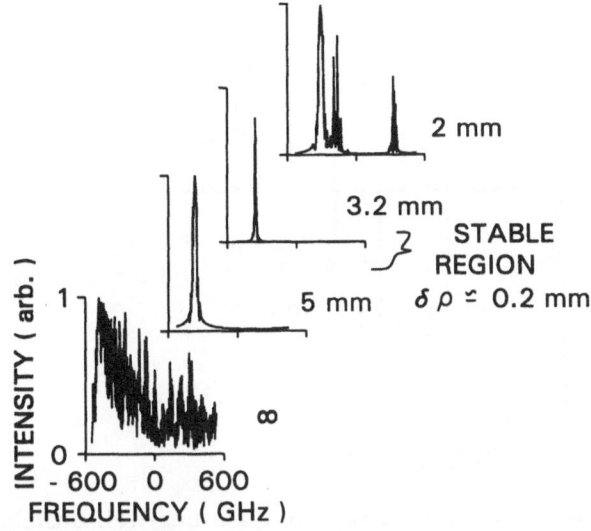

Fig. 10-35. Laser spectra for the bulk GaAs unstable resonator laser. Each spectrum is labeled by the curve mirror radii of curvature. The lasers are operating with an excitation of $N/N_{th} = 2$ and the gain medium parameters are $\gamma = 10^{13}/s$, $n = 3.5$, $D_f = 30cm^2/s$, $\gamma_{sp} = 3 \times 10^8/s$, $\Gamma = 0.3$ and $\alpha_{abs} = 10/cm$. (From Ru et al. 1993)

we use the general nonlinear functions $g(N)$ and $\delta n_g(N)$. As discussed earlier chapters studies have shown that a linearized gain model can lead to errors in predicting the dynamical behavior of quantum-well lasers. Another result is that with the proper choice of a strained quantum-well gain structure, one can further improve unstable resonator laser performance. However, since the carrier dependence of the gain and carrier-induced refractive index can vary considerably with the amount and type of strain (Chap. 6), one has to be careful when designing the strained quantum-well structure. Finally, the results discussed in this chapter would be difficult to obtain experimentally because fabrication uncertainties from laser to laser make it difficult to distinguish gain medium effects from resonator effects.

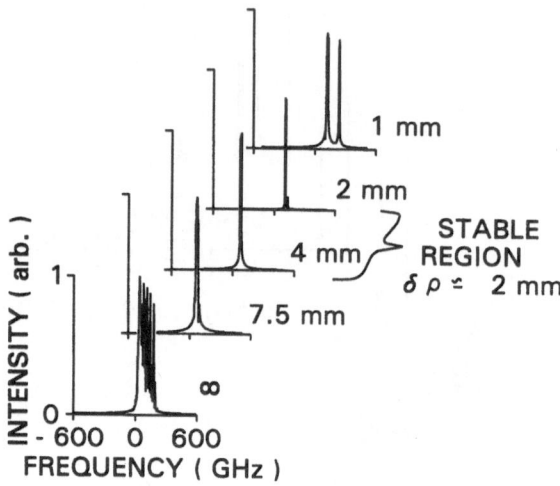

Fig. 10-36. Laser spectra for the quantum-well unstable resonator laser. All notation and parameters are the same as those in Fig. 10-35, except $\Gamma = 0.03$ and $\alpha_{abs} = 2/cm$. [From Ru *et al.* (1993)].

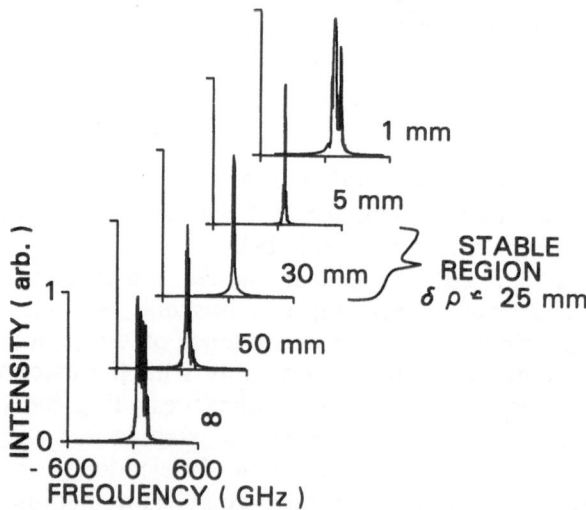

Fig. 10-37. Laser spectra for the strained quantum-well unstable resonator laser. All notation and parameters are the same as those in Fig. 10-36.

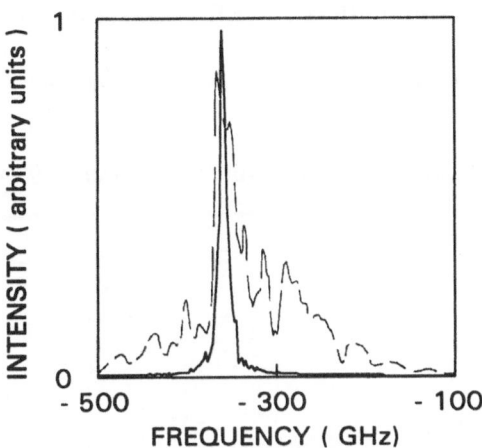

Fig. 10-38. Bulk GaAs laser spectra for excitation, N/N_{th} = 2 (solid curve) and 3 (dashed curve). The mirror curvature is ρ = 3.2mm, which is in the middle of the stable region. [From Ru *et al.* (1993)].

References

For discussion of semiconductor amplifiers see, e.g. the review by

Mukai, T., Y. Yamamoto, and T. Kimura (1985), "Optical amplification by semiconductor lasers," Chap. 3 in *Semiconductors and Semimetals* 22, Part E.

For more details on the Rigrod analysis see

Rigrod, W. W. (1963), J. Appl. Phys. 36, 2487.

For filamentation see, e.g.

Kirkby, P., A. Goodwin, G. Thompson, and P. Selway (1977), IEEE J. Quantum Electron. QE-13, 705.

Some of the figures in this chapters are taken from

Chow, W.W. and D. Depatie (1988), IEEE J. Quantum Electron. 24, 1297;

Chow, W.W. and R. Craig (1990), IEEE J. Quantum Electron. 26, 1363;

Chow, W.W. and R. Craig (1991), IEEE J. Quantum Electron. 27, 2267;

Chow, W.W., S.W. Koch, and M. Sargent (1990), IEEE J. Quantum Electron. **26**, 1052;

Ru, P., W.W. Chow, J.V. Moloney, and S.W. Koch (1993), submitted.

Chapter 11
BEYOND QUASIEQUILIBRIUM THEORY

In this final chapter we discuss results for semiconductor lasers and amplifiers that go beyond the quasiequilibrium assumption made in the previous chapters. First, in Sec. 11-1 we outline a microscopic theory for semiconductor lasers that treats both the light field and the carrier system quantum mechanically. This approach applies nonequilibrium quantum mechanical (Green's function) techniques to derive a set of coupled equations for the light field and carrier distribution functions. It allows us to analyze the stationary and dynamical responses of the interacting photon and electron-hole systems in semiconductor lasers and amplifiers including many-body effects of the electron-hole plasma and the photon system.

In Sec. 11-2 the theory is applied to the case of single-mode microcavity lasers, the VCSELs discussed in Sec. 1-5. The almost ideal design characteristics of these microlasers should make it possible to perform a quantitative experimental analysis of laser and electron-hole-plasma characteristics, such as laser linewidth, spectral and kinetic hole burning, spontaneous emission effects, photon correlations, and much more. To analyze the intricate dynamical interaction between the laser photons and the semiconductor electron-hole plasma, it is necessary to apply a sophisticated microscopic nonequilibrium theory. This theory not only allows us to determine the laser response for the full range of parameter variations, but it also makes it possible to define the regimes where the simpler quasiequilibrium theory of the previous chapters can be safely applied.

In Sec. 11-3 we present an analysis of amplification and absorption in semiconductor amplifiers. To stress the nonequilibrium aspects, we discuss the case of ultrashort pulse propagation. The calculations are based on the semiconductor Bloch equations (see Sec. 4-3) coupled to Maxwell's equation for plane-wave pulse propagation. The results show that the propagating pulse is not simply amplified, as commonly expected for light injected into an amplifier medium. In addition to the amplification of weak pulses, we also find lossless propagation and even pulse absorption for increasing pulse intensities.

11-1. Nonequilibrium Laser Theory

To provide a consistent microscopic analysis of the nonequilibrium many-body effects in semiconductor lasers we use an approach that is based on the direct investigation of correlation functions for the carrier distributions and the quantized light field. Using nonequilibrium Green's functions we compute the spectrally resolved laser intensity from a second-order photon correlation function, which is obtained from the solution of the kinetic equation for the photon propagator. This photon kinetic equation, together with kinetic equations for electrons and holes, describes the carrier and laser dynamics under nonequilibrium conditions.

The details of the nonequilibrium Green's functions technique are presented in the literature listed at the end of this chapter. Here we only discuss the resulting equations for the carrier distribution functions and the frequency dependent laser intensity. This presentation follows more or less the paper by Jahnke, Henneberger, Schäfer and Koch (1993).

Generally, in the Green's function theory, one works with coupled sets of equations for correlation functions rather than with equations for the operators directly, as in Chap. 9. As an example of a relevant correlation function, we introduce the time and frequency dependent laser intensity $I_{\mathbf{q}}(\omega, t)$, which is given as the Fourier transform of the field correlation function

$$I_{\mathbf{q}}(\omega, t) = \int d\tau \, e^{i\omega\tau} \, I_{\mathbf{q}}(\tau, t) \, , \tag{1}$$

where

$$I_{\mathbf{q}}(\tau, t) = \langle \mathbf{A}_{\mathbf{q}}(t+\tau/2) \cdot \mathbf{A}_{\mathbf{q}}(t-\tau/2) \rangle \, . \tag{2}$$

Here $\mathbf{A}$ is the field operator for the vector potential, which is related to the photon creation and destruction operators, $a^{\dagger}$ and a, of Chap. 9 via

$$\mathbf{A}(\mathbf{r}, t) = \sum_{\mathbf{q}} \sum_{\lambda=1,2} [\, a_{\lambda\mathbf{q}} \, \mathbf{u}_{\lambda\mathbf{q}}(\mathbf{r}) + a^{\dagger}_{\lambda\mathbf{q}} \, \mathbf{u}_{\lambda\mathbf{q}}{}^{*}(\mathbf{r}) \,] \, , \tag{3}$$

where $\mathbf{u}_{\lambda\mathbf{q}}(\mathbf{r})$ are the products of the polarization unit vector $\mathbf{e}_{\lambda\mathbf{q}}$ and the properly normalized cavity eigenfunctions.

The analysis starts from the many-body Hamiltonian of Eq. (4.2) plus the quantum dipole interaction Hamiltonian Eq. (9.15) expressed in terms of the vector potential (3). Applying the functional derivative technique of nonequilibrium Green's functions theory, one can obtain the kinetic equation for $I_{\mathbf{q}}(\omega, t)$

$$\frac{\partial}{\partial t} I_{\mathbf{q}}(\omega) = [F_{\mathbf{q}} g(\omega) - \kappa] I_{\mathbf{q}}(\omega) + F_{\mathbf{q}} W(\omega) S_{\mathbf{q}}(\omega).$$ (4)

This equation describes the changes of the laser intensity according to the balance between stimulated emission (modal gain), cavity loss, and the spontaneous emission into the respective cavity modes. The function $g(\omega)$ is the frequency dependent gain rate, κ is the cavity loss rate, and $W(\omega)$ is the rate of spontaneous emission. The function $F_{\mathbf{q}}$ is the cavity function, which describes the cavity resonance structure for the light modes with wavevector q

$$F_{\mathbf{q}} = \frac{n_{eff}}{2} \left[\frac{1}{n_{eff}^2 \cos^2(qL/2) + \sin^2(qL/2)} + \frac{1}{n_{eff}^2 \sin^2(qL/2) + \cos^2(qL/2)} \right].$$ (5)

Here L is the cavity length and, for the case of a cavity with amplitude mirror reflectivity R, the effective index of refraction is given by

$$n_{eff} = \frac{1+R}{1-R}.$$ (6)

In Eq. (4) the cavity function (5) modulates the gain and the spontaneous emission.

The gain rate $g(\omega)$ has the same formal structure as the gain in quasiequilibrium theory [real part of Eq. (5.9)]

$$g(\omega) = \alpha_0 \sum_{\mathbf{k}} |\mu_{\mathbf{k}}|^2 [n_{e\mathbf{k}} + n_{h\mathbf{k}} - 1] G_{eh}(\mathbf{k}, \omega) [1 + 2w(\mathbf{k}, \omega)].$$ (7)

However instead of the quasiequilibrium carrier distributions in Eq. (5.9), the actual nonequilibrium distributions appear in Eq. (7). The factor $[1 + 2w(\mathbf{k})]$, in which

$$w(\mathbf{k},\omega) = \sum_{\mathbf{k}'} V_{s,|\mathbf{k}-\mathbf{k}'|}[1-n_{e\mathbf{k}'}-n_{h\mathbf{k}'}] \frac{\varepsilon_g+\varepsilon_{e\mathbf{k}'}+\varepsilon_{h\mathbf{k}'}-\hbar\omega}{[\varepsilon_g+\varepsilon_{e\mathbf{k}'}+\varepsilon_{h\mathbf{k}'}-\hbar\omega]^2 + \gamma_{\mathbf{k}'}^2} , \tag{8}$$

is the interband Coulomb enhancement factor. In the quasiequilibrium Padé approximation of Eq. (5.9) the Coulomb enhancement is contained in the factor $1/(1 - q(\mathbf{k}))$. The lineshape function in Eq. (7) is

$$G_{eh}(\mathbf{k},\omega) = \frac{2\gamma_{\mathbf{k}}}{[\varepsilon_g+\varepsilon_{e\mathbf{k}}+\varepsilon_{h\mathbf{k}}-\hbar\omega]^2 + \gamma_{\mathbf{k}}^2} , \tag{9}$$

where the renormalized energies and the dephasing rate are determined by the actual distribution functions $n_{\mathbf{k}}$.

The rate of spontaneous emission in Eq. (4) is

$$W(\omega) = \alpha_0 \sum_{\mathbf{k}} |\mu_{\mathbf{k}}|^2 n_{e\mathbf{k}} n_{h\mathbf{k}} G_{eh}(\mathbf{k},\omega) [1 + 2w(\mathbf{k},\omega)] . \tag{10}$$

In Eq. (4) the spectral function of the photons, $S_{\mathbf{q}}(\omega)$, appears as prefactor of the spontaneous emission contribution. This spectral function can be written as

$$S_{\mathbf{q}}(\omega) = \frac{2\chi_{\mathbf{q}}(\omega)}{[(q^2 - \omega^2/c^2)/F_{\mathbf{q}}]^2 + \chi_{\mathbf{q}}^2(\omega)} , \tag{11}$$

with

$$\chi_{\mathbf{q}}(\omega) = g(\omega) - \frac{\kappa}{F_{\mathbf{q}}} . \tag{12}$$

Eq. (11) shows that the effective damping of the photons in the laser cavity is given by the degree of gain - loss compensation. For $\chi \to 0$, Eq. (11) approaches a δ-function describing the usual photon dispersion $\omega = cq$. However, in general $S_{\mathbf{q}}(\omega)$ is broadened by the frequency and momentum dependent photon damping $\chi_{\mathbf{q}}(\omega)$.

In order to obtain the frequency dependent laser intensity, Eq. (4) is summed over $\mathbf{q}$. For the analysis of microcavity lasers we may restrict the summation to longitudinal modes, so that we can deal with wavevectors in one dimension. The finite geometry of the cavity in the other two dimensions can be treated by introducing the geometrical factor S_C. Using polar coordinates, this factor can be traced back to a beam divergence angle $\theta_{max} \simeq d/2L$ where d is the beam diameter. Hence the q-summation yields

$$S(\omega) = \frac{|\mu|^2}{V} \sum_{\mathbf{q}} S_{\mathbf{q}}(\omega) = \frac{|\mu|^2}{(2\pi)^3} \int_{-\infty}^{\infty} dq \, q^2 \int_{0}^{2\pi} d\phi \int_{0}^{d/2L} d\theta \, \sin\theta \, S_{\mathbf{q}}(\omega)$$

$$= S_C \frac{|\mu|^2}{2\pi^2} \int_{-\infty}^{\infty} dq \, q^2 \, S_q(\omega) . \tag{13}$$

For the laser intensity spectrum we have

$$I(\omega) = \frac{|\mu|^2}{V} \sum_{\mathbf{q}} I_{\mathbf{q}}(\omega) , \tag{14}$$

and no prefactor S_c appears since $I(\omega)$ is the *modal intensity spectrum*, which gives the total intensity for the case of a single-mode laser [see Eq. (24)]. Generally, the geometrical factor S_C appears in front of the spontaneous emission contribution, since the spontaneously emitted light goes into all available photon modes, however, only that part which goes into the laser mode acts as source term in the intensity equation. After the q-summation, we can write the final equation for the spectrally resolved laser intensity as

$$\boxed{\frac{\partial}{\partial t} I(\omega) = [\, F(\omega) \, g(\omega) - \kappa \,] \, I(\omega) + S_C \, F(\omega) \, W(\omega) \, S(\omega)} . \tag{15}$$

Through the microscopic expressions (7) and (10) for gain and spontaneous emission, the laser intensity is coupled to the carrier distributions $n_{\alpha\mathbf{k}}$. These distribution functions have to be computed self-consistently with the instantaneous light intensity. The nonequilibrium Green's function theory yields a kinetic equation for the carrier distribution functions, which can be regarded as a generalization of Eq. (3.15)

$$\boxed{\dot{n}_{\alpha\mathbf{k}} = \dot{n}_{\alpha\mathbf{k}}\big|_{pump} + \dot{n}_{\alpha\mathbf{k}}\big|_{relax} + \dot{n}_{\alpha\mathbf{k}}\big|_{photon} \, , \quad \alpha = e \text{ or } h} , \tag{16}$$

where

$$\dot{n}_{\alpha\mathbf{k}}\big|_{pump} = \Lambda_{\alpha\mathbf{k}} \qquad\qquad (17) ,$$

and $\Lambda_{\alpha\mathbf{k}}$ given by Eq. (3.20). The carrier relaxation rate describes the redistribution of carriers due to carrier-carrier and carrier-phonon scattering

$$\dot{n}_{\alpha\mathbf{k}}\big|_{relax} = \dot{n}_{\alpha\mathbf{k}}\big|_{col} + \dot{n}_{\alpha\mathbf{k}}\big|_{\alpha-p} , \qquad\qquad (18)$$

with $\dot{n}_{\alpha\mathbf{k}}\big|_{col}$ and $\dot{n}_{\alpha\mathbf{k}}\big|_{\alpha-p}$ given by Eqs. (4.76) and (4.89), respectively.

The change of the carrier distributions due to the coupling to the photons is given by the sum of a stimulated and a spontaneous contribution

$$\dot{n}_{\alpha\mathbf{k}}\big|_{photon} = \dot{n}_{\alpha\mathbf{k}}\big|_{stim} + \dot{n}_{\alpha\mathbf{k}}\big|_{spont} , \qquad\qquad (19)$$

where

$$\dot{n}_{\alpha\mathbf{k}}\big|_{stim} = \frac{1 - n_{e\mathbf{k}} - n_{h\mathbf{k}}}{\tau_{\mathbf{k},stim}} \qquad\qquad (20)$$

$$\dot{n}_{\alpha\mathbf{k}}\big|_{spont} = \frac{n_{e\mathbf{k}}\, n_{h\mathbf{k}}}{\tau_{\mathbf{k},spont}} , \qquad\qquad (21)$$

and the stimulated and spontaneous carrier "lifetimes" are defined by

$$1/\tau_{\mathbf{k},stim} = \int \frac{d\omega}{2\pi}\, G_{eh}(\mathbf{k},\omega)\, I(\omega)\, [1 + w(\mathbf{k},\omega)] \qquad\qquad (22)$$

$$1/\tau_{\mathbf{k},spont} = \int \frac{d\omega}{2\pi}\, S(\omega)\, [1 + w(\mathbf{k},\omega)] . \qquad\qquad (23)$$

These relations show that the electron-hole-pair lifetime is shortened when the rate of stimulated or spontaneous emission increases. As soon as the stimulated recombination time becomes comparable to either one of the relevant relaxation times in semiconductor lasers (see the discussion in Sec. 4-6), we can expect deviations from quasiequilibrium conditions. We come back to this point in the next section.

11-2. Numerical Results for VCSELs

In this section we summarize some of the results which were obtained by F. Jahnke, who succeeded in numerically solving the coupled Eqs. (15) and (16) for the case of a microlaser with a cavity length of $1\mu m$ and a mirror reflectivity of 99%. The numerical task is quite involved because of the large difference between the characteristic time scales for the various microscopic processes. At each time step, one has to compute the changes of the carrier occupation due to carrier-photon, carrier-carrier and carrier-phonon scattering, the corresponding new gain including Coulomb renormalization and screening, spontaneous and stimulated emission and the changes of the cavity properties due to index changes. To give a benchmark for the numerical complexity, we mention that a single calculation for laser switch-on up to 1 ns takes about 12 CPU hours on a CONVEX vector computer.

Solving Eqs. (15) and (16) for different pump rates allows one to obtain the stationary laser intensity

$$I = \int \frac{d\omega}{2\pi} I(\omega) \tag{24}$$

and the total carrier density. Figure 11-1a displays examples of the computed laser intensity and carrier density for different spontaneous emission coupling into the lasing mode. For small S_C, which is characteristic for conventional semiconductor lasers, Fig. 1l-1 shows a pronounced threshold for the stationary laser intensity. This threshold gradually vanishes with increasing spontaneous emission coupling S_C. Basically, the magnitude of the "jump" of the output intensity from below to above the threshold is given by the inverse of S_C.

Figure 1l-1b shows that the carrier density increases with increasing pump rate, until the laser threshold is reached. For pump intensities just above the threshold, carrier clamping is observed, pretty much as predicted by the semiclassical quasiequilibrium semiconductor laser theory of Sec. 7-2. However in contrast to the expected density clamping for all pump rates above threshold, we see in Fig. 11-1 that clamping occurs only for pump rates up to about five times the threshold pump rate. For stronger pumping the total carrier density increases again. We show below that this increase in the total carrier density is related to deviations of the carrier distributions from quasiequilibrium Fermi-Dirac distributions.

To analyze VCSEL laser operation and to get a feeling for the range of validity of the quasiequilibrium theory, we investigate the system for different pumping levels. To obtain the steady state results for a given pump

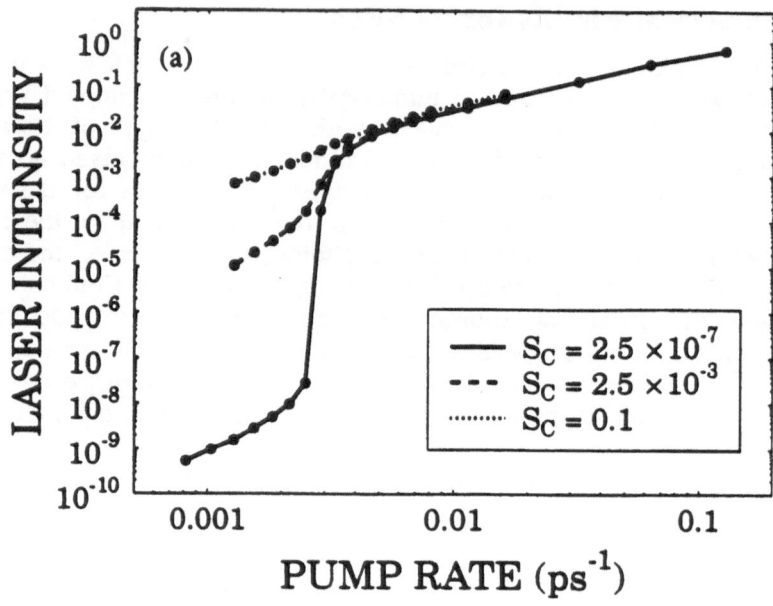

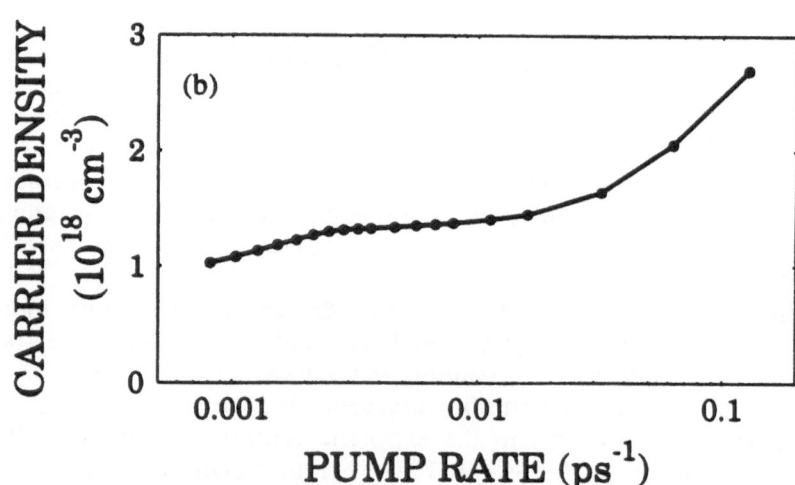

Fig. 11-1. (a) Stationary laser intensity as function of the pump rate for various spontaneous emission coupling into the lasing mode S_C. (b) Stationary carrier density as function of the pump rate. Since the density exhibits basically no dependence on the spontaneous emission coupling, only the curve for $S_C = 2.5 \times 10^{-7}$ is shown. [From Jahnke *et al.* (1993)].

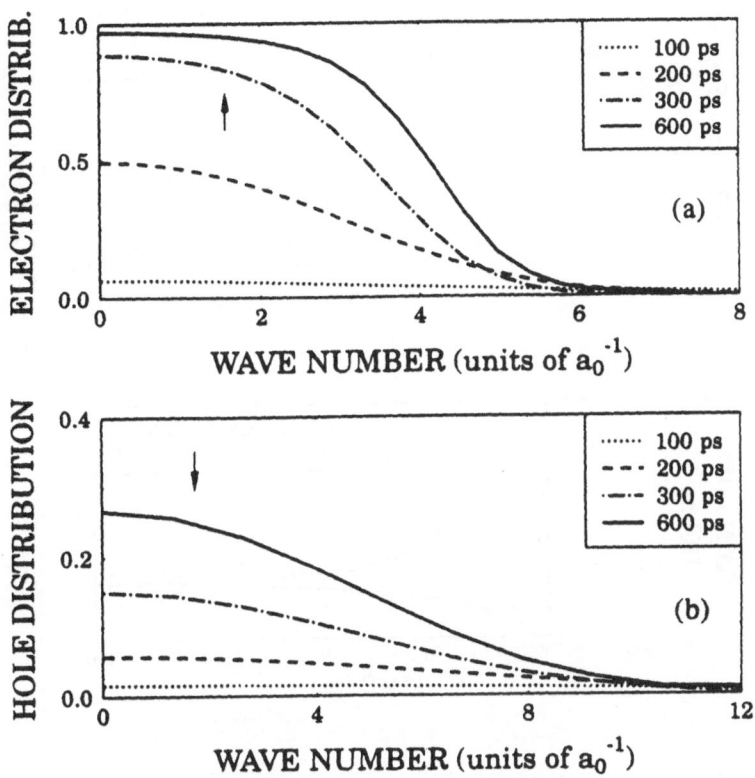

Fig. 11-2. Electron (*a*) and hole (*b*) distribution functions for various times after the beginning of the pump process. The pump rate is 0.008 ps^{-1} and the spontaneous emission coupling is $S_C = 2.5 \times 10^{-3}$. The full lines correspond to the stationary result. The arrows indicate the wavenumber corresponding to the stationary laser frequency. [From Jahnke *et al.* (1993)].

intensity, we start with a passive system and ramp up the pump rate within 1 *ps* to the stationary value. For a pump rate slightly above threshold, we obtain the results in Figs. 11-2 and 11-3. Figure 11-2 shows the gradual increase of the electron and hole populations after the onset of pumping. The total carrier density saturates after about 600 *ps*, Fig. 11-3*a*, when the laser has switched on. We see in Fig. 11-3*c* that the spectral laser intensity is initially a broad line which becomes increasingly sharp in time as a consequence of the increasing gain-loss compensation. In addition to the sharpening of the laser line, we note a small shift of the lasing frequency,

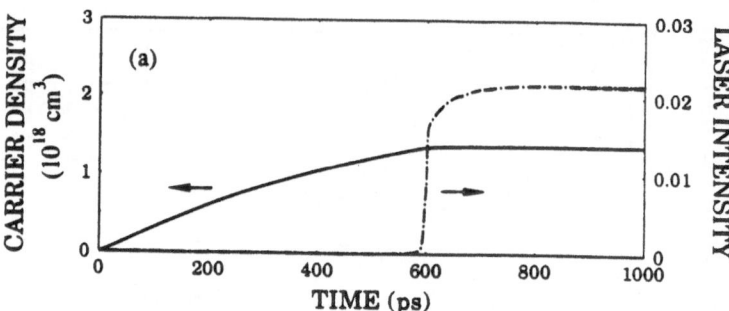

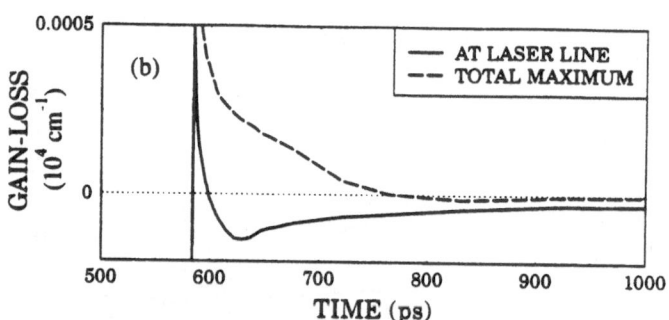

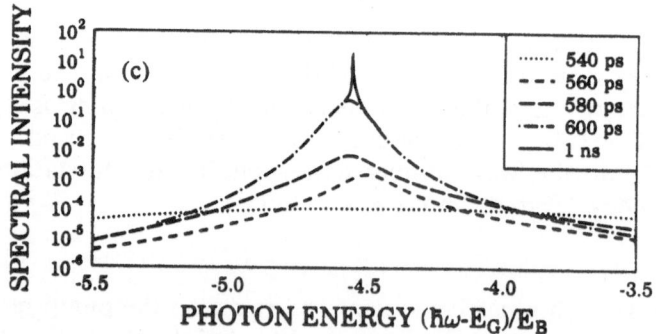

Fig. 11-3. (a) Time resolved carrier density and laser intensity for the same parameters as in Fig. 11-2. (b) Dynamic evolution of the modal gain at the laser line (full line) and at the total maximum of the frequency resolved modal gain (dashed line). (c) Spectrally resolved laser intensity for various times after the beginning of the pump process. The photon energy $\hbar\omega$ is given as detuning from the unrenormalized band-gap energy, normalized to the excitonic Rydberg energy. [From Jahnke *et al.* (1993)].

which is a consequence of the index changes caused by the carrier density changes.

Analyzing the stationary carrier distribution functions in Fig. 11-2, we find only very small deviations from quasiequilibrium Fermi-Dirac distributions. However, the temperature of these carrier distributions is almost 10% above the lattice temperature (300 K). We come back to this carrier heating effect later in this section. Generally, the amount by which the stationary carrier distribution in the running laser deviates from Fermi-Dirac distributions depends on the competition between frequency selective stimulated carrier recombination in the spectral vicinity of the running laser mode and the intraband carrier relaxation through carrier-carrier and carrier-phonon scattering.

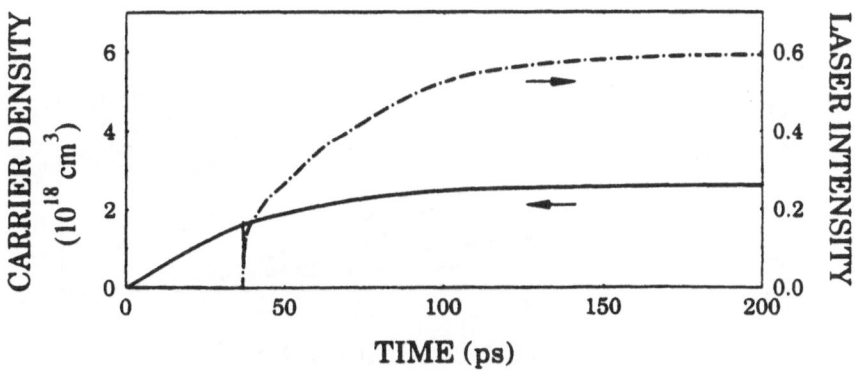

Fig. 11-4. Time resolved carrier density and laser intensity for the pump rate $p = 0.13 \, ps^{-1}$ and a spontaneous emission coupling $S_C = 2.5 \, 10^{-3}$. [From Jahnke et al. (1993)].

The nonequilibrium effects are more pronounced for higher pump rates. In Fig. 11-4, we show the temporal evolution of the total carrier density and the laser intensity for a higher pump rate in the region, where the total carrier density in Fig. 11-1b is no longer clamped. We can see that also for high pumping the carrier density increases rapidly until lasing sets in (in this case around 40 ps after pump begin). However, in contrast to the low-pumping case in Fig. 11-3, for high pumping the carrier density continues to increase at a slower pace even after the onset of lasing. Furthermore, the laser intensity goes through a brief switch-on oscillation and then continues to gradually increase for about 150 ps after lasing begins.

For the same conditions as in Fig. 11-4 we show in Fig. 11-5 the dynamical evolution of the carrier distribution functions during the laser

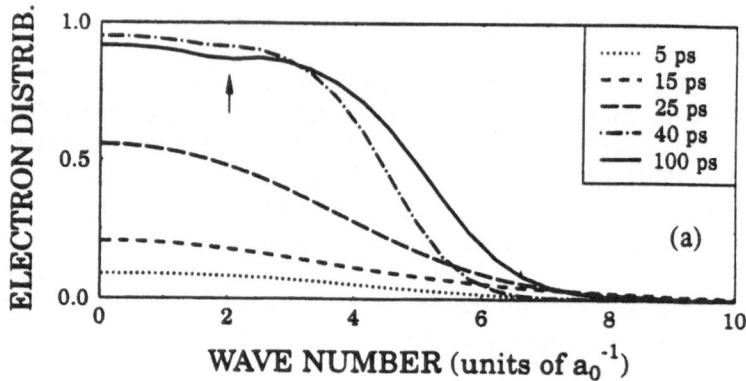

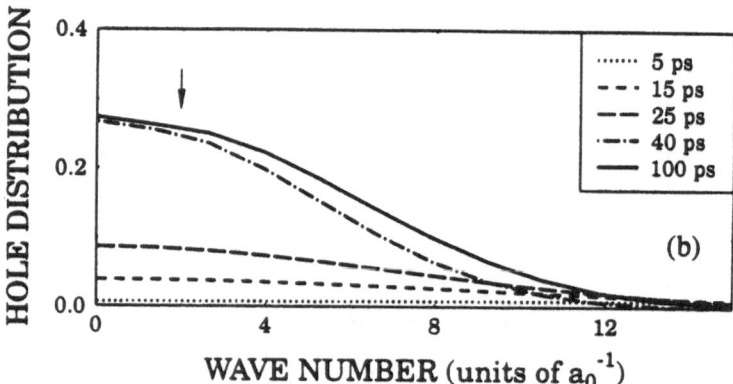

Fig. 11-5. Electron (*a*) and hole (*b*) distribution functions for various times after the beginning of the pump process using the same parameters as in Fig. 11-4. The full line corresponds to the stationary result. The arrows indicate the wavenumber corresponding to the stationary laser frequency. [From Jahnke *et al.* (1993)].

switch-on. As in the low pumping case [see Fig. 11-3*a*], we again see the gradual build-up of the carrier populations. However, after the approach of quasi-stationary distributions ($t \geq 100\ ps$), in particular the electron distribution function exhibits pronounced deviations from quasiequilibrium. The distribution function is deformed due to kinetic hole burning, which occurs as a consequence of the wavenumber selective carrier removal by the stimulated emission. The width of the kinetic hole is determined by the balance between stimulated carrier removal and carrier Coulomb scattering. Equation (9) shows that the optical transitions at a certain frequency are connected with a wavenumber range the width of which is governed by the dephasing rate γ. Hence the carrier depletion due to the lasing process occurs over a relatively wide range of wavenumbers, leading to a reduction of the entire population of the low momentum states.

In Fig. 11-6 we show the dynamic evolution of the frequency resolved laser intensity which reflects the dynamic interplay between carrier density and laser intensity. At early times, up to 35 *ps* after the onset of pumping, where the system is still absorbing or has small gain, Fig. 11-6*a* shows only enhanced spontaneous emission, which is weakly modulated by the cavity due to the strong photon damping. After approximately 40*ps* the gain overcompensates the loss and lasing begins. During the switch-on period the spectral intensity represents a rather broad amplification of spontaneous emission since the gain exceeds the loss in a broad spectral region. With increasing gain-loss compensation the spectral region of gain exceeding the loss decreases substantially, leading to a narrowing of the laser line, which then becomes significantly sharper than the passive cavity linewidth.

The delicate compensation process between gain, loss and spontaneous emission does not necessarily lead to a continuous increase of the spectral peak intensity. When the laser line begins to grow initially, the growing intensity reduces the gain, and the initial laser line goes down (at $t \simeq 60$ *ps* in Fig. 11-6*b*). When the gain begins to increase again, it produces a shifted new line instead of supporting the initial laser line. This line shift occurs because the density continues to change during the intensity switch-on oscillation and the corresponding index change causes a slight tuning of the cavity. We note in Fig. 11-6*c* that even though the total intensity has already reached the stationary value after about 200 *ps*, the sharpening of the laser line continues well into the nanosecond region.

Hence we can conclude at this point that the VCSEL output characteristics for low and medium pump rates are close to results obtained using quasiequilibrium carrier distributions, whereas for high pump rates the quasiequilibrium assumption clearly fails. However even though the carrier distributions are described by Fermi-Dirac distributions, the carrier temperature is not necessarily that of the lattice.

To understand the origin of these carrier heating effects, we recall the modelling of carrier generation through injection pumping. As discussed in Sec. 3-1, we assume that the injected carriers arrive at the active region in a quasiequilibrium Fermi-Dirac distribution $f_{\alpha k0}$, which is at the lattice temperature. However since each carrier state can be occupied only once, the momentum-dependent pump rate

$$\Lambda_{\alpha k} = \frac{\eta_{tr} J}{e d N_0} f_{\alpha k0} \left(1 - n_{\alpha k}\right) , \qquad (3.20)$$

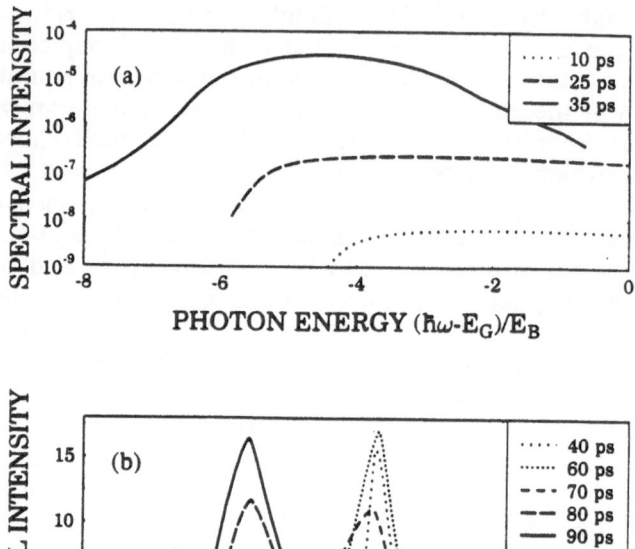

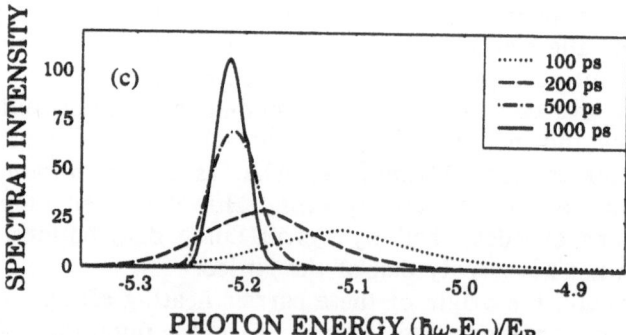

Fig. 11-6. Spectral resolved laser intensity for various times after the beginning of the pump process for the same parameters as in Fig. 11-4. (a) between 10 *ps* and 35 *ps*, (b) between 40 *ps* and 90 *ps*, and (c) between 100 *ps* and 1 *ns*. Note the different frequency ranges in parts (a)-(c). [From Jahnke *et al.* (1993)].

contains the Pauli blocking factor $1 - n_{\alpha k}$. As shown in Figs. 3-1 and 3-2, this Pauli blocking becomes more and more effective for increasing carrier population in the active region. Consequently, the pumping is gradually blocked and the quantum efficiency decreases. Moreover since the Pauli blocking is more efficient for energetically lower states, the main kinetic energy, and therefore the average "temperature" of the injected carriers increases.

During the pump process the carriers in the laser region are constantly redistributed into thermal distributions by the very fast carrier-carrier Coulomb scattering which leads to scattering times around 50 fs (Sec. 4-6). However, since these elastic scattering processes do not dissipate kinetic energy, the carrier system equilibrates at a temperature which is higher than that of the lattice. Energy from the carrier system is transferred to the lattice only through the carrier - phonon coupling. Hence, we have to study the balance between the different contributions in the carrier kinetic equation (16) to obtain the stationary carrier distributions (Jahnke and Koch, 1993).

In Fig. 11-7 we plot pump distribution functions, effective pump rate and carrier distributions functions in the active laser region computed for stationary laser conditions. In each case the pump rate was switched on in 1 ps and the coupled Eqs. (15) and (16) were solved until the system has evolved into a steady state. The injected carrier are always in Fermi-Dirac distributions at the lattice temperature ($T = 300\ K$). The general behavior of the results in Fig. 11-7 is similar to those in Fig. 3-1. Again, we clearly see the effects of pump blocking due to the carrier population inside the active region. Additionally, we notice that for higher pump densities the peak of the effective pump rate, $f_{ek0}\ (1-n_{ek})$, shifts to larger wavenumbers. This shift indicates the preferred pumping of energetically higher states. Correspondingly, we find a strong deviation of the actual carrier occupation n_{ek} from a lattice temperature Fermi-Dirac distribution with the same carrier density. For low pump density [see Fig. 11-7c], the carrier injection is restricted to the region of effective blocking. This results in a lower carrier density inside the active region due to less effective pumping.

In the presence of a running laser mode, the stimulated recombination of carriers due to the lasing process occurs in the spectral region where the carrier system is inverted. Since this inversion region is located just above the (renormalized) band gap, the stimulated emission removes electrons and holes with below average kinetic energy, which further increases the carrier temperature. The combined influence of the removal of "cold" carriers in the wavevector region around the laser mode (marked by arrows in Fig. 11-7) and the injection of "hot" carriers by the pumping leads to a substantial carrier heating. Fitting the actual carrier distributions n_{ek} to Fermi-Dirac distributions with the same carrier density, we obtain the effective

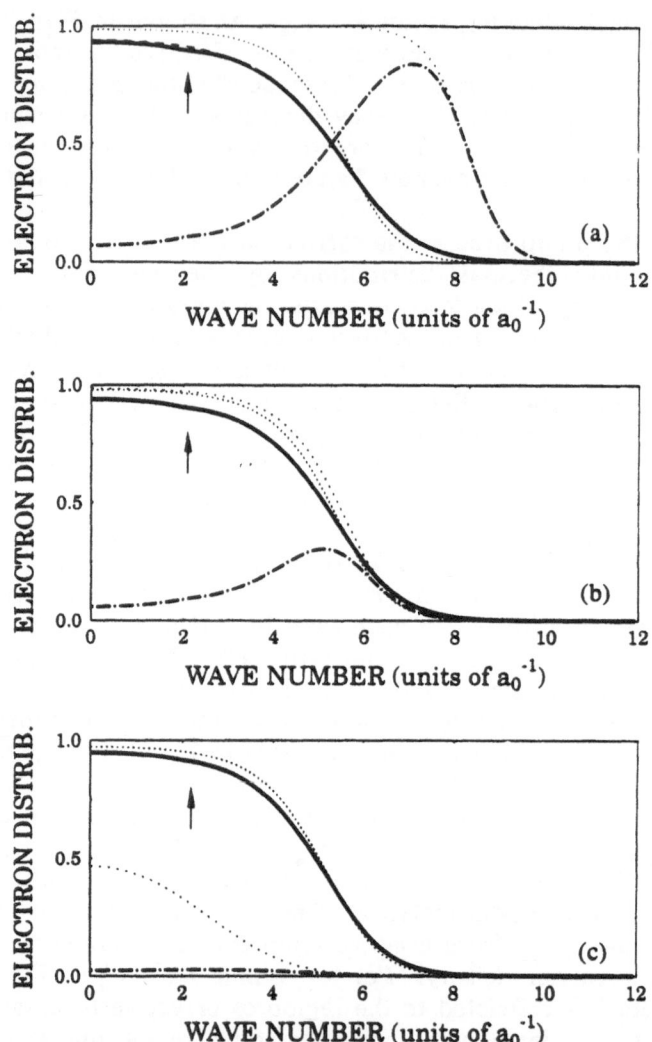

Fig. 11-7. Steady state electron distribution functions (full lines) for a VCSEL pumped at the densities $N_0 = 10^{19}$ cm^{-3} (a), $N_0 = 3 \times 10^{18}$ cm^{-3} (b), $N_0 = 3 \times 10^{17}$ cm^{-3} (c). The dotted line shows $f_{e k 0}$, the dashed dotted line is $f_{e k 0}$ $(1-n_{e k})$, and the short dotted line shows a Fermi-Dirac distribution function with the same density as $n_{e k}$ and lattice temperature $T = 300K$. The effective carrier temperatures are 435 K (a), 390 K (b) and 355 K (c). [From Jahnke and Koch (1993)]

carrier temperatures 435 K, 390 K, and 355 K for Figs. 11-7a, 11-7b, and 11-7c, respectively.

11-3. Pulse Propagation in Semiconductor Amplifiers

To provide some additional insight into electron-hole nonequilibrium effects in semiconductor laser media we discuss in this section aspects of light propagation in semiconductor amplifiers. In particular, we study the propagation of femtosecond pulses whose frequency spectrum is fully inside the semiconductor gain region. For this purpose we start from the semiconductor Bloch equations

$$\dot{p}_{\mathbf{k}} = - i\omega_{\mathbf{k}} p_{\mathbf{k}} - i\Omega_{\mathbf{k}} [n_{e\mathbf{k}} + n_{h\mathbf{k}} - 1] + \frac{\partial p_{\mathbf{k}}}{\partial t}\bigg|_{col} \tag{4.42}$$

$$\dot{n}_{e\mathbf{k}} = i[\Omega_{\mathbf{k}} p_{\mathbf{k}}^* - \Omega_{\mathbf{k}}^* p_{\mathbf{k}}] + \frac{\partial n_{e\mathbf{k}}}{\partial t}\bigg|_{col} \tag{4.43}$$

$$\dot{n}_{h\mathbf{k}} = i[\Omega_{\mathbf{k}} p_{\mathbf{k}}^* - \Omega_{\mathbf{k}}^* p_{\mathbf{k}}] + \frac{\partial n_{h\mathbf{k}}}{\partial t}\bigg|_{col} \tag{4.44}$$

with

$$\Omega_{\mathbf{k}}(\mathbf{r},t) = \frac{\mu_{\mathbf{k}} \cdot \mathbf{E}(\mathbf{r},t)}{\hbar} + \frac{1}{\hbar} \sum_{\mathbf{k}' \neq \mathbf{k}} V|_{\mathbf{k}-\mathbf{k}'}|p_{\mathbf{k}'} . \tag{4.46}$$

In these equations the interband polarization $p_{\mathbf{k}}$ and the electric field E represent the slowly varying envelopes of the fields with respect to the laser frequency ω_L.

For simplicity, we restrict the propagation studies to the case of plane waves traveling in z-direction

$$\mathbf{E}(\mathbf{r},t) = \frac{1}{2} E(z,t) \exp[i(k_L z - \omega_L t)] + \text{c.c.} , \tag{25}$$

where k_L and ω_L are related by the linear dispersion relation for the medium. Then in the slowly varying envelope approximation, the reduced wave equation becomes

$$\frac{\partial E(\xi,\eta)}{\partial \xi} = i \ \frac{4\pi}{c^2} \ \frac{\omega_L^2}{k_L} \ P(\xi,\eta) \ , \tag{26}$$

where a moving coordinate frame $(\xi,\eta) = (z, t-z/v_g)$, v_g being the group velocity of the pulse, has been introduced. The total optical polarization $P = 2 \ \Sigma_{\mathbf{k}} \ \mu_{\mathbf{k}} \ p_{\mathbf{k}}$ is obtained by summing over all $\mathbf{k}$. The Maxwell wave equation (26) coupled to the semiconductor Bloch equations (4.42) - (4.46) constitute the simplest case of the *Semiconductor Maxwell Bloch Equations* (SMBE).

Knorr *et al.* (1993) evaluate the SMBE for parameters which are typical for a GaAs amplifier at room temperature. They assume that the carrier density is elevated to $N = 2 \ \Sigma_{\mathbf{k}} \ n_{\mathbf{k}} = 2.5 \times 10^{18} \ cm^{-3}$ by the pump in the absence of the light field. For such a plasma density carrier-carrier Coulomb scattering is the dominant dephasing mechanism. In this case one can use the dephasing approximation (4.83) for the interband polarization. The input pulses for the propagation studies were taken as *sech* shaped in time with a carrier frequency ω_L centered around peak gain and a pulse duration of $150 \, fs$ (intensity FWHM). For these parameters the spectral pulse width is substantially less than the gain bandwidth. The sample length L is chosen to be one linear gain length, $L = g_0^{-1}$ where g_0 is the linear gain, so the transmitted energy is amplified by a factor e^1 for a vanishingly small input pulse energy. In the present case g_0 is about 5 μm.

Figure 11-8 shows examples of the numerical results for various peak input intensities. The input (solid lines) and transmitted (dashed lines) pulse profiles for low ($I_0/4$), intermediate ($I_0 = 1.8 \ GW/cm^2$), and high ($4I_0$) peak input intensities are shown in Fig. 11-8a. The corresponding variation of the pulse energy, normalized to the input pulse energy, with scaled propagation distance $g_0\xi$ is shown in Fig. 11-8b. For the low input intensity, amplification of the pulse energy is clearly seen. In contrast, for the intermediate intensity case the pulse energy propagates essentially unchanged, and for the high intensity case the transmitted pulse suffers a net absorption.

To further highlight these features, Fig. 11-9a displays the temporal profile of the carrier density $N = 2 \ \Sigma_{\mathbf{k}} \ n_{\mathbf{k}}$ at the input face ($\xi = 0$) for the three cases studied in Fig. 11-8. For the low intensity pulse we obtain a decrease of the carrier density consistent with the pulse amplification by stimulated recombination of carriers. However for the intermediate and high-intensity pulses we see that the carrier density initially decreases and then increases. In particular, for the intermediate input intensity I_0, the carrier-density decrease and subsequent increase more or less cancel each other. There is no net change, consistent with the fact that the pulse energy remains constant under propagation. In contrast, for the high pulse-energy intensity there is a net increase in the carrier density after the

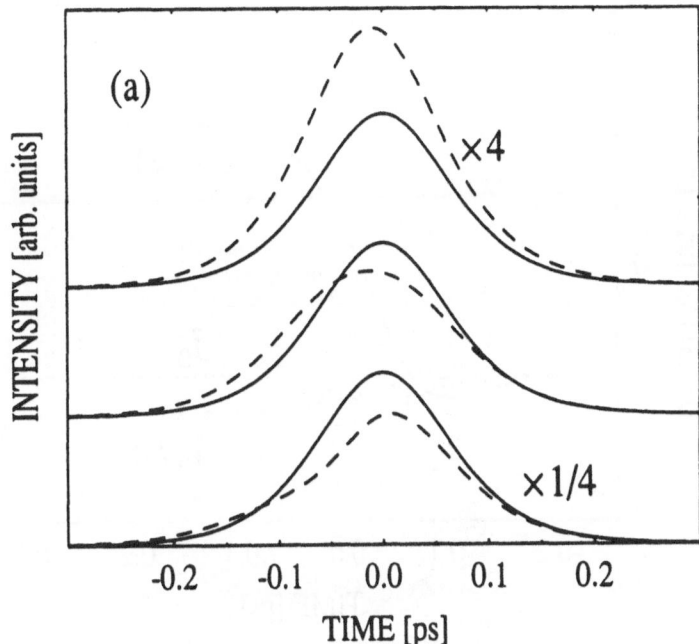

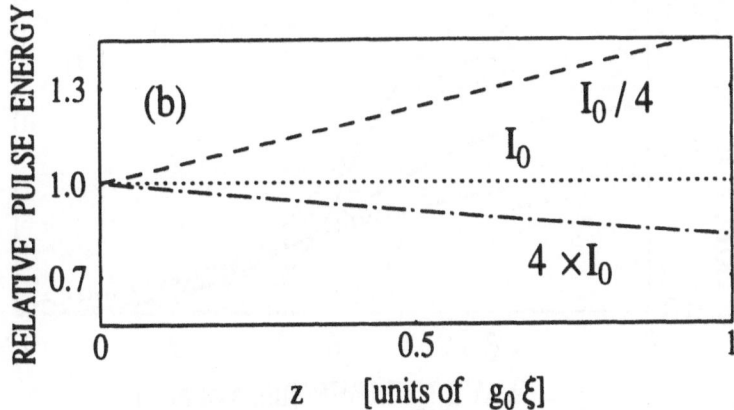

Fig. 11-8. (a) Input (solid) and output (dashed) pulse intensity profiles for propagation over one linear amplification length g_0^{-1} and three different peak input intensities $I_0/4$ (top), $I_0 = 1.8 GW/cm^2$ (middle), and $I_0 = 4I_0$ (bottom), and (b) pulse energy transmission (pulse energy divided by the input pulse energy). [From Knorr *et al.* (1993)].

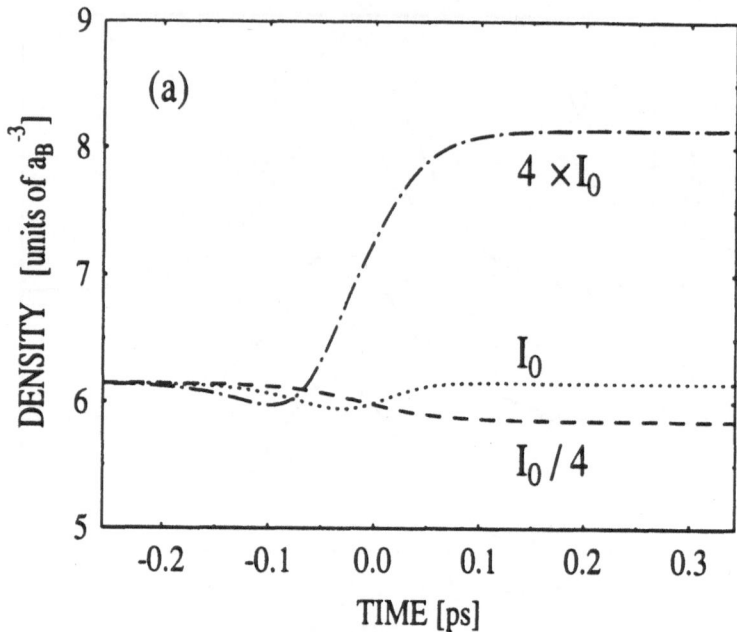

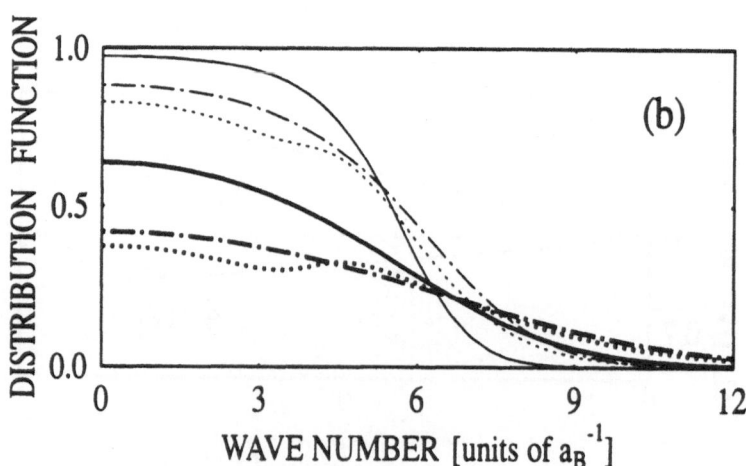

Fig. 11-9. (a) Temporal variation of the carrier density at the front face of the sample under the same conditions as Fig. 11-8. (b) Electron distributions $n_{e,\mathbf{k}}$ (thin lines) and hole distributions $n_{h,\mathbf{k}}$ (thick lines) at the front face, before the pulse ($t = -\infty$, solid line), during the pulse ($t = 0$, dotted line), and after the pulse ($t = \infty$, dashed line), for the example showing net absorption ($I = 4I_0$). [From Knorr et al. (1993)].

pulse, which forces the pulse energy to decrease.

To understand the pulse propagation phenomena one we have to realize that for these ultrafast time scales transient and nonequilibrium effects become important. Furthermore, we have to consider the fact that prior to the pulse, there is a continuum of electronic states in an inverted semiconductor, which are partially populated by the electron-hole plasma according to Fermi-Dirac distributions. The states are only inverted in the spectral region between the renormalized bandgap up to the chemical potential, which is determined by the total carrier density. Hence the spectral gain region is always adjacent to a region of optical absorption above the chemical potential.

The analysis of Knorr et al. (1993) shows that the results in Figs. 11-8 and 11-9 result from the interplay of the states in the gain and absorption regions. To appreciate this feature, one has to realize, that even when tuned into the gain region, the optical pulse also interacts with absorbing states, since these states have a finite linewidth. For high-intensity pulses, when gain saturation becomes appreciable, this absorption can exceed the amplification since the absorbing states are weighted by the high k^2 density-of-states factor.

Figure 11-9b shows the variation of the carrier distributions $n_{\alpha k}(0, t)$, $\alpha = e, h$, at the front face of the sample, before, during and after the pulse for the example showing net absorption ($I = 4I_0$). Prior to the pulse the electrons and holes are both in Fermi-Dirac distributions, the corresponding temperature being that of the lattice ($T = 300K$). During the pulse the distributions are in a highly nonequilibrium state, whereas after the pulse they rapidly relax toward a Fermi-Dirac distribution, but at a greatly elevated temperature ($T > 600K$). This temperature increase results from the absorption into high-lying k-states with above average kinetic energy. A Fermi-Dirac distribution with elevated temperature exhibits a reduced occupation of the lower k-states, which corresponds to gain saturation. This nonresonant carrier generation in high-lying k-states with simultaneous carrier recombination in low-lying k-states thus comprises a carrier heating mechanism, similar in some ways to the Pauli-blocked injection-current pumping in semiconductor lasers [see in Sec. 11-2].

We summarize the explanation of the results shown in Figs. 11-8 and 11-9 as follows: at low input intensity ($I < I_0$) the amplifying states in the spectral gain region dominate over the nonresonant absorbing states since the latter are only weakly populated by the pulse. Hence a net amplification of the input pulse occurs due to stimulated recombination of carriers [dashed line, Fig. 11-8b], and the carrier density decreases [dashed line, Fig. 11-9a]. In contrast, for high peak input intensity ($I > I_0$), the nonresonant absorbing states become populated [dash-dotted line, Fig. 11-9b] and dominate over the amplifying states giving rise to a net absorption of the pulse (dash-dotted line, Fig. 11-8a) and an increase in carrier density (dash–dotted line, Fig. 11-9a). The intermediate intensity ($I = I_0$) case shown in Figs. 11-8 and 11-9 represents the special situation in which the amplifying states and the nonresonant absorbing states contribute equally and the density returns to its initial value after the pulse. This means that there is no net absorption or amplification, which is reflected by the fact that the transmitted pulse energy is almost identical to that of the input pulse [dotted line, Fig. 11-9a]. Under these conditions the pulse therefore undergoes lossless propagation.

References

Nonequilibrium Green's functions are discussed in:

Kadanoff, L.P., and G. Baym (1962), *Quantum Statistical Mechanics*, Benjamin, New York.

Keldysh, L.V. (1965), Sov. Phys. JETP **20**, 4;

Korenman, V. (1966), Annals of Physics **39**, 72;

DuBois, D.F. (1967), p.469 in *Lectures in Theoretical Physics*, Vol. IX, edited by W.E. Brittin and A.O. Barut, Gordon and Breach, New York.

Application of nonequilibrium Green's function theory to semiconductor lasers can be found in:

Henneberger, K., F. Herzel, S.W. Koch, R. Binder, A.E. Paul, and D. Scott (1992), Phys. Rev. **A45**, 1853;

Henneberger, K., F. Jahnke and F. Herzel (1992), phys. stat. sol. (*b*) **173**, 423;

Jahnke, F., S.W. Koch and K. Henneberger (1993), Appl. Phys. Lett. **62**, 2313;

Jahnke, F. and S.W. Koch (1993), Opt. Lett. **18**, 1438;

Jahnke, F., K. Henneberger, F. Schäfer, and S.W. Koch (1993), Journ. Opt. Soc. Am. **10**.

A rate equation analysis of the effects of spontaneous emission coupling on semiconductor laser operation can be found in

Yamamoto, Y., S. Machida, and G. Bjork (1991), Phys. Rev. **A44**, 657.

The discussion of the pulse propagation studies in Sec. 11-3 closely follows

Knorr, A., R. Binder, E.M. Wright, and S.W. Koch (1993), Opt. Lett. **18**, 1538.

See also

Knorr, A., R. Binder, M. Lindberg, and S.W. Koch (1992), Phys. Rev. **A46**, 7179.

Koch, S.W., A. Knorr, R. Binder, and M. Lindberg (1992), phys. stat. sol. (*b*) **173**, 177.

Appendix A
TWO-LEVEL SYSTEMS AND RATE EQUATIONS

Chapters 2 and 3 show that for semiconductor laser purposes, the interaction of light with a semiconductor medium can often be modeled in terms of electronic transitions between a valence and a conduction band. A spread of transition energies occur that depend on the value of the carrier momentum **k**. Elsewhere in physics such a range of transitions is known to occur, namely in ensembles of inhomogeneously broadened two-level systems. These appear to good approximation in the interaction of light with atoms and with the magnetic dipoles of nuclear magnetic resonance. Hence people have been led to model semiconductor laser media using the theory of two-level systems. In appropriate limits, the two-level approaches reduce to rate equation theory, also very popular in modeling aspects of semiconductor laser operation. On the other hand, Sec. 3-1 reveals that the semiconductor medium is at the very least an inhomogeneously broadened four-level medium, all of whose levels have appreciable probability in a gain medium. Two of the four levels correspond to the levels in a two-level medium, but the other two are absent in the two-level medium. Hence the degree to which a two-level model can describe semiconductor response is disturbingly uncertain. One is really better off using a real semiconductor model, for which the approximations are well defined. Nevertheless much of the physics of the two-level model has counterparts in the semiconductor models and one can study this physics in a relatively simple context. This appendix seeks to present the two-level model in a way well suited to people primarily interested in semiconductor media. More complete discussions can be found, for example, in Meystre and Sargent (1991).

Section A-1 reviews the physics of two-level systems, starting with the wave function and proceeding to the density matrix. The corresponding equations of motion are often called the optical Bloch equations. Section A-2 derives simple rate equations and solves for the gain and index of homogeneously and inhomogeneously broadened two-level media. As Sec. 1-9 shows, the simple semiconductor formalism is very similar to the homogeneously broadened two-level formalism. At first this might be surprising, since the semiconductor obviously involves a wide range of transition frequencies, i.e., is inhomogeneously broadened. The resolution of this apparent contradiction lies in rapid carrier-carrier scattering, which for sufficiently small laser fields allows the fields to experience gain from a large part of the carrier distribution, rather than from just in the vicinity of the laser frequency.

The material in this appendix is not necessary for an understanding of semiconductor lasers. However, it helps both in defining many common variables and nomenclature, as well as in talking with people about semiconductor lasers, since they may tend to think in terms of two-level media. After reading this appendix, the reader should read the end of Sec. 3-2, which discusses the relationship between two-level systems and semiconductor media in greater detail.

A-1. Two-Level Physics

In this section, we start with the wave function and its equation of motion, the Schrödinger equation. We expand this function in terms of energy eigenfunctions, and simplify the treatment to two levels. We solve for the time evolution of the level probability amplitudes under the influence of light. We then define the corresponding density matrix, whose diagonal elements are level probabilities and off-diagonal elements are proportional to induced electric-dipole moments. The latter give the polarization of the medium that acts as a source in Maxwell's equations to amplify or absorb light. The density matrix elements obey equations of motion that reduce to rate equations in appropriate limits. Think of this section as a crash course in the basics of two-level theory. More detailed discussions are available in many textbooks, such as Sargent, Scully, and Lamb (1977) and Meystre and Sargent (1991).

According to the postulates of quantum mechanics, the best possible knowledge about a quantum mechanical system is given by its wave function $\psi(\mathbf{r}, t)$. Although $\psi(\mathbf{r}, t)$ itself has no direct physical meaning, it allows us to calculate the *expectation values* of all *observables* of interest. This is due to the fact that the quantity

$$\psi(\mathbf{r}, t)^* \psi(\mathbf{r}, t) \, d^3 r$$

is the probability of finding the system in the volume element $d^3 r$. Since the system described by $\psi(\mathbf{r}, t)$ is assumed to exist, its probability of being somewhere has to equal 1. This gives the normalization condition

$$\int d^3r \ \psi(\mathbf{r},t)^* \psi(\mathbf{r},t) = 1 \ , \tag{1}$$

where the integration is taken over all space.

An observable is represented by a *Hermitian* operator $\mathcal{O}$ and its expectation value is given in terms of $\psi(\mathbf{r},t)$ by

$$\langle \mathcal{O} \rangle = \int d^3r \ \psi(\mathbf{r},t)^* \mathcal{O}\psi(\mathbf{r},t) \ . \tag{2}$$

Experimentally this expectation value is given by the average value of the results of many measurements of the observable $\mathcal{O}$ acting on identically prepared systems. The accuracy of the experimental value for $\langle \mathcal{O} \rangle$ typically depends on the number of measurements performed. Hence enough measurements should be made so that the value obtained for $\langle \mathcal{O} \rangle$ doesn't change significantly when still more measurements are performed. The reason observables, such as position, momentum, energy, and dipole moment, are represented by Hermitian operators is that the expectation values (2) must be real. Denoting by (ϕ, ψ) the inner or scalar product of two vectors ϕ and ψ, we say that a linear operator $\mathcal{O}$ is Hermitian if the equality

$$(\phi, \mathcal{O}\psi) = (\mathcal{O}\phi, \psi)^* . \tag{3}$$

holds for all ϕ and ψ.

An important observable in the interaction of light with electrons is the electric dipole $e\mathbf{r}$. This operator provides the bridge between the quantum mechanical description of a system and the polarization of the medium $\mathbf{P}$ used as a source in Maxwell's equations for the electromagnetic field. According to Eq. (2), the expectation value of $e\mathbf{r}$ is

$$\langle e\mathbf{r} \rangle = \int d^3r \ \mathbf{r} \ e|\psi(\mathbf{r},t)|^2 \ , \tag{4}$$

where we can move $e\mathbf{r}$ to the left of $\psi(\mathbf{r},t)^*$ since the two commute (an operator like ∇ cannot be so moved). Here we see that the dipole–moment expectation value has the same form as the classical value if we identify $\rho = e|\psi(\mathbf{r},t)|^2$ as the charge density.

In nonrelativistic quantum mechanics, the evolution of $\psi(\mathbf{r},t)$ is governed by the Schrödinger equation

$$\boxed{i\hbar \frac{\partial}{\partial t}\psi(\mathbf{r},t) = \mathcal{H}\psi(\mathbf{r},t)} \quad, \tag{5}$$

where $\mathcal{H}$ is the Hamiltonian for the system and $\hbar = 1.054 \times 10^{-34}$ joule-seconds is Planck's constant divided by 2π. The Hamiltonian of an unperturbed system, for instance an atom not interacting with light, is the sum of its potential and kinetic energies

$$\mathcal{H} = \frac{p^2}{2m} + V(\mathbf{r}) \quad, \tag{6}$$

where p is the system momentum, m is the system mass, and $V(\mathbf{r})$ the potential energy. We suppose that the time and space dependencies in Eq. (5) separate as

$$\psi_n(\mathbf{r},t) = u_n(\mathbf{r})\, e^{-i\omega_n t} \tag{7}$$

for which the $u_n(\mathbf{r})$ satisfy the energy eigenvalue equation

$$\mathcal{H} u_n(\mathbf{r}) = \hbar\omega_n\, u_n(\mathbf{r}) \quad. \tag{8}$$

The eigenfunctions $u_n(\mathbf{r})$ can be shown to be orthonormal and complete so that any function can be written as a superposition of the $u_n(\mathbf{r})$. In particular the wave function $\psi(\mathbf{r},t)$ itself can be written as the superposition

$$\psi(\mathbf{r},t) = \sum_n C_n(t)\, u_n(\mathbf{r})\, e^{-i\omega_n t} \quad. \tag{9}$$

Here the expansion coefficients $C_n(t)$ are constants for problems described by a Hamiltonian satisfying the eigenvalue equation (8). We include the time dependence in anticipation of adding an interaction energy to the Hamiltonian. Such a modified Hamiltonian wouldn't quite satisfy Eq. (8), thereby causing the $C_n(t)$ to change in time.

Substituting Eq. (9) into the normalization condition (1) and using the orthonormality of the $u_n(\mathbf{r})$, we find

$$\sum_n |C_n|^2 = 1 .$$

(10)

The $|C_n|^2$ can be interpreted as the probability that the system is in the nth energy state. The C_n are complex probability amplitudes and completely determine the wave function. In terms of the $C_n(t)$, the expectation value (2) of the operator $\mathcal{O}$ is given by

$$\langle \mathcal{O} \rangle = \sum_n \sum_m C_n(t) C_m(t) e^{-i\omega_{nm} t} \mathcal{O}_{mn} ,$$

(11)

where $\mathcal{O}_{mn}$ is the matrix element $\int d^3 r \, u_m{}^*(\mathbf{r}) \mathcal{O} u_n(\mathbf{r})$.

We are primarily interested in the interaction of a medium with light. To treat such interactions, we add the appropriate interaction energy to the Hamiltonian, that is

$$\mathcal{H} = \mathcal{H}_0 + \mathcal{V}.$$

(12)

If we expand the wave function in terms of the eigenfunctions of the "unperturbed Hamiltonian" $\mathcal{H}_0$, rather than those of the total Hamiltonian $\mathcal{H}$, the probability amplitudes $C_n(t)$ change in time. To find out just how, we substitute the wave function (9) and the Hamiltonian (12) into Schrödinger's equation (5) to find

$$\boxed{\dot{C}_n(t) = -\frac{i}{\hbar} \sum_m \langle n|\mathcal{V}|m\rangle \, e^{i\omega_{nm} t} \, C_m(t)} ,$$

(14)

where the matrix element

$$\langle n|\mathcal{V}|m\rangle = \int d^3 r \, u_n{}^*(\mathbf{r}) \, \mathcal{V} \, u_m(\mathbf{r}) ,$$

(15)

and the frequency difference $\omega_{nm} = \omega_n - \omega_m$.

We can solve Eq. (14) approximately by using first-order perturbation theory. Since we are interested in interactions with light, we suppose that the interaction energy matrix element has the time-dependent form

$$\langle n|\mathcal{V}|m\rangle = \mathcal{V}_{nm}(0)\cos\nu t \, , \tag{16}$$

where ν is an optical frequency. Starting at time $t = 0$ with the system in the initial level i ($C_i(0) = 1$, $C_{m\neq i}(0) = 0$)), we integrate Eq. (14) with $C_m(t) \simeq C_m(0)$ to find

$$C_n(t) \simeq C_n^{(1)}(t) = -i \, \frac{\mathcal{V}_{ni}(0)}{2\hbar} \left[\frac{e^{i(\omega_{ni}+\nu)t} - 1}{i(\omega_{ni}+\nu)} + \frac{e^{i(\omega_{ni}-\nu)t} - 1}{i(\omega_{ni}-\nu)} \right]. \tag{17}$$

Here the superscript (1) says that the interaction energy has been applied 1 time, i.e., to first order. For the sake of definiteness, consider the case $\omega_{ni} > 0$. Then the denominator $\omega_{ni}+\nu$ is always positive and larger than ω_{ni}. This is not true for the denominator $\omega_{ni}-\nu$, which vanishes if the resonance condition

$$\nu \simeq \omega_{ni} \tag{18}$$

is satisfied. For interactions near resonance, the term with the relatively small denominator $\omega_{ni}-\nu$ is much larger than that with the $\omega_{ni}+\nu$, allowing us to neglect the latter, an approximation known as the *rotating-wave approximation*. The origin of this name is due to thinking about the spin-½ case in nuclear magnetic resonance, where the two waves with depencencies $e^{\pm i\nu t}$ actually correspond to wave rotating in opposite directions in space. The wave kept in the rotating wave approximation is the one that rotates in the same direction as the precessing spin, while that rotating in the opposite direction is dropped.

Two-Level Approximation

Similarly we can probably neglect transitions to levels with energies very different from $\hbar\nu$. In fact we may be able to neglect all except two levels, namely those satisfying Eq. (18). This observation is used to justify the *two-level approximation* in an atom, which in general has an infinite number of bound states. In this limit, the wave function of Eq. (9) reduces to

$$\psi(\mathbf{r},t) = C_a(t)e^{-i\omega_a t} \, u_a(\mathbf{r}) + C_b(t)e^{-i\omega_b t} \, u_b(\mathbf{r}) \tag{19}$$

According to Eq. (14), the probability amplitudes obey the equations of motion

$$\dot{C}_a = -\frac{i}{\hbar} \mathcal{V}_{ab} e^{i\omega t} C_b ,\qquad (20)$$

$$\dot{C}_b = -\frac{i}{\hbar} \mathcal{V}_{ba} e^{-i\omega t} C_a ,\qquad (21)$$

where the transition frequency $\omega \equiv \omega_{ab}$.

Electric–Dipole Interactions

The polarization of the medium acts as a source in Maxwell's equations, amplifying or absorbing light incident on the medium. This polarization is given by the expectation value of the electric-dipole operator multiplied by the number of systems per unit volume. To compute this value, we note first that the matrix element of the dipole operator $e\mathbf{r}$ between a level and itself vanishes unless the system has a permanent dipole moment (like H_2O), since $|u_n(\mathbf{r})|^2$ is inevitably a symmetrical function of $\mathbf{r}$ and $\mathbf{r}$ itself is antisymmetrical. Matrix elements of $\mathbf{r}$ between different states can also vanish, but we are interested in two levels a and b between which the matrix element does not vanish. For this case, we substitute Eq. (19) into Eq. (4) to find the electric-dipole expectation value

$$\langle e\mathbf{r} \rangle = e\mathbf{r}_{ba} C_a(t) C_b{}^*(t) e^{-i\omega t} + \text{c.c.},\qquad (22)$$

where $\mathbf{r}_{ab}$ is the complex electric-dipole matrix element

$$e\mathbf{r}_{ab} = \int d^3r\, u_a{}^*(\mathbf{r})\, e\mathbf{r}\, u_b(\mathbf{r}).\qquad (23)$$

Hence we don't need C_a and C_b separately, but only in the *bilinear* combination $C_a(t) C_b{}^*(t)$.

In addition to providing the link between Maxwell's equations and the medium, the electric-dipole operator also yields the light-matter interaction energy $\mathcal{V}$. Specifically, this interaction energy is

$$\mathcal{V} = -e\mathbf{r} \cdot \mathbf{E}(\mathbf{R}, t) ,\qquad (24)$$

where $\mathbf{R}$ is the position of the center of mass of the dipole. This gives a higher energy for the dipole $e\mathbf{r}$ aligned against the field $\mathbf{E}$ than along it, as it should. In Eq. (24), we have approximated $\mathbf{E}(\mathbf{r}, t)$ by $\mathbf{E}(\mathbf{R}, t)$, since we are interested in electromagnetic fields with wavelengths much larger than atomic dimensions. We can therefore approximate the electric field by a constant over the dimensions of the dipole. This is called the *dipole approximation*. Typically we are also interested in plane waves, for which we write the electric field simply as $E(z,t)$, where z is the axis of propagation. The dipole traditionally is written as the positive charge value times the distance vector pointing from the negative to the positive charge. This gives the same answer as $e\mathbf{r}$, which is the negative charge value times the distance vector pointing from the positive charge to the negative charge. We can then write the electric-dipole interaction energy matrix element as

$$\mathcal{V}_{ab} = - \wp E(\mathbf{R},t), \tag{25}$$

where $\wp$ (pronounced "squiggle") is the component of $e\mathbf{r}_{ab}$ along $\mathbf{E}$.

For the sake of simplicity, we ignore the spatial dependence altogether in the remainder of this section, and use the electric field

$$E(t) = E_0 \cos \nu t , \tag{26}$$

which gives the interaction energy matrix element

$$\mathcal{V}_{ab} = - \wp E_0 \cos \nu t . \tag{27}$$

Examining the interaction energy (16) and Eq. (17), we see that in the rotating-wave approximation we keep the $e^{-i\nu t}$ term for $\omega_{ni} > 0$. In the present case, the transition frequency $\omega \equiv \omega_{ab} > 0$, and hence in the rotating-wave approximation we keep only

$$\mathcal{V}_{ab} \simeq - \frac{1}{2} \wp E_0 e^{-i\nu t} , \tag{28}$$

and $\mathcal{V}_{ba} = \mathcal{V}_{ab}{}^*$.

We can very quickly discover some basic physics by considering exact resonance, i.e., $\nu = \omega$. Substituting the complex conjugate of Eq. (28) into Eq. (21), differentiating with respect to time, and substituting Eq. (20), we find

$$\ddot{C}_b = -\tfrac{1}{4}\mathcal{R}_0{}^2 C_b , \tag{29}$$

where $\mathcal{R}_0$ is the *Rabi flopping frequency*

$$\mathcal{R}_0 \equiv \frac{\wp E_0}{\hbar} \tag{30}$$

after Rabi (1936), who studied the the spin-$\frac{1}{2}$ magnetic dipole in nuclear magnetic resonance. Equation (29) is the differential equation for sinusoids. In particular if at time $t=0$ the system is in the lower state ($C_b(0) = 1$, $C_a(0) = 0$), then

$$C_b(t) = \cos\tfrac{1}{2}\mathscr{R}_0 t ,\tag{31}$$

which from Eq. (31) gives

$$C_a(t) = i\sin\tfrac{1}{2}\mathscr{R}_0 t .\tag{32}$$

The probability that the system is in the lower level $|C_b(t)|^2 = \cos^2\tfrac{1}{2}\mathscr{R}_0 t = (1 + \cos\mathscr{R}_0 t)/2$, while $|C_a|^2 = \sin^2\tfrac{1}{2}\mathscr{R}_0 t = (1 - \cos\mathscr{R}_0 t)/2$. Hence the wave function oscillates between the lower and upper states sinusoidally at the frequency $\mathscr{R}_0$, a phenomenon known as Rabi flopping.

The Density Matrix

So far we have dealt with a single quantum system described by a wave function. In treating laser media, we need to consider ensembles of such systems created by pump processes and destroyed by damping processes. These processes are examples of reservoir interactions that by nature are irreversible, unlike processes whose dynamics are determined by the Schrödinger equation. Although some limited examples of decay processes can be described using a wave function, two important kinds cannot: decay from an upper level to a lower level and more rapid decay of the electric dipole than the average level decay rate. Both of these decay processes occur in semiconductors. Furthermore most pump processes are incoherent, so that only the level probabilities are pumped. For these damping and pump mechanisms, we need a more general description than can be provided by a single wave function. Specifically, the pump and decay processes tend to wash out the coherent superposition of states implied by the wave function, reducing the size of electric dipole moment given by the bilinear product $C_a(t)C_b{}^*(t)$ in Eq. (22). The density matrix provides a convenient way to describe the resulting partially incoherent superposition.

The density matrix elements ρ_{ij} corresponding to the wave function of Eq. (29) are given by the bilinear products

$$\rho_{aa} = C_a C_a{}^* \text{, probability of being in upper level}$$

$$\rho_{ab} = C_a C_b{}^* \exp(-i\omega t) \text{ , dimensionless complex dipole moment}$$

$$\rho_{ba} = C_b C_a{}^* = \rho_{ab}{}^* ,$$

$$\rho_{bb} = C_b C_b{}^* \text{, probability of being in lower level .}$$

that is,

$$\rho = \begin{pmatrix} C_a C_a{}^* & C_a C_b{}^* \exp(-i\omega t) \\ C_b C_a{}^* \exp(i\omega t) & C_b C_b{}^* \end{pmatrix} = \begin{pmatrix} \rho_{aa} & \rho_{ab} \\ \rho_{ba} & \rho_{bb} \end{pmatrix}. \quad (33)$$

In terms of this 2×2 density matrix, the expectation value (2) of an operator $\mathcal{O}$ is given by

$$\langle \mathcal{O} \rangle = \rho_{aa}\, \mathcal{O}_{aa} + \rho_{ab}\, \mathcal{O}_{ba} + \rho_{ba}\, \mathcal{O}_{ab} + \rho_{bb}\, \mathcal{O}_{bb}. \quad (34)$$

In particular the dipole moment is given in the u_a, u_b basis by

$$\langle er \rangle = \wp \rho_{ab} + \text{c.c.} \quad (35)$$

The density matrix element $\rho_{nm}(t)$ corresponding to the many-level wave function of Eq. (9) is defined by

$$\rho_{nm}(t) = C_n(t) C_m(t)^* e^{-i\omega_{nm} t}. \quad (36)$$

In terms of this, the expectation value (11) is the trace of the matrix product $\rho \mathcal{O}$:

$$\langle \mathcal{O} \rangle = \sum_n \sum_m \rho_{nm}\, \mathcal{O}_{mn} = \sum_n (\rho \mathcal{O})_{nn} = \text{tr}\{\rho \mathcal{O}\}. \quad (37)$$

The expectation value of an operator is given by Eq. (37) even when the system is described by the most general density matrix.

Using the upper-level probability-amplitude equation of motion (20), we find the equation of motion for ρ_{aa} as

$$\begin{aligned} \dot{\rho}_{aa} &= \dot{C}_a C_a{}^* + C_a \dot{C}_a{}^* \\ &= -i\hbar^{-1} \mathcal{V}_{ab} e^{i\omega t} C_b C_a{}^* + i\hbar^{-1} \mathcal{V}_{ba} e^{-i\omega t} C_a C_b{}^* \\ &= -i\hbar^{-1} \mathcal{V}_{ab} \rho_{ba} + \text{c.c.} \end{aligned} \quad (38)$$

Here we expect to find the complex conjugate since ρ_{aa} is a probability and probabilites are real. Similarly we find

$$\dot{\rho}_{bb} = +i\hbar^{-1} \mathcal{V}_{ab} \rho_{ba} + \text{c.c.} = -\dot{\rho}_{aa}, \quad (39)$$

i.e., the probability that leaves the upper level due to stimulated emission enters the lower level. Using both equations of motion (20) and (21), we find

$$\dot{\rho}_{ab} = -i\omega\rho_{ab} + \frac{i}{\hbar}\mathcal{V}_{ab}(\rho_{aa} - \rho_{bb}) .$$ (40)

Using Eq. (14), we find the corresponding equations of motion for the many-level wave function to be

$$\dot{\rho}_{nm} = -i\omega_{nm}\rho_{nm} - \frac{i}{\hbar}[\mathcal{V}_{nm}\rho_{mn} - \rho_{nm}\mathcal{V}_{mn}] .$$ (41)

This equation can be written most simply as

$$\dot{\rho} = -\frac{i}{\hbar}[\mathcal{H}, \rho] ,$$ (42)

where the commutator $[\mathcal{H}, \rho] \equiv \mathcal{H}\rho - \rho\mathcal{H}$.

In this section, we have blazed a bee-line path from the basic elements of quantum theory to the equations of motion (38) through (40) for the two-level density matrix. In the next section, we add pump and decay processes to find the polarization of the two-level medium. Combined with the electric-field self-consistency equations (2.38) and (2.39), we have a simple laser theory that describes some features of semiconductor laser operation. This is surprising since the theory assumes *homogeneous* broadening, that is, there's only one transition frequency, while the band diagram of Fig. 2-1 shows a whole range. This is due to the establishment of quasiequilibrium, a situation that we can simply postulate or, with a more advanced theory, which we can predict.

A-2. Two-Level Laser Theory

To derive the polarization of a medium $P(z,t)$, it is convenient to define the population matrix, or more precisely, the population density matrix. This matrix is intimately related to the traditional density matrix and unlike the density matrix actually has units of 1/volume. We define the population matrix by

$$\rho(z,t) = \sum_{\alpha=a,b} \int_{-\infty}^{t} dt_0 \, \lambda_\alpha(z,t_0) \, \rho(\alpha,t_0,z,t) \,. \tag{43}$$

Here $\lambda_a(z,t_0)$ is the pump rate per unit volume to the level α ($\alpha=a$ or b) for a homogeneously broadened medium, t_0 is the time of excitation, and $\rho(\alpha,t_0,z,t)$ is the density matrix describing a system excited to the level α at the time t_0. $\rho(z,t)$ has diagonal elements giving the population densities (rather than probabilities) of the levels, and hence the name population matrix. The population matrix formalism leads directly to the popular rate equations (Eqs. (15) and (16)) for the population densities for sufficiently rapidly decaying dipole moments. It furthermore bypasses some of the algebra in summing contributions from all systems at the place z at the time t regardless of their initial times and levels of excitation. In terms of $\rho(z,t)$ the polarization is given by

$$P(z,t) = \wp \rho_{ab}(z,t) + \text{c.c.} \tag{44}$$

Equation (14) shows that $\rho_{ab} \propto e^{i(Kz - \nu t)}$, so that we can combine Eqs. (2) and (7) to find

$$\mathscr{P}(z) = 2\wp \, e^{-i(Kz - \nu t)} \, \rho_{ab}. \tag{45}$$

Hence we need to find $\rho_{ab}(z,t)$ as it evolves under the influence of the electric-dipole interaction energy.

To do this we calculate the equation of motion of $\rho(z,t)$ by differentiating (6) with respect to time. Two time dependencies exist: that of the upper limit of integration over the excitation time t_0 and that of the single-atom density matrix. We have

$$\frac{d\rho(z,t)}{dt} = \sum_\alpha \lambda_\alpha(z,t)\rho(\alpha,z,t,t) + \sum_\alpha \int_{-\infty}^{t} dt_0 \, \lambda_\alpha(z,t_0) \, \dot{\rho}(\alpha,z,t_0,t).$$

Assuming no off-diagonal excitation, we have by construction $\rho_{ij}(\alpha,z,t,t) = \delta_{i\alpha}\delta_{j\alpha}$, and the first term can be replaced by the operator with the matrix representation

$$\begin{bmatrix} \lambda_a(z,t) & 0 \\ 0 & \lambda_b(z,t) \end{bmatrix}.$$

The second term has components identical to the right-hand sides of the equations of motion for the single-atom density matrix [see Eqs. (38) through (40)]. This result follows because the Hamiltonian does not depend on a or b and the pump rates $\lambda_\alpha(z, t_0)$ are assumed to vary slowly enough to be evaluated at the time t. To model damping processes, we include the phenomenological decay rates $-\gamma_\alpha \rho_{\alpha\alpha}(z,t)$ in the $\rho_{\alpha\alpha}$ equations of motion and $-\gamma \rho_{ab}(z,t)$ in the ρ_{ab} equation of motion. In component form, the equations of motion for the population matrix $\rho(z,t)$ become

$$\dot\rho_{ab}(z,t) = -(i\omega + \gamma)\rho_{ab}(z,t) + i\hbar^{-1}\mathcal{V}_{ab}(z,t)D(z,t) , \tag{46}$$

$$\dot\rho_{aa}(z,t) = \lambda_a(z,t) - \gamma_a\rho_{aa}(z,t) - [i\hbar^{-1}\mathcal{V}_{ab}\rho_{ba}(z,t) + \text{c.c.}] , \tag{47}$$

$$\dot\rho_{bb}(z,t) = \lambda_b(z,t) - \gamma_b\rho_{bb}(z,t) + [i\hbar^{-1}\mathcal{V}_{ab}\rho_{ba}(z,t) + \text{c.c.}] , \tag{48}$$

where $D(z,t) = \rho_{aa}(z,t) - \rho_{bb}(z,t)$ is the population (density) difference.

These equations of motion have been used to model laser media with both active levels excited, such as in the He-Ne laser. For the purposes of modeling something like a semiconductor medium, the lower level is the ground state and the pump mechanism transfers systems from the lower state to the upper state. This situation is characteristic also of the ruby laser transition. The desired pump rate is given by

$$\lambda_a(z,t) = \Lambda\rho_{bb}(z,t) . \tag{49}$$

The lower level is "pumped" by decays from the upper level, that is,

$$\lambda_b(z,t) = \Gamma\rho_{aa}(z,t) . \tag{50}$$

Accordingly using these values in Eqs. (47) and (48), replacing γ_a by Γ and setting $\gamma_b = 0$, we have the upper-to-ground-lower-level equation of motion

$$\dot\rho_{aa}(z,t) = -\dot\rho_{bb}(z,t)$$
$$= \Lambda(z,t)\rho_{bb} - \Gamma\rho_{aa}(z,t) - [i\hbar^{-1}\mathcal{V}_{ab}\rho_{ba}(z,t) + \text{c.c.}] . \tag{51}$$

Since the total number of systems $N'(z,t)$ is distributed between levels a and b, we have

$$N'(z,t) = \rho_{aa}(z,t) + \rho_{bb}(z,t) . \tag{52}$$

Combining Eqs. (51) and (52), we find that the population difference $D(z,t)$ obeys the equation of motion

$$\dot{D}(z,t) = -\frac{D(z,t) - D_0(z,t)}{T_1} - 2[i\hbar^{-1}\mathcal{V}_{ab}\rho_{ba}(z,t) + \text{c.c.}] , \qquad (53)$$

where T_1 is the population difference decay time

$$T_1 = \frac{1}{\Lambda + \Gamma} , \qquad (54)$$

and D_0 is the zero-field steady-state population difference

$$D_0(z,t) = N'(z,t)(\Lambda - \Gamma)T_1 . \qquad (55)$$

Our working equations of motion are (46) for $\rho_{ab}(z,t)$ coupled to (53) for $D(z,t)$. These equations are often called the *optical Bloch equations*, although technically Bloch considered the real and imaginary parts of the polarization element ρ_{ab} separately so that one can define a Bloch vector. This vector approach has some pictorial advantages, but for most calculations we're better off letting complex arithmetic do the work for us.

Rate Equation Approximation

Often the dipole decay rate constant γ greatly exceeds the level decay rate constant Γ. This is due to elastic collisions that destroy the phase relationship between the applied electromagnetic field and the induced dipole, but are not severe enough to cause appreciable energy change. In semiconductors, for example, γ is due to carrier-carrier interactions on the order of 100 femtoseconds, while the decay of the total carrier density is due to electron-hole recombination on the order of nanoseconds. This case is quite different from the pure radiative decay observed in a low pressure sodium discharge, for which $\Gamma = 2\gamma$. In general we can integrate Eq. (46) formally to find

$$\rho_{ab}(z,t) = \frac{i}{\hbar}\int_{-\infty}^{t} dt' \, e^{-(i\omega+\gamma)(t-t')} \, \mathcal{V}_{ab}(z,t') \, D(z,t') . \qquad (56)$$

We now assume that the dipole decay time $T_2 \equiv 1/\gamma$ is much smaller than times for which the population difference $D(z,t)$ or the field envelope can change. For steady state, this approximation is exact since the population difference and field envelope are constant. Using Eq. (28) for $\mathcal{V}_{ab}$, we can then factor both the population difference and the field envelope outside the t' integration, perform the integral over exponentials, and find

$$\rho_{ab}(z,t) = -i(\wp\mathcal{E}/2\hbar)e^{i(Kz - \nu t)}\mathcal{D}(\omega-\nu)D(z,t) \; , \tag{57}$$

where the complex Lorentzian denominator

$$\mathcal{D}(\omega-\nu) = \frac{1}{\gamma + i(\omega-\nu)} \; . \tag{58}$$

Substituting Eq. (57) into the population difference equation of motion (53), we find the rate equation

$$\dot{D}(z,t) = \frac{D_0(z,t) - [1 + I\mathcal{L}(\omega-\nu)/I_s]D(z,t)}{T_1} \; , \tag{59}$$

where in mks units I is the intensity $c\epsilon_0|\mathcal{E}|^2$, I_s is the *saturation intensity*

$$I_s = c\epsilon_0|\hbar/\wp|^2/T_1 T_2 \; , \tag{60}$$

$T_2 \equiv 1/\gamma$ is the dipole decay time, and $\mathcal{L}(\omega-\nu)$ is the dimensionless Lorentzian $\mathcal{L}(\omega-\nu)$

$$\mathcal{L}(\omega - \nu) = \frac{\gamma^2}{\gamma^2 + (\omega-\nu)^2} \; . \tag{61}$$

Substituting Eq. (57) for ρ_{ab} into (45) for $\mathcal{P}$, we have the complex polarization

$$\mathcal{P}(z,t) = -\frac{i\wp^2\mathcal{E}}{\hbar} \mathcal{D}(\omega-\nu)D(z,t) \; . \tag{62}$$

Using Eq. (2.28), we have the susceptibility

$$\chi = -\frac{i\chi_0\gamma\mathcal{D}(\omega-\nu)D}{D_0} \; , \tag{63}$$

where χ_0 is given by Eq. (2.34). This susceptibility can be used either in the propagation equations (2.29) and (2.30) or in the cavity Eqs. (2.38) through (2.39). In particular substituting it into the intensity equation of motion (2.40), we have

$$\frac{dI(t)}{dt} = -\frac{\nu}{Q}\left[1 - \frac{\chi_0 Q\mathcal{L}(\omega-\nu)D}{D_0}\right]I(t) \; . \tag{64}$$

Equations (64) and (59) are a coupled set of rate equations describing how the laser field and the population difference are coupled together in the rate equation approximation. As we see in Sec. xxx, they exhibit relaxation oscillations in their approach to their steady-state values.

Steady–State Intensity and Oscillation Frequency

Solving Eq. (59) in steady state ($\dot{D} = 0$), we find the saturated population difference

$$D(z) = \frac{D_0(z)}{1 + I\mathscr{L}(\omega-\nu)/I_s} , \tag{65}$$

which gives the susceptibility

$$\chi = -\frac{i\chi_0\gamma\mathscr{D}(\omega-\nu)}{1 + I\mathscr{L}(\omega-\nu)/I_s} . \tag{66}$$

This shows that the response *saturates* or *bleaches* as $1/(1 + I\mathscr{L}/I_s)$. The half value is reached when $I\mathscr{L} = I_s$, i.e., for central tuning ($\nu = \omega$) when the intensity equals the saturation intensity. Expanding Eq. (66) keeping only the first two terms of the geometric series, we find

$$\chi = \chi^{(1)} + \chi^{(3)} I , \tag{67}$$

where the linear response is given by

$$\chi^{(1)} = -i\chi_0\gamma\mathscr{D}(\omega-\nu) , \tag{68}$$

and the lowest-order nonlinear response is given by

$$\chi^{(3)} = i\chi_0\gamma\mathscr{D}(\omega-\nu)\mathscr{L}(\omega-\nu)/I_s . \tag{69}$$

We see that Eq. (66) contains all terms in the perturbation series, and is valid even when the series fails to converge, i.e., for $I\mathscr{L}(\omega-\nu)/I_s > 1$. Expanding the $\mathscr{D}$ and $\mathscr{L}$ functions in Eq. (66), we find

$$\chi = -\frac{i\chi_0\gamma[\gamma - i(\omega-\nu)]}{\gamma^2(1 + I/I_s) + (\omega-\nu)^2} . \tag{70}$$

This formula reveals a *power-broadened* response, since the full-width at half maximum value of χ'' is $2\gamma(1 + I/I_s)$.

Substituting Eq. (65) into Eq. (64) and solving for the steady-state intensity ($I = 0$), we find

$$I = I_s \left[\chi_0 Q - \frac{1}{\mathscr{L}} \right] . \tag{71}$$

Since χ_0 is proportional to the pump rate Λ minus Γ, when Λ exceeds a threshold value, the laser output intensity increases from 0 linearly with pumping.

Substituting the real part of the susceptibility into Eq. (2.39) and using the steady-state intensity of Eq. (71), we find the laser oscillation frequency

$$\nu = \frac{\Omega \gamma + \omega(\nu/2Q)}{\gamma + \nu/2Q} . \tag{73}$$

This formula can be interpreted as a center of mass formula in which the oscillation frequency ν assumes a weighted average value of the passive cavity frequency Ω and the atomic line-center frequency ω with weights γ and $\nu/2Q$, respectively. For high Q cavities, $\nu/2Q \ll \gamma$, and ν is pulled slightly from Ω toward ω.

Inhomogeneous Broadening

From Eq. (1.3), we see that the transition frequency $\omega(\mathbf{k})$ in a semiconductor is a function of the momentum $\hbar\mathbf{k}$. This yields a whole range of transition frequencies, rather than the single frequency assumed so far in this section. A range of transition frequencies is called inhomogeneous broadening, and is well known also in gases due to Doppler shifts and in low-temperature crystals due to nonuniform strains. We can add inhomogeneous broadening to the homogeneous broadening case considered so far by mulitplying the susceptibility χ of Eq. (66) by a distribution function $W(\omega')$ and summing over ω'. This gives

$$\chi = -i\chi_0 \int_{-\infty}^{\infty} W(\omega') \frac{\gamma \mathscr{D}(\omega-\nu)}{1 + I \mathscr{L}(\omega-\nu)/I_s} . \tag{74}$$

This formula reveals the phenomenon of *spectral hole burning*, namely that a broad distribution $W(\omega')$ is selectively bleached by a relatively sharp Lorentzian $\mathcal{L}(\omega'-\nu)$. Alternatively, we can expand the $\mathcal{L}$ and $\mathcal{D}$ as in Eq. (70) to find

$$\chi = -i\chi_0 \int_{-\infty}^{\infty} W(\omega') \frac{\gamma[\gamma - i(\omega'-\nu)]}{\gamma^2(1 + I/I_s) + (\omega'-\nu)^2}. \tag{75}$$

Although this formula is solvable in general numerically, we note that if the width of the distribution $W(\omega')$ is much greater than γ, we can evaluate $W(\omega')$ at ν and take it outside the integral. This approximation is called the inhomogeneous broadening limit. The term $\omega' - \nu$ is antisymmetric about ν and thereby integrates to 0. Thus Eq. (75) reduces to

$$\chi = -i\frac{\wp^2\mathcal{E}N}{\hbar}W(\nu) \int_{-\infty}^{\infty} \frac{dx}{1 + x^2 + I/I_s} = -\frac{i\pi\wp^2 N W(\nu)\mathcal{E}}{\hbar(1+I/I_s)^{1/2}}. \tag{76}$$

Comparing this to the homogeneous-broadening case of Eq. (66), we see that the tuning dependence is gone, and that the saturation is weaker, i.e., Eq. (76) is proportional to $(1+I/I_s)^{-1/2}$, while Eq. (66) is proportional to $(1+I\mathcal{L}/I_s)^{-1}$. The reduced saturation of the inhomogeneously broadened case is due to the fact that contributions from detuned (and therefore less saturated) atoms are included in the average over ω'.

References

Rabi, I. I. (1936), Phys. Rev. 49, 324; (1937), Phys. Rev. 51, 652.

Meystre, M. and M. Sargent III, *Elements of Quantum Optics*, Springer-Verlag, Heidelberg (1990).

Sargent M. III, M. O. Scully and W. E. Lamb, Jr., *Laser Physics*, Addison-Wesley Publishing Co., Reading, Mass. 3rd Printing (1977).

Appendix B
k·p THEORY

In this Appendix we derive the results of the k·p theory used in Chap. 6. We begin by applying perturbation theory to solve the Schrödinger equation for the lattice periodic functions of a bulk semiconductor. The cases where the unperturbed (zone center) eigenstates are nondegenerate and degenerate are treated separately. Spin-orbit coupling, which was initially neglected, is then taken into account by switching to basis states that are eigenstates of the total (orbital and spin) angular momentum. For the case of the top valence bands, i.e., those with $j = 3/2$, degenerate perturbation theory leads to the Luttinger Hamiltonian. The Luttinger Hamiltonian may be reduced to a block-diagonal form by a unitrary transformation. Diagonalization of the block-diagonal Luttinger Hamiltonian gives a valence band structure that consists of two degenerate heavy-hole bands and two degenerate light-hole bands. All four bands are degenerate at the zone center.

B-1. Nondegenerate Perturbation Theory

Equations (6.1) - (6.5) show that the Schrödinger equation for the lattice periodic functions may written as

$$(\mathcal{H}_0 + \lambda \mathcal{H}_1) \left| n \; \mathbf{k} \right\rangle = W_{n\mathbf{k}} \left| n \; \mathbf{k} \right\rangle , \tag{1}$$

where

$$\mathcal{H}_0 = \frac{p^2}{2m_0} + V_0 \tag{2}$$

$$\mathcal{H}_1 = \frac{\hbar}{m_0} \mathbf{k} \cdot \mathbf{p} \tag{3}$$

$$W_{n\mathbf{k}} = \varepsilon_{n\mathbf{k}} - \frac{\hbar^2 k^2}{2m_0} . \tag{4}$$

In these equations, m_0 is the free electron mass and V_0 is an effective periodic potential due to the lattice. We treat $\mathcal{H}_1$ as a perturbation and use λ to keep track of the order of the terms in the perturbation expansion. The unperturbed eigenstates are the zone center eigenstates

$$|n \text{ k=0}\rangle \equiv |n\rangle \quad \text{and} \quad \varepsilon_{n,\text{k=0}} \equiv \varepsilon_n , \tag{5}$$

where

$$\mathcal{H}_0 |n\rangle = \varepsilon_n |n\rangle . \tag{6}$$

Ignoring spin-orbit coupling, the states $|m\rangle$ are given by $|l \ m_l\rangle$, where l and m_l are the orbital angular momentum and z-component angular momentum quantum numbers, respectively. For the conduction band, we have the nondegenerate eigenstate, $|S\rangle = |0 \ 0\rangle$, and for the valence band, we have the degenerate eigenstates $|1 \ \pm1\rangle$ and $|1 \ 0\rangle$, or $|X\rangle$, $|Y\rangle$ and $|Z\rangle$. Let us first treat the nondegenerate case.

Suppose we wish to find $|n \text{ k}\rangle$, which is the state that reduces to $|n\rangle$ at $\text{k} = 0$. First, we write $|n \text{ k}\rangle$ and $W_{n\text{k}}$ as the expansions

$$|n \text{ k}\rangle = |n \text{ k}\rangle_0 + \lambda|n \text{ k}\rangle_1 + \lambda^2|n \text{ k}\rangle_2 + \cdots \tag{7}$$

$$W_{n\text{k}} = W_{n\text{k}}^{(0)} + \lambda W_{n\text{k}}^{(1)} + \lambda^2 W_{n\text{k}}^{(2)} + \cdots . \tag{8}$$

Substituting Eqs. (7) and (8) into (1) and collecting terms with equal powers of λ, we get

$$(\mathcal{H}_0 - W_{n\text{k}}^{(0)})|n \text{ k}\rangle_0 = 0 , \tag{9a}$$

$$(\mathcal{H}_0 - W_{n\text{k}}^{(0)})|n \text{ k}\rangle_1 = (W_{n\text{k}}^{(1)} - \mathcal{H}_1)|n \text{ k}\rangle_0 , \tag{9b}$$

$$(\mathcal{H}_0 - W_{n\text{k}}^{(0)})|n \text{ k}\rangle_2 = (W_{n\text{k}}^{(1)} - \mathcal{H}_1)|n \text{ k}\rangle_1 + W_{n\text{k}}^{(2)}|n \text{ k}\rangle_0 \tag{9c}$$

and so on. Equation (9a) gives

$$W_{n\text{k}}^{(0)} = \varepsilon_n \quad \text{and} \quad |n \text{ k}\rangle_0 = |n\rangle . \tag{10}$$

Taking the scalar product of Eq. (9b) with $\langle n|$, we find

$$W_{n\mathbf{k}}^{(1)} = \langle n|\mathcal{H}_1|n\rangle = 0 , \tag{11}$$

which vanishes because $\mathcal{H}_0$ has inversion symmetry, so that $|n\rangle$ has definite parity. Consequently, all diagonal matrix elements of $\mathbf{p}$ vanish, since $\mathbf{p}$ changes sign under space inversion :

$$\langle n|\mathbf{p}|n\rangle = \int d^3r \, \langle n|\mathbf{r}\rangle \frac{\hbar}{i}\nabla\langle\mathbf{r}|n\rangle = 0 , \tag{12}$$

and we have no first-order correction to the energy. If we take the scalar product of Eq. (9b) with $\langle m|$, where $m \neq n$, then

$$\langle m|n\,\mathbf{k}\rangle_1 = \frac{\langle m|\mathcal{H}_1|n\rangle}{\varepsilon_n - \varepsilon_m} , \tag{13}$$

provided $\varepsilon_n \neq \varepsilon_m$. Then to first order in the perturbation, the eigenstate is

$$|n\,\mathbf{k}\rangle_1 = \sum_m |m\rangle\langle m|n\,\mathbf{k}\rangle_1 = \sum_{m\neq n} \frac{\langle m|\mathcal{H}_1|n\rangle}{\varepsilon_n - \varepsilon_m} |m\rangle . \tag{14}$$

Note that we could add a term containing $|n\rangle$ to $|n\,\mathbf{k}\rangle_1$ and still satisfy Eq. (9b). However, we choose not to do so in order to make the unperturbed eigenstate orthogonal to all the higher-order corrections, that is

$$\langle n|n\,\mathbf{k}\rangle_j = 0 \tag{15}$$

for $j \neq 0$. Taking the scalar product of Eq. (9c) with $\langle n|$ and using Eq. (13), we find

$$W_{n\mathbf{k}}^{(2)} = \sum_{m\neq n} \frac{|\langle n|\mathcal{H}_1|m\rangle|^2}{\varepsilon_n - \varepsilon_m} , \tag{16}$$

From Eqs. (4) and (16), we have

$$\varepsilon_{n\mathbf{k}} = \varepsilon_n + \frac{\hbar^2 k^2}{2m_0} + \frac{\hbar^2}{m_0^2} \sum_{m \neq n} \frac{|\langle n | \mathbf{k} \cdot \mathbf{p} | m \rangle|^2}{\varepsilon_n - \varepsilon_m}. \tag{17}$$

Given that the eigenstates have definite parity, one can readily see that

$$\langle n | p_\alpha | m \rangle \langle m | p_\beta | n \rangle = 0 \tag{18}$$

for $\alpha \neq \beta$. Furthermore, for bulk III-V semiconductors,

$$|\langle n | p_x | m \rangle|^2 = |\langle n | p_y | m \rangle|^2 = |\langle n | p_z | m \rangle|^2 \tag{19}$$

because of the symmetry about the zone center. Then, we can write

$$\varepsilon_{n\mathbf{k}} = \varepsilon_n + \frac{\hbar^2 k^2}{2m_{n,eff}}, \tag{20}$$

where the *effective mass* $m_{n,eff}$ is defined by

$$\frac{1}{m_{n,eff}} = \frac{1}{m_0} \left[1 + \frac{2}{m_0} \sum_{m \neq n} \frac{|\langle n | p_x | m \rangle|^2}{\varepsilon_n - \varepsilon_m} \right] \tag{21}$$

B-2. Degenerate Perturbation Theory

To treat the situation when the bands of interest have degenerate unperturbed (zone center) eigenstates, let us suppose the first N eigenstates of $\mathcal{H}_0$, $|1\rangle, |2\rangle, |3\rangle, \cdots, |N\rangle$, are degenerate, so that

$$\varepsilon_1 = \varepsilon_2 = \varepsilon_3 = \cdots = \varepsilon_N. \tag{22}$$

The procedure involves finding N new orthonormal eigenstates that evolve continuously into nondegenerate eigenstates as we leave the zone center. We write these new eigenstates as linear superpositions of the old ones,

$$|n\rangle' = \sum_{m=1}^{N} |m\rangle \langle m | n \rangle', \tag{23}$$

where $1 \leq n \leq N$ and $\langle m | n \rangle' = 0$ for $m > N$. The scalar product of $\langle j |$ with

Eq. (9*b*) gives

$$\sum_{m=1}^{N} \langle j|\mathcal{H}_1|m\rangle\langle m|n\rangle' - W_{n\mathbf{k}}^{(1)}\langle j|n\rangle' = 0 \qquad \text{for } 1 \le j \le N \quad (24)$$

$$\langle j|n\rangle' = \sum_{m=1}^{N} \frac{\langle j|\mathcal{H}_1|m\rangle\langle m|n\rangle'}{\varepsilon_1 - \varepsilon_m} \qquad \text{for } j > N . \quad (25)$$

Equation (24) provides us with N coupled equations that can be solved for the N new eigenstates. However, if $\langle j|\mathcal{H}_1|m\rangle = 0$ for all j and m between 1 and N, then $W_{n\mathbf{k}}^{(1)} = 0$ and we have to move to the next higher order in the perturbation. This turns out to be the case for every group of degenerate states, since all states within each group have the same orbital angular momentum. Repeating the above procedure with the second-order equation (9*c*), we find

$$\sum_{l=1}^{N} \sum_{m>N} \frac{\langle j|\mathcal{H}_1|m\rangle\langle m|\mathcal{H}_1|l\rangle}{\varepsilon_1 - \varepsilon_m} \langle l|n\rangle' - W_{n\mathbf{k}}^{(2)}\langle j|n\rangle' = 0 ,$$

$$(26)$$

for $1 \le j \le N$. Equation (26) gives a set of N coupled homogeneous equations that can be solved for $W_{n\mathbf{k}}^{(2)}$ and $|n\rangle'$.

B-3. Luttinger Hamiltonian

In this section, we apply the results of the previous section to treat the valence bands. The degenerate eigenstates at $k = 0$ are the three $l = 1$, $m_l = 0, \pm 1$ eigenstates, see Eqs. (6.8) and (6.9). To second order in the perturbation, the energy of the nth band is

$$\varepsilon_{n\mathbf{k}} = \varepsilon_n + \frac{\hbar^2 k^2}{2m_0} + W_{n\mathbf{k}}^{(2)} , \qquad (27)$$

where $W_{n\mathbf{k}}^{(2)}$ is given by Eq. (26). Hence we have the set of three equations

$$\sum_{l=1}^{3} \left[{\sum_{m}}' \frac{\langle j|\mathcal{H}_1|m\rangle \langle m|\mathcal{H}_1|l\rangle}{\varepsilon_1 - \varepsilon_m} + \left[\varepsilon_1 + \frac{\hbar^2 k^2}{2m_0} - \varepsilon_{n\mathbf{k}} \right] \delta_{j,l} \right] \langle l|n\ \mathbf{k}\rangle = 0 \ ,$$

$$(28)$$

where the primed summation sign indicates the exclusion of the $l = 1$ states from the summation, or in other words, the summation involves only the remote bands. Nontrivial solutions of this set of N coupled homogeneous equations occur only if

$$\det(\underline{\mathcal{H}} - \varepsilon_{n\mathbf{k}} \underline{I}) = 0 \ , \qquad (29)$$

where $\underline{I}$ is the identity matrix and $\underline{\mathcal{H}}$ is a 3×3 matrix, whose elements are

$$\mathcal{H}_{jl} = \left[\varepsilon_1 + \frac{\hbar^2 k^2}{2m_0} \right] \delta_{j,l} + {\sum_{m}}' \frac{\langle j|\mathcal{H}_1|m\rangle \langle m|\mathcal{H}_1|l\rangle}{\varepsilon_1 - \varepsilon_m} \ . \qquad (30)$$

The computation of the matrix elements is straightforward. For example, if we order the basis states so that $|j\rangle$ for $j = 1, 2$ and 3 are $|X\rangle$, $|Y\rangle$ and $|Z\rangle$, then

$$\mathcal{H}_{11} \equiv \langle X|\mathcal{H}|X\rangle = \varepsilon_1 + \frac{\hbar^2 k^2}{2m_0} + {\sum_{m}}' \frac{|\langle X|\mathcal{H}_1|m\rangle|^2}{\varepsilon_1 - \varepsilon_m} \ . \qquad (31)$$

Using Eq. (6.9), we can convince ourselves that

$$\frac{m_0^2}{\hbar^2} |\langle X|\mathcal{H}_1|m\rangle|^2 = |\langle X|p_x|m\rangle|^2 k_x^2 + |\langle X|p_y|m\rangle|^2 k_y^2 + |\langle X|p_z|m\rangle|^2 k_z^2 \ ,$$

$$(32)$$

so that

$$\mathcal{H}_{11} = \varepsilon_1 + \sum_{j=x,y,z} \left[\frac{\hbar^2}{2m_0} + \frac{\hbar^2}{m_0^2} {\sum_{m}}' \frac{|\langle X|p_j|m\rangle|^2}{\varepsilon_1 - \varepsilon_m} \right] k_j^2 \ . \qquad (33)$$

Due to the symmetry of III-V compound semiconductors at $k = 0$,

$$|\langle X|p_y|m\rangle|^2 = |\langle X|p_z|m\rangle|^2 , \tag{34}$$

which reduces Eq. (33) to

$$\mathcal{H}_{11} = \varepsilon_1 + Ak_x^2 + B(k_y^2 + k_z^2) , \tag{35}$$

where

$$A = \frac{\hbar^2}{2m_0} + \frac{\hbar^2}{m_0^2} \sum_j {}' \frac{|\langle X|p_x|j\rangle|^2}{\varepsilon_1 - \varepsilon_j} \tag{36}$$

$$B = \frac{\hbar^2}{2m_0} + \frac{\hbar^2}{m_0^2} \sum_j {}' \frac{|\langle X|p_y|j\rangle|^2}{\varepsilon_1 - \varepsilon_j} . \tag{37}$$

The same procedure can be used for the remaining matrix elements, resulting in

$$\mathcal{H} = \begin{bmatrix} \mathcal{E}_1 + Ak_x^2 + B(k_y^2 + k_z^2) & Ck_x k_y & Ck_x k_z \\ Ck_x k_y & \mathcal{E}_1 + Ak_y^2 + B(k_x^2 + k_z^2) & Ck_y k_z \\ Ck_x k_z & Ck_y k_z & \mathcal{E}_1 + Ak_z^2 + B(k_x^2 + k_y^2) \end{bmatrix} , \tag{38}$$

where

$$C = \frac{\hbar^2}{m_0^2} \sum_j {}' \frac{\langle X|p_x|j\rangle\langle j|p_y|Y\rangle + \langle X|p_y|j\rangle\langle j|p_x|Y\rangle}{\varepsilon_1 - \varepsilon_j} . \tag{39}$$

As discussed in Sec. 6-2, to include the effects of spin-orbit coupling, we need to use the eigenstates for the total (orbit and spin) angular momentum, $|j \; m_j\rangle$. In general, the Clebsch-Gordan coefficients relate the old states to the new ones by

$$|l \; s; j \; m_j\rangle = \sum_l^m \sum_s^m |l \; s \; m_l \; m_s\rangle\langle l \; s \; m_l \; m_s|l \; s; j \; m_j\rangle . \tag{40}$$

For $s = \frac{1}{2}$, the Clebsch-Gordan coefficients are given by the table (note that $m_l = m_j - m_s$)

$$\langle l \ s=\tfrac{1}{2} \ m_l \ m_s | l \ s=\tfrac{1}{2}; \ j \ m_j \rangle$$

m_s\j	$l + \tfrac{1}{2}$	$l - \tfrac{1}{2}$
$\tfrac{1}{2}$	$\sqrt{\dfrac{l + m_j + \tfrac{1}{2}}{2l + 1}}$	$-\sqrt{\dfrac{l - m_j + \tfrac{1}{2}}{2l + 1}}$
$-\tfrac{1}{2}$	$\sqrt{\dfrac{l - m_j + \tfrac{1}{2}}{2l + 1}}$	$\sqrt{\dfrac{l + m_j + \tfrac{1}{2}}{2l + 1}}$

Using these coefficients, we find the total angular momentum states for the heavy and light-hole valence bands of GaAs as given by Eq. (6.17).

The spin-orbit potential

$$\frac{1}{2m_0{}^2 c^2} (S \times \nabla V) \cdot p = \xi \ S \cdot L = \frac{\xi}{2} (J^2 - L^2 + S^2) , \qquad (41)$$

removes the degeneracy between states with different total angular momenta. For the $j = 3/2$ and $j = 1/2$ states the energy separation is

$$\Delta \simeq 9 \langle \xi \rangle /8 , \qquad (42)$$

where ξ is the spin-orbit function and $\langle \xi \rangle$ is calculated with the eigenstates of the electrostatic problem. Since the spin-orbit energy is of the order of several hundreds meV, the split-off bands (i.e., the $j = 1/2$ bands) are usually too far removed from the band edge to be directly involved in the optical laser transitions. Therefore we can limit our discussion to the $j = 3/2$ states, which may be written as

$$|\tfrac{3}{2} \ \tfrac{3}{2}\rangle = -2^{-1/2} [|X \uparrow\rangle + i|Y\uparrow\rangle] , \qquad (43a)$$

$$|\tfrac{3}{2} \ \tfrac{1}{2}\rangle = -6^{-1/2} [|X \downarrow\rangle + i|Y\downarrow\rangle] + \sqrt{2/3} \ |Z \uparrow\rangle , \qquad (43b)$$

$$|\tfrac{3}{2} \ -\tfrac{1}{2}\rangle = 6^{-1/2} [|X \uparrow\rangle - i|Y\uparrow\rangle] + \sqrt{2/3} \ |Z \downarrow\rangle , \qquad (43c)$$

$$|\tfrac{3}{2} \ -\tfrac{3}{2}\rangle = 2^{-1/2} [|X \downarrow\rangle - i|Y\downarrow\rangle] , \qquad (43d)$$

where

$$|X \uparrow\rangle = |X\rangle|\uparrow\rangle \, ,$$

etc., and $|X\rangle$, $|Y\rangle$, $|Z\rangle$ are defined by Eq. (6.8). Furthermore, in Eq. (43) we simplified $|l \, s; \, j \, m_j\rangle$ to $|j \, m_j\rangle$ since it is understood that $l = 1$ and $s = 1/2$, and $\uparrow$ indicates $m_s = 1/2$ and $\downarrow$ indicates $m_s = -1/2$, respectively.

With the addition of spin, and taking into account only the $j = 3/2$ states, $\mathcal{H}$ in Eq. (29) becomes a 4×4 matrix, where the matrix elements are computed using Eq. (38). For example, if we arrange the basis states so that $|j\rangle$ for $j = 1, 2, 3$ and 4 is $|\frac{3}{2} \, \frac{3}{2}\rangle$, $|\frac{3}{2} \, -\frac{1}{2}\rangle$, $|\frac{3}{2} \, \frac{1}{2}\rangle$, and $|\frac{3}{2} \, -\frac{3}{2}\rangle$, respectively, then

$$\mathcal{H}_{11} \equiv \langle \tfrac{3}{2} \, \tfrac{3}{2} | \mathcal{H} | \tfrac{3}{2} \, \tfrac{3}{2} \rangle$$

$$= \tfrac{1}{2}(\langle X \uparrow | \mathcal{H} | X \uparrow \rangle + \langle Y \uparrow | \mathcal{H} | Y \uparrow \rangle$$

$$+ \, i \langle X \uparrow | \mathcal{H} | Y \uparrow \rangle - i \langle Y \uparrow | \mathcal{H} | X \uparrow \rangle)$$

$$= \varepsilon_1 + \frac{A}{2}(k_x^2 + k_y^2) + \frac{B}{2}(k_x^2 + k_y^2 + k_z^2) \, . \qquad (44)$$

Here we use the fact that the momentum operator $\mathbf{p}$ does not couple states with different spin orientations, so that matrix elements like $\langle X \uparrow | \mathcal{H} | X \downarrow \rangle$ vanish. In practice, instead of computing the matrix elements $\langle n | \mathbf{p} | m \rangle$ from first principles, one replaces them with experimentally determined parameters called *Luttinger parameters*. There are three Luttinger parameters

$$\gamma_1 = - \, 2m_0(A + 2B)/3\hbar^2 \, , \qquad (45)$$

$$\gamma_2 = - \, m_0(A - B)/3\hbar^2 \, , \qquad (46)$$

$$\gamma_3 = - \, m_0 C/3\hbar^2 \, . \qquad (47)$$

In terms of the Luttinger parameters,

$$\mathcal{H}_{11} = \varepsilon_1 - \frac{\hbar^2 k_z^2}{2m_0}(\gamma_1 - 2\gamma_2) - \frac{\hbar^2(k_x^2 + k_y^2)}{2m_0}(\gamma_1 + \gamma_2) \, . \qquad (48)$$

The bandstructure calculated using the Luttinger parameters is the hole bandstructure because the experiments performed to measure the Luttinger parameters kept track of the hole in an otherwise filled valence band.

Changing to the convention where the hole energy is positive, Eq. (48) becomes

$$\mathcal{H}_{hh} \equiv \mathcal{H}_{11} = -\varepsilon_1 + \frac{\hbar^2 k_z^2}{2m_0}(\gamma_1 - 2\gamma_2) + \frac{\hbar^2(k_x^2 + k_y^2)}{2m_0}(\gamma_1 + \gamma_2) , \quad (49)$$

where the zero-energy reference for $\mathcal{H}_{hh}$ is usually defined such that $\varepsilon_1 = 0$. Repeating the above calculation for the other matrix elements, we find the Luttinger Hamiltonian

$$\begin{array}{cccc} |\tfrac{3}{2}\,\tfrac{3}{2}\rangle & |\tfrac{3}{2}\,-\tfrac{1}{2}\rangle & |\tfrac{3}{2}\,\tfrac{1}{2}\rangle & |\tfrac{3}{2}\,-\tfrac{3}{2}\rangle \end{array}$$

$$\mathcal{H} = \begin{pmatrix} \mathcal{H}_{hh} & -c & -b & 0 \\ -c^* & \mathcal{H}_{lh} & 0 & b \\ -b^* & 0 & \mathcal{H}_{lh} & -c \\ 0 & b^* & -c^* & \mathcal{H}_{hh} \end{pmatrix} \begin{array}{c} |\tfrac{3}{2}\,\tfrac{3}{2}\rangle \\ |\tfrac{3}{2}\,-\tfrac{1}{2}\rangle \\ |\tfrac{3}{2}\,\tfrac{1}{2}\rangle \\ |\tfrac{3}{2}\,-\tfrac{3}{2}\rangle \end{array} \quad (50)$$

Luttinger Hamiltonian

where

$$\mathcal{H}_{lh} = \frac{\hbar^2 k_z^2}{2m_0}(\gamma_1 + 2\gamma_2) + \frac{\hbar^2(k_x^2 + k_y^2)}{2m_0}(\gamma_1 - \gamma_2) , \quad (51)$$

$$c = \frac{\sqrt{3}\hbar^2}{2m_0}[\gamma_2(k_x^2 - k_y^2) - 2i\gamma_3 k_x k_y] , \quad (52)$$

$$b = \frac{\sqrt{3}\hbar^2}{m_0}\gamma_3 k_z(k_x - ik_y) . \quad (53)$$

In the vicinity of $k = 0$, one may use the *axial approximation*, where the Luttinger parameters γ_2 and γ_3 in Eq. (52) are replaced by an effective Luttinger parameter

$$\bar{\gamma} = \tfrac{1}{2}(\gamma_2 + \gamma_3) . \quad (54)$$

The function c then simplifies to

$$c \simeq \frac{\sqrt{3}\hbar^2\bar{\gamma}}{2m_0} (k_x - ik_y)^2 \ . \tag{55}$$

In the axial approximation, the Luttinger Hamiltonian (50) can be transformed into a block diagonal form by a unitary transformation [Broido and Sham (1985)]

$$\mathcal{H}' = \mathcal{U} \, \mathcal{H} \, \mathcal{U}^\dagger \ , \tag{56}$$

where

$$\mathcal{U} = \begin{pmatrix} v^* & 0 & 0 & -v \\ 0 & w^* & -w & 0 \\ 0 & w^* & w & 0 \\ v^* & 0 & 0 & v \end{pmatrix} \tag{57}$$

$$v = 2^{-1/2} \, e^{i(3\pi/4 - 3\xi/2)} \ , \tag{58}$$

$$w = 2^{-1/2} \, e^{i(-\pi/4 + \xi/2)} \ , \tag{59}$$

$$\xi = \mathrm{atan}(k_y / k_x) \ . \tag{60}$$

Evaluating Eq. (56), we find the block-diagonal Luttinger Hamiltonian

$$\mathcal{H}' = \begin{pmatrix} \mathcal{H}^U & 0 \\ 0 & \mathcal{H}^L \end{pmatrix} , \tag{61}$$

where

$$\mathcal{H}^U = \begin{pmatrix} \mathcal{H}_{hh} & R \\ R^* & \mathcal{H}_{lh} \end{pmatrix} , \tag{62}$$

$$\mathcal{H}^L = \begin{pmatrix} \mathcal{H}_{lh} & R \\ R^* & \mathcal{H}_{hh} \end{pmatrix} , \tag{63}$$

$$R = |c| - i|b| \ . \tag{64}$$

The block-diagonal basis is

$$|1\rangle = v^* \left| \tfrac{3}{2} \ \ \tfrac{3}{2} \right\rangle - v \left| \tfrac{3}{2} \ -\tfrac{3}{2} \right\rangle, \tag{65a}$$

$$|2\rangle = w^* \left| \tfrac{3}{2} \ -\tfrac{1}{2} \right\rangle - w \left| \tfrac{3}{2} \ \ \tfrac{1}{2} \right\rangle, \tag{65b}$$

$$|3\rangle = w^* \left| \tfrac{3}{2} \ -\tfrac{1}{2} \right\rangle + w \left| \tfrac{3}{2} \ \ \tfrac{1}{2} \right\rangle, \tag{65c}$$

$$|4\rangle = v^* \left| \tfrac{3}{2} \ \ \tfrac{3}{2} \right\rangle + v \left| \tfrac{3}{2} \ -\tfrac{3}{2} \right\rangle. \tag{65d}$$

Appendix C
ENVELOPE FUNCTION APPROACH

In this appendix we summarize the technical details of the envelope function approximation for semiconductor quantum wells. Furthermore, we evaluate the dipole matrix elements for the light propagating along the x-y plane assuming polarization perpendicular to the plane (TM mode) or in the plane (TE mode), respectively.

C-1. Envelope Function Approximation

The time-independent Schrödinger equation for a quantum-well structure is

$$\left[\frac{p^2}{2m_0} + V_0 + V_{con}\right]|\phi_\lambda^{QW}\rangle = \varepsilon_\lambda \, |\phi_\lambda^{QW}\rangle \, , \tag{1}$$

where V_{con} is the potential due to the heterostructure and λ represents the combination of quantum numbers to be specified later for identifying the quantum well electronic states. Assuming that the effect of V_{con} is to mix the different k states of the bulk material, we write

$$|\phi_\lambda^{QW}\rangle = \sum_{n,\mathbf{k}} |\phi_{n\mathbf{k}}\rangle\langle\phi_{n\mathbf{k}}|\phi_\lambda^{QW}\rangle \, . \tag{2}$$

Here $|\phi_{n\mathbf{k}}\rangle$ is the bulk eigenstate statisfying

$$\left[\frac{p^2}{2m_0} + V_0\right]|\phi_{n\mathbf{k}}\rangle = \varepsilon_{n\mathbf{k}}|\phi_{n\mathbf{k}}\rangle \, , \tag{3}$$

and $\langle\phi_{n\mathbf{k}}|\phi_\lambda^{QW}\rangle$ is the probability amplitude for finding the quantum-well eigenstate $|\phi_\lambda^{QW}\rangle$ in the bulk eigenstate $|\phi_{n\mathbf{k}}\rangle$. In the coordinate representation,

$$\langle \mathbf{r}|\phi_\lambda^{QW}\rangle = \sum_{n,\mathbf{k}} \langle \mathbf{r}|\phi_{n\mathbf{k}}\rangle \langle \phi_{n\mathbf{k}}|\phi_\lambda^{QW}\rangle = \sum_{n,\mathbf{k}} e^{i\mathbf{k}\cdot\mathbf{r}} \langle \mathbf{r}|n\ \mathbf{k}\rangle \langle \phi_{n\mathbf{k}}|\phi_\lambda^{QW}\rangle \ , \quad (4)$$

where we used the Bloch theorem, and $|n\ \mathbf{k}\rangle$ is a lattice periodic eigenstate of the bulk material. Expanding $|n\ \mathbf{k}\rangle$ in terms of the eigenstates at $k = 0$, we find

$$\langle \mathbf{r}|\phi_\lambda^{QW}\rangle = \sum_{m,n,\mathbf{k}} e^{i\mathbf{k}\cdot\mathbf{r}} \langle \mathbf{r}|m\rangle \langle m|n\ \mathbf{k}\rangle \langle \phi_{n\mathbf{k}}|\phi_\lambda^{QW}\rangle$$

$$= \sum_m \left[\sum_{\mathbf{k}} e^{i\mathbf{k}\cdot\mathbf{r}} W_{\lambda m \mathbf{k}} \right] \langle \mathbf{r}|m\rangle \ , \quad (5)$$

where

$$W_{\lambda m \mathbf{k}} = \sum_n \langle m|n\ \mathbf{k}\rangle \langle \phi_{n\mathbf{k}}|\phi_\lambda^{QW}\rangle \ . \quad (6)$$

We note that Eq. (4) may be written as

$$\langle \mathbf{r}|\psi_\lambda^{QW}\rangle = \sum_m W_{\lambda m}(\mathbf{r}) \langle \mathbf{r}|m\rangle \ , \quad (7)$$

where

$$W_{\lambda m}(\mathbf{r}) = \sum_{\mathbf{k}} e^{i\mathbf{k}\cdot\mathbf{r}} W_{\lambda m \mathbf{k}} \ , \quad (8)$$

plays the role of an *envelope function*.

Substituting Eq. (4) into Eq. (1), multiplying the result by $e^{-i\mathbf{k}'\cdot\mathbf{r}} \langle m|\mathbf{r}\rangle$ and integrating over the volume of the crystal, we find

$$\sum_{n,\mathbf{k}'} W_{\lambda n \mathbf{k}} \frac{1}{V} \int_V d^3r \; e^{i(\mathbf{k} - \mathbf{k}') \cdot \mathbf{r}} \langle m | \mathbf{r} \rangle$$

$$\times \left[\frac{\hbar^2 k^2}{2m_0} - \varepsilon_n - \varepsilon_\lambda - \frac{i\hbar^2}{m_0} \mathbf{k} \cdot \nabla + V_{con}(z) \right] \langle r | n \rangle = 0 \; . \tag{9}$$

The spatial integral may be evaluated by writing $\mathbf{r} = \mathbf{R}_l + \rho$, where $\mathbf{R}_l$ is a lattice vector and ρ lies within the unit cell. For example,

$$\frac{1}{V} \int_V d^3r \; e^{i(\mathbf{k} - \mathbf{k}') \cdot \mathbf{r}} \langle m | \mathbf{r} \rangle \langle \mathbf{r} | n \rangle$$

$$= \frac{1}{N} \sum_{\nu=1}^{N} e^{i(\mathbf{k} - \mathbf{k}') \cdot \mathbf{R}_\nu} \frac{1}{v} \int_v d^3\rho \; \exp(i(\mathbf{k} - \mathbf{k}') \cdot \rho) \langle m | \rho \rangle \langle \rho | n \rangle$$

$$= \delta_{\mathbf{k},\mathbf{k}'} \frac{1}{v} \int_v d^3\rho \; \langle m | \rho \rangle \langle \rho | n \rangle = \delta_{\mathbf{k},\mathbf{k}'} \delta_{j,m} \; . \tag{10}$$

Here we note that $\langle \mathbf{R}_\nu - \rho | n \rangle = \langle \rho | n \rangle$ and the crystal volume, $V = Nv$ [compare Eqs. (6.45) – (6.46)]. Similarly,

$$\frac{1}{V} \int_V d^3r \; e^{i(\mathbf{k}-\mathbf{k}') \cdot \mathbf{r}} \langle m | \mathbf{r} \rangle \left(-\frac{i\hbar}{m_0} \mathbf{k} \cdot \nabla \right) \langle r | n \rangle = \delta_{\mathbf{k},\mathbf{k}'} \frac{\hbar}{m_0} \mathbf{k} \cdot \mathbf{p}_{mn} \; , \tag{11}$$

and

$$\frac{1}{V} \int_V d^3r \; e^{i(\mathbf{k} - \mathbf{k}') \cdot \mathbf{r}} \langle m | \mathbf{r} \rangle V_{con}(z) \langle r | n \rangle$$

$$\simeq \frac{1}{N} \sum_{\nu=1}^{N} e^{i(\mathbf{k} - \mathbf{k}') \cdot \mathbf{R}_\nu} V_{con}(Z^\nu) \frac{1}{v} \int_v d^3\rho \; e^{i(\mathbf{k}-\mathbf{k}') \cdot \mathbf{r}} \langle m | \rho \rangle \langle \rho | n \rangle$$

$$\simeq V_{con,\mathbf{k}'-\mathbf{k}} \; \delta_{j,m} \; . \tag{12}$$

Here we abbreviated $\mathbf{p}_{mn} = \langle m|\mathbf{p}|n\rangle$ and

$$V_{con,\mathbf{k'}-\mathbf{k}} = \frac{1}{N} \sum_{\nu=1}^{N} e^{i(\mathbf{k} - \mathbf{k'})\cdot\mathbf{R}_{\nu}} V_{con}(Z^{\nu}) . \tag{13}$$

In getting Eq. (12), we assume that the external potential varies little in a unit cell so that $V_{con}(z) \simeq V_{con}(Z^{\nu})$ and $\exp[i(\mathbf{k'} - \mathbf{k})\cdot\mathbf{r}] \simeq 1$, which is a good approximation in the case when the electronic state mixing is such that $|\mathbf{k} - \mathbf{k'}|$ is much smaller than the reciprocal lattice vector. With these results, Eq. (8) becomes

$$\left[\frac{\hbar^2 k^2}{2m_0} + \varepsilon_m\right] W_{\lambda m \mathbf{k}} + \frac{\hbar}{m_0} \sum_{n} \mathbf{k}\cdot\mathbf{p}_{mn} W_{\lambda n \mathbf{k}} + \sum_{\mathbf{k'}} V_{con,\mathbf{k}-\mathbf{k'}} W_{\lambda m \mathbf{k'}} = \varepsilon_{\lambda} W_{\lambda m \mathbf{k}} . \tag{14}$$

To solve the set of equations given by Eq. (14), let us first consider the case where the band of interest is nondegenerate at $k = 0$. We label that band with the subscript n. Equation (14) for all other bands may be approximated by

$$\left[\frac{\hbar^2 k^2}{2m_0} + \varepsilon_j\right] W_{\lambda j \mathbf{k}} + \frac{\hbar}{m_0} \mathbf{k}\cdot\mathbf{p}_{jn} W_{\lambda n \mathbf{k}} = \varepsilon_{\lambda} W_{\lambda j \mathbf{k}} , \tag{15}$$

where we ignore the effects of V_{con} and consider only the coupling to the nth band. Approximating $\varepsilon_{\lambda} \simeq \varepsilon_n + \hbar^2 k^2/2m_0$, we have

$$W_{\lambda j \mathbf{k}} = \frac{\hbar}{m_0} \frac{\mathbf{k}\cdot\mathbf{p}_{jn}}{\varepsilon_n - \varepsilon_j} W_{\lambda n \mathbf{k}} . \tag{16}$$

Substituting Eq. (16) into Eq. (14) for the nth band gives

$$\left[\frac{\hbar^2 k^2}{2m_0} + \frac{\hbar^2}{m_0^2} \sum_{m} \frac{|\mathbf{k}\cdot\mathbf{p}_{mn}|^2}{\varepsilon_n - \varepsilon_m}\right] W_{\lambda n \mathbf{k}} + \sum_{\mathbf{k'}} V_{con,\mathbf{k}-\mathbf{k'}} W_{\lambda n \mathbf{k'}} = \varepsilon_{\lambda} W_{\lambda n \mathbf{k}} , \tag{17}$$

which can be written in the form

$$\frac{\hbar^2 k^2}{2m_n} W_{\lambda n \mathbf{k}} + \sum_{\mathbf{k}'} V_{con,\mathbf{k}-\mathbf{k}'} W_{\lambda n \mathbf{k}'} = \varepsilon_\lambda W_{\lambda n \mathbf{k}} , \tag{18}$$

where m_n is the effective mass defined in Eq. (6.26). Taking the Fourier transform of Eq. (18) gives

$$\left[-\frac{\hbar^2 \nabla^2}{2m_n} + V_{con}(Z) \right] W_{\lambda n}(\mathbf{R}) = \varepsilon_\lambda W_{\lambda n}(\mathbf{R}) , \tag{19}$$

which we recognize as the equation for a particle in a one-dimensional potential $V_{con}(Z)$. In Eq. (19),

$$W_{\lambda n}(\mathbf{R}) = \sum_{\mathbf{k}} e^{i\mathbf{k}\cdot\mathbf{R}} W_{\lambda n \mathbf{k}} . \tag{20}$$

We now consider the case where the bands of interest are degenerate at $k = 0$. Let us label these bands with the subscripts 1 to N. Then repeating the steps used for the nondegenerate case give

$$W_{\lambda j \mathbf{k}} = \frac{\hbar}{m_0(\varepsilon_1 - \varepsilon_j)} \sum_{n=1}^{N} \mathbf{k}\cdot\mathbf{p}_{jn} W_{\lambda n \mathbf{k}} , \tag{21}$$

where $j > N$. Substituting Eq. (21) into Eq. (14) for the N degenerate bands gives

$$\left[\varepsilon_1 + \frac{\hbar^2 k^2}{2m_0} \right] W_{\lambda n \mathbf{k}} + \frac{\hbar^2}{m_0{}^2} \sum_{m=1}^{N} \sum_{j>N} \frac{\mathbf{k}\cdot\mathbf{p}_{nj}\,\mathbf{k}\cdot\mathbf{p}_{jm}}{\varepsilon_1 - \varepsilon_j} W_{\lambda m \mathbf{k}}$$

$$+ \sum_{\mathbf{k}'} V_{con,\mathbf{k}-\mathbf{k}'} W_{\lambda n \mathbf{k}'} = \varepsilon_\lambda W_{\lambda n \mathbf{k}} , \tag{22}$$

where we assume that the degenerate states have equal parity so that the matrix element of $\mathbf{p}$ between any two degenerate states vanishes. With the exception of the terms containing V_{con}, Eq. (22) is similar to Eq. (6.29) – (6.30), respectively (B.26). If we repeat the degenerate perturbation theory as outlined in App. B, we get for the holes

$$\mathcal{H}_{hh} W_{\lambda 1\mathbf{k}} + R W_{\lambda 2\mathbf{k}} + \sum_{\mathbf{k}'} V_{con,\mathbf{k}-\mathbf{k}'} W_{\lambda 1\mathbf{k}'} = \varepsilon_\lambda W_{\lambda 1\mathbf{k}} , \qquad (23a)$$

$$\mathcal{H}_{lh} W_{\lambda 2\mathbf{k}} + R^* W_{\lambda 1\mathbf{k}} + \sum_{\mathbf{k}'} V_{con,\mathbf{k}-\mathbf{k}'} W_{\lambda 2\mathbf{k}'} = \varepsilon_\lambda W_{\lambda 2\mathbf{k}} , \qquad (23b)$$

$$\mathcal{H}_{lh} W_{\lambda 3\mathbf{k}} + R W_{\lambda 4\mathbf{k}} + \sum_{\mathbf{k}'} V_{con,\mathbf{k}-\mathbf{k}'} W_{\lambda 3\mathbf{k}'} = \varepsilon_\lambda W_{\lambda 3\mathbf{k}} , \qquad (23c)$$

$$\mathcal{H}_{hh} W_{\lambda 4\mathbf{k}} + R^* W_{\lambda 3\mathbf{k}} + \sum_{\mathbf{k}'} V_{con,\mathbf{k}-\mathbf{k}'} W_{\lambda 4\mathbf{k}'} = \varepsilon_\lambda W_{\lambda 4\mathbf{k}} , \qquad (23d)$$

where we used the states given by Eq. (6.76) and $\mathcal{H}_{hh}$, $\mathcal{H}_{lh}$, and R are defined in Eqs. (6.54), (6.55), and (6.67), respectively. The symmetrized Fourier transform of Eqs. (23) is

$$\left[-\frac{\partial}{\partial Z} \frac{\hbar^2}{2m_{hhZ}} \frac{\partial}{\partial Z} - \frac{\hbar^2}{2m_{hh\perp}} \left(\frac{\partial^2}{\partial X^2} + \frac{\partial^2}{\partial Y^2} \right) + V_{con}(Z) \right] W_{\lambda n}(\mathbf{R})$$

$$+ \frac{\sqrt{3}\hbar^2}{2m_0} k_\perp \left[\gamma_2 k_\perp - 2\gamma_3 \frac{d}{dZ} \right] W_{\lambda m}(\mathbf{R}) = \varepsilon_{\lambda 1} W_{\lambda n}(\mathbf{R}) . \qquad (24a)$$

for $n,m = 1,2$ and $4,3$ and

$$\left[-\frac{\partial}{\partial Z} \frac{\hbar^2}{2m_{lhZ}} \frac{\partial}{\partial Z} - \frac{\hbar^2}{2m_{lh\perp}} \left(\frac{\partial^2}{\partial X^2} + \frac{\partial^2}{\partial Y^2} \right) + V_{con}(Z) \right] W_{\lambda n}(\mathbf{R})$$

$$+ \frac{\sqrt{3}\hbar^2}{2m_0} k_\perp \left[\gamma_2 k_\perp - 2\gamma_3 \frac{\partial}{\partial Z} \right] W_{\lambda m}(\mathbf{R}) = \varepsilon_{\lambda 2} W_{\lambda n}(\mathbf{R}) , \qquad (24b)$$

for $n,m = 2,1$ and $3,4$. The effective masses are

$$m_{hhz} = m_0/(\gamma_1 - 2\gamma_2) , \qquad (25a)$$

$$m_{lhz} = m_0/(\gamma_1 + 2\gamma_2) \,, \tag{25b}$$

$$m_{hh\perp} = m_0/(\gamma_1 + \gamma_2) \,, \tag{25c}$$

$$m_{lh\perp} = m_0/(\gamma_1 - \gamma_2) \,. \tag{25d}$$

C-2. Dipole Matrix Elements

In this section we summarize the details of the calculations needed to obtain the dipole matrix elements for TE and TM polarization. The wavefunctions for the conduction bands are

$$\langle \mathbf{r} \mid \phi^e_{l\mathbf{k}_\perp} \rangle = e^{i\mathbf{k}_\perp \cdot \mathbf{R}_\perp} C_l(Z) \langle \mathbf{r} \mid S\uparrow \rangle \tag{26a}$$

$$\langle \mathbf{r} \mid \phi^e_{l\mathbf{k}_\perp} \rangle = e^{i\mathbf{k}_\perp \cdot \mathbf{R}_\perp} C_l(Z) \langle \mathbf{r} \mid S\downarrow \rangle \,, \tag{26b}$$

for the states with spin $\tfrac{1}{2}$ and $-\tfrac{1}{2}$, respectively. Correspondingly, the wavefunctions for the hole states are

$$\langle \mathbf{r} \mid \phi^h_{n\mathbf{k}_\perp} \rangle = e^{i\mathbf{k}_\perp \cdot \mathbf{R}_\perp} \sum_{m=1}^{2} \sum_{n_m=1}^{N_m} A_{n_m m}(Z) \langle \mathbf{r} \mid m \rangle \,, \tag{27a}$$

for the upper block and

$$\langle \mathbf{r} \mid \phi^h_{n\mathbf{k}_\perp} \rangle = e^{i\mathbf{k}_\perp \cdot \mathbf{R}_\perp} \sum_{m=3}^{4} \sum_{n_m=1}^{N_m} A_{n_m m}(Z) \langle \mathbf{r} \mid m \rangle \,, \tag{27b}$$

for the lower block of the Luttinger Hamiltonian (6.64).

For the TM mode we have to evaluate the matrix element

$$\mu_{TM} = \langle \phi^e_{l\mathbf{k}_\perp} \mid ez \mid \phi^h_{n\mathbf{k}_\perp} \rangle$$

$$= \frac{1}{N} \sum_\nu \sum_{m=1}^{4} \sum_{n_m=1}^{N_m} C^*_l(Z^\nu) A_{n_m m}(Z^\nu) \int_\nu \frac{d^3 r}{v} \langle Sm_s \mid r \rangle \, ez \, \langle r \mid m \rangle \,, \tag{28}$$

where we split the integral over all space into the sum over unit cells ν and the integral within the unit cells, Eq. (6.45). The m-summation in Eq. (28) runs over the four $j=3/2$ bulk hole states and the n_m summation runs over the number of confined quantum well states, respectively. As shown in App. B, Eq. (B.66), the hole states $\langle r|m \rangle$ of the block diagonal Luttinger Hamiltonian can be expressed as

$$\langle r|m \rangle = \sum_{m_j=-3/2}^{3/2} a_{m,m_j} \langle r|3/2\, m_j \rangle \,, \tag{29}$$

where a_{m,m_j} is either v, v^*, w, w^*, see Eqs. (B.59), (B.60), and (B.66). The states $|3/2\, m_j \rangle$ in turn are related to the states $|q\, s_z \rangle$ as given by Eq. (B.44),

$$\langle r|3/2\, m_j \rangle = \sum_{q=X,Y,Z} \sum_{s_z=-1/2}^{1/2} \alpha_{m_j,q,s_z} \langle r|q,s_z \rangle \,, \tag{30}$$

where α_{m_j,q,s_z} are the Clebsh-Gordan coefficients evaluated in Eq. (B.44). Using Eqs. (29) and (30) in (28), we obtain

$$\mu_{TM} = \frac{1}{N} \sum_{\nu} \sum_{m} \sum_{n_m} C_l^*(Z^\nu)\, A_{n_m,m}(Z^\nu) \sum_{j}^{m} \sum_{q} \sum_{s'}^{m} \langle s\, m_s|ez|q\, m_{s'} \rangle \,, \tag{31}$$

where we defined

$$\langle S\, m_s|ez|q\, m_{s'} \rangle = \int_\nu \frac{d^3r}{\nu} \langle S m_s|r \rangle\, ez \langle r|q,m_{s'} \rangle \,. \tag{32}$$

For the case of conduction band spin ↑ the only non-vanishing matrix element is $\langle S\uparrow|ez|Z\uparrow \rangle$, as can be verified using the representation (6.9) and the fact that the state $|S\uparrow \rangle$ has spherical symmetry. From Eq. (B.44b) we see that

$$\alpha_{1/2,Z,\uparrow} = \sqrt{\frac{2}{3}} \,, \tag{33}$$

and from Eq. (B.66) we obtain

$$a_{2,1/2} = - w \quad \text{and} \quad a_{3,1/2} = w \,, \tag{34}$$

respectively. Hence we get

$$\mu_{TM} = - w \sqrt{\frac{2}{3}} \frac{1}{N} \sum_{\nu} \sum_{n_2} C_l^*(Z^\nu) A_{n_2,2}(Z^\nu) \langle S \uparrow | ez | Z \uparrow \rangle \tag{35a}$$

for the hole state $|2\rangle$ and

$$\mu_{TM} = w \sqrt{\frac{2}{3}} \frac{1}{N} \sum_{\nu} \sum_{n_3} C_l^*(Z^\nu) A_{n_3,3}(Z^\nu) \langle S \uparrow | ez | Z \uparrow \rangle \tag{35b}$$

for the hole state $|3\rangle$. To compute the quantum-well gain and refractive index we always need the absolute square of the respective matrix element. Using Eq. (35) we get

$$|\mu_{TM}|^2 = \frac{1}{3} \left| \sum_{n_m=1}^{N_m} \langle A_{n_m} | C_l \rangle \right|^2 |\langle S \uparrow | ez | Z \uparrow \rangle|^2 \tag{36}$$

for the conduction band with spin $\uparrow$, and $m = 2, 3$, where we used Eq. (B.60) to see that

$$|w|^2 = \tfrac{1}{2} \,, \tag{37}$$

$$\langle A_{n_m} | C_l \rangle = \frac{1}{N} \sum_{\nu} C_l^*(Z^\nu) A_{n_m m}(Z^\nu) \,. \tag{38}$$

For the TE mode we have to evaluate

$$\mu_{TE} = \langle \phi_{l,\mathbf{k}_\perp}^e | ex | \phi_{n\mathbf{k}_\perp}^h \rangle \,. \tag{39}$$

Using the same arguments as before, we can convince ourselves that for the case of conduction band spin ↑ the only nonvanishing matrix element is $\langle S\uparrow|ex|X\uparrow\rangle$. Furthermore, we use the results from App. B to get

$$\alpha_{3/2,X,\uparrow} = -\sqrt{\tfrac{1}{2}} \quad \text{and} \quad \alpha_{-1/2,X,\uparrow} = \sqrt{\tfrac{1}{6}} . \tag{40}$$

From Eq. (B.66) we obtain

$$a_{1,3/2} = v^* = a_{4,3/2} \quad \text{for } m_j = \tfrac{3}{2}$$

$$a_{2,-1/2} = w^* = a_{3,-1/2} \quad \text{for } m_j = -\tfrac{1}{2} . \tag{41}$$

Therefore,

$$\mu_{TE} = \frac{1}{N} \sum_{\nu} c_l^*(Z^\nu) \left[-\sqrt{\tfrac{1}{2}} \sum_{n_1} A_{n_1,1}(Z^\nu)\, v^* \right.$$

$$\left. + \sqrt{\tfrac{1}{6}} \sum_{n_2} A_{n_2,2}(Z^\nu)\, w^* \right] \langle S\uparrow|ex|X\uparrow\rangle \tag{42a}$$

for the upper block and

$$\mu_{TE} = \frac{1}{N} \sum_{\nu} c_l^*(Z^\nu) \left[-\sqrt{\tfrac{1}{2}} \sum_{n_4} A_{n_4,4}(Z^\nu)\, v^* \right.$$

$$\left. + \sqrt{\tfrac{1}{6}} \sum_{n_3} A_{n_3,3}(Z^\nu)\, w^* \right] \langle S\uparrow|ex|X\uparrow\rangle \tag{42b}$$

for the lower block, respectively. For the square of the TE dipole matrix element we get

$$|\mu_{TE}|^2 = \frac{1}{4} |\langle S \uparrow |ex| X \uparrow \rangle|^2 \left[\left| \sum_{n_1=1}^{N_1} \langle A_{n_1} | C_l \rangle \right|^2 + \frac{1}{3} \left| \sum_{n_2=1}^{N_2} \langle A_{n_2} | C_l \rangle \right|^2 \right.$$

$$\left. + \frac{2}{\sqrt{3}} \left(\sum_{n_1=1}^{N_1} \langle C_l | A_{n_1} \rangle \right) \left(\sum_{n_2=1}^{N_2} \langle A_{n_2} | C_l \rangle \right) \cos(2\phi) \right] . \qquad (43)$$

Here we used

$$v^* w = -\frac{1}{2} e^{i2\phi} , \qquad (44)$$

where ϕ is the angle between $\mathbf{k}$ and $\mathbf{x}$. For the lower block of the Luttinger Hamiltonian we have to replace $1 \rightarrow 4$ and $2 \rightarrow 3$.

Appendix D
STRAIN EFFECTS

Restricting ourselves to the ideal case of a single quantum well with cubic symmetry and bulk lattice constant a_w, which grows under elastic strain in the x-y plane, between barriers of lattice constant a_b, the in-plane components of the strain tensor e_{ij} are given by the lattice mismatch

$$e_{xx} = e_{yy} = \frac{a_b - a_w}{a_b} \equiv e_0 .$$ (1)

As a consequence of the structural geometry, all strain components

$$e_{\alpha\beta} = 0 \quad \text{for} \quad \alpha \neq \beta ,$$ (2)

so that the only other non-vanishing component is e_{zz}. To determine e_{zz} we make use of the fact that there is no net force acting pependicular to the quantum-well plane. This force is

$$Z_z = C_{12} e_{xx} + C_{12} e_{yy} + C_{11} e_{zz} ,$$ (3)

where the capital Z denotes the force acting in z-direction and the index is the normal of the plane, which in this case is also in the z-direction (see e.g. Kittel, 1971, Chap. 4). The quantities C_{ij} are the *elastic moduli* or *elastic stiffness constants*. Without going into detail, we just note that cubic symmetry requires that only three, i.e. C_{11}, C_{12}, and C_{44}, of the generally possible twenty-one C_{ij} are unequal to zero. The values of these material dependent constants can be found e.g. in Landolt-Börnstein (1982). For the condition that $Z_z = 0$, Eq. (3) immediately yields

$$e_{zz} = -2 \frac{C_{12}}{C_{11}} e_0 ,$$ (4)

where (1) has been used.

Knowing the components of the strain tensor, we now proceed with the analysis of the bandstructure modifications. Clearly, in the strained material an equation like Eq. (6.5) holds, where however all space variables are now in the strained system. Since we always assume small amounts of strain, we can expand the functions of the new variables in terms of the old, unstrained variables. In terms of the vector $\mathbf{r}$ in the unstrained lattice, the component α of the vector $\mathbf{r}'$ of the strained lattice is

$$r'_\alpha = \sum_\beta (\delta_{\alpha\beta} + e_{\alpha\beta}) \, r_\beta \; . \tag{5}$$

Ignoring all terms of order $\mathcal{O}(e^2)$ or higher, we can write

$$\frac{\partial}{\partial r'_\alpha} = \sum_\beta \frac{\partial r_\beta}{\partial r'_\alpha} \frac{\partial}{\partial r_\beta} = \sum_j \frac{1}{\delta_{\alpha\beta} + e_{\alpha\beta}} \frac{\partial}{\partial r_\beta} \simeq \sum_\beta (\delta_{\alpha\beta} - e_{\alpha\beta}) \frac{\partial}{\partial r_\beta} \; , \tag{6}$$

and correspondingly

$$\frac{\partial^2}{\partial r'^2_\alpha} \simeq \frac{\partial^2}{\partial r^2_\alpha} - 2 \sum_\beta e_{\alpha\beta} \frac{\partial^2}{\partial r_\alpha \partial r_\beta} \; , \tag{7}$$

$$k_\alpha p'_\alpha \simeq k_\alpha p_\alpha - \sum_\beta k_\alpha \, e_{\alpha\beta} \, p_\beta \; , \tag{8}$$

and

$$V_0(r') \simeq V_0(r) + \sum_{\alpha\beta} V_{\alpha\beta} \, e_{\alpha\beta} \; , \tag{9}$$

where $V_{\alpha\beta}$ is the derivative of the lattice periodic potential $V_0(\mathbf{r}')$ with respect to $e_{\alpha\beta}$.

Inserting these expansions into Eq. (6.5), we obtain

$$\left[- \frac{\hbar^2}{2m_0} \nabla^2 + V_0(\mathbf{r}) + \frac{\hbar}{m_0} \mathbf{k} \cdot \mathbf{p} + \sum_{\alpha\beta} \mathcal{S}^{\alpha\beta} e_{\alpha\beta} \right] |n\,\mathbf{k}\rangle$$

$$= \left[\varepsilon_{n\mathbf{k}} - \frac{\hbar^2 k^2}{2m_0} \right] |n\,\mathbf{k}\rangle , \qquad (10)$$

where

$$\mathcal{S}^{\alpha\beta} = \frac{\hbar}{m_0} k_\alpha p_\beta + \frac{\hbar^2}{m_0} e_{\alpha\beta} \frac{\partial^2}{\partial r_\alpha \partial r_\beta} + V_{\alpha\beta} . \qquad (11)$$

The comparison shows that Eqs. (11) and (6.5) differ only by the additional term proportional to the strain tensor elements. Hence, we can now repeat all steps of the previous sections including the additional strain terms. For example, in the $\mathbf{k} \cdot \mathbf{p}$ theory of Sec. 6-3 we obtain the modified Eq. (C.23) as

$$\sum_j W_{nj}(\mathbf{k}) \left[\left\{ \frac{\hbar^2 k^2}{2m_0} + \mathcal{E}_j(0) - E_n(\mathbf{k}) \right\} \delta_{ij} + \frac{\hbar}{m_0} \mathbf{k} \cdot \mathbf{p}_{ij} + \mathcal{S}_{ij} \right] = 0 , \qquad (12)$$

where

$$\mathcal{S}_{ij} = \sum_{\alpha\beta} \mathcal{S}^{\alpha\beta}_{ij} e_{\alpha\beta} \quad \text{and} \quad \mathcal{S}^{\alpha\beta}_{ij} = \int \frac{d^3 r}{L^3} u_i^*(0,\mathbf{r}) \, \mathcal{S}^{\alpha\beta} \, u_j(0,\mathbf{r}) . \qquad (13)$$

In the subsequent perturbation theory we keep terms of order k, k^2, as in Sec. 6-3, but now we include also terms linear in the strain tensor, $\mathcal{O}(e)$. We ignore all contributions containing products of k and strain tensor, or higher orders. This way, the effective $\mathbf{k} \cdot \mathbf{p}$ Hamiltonian is generalized as

$$\mathcal{H}_{ij} = \mathcal{E}_0 \, \delta_{ij} + \frac{\hbar}{m_0} \mathbf{k} \cdot \mathbf{p}_{ij} + \sum_{\alpha,\beta = x,y,z} (D^{\alpha\beta}_{ij} k_\alpha k_\beta + \mathcal{S}^{\alpha\beta}_{ij} e_{\alpha\beta}) .$$

$$(14)$$

Equation (14) shows a one-to-one correspondence between the terms proportional to $k_\alpha k_\beta$ and to $e_{\alpha\beta}$. Hence, we can proceed to generalize the Luttinger Hamiltonian simply by adding the proper $e_{\alpha\beta}$ terms to the $k_\alpha k_\beta$ ones. As in the unstrained case, where we introduced the empirical Luttinger parameters, we do not attempt to explicitly compute the matrix elements entering the strain part of the Hamiltonian. Instead, we introduce new parameters, the so-called *hydrostatic* and *shear deformation potentials*, which are obtained from experiments.

The strain induced addition to the term b, Eq. (B.53), vanishes for the present example,

$$b_{strain} \propto (e_{zx} - ie_{zy}) = 0 , \tag{15}$$

since $e_{zx} = e_{zy} = 0$, see Eq. (2). The correction to the term c, Eq. (B.52), vanishes as well,

$$c_{strain} = c_1 (e_{xx} - e_{yy}) - c_2 e_{xy} = 0 , \tag{16}$$

since $e_{xx} = e_{yy}$ and $e_{xy} = 0$. Hence, we have only strain corrections to the diagonal terms in the Luttinger Hamiltonian

$$\mathcal{H}_{total} = \mathcal{H} + \mathcal{H}_{strain} , \tag{17}$$

where

$$\mathcal{H}_{strain} = \begin{bmatrix} \mathcal{H}_{strain,hh} & 0 & 0 & 0 \\ 0 & \mathcal{H}_{strain,lh} & 0 & 0 \\ 0 & 0 & \mathcal{H}_{strain,lh} & 0 \\ 0 & 0 & 0 & \mathcal{H}_{strain,hh} \end{bmatrix} . \tag{18}$$

Using the one-to-one correspondence between the terms proportional to $k_\alpha k_\beta$ and to $e_{\alpha\beta}$, and replacing the parameter combinations $\hbar^2\gamma_1/2m_0$ and $\hbar^2\gamma_2/2m_0$ in Eqs. (B.49) and (B.51) by a_1 and a_2, respectively, we write the strain corrections $\mathcal{H}_{strain,hh}$ and $\mathcal{H}_{strain,lh}$ in complete symmetry to the respective unstrained parts:

$$\mathcal{H}_{strain,hh} = - e_{zz} (a_1 - 2a_2) - (e_{xx} + e_{yy}) (a_1 + a_2) \tag{19}$$

and

$$\mathcal{H}_{strain,lh} = - e_{zz} (a_1 + 2a_2) - (e_{xx} + e_{yy}) (a_1 - a_2) . \tag{20}$$

Reordering the terms and using Eqs. (6.97) and (6.98), we obtain

$$\mathcal{H}_{strain,hh} = - a_1 (e_{xx} + e_{yy} + e_{zz}) - a_2 (e_{xx} + e_{yy} - 2e_{zz})$$

$$= - 2a_1 e_0 \frac{C_{11} - C_{12}}{C_{11}} - 2a_2 e_0 \frac{C_{11} + 2C_{12}}{C_{11}}$$

$$\equiv - \delta\varepsilon_H - \frac{1}{2} \delta\varepsilon_S , \qquad (21)$$

and

$$\mathcal{H}_{strain,lh} = - \delta\varepsilon_H + \frac{1}{2} \delta\varepsilon_S , \qquad (22)$$

respectively.

REFERENCES

Kittel, C. (1971), *Introduction to Solid State Physics*, Wiley & Sons, New York; Kittel, C. (1967) *Quantum Theory of Solids*, Wiley & Sons, New York.

Landolt-Börnstein (1982), *Numerical Data and Functional Relationships in Science and Technology*, ed. K.H. Hellwege, Vol. 17 Semiconductors, edited by O. Madelung, M. Schulz and H. Weiss, Springer Verlag, Berlin.

Appendix E
SOME LANGEVIN GOODIES

This appendix derives diffusion coefficients and other quantities needed for the noise spectra discussions of Sec. 9-5.

We transform the diffusion coefficients using Eq. (9.76). For $\langle D_{EE} \rangle$, we compute

$$\langle D_{EE} \rangle = \frac{\partial E}{\partial A^\dagger} \frac{\partial E}{\partial A} \langle D_{A^\dagger A} \rangle + \frac{\partial E}{\partial A} \frac{\partial E}{\partial A^\dagger} \langle D_{A A^\dagger} \rangle$$

$$= \frac{\mathscr{E}^4}{4E^2} [A A^\dagger \langle D_{A^\dagger A} \rangle + A^\dagger A \langle D_{A A^\dagger} \rangle]$$

$$\simeq \tfrac{1}{4} \mathscr{E}^2 [\langle D_{A^\dagger A} \rangle + \langle D_{A A^\dagger} \rangle] . \tag{1}$$

Using Eqs. (9.95) and (9.97), we find

$$\langle D_{EE} \rangle = \frac{\mathscr{E}^2}{4} \left[\frac{\nu}{Q} [\bar{n}(\nu) + \tfrac{1}{2}] + \tfrac{1}{2}(R_{sp} + R_{abs}) \right] .$$

We can simplify this expression by using Eq. (9.100) in which we neglect the noise terms, i.e., use the steady-state semiclassical oscillation condition $\nu/2Q = \tfrac{1}{2}(R_{sp} - R_{abs})$, and we neglect $\bar{n}(\nu)$, since we're interested only in optical frequencies. This gives

$$\langle D_{EE} \rangle = \tfrac{1}{4} \mathscr{E}^2 R_{sp} . \tag{2}$$

In Sec. 9-5, we calculate the Relative Intensity Noise (RIN), for which we need the intensity diffusion coefficient $\langle D_{II} \rangle$, where $I = E^2 = \mathscr{E}^2 \langle A^\dagger A \rangle$. Using I instead of E in Eq. (1), we find

$$\langle D_{II} \rangle = \mathscr{E}^2 I [\langle D_{A^\dagger A} \rangle + \langle D_{A A^\dagger} \rangle] = \mathscr{E}^2 I R_{sp} . \tag{3}$$

To find the phase diffusion coefficient, we use

$$\frac{\partial \phi}{\partial A} = \frac{i}{2A} , \quad \frac{\partial \phi}{\partial A\dagger} = -\frac{i}{2A\dagger} , \tag{4}$$

together with the simplifications leading to Eq. (2) to find

$$\langle D_{\phi\phi} \rangle = \frac{1}{4A\dagger A}\langle D_{A A\dagger} \rangle + \frac{1}{4A A\dagger}\langle D_{A\dagger A} \rangle \simeq \frac{\mathcal{E}^2 R_{sp}}{4E^2} . \tag{5}$$

Furthermore, we also see that

$$\langle D_{E\phi} \rangle = \frac{\partial E}{\partial A\dagger}\frac{\partial \phi}{\partial A}\langle D_{A\dagger A} \rangle + \frac{\partial E}{\partial A}\frac{\partial \phi}{\partial A\dagger}\langle D_{A A\dagger} \rangle$$

$$= \frac{i\mathcal{E}A}{4EA}\langle D_{A\dagger A} \rangle - \frac{i\mathcal{E}A\dagger}{4EA\dagger}\langle D_{A A\dagger} \rangle = \frac{i\mathcal{E}}{4E}[\langle D_{A\dagger A} \rangle - \langle D_{A A\dagger} \rangle]$$

$$= \frac{i\mathcal{E}}{4E}[R_{sp} - R_{abs} - \nu/Q] \simeq 0 . \tag{6}$$

The final values of the diffusion coefficients in Eqs. (2) and (5) are the ones used in much of the semiconductor laser literature. These values are appropriate for fluctuations about the steady-state operating point.

Using the results of Sec. 9-5, we find the autocorrelation of $F_N(t)$ to be

$$\langle F_N(t)F_N(t') \rangle = \frac{1}{V^2} \sum_{\mathbf{k}} \frac{|g_{\mathbf{k}}|^2}{\gamma^2}\mathcal{L}_{\mathbf{k}}\left[A\dagger A \langle F_{\sigma_{\mathbf{k}}}(t)F_{\sigma_{\mathbf{k}}}^{\dagger}(t') \rangle + A A\dagger \langle F_{\sigma_{\mathbf{k}}}^{\dagger}(t)F_{\sigma_{\mathbf{k}}}(t') \rangle \right]$$

$$+ \frac{1}{V^2} \sum_{\mathbf{k}} \langle F_{e\mathbf{k}}(t)F_{e\mathbf{k}}(t') \rangle$$

$$= \frac{1}{V^2}\left[A\dagger A[R_{sp} + R_{abs}] + 2\sum_{\mathbf{k}} \langle D_{ee} \rangle \right]\delta(t - t') ,$$

that is,

$$2\langle D_{NN}\rangle = \frac{I}{\mathscr{E}^2 V^2}[R_{sp} + R_{abs}] + \frac{2}{V^2}\sum_{\mathbf{k}}\langle D_{ee}\rangle \; . \tag{7}$$

The carrier density N and the annihilation operator are correlated through their common dependence on the dipole noise operator. Using Eqs. (9.93) and (9.104), we find

$$\langle F_N(t)F_A(t')\rangle = -\frac{1}{V}\sum_{\mathbf{k}}\frac{|g_{\mathbf{k}}|^2}{\gamma^2}A\mathscr{L}_{\mathbf{k}}\langle F_{\sigma_{\mathbf{k}}}^{\dagger}(t)F_{\sigma_{\mathbf{k}}}(t')\rangle = -\frac{AR_{sp}}{V}\delta(t - t') \; ,$$

which gives

$$2\langle D_{NA}\rangle = -AR_{sp}/V \; . \tag{8}$$

Similarly

$$\langle F_N(t)F_A^{\dagger}(t')\rangle = -\frac{1}{V}\sum_{\mathbf{k}}\frac{|g_{\mathbf{k}}|^2}{\gamma^2}A^{\dagger}\mathscr{L}_{\mathbf{k}}\langle F_{\sigma_{\mathbf{k}}}(t)F_{\sigma_{\mathbf{k}}}^{\dagger}(t')\rangle$$

$$= -\frac{A^{\dagger}R_{abs}}{V}\delta(t - t') \; ,$$

which gives

$$2\langle D_{NA^{\dagger}}\rangle = -A^{\dagger}R_{abs}/V \; . \tag{9}$$

Similarly, $2\langle D_{A^{\dagger}N}\rangle = -A^{\dagger}R_{sp}/V$ and $2\langle D_{AN}\rangle = -AR_{abs}/V$. Using Eq. (9.82), we have

$$\langle D_{NE}\rangle = \frac{\partial N}{\partial N}\frac{\partial E}{\partial A}\langle D_{NA}\rangle + \frac{\partial N}{\partial N}\frac{\partial E}{\partial A^{\dagger}}\langle D_{NA^{\dagger}}\rangle = -\frac{\mathscr{E}^2 A^{\dagger}A}{4E}R_{sp} - \frac{\mathscr{E}^2 AA^{\dagger}}{4E}R_{abs}$$

$$\simeq -\frac{E}{4V}[R_{sp} + R_{abs}] = \langle D_{EN}\rangle \; . \tag{10}$$

Alternatively for $I = \mathscr{E}^2 A^{\dagger}A$, we find

$$\langle D_{NI} \rangle \simeq -\frac{I}{2V}[R_{sp} + R_{abs}] = \langle D_{IN} \rangle . \tag{11}$$

Finally, we have using Eqs. (9.89), (9.102), (9.104), and (4)

$$\langle D_{N\phi} \rangle = \frac{\partial N}{\partial N}\frac{\partial \phi}{\partial A}\langle D_{NA} \rangle + \frac{\partial N}{\partial N}\frac{\partial \phi}{\partial A\dagger}\langle D_{NA\dagger} \rangle$$

$$= -\frac{i}{4VA}AR_{sp} + \frac{i}{4VA\dagger}A\dagger R_{abs} = -\frac{i}{4V}[R_{sp} - R_{abs}] = -i\frac{\nu}{4QV} . \tag{12}$$

In the literature, $\langle D_{NI} \rangle$ is usually taken equal to $-IR_{sp}/V$ and $\langle D_{N\phi} \rangle = 0$, but the arguments we've found leading to these values contain more intuition than one would hope for.

Phase Diffusion for Gaussian Noise

To show for a Gaussian noise process that $\langle e^{i\Delta\phi(t)} \rangle = \exp\{-\frac{1}{2}\langle[\Delta\phi(t)]^2\rangle\}$, we expand the phase exponential as

$$\langle e^{i\Delta\phi(t)} \rangle = \sum_{n=0}^{\infty} \frac{i^n \langle[\Delta\phi(t)]^n\rangle}{n!} . \tag{13}$$

We assume the phase fluctuations are due to a Gaussian random process with a zero mean, which has the property that the average of an nth-order product can be written as the sum of the products of all pairs of terms. Odd power expressions always include the average of one unpaired term, which has a zero average, so Eq. (13) reduces to

$$\langle e^{i\Delta\phi(t)} \rangle = \sum_{n=0}^{\infty} \frac{(-1)^n}{2n!}\langle[\Delta\phi(t)]^{2n}\rangle . \tag{14}$$

The number of combinations of $2n$ terms in pairs is given by $\binom{2n}{2}$, that for the remaining $2n-2$ terms is $\binom{2n-2}{2}$, and so forth. Hence the total number of distinguishable ways of breaking $2n$ terms into products of n pairs is

$$\frac{1}{n!}\binom{2n}{2}\binom{2n-2}{2}\cdots\binom{2}{2} = \frac{(2n)!}{n!2^n} .$$

Substituting this into Eq. (14), we have

$$\langle e^{i\Delta\phi(t)}\rangle = \sum_{n=0}^{\infty} \frac{\{-\frac{1}{2}\langle[\Delta\phi(t)]^2\rangle\}^n}{n!} = \exp\{-\frac{1}{2}\langle[\Delta\phi(t)]^2\rangle\} \tag{15}$$

as needed in Eq. (9.123).

Contour Evaluation of Phase Diffusion

Equation (9.126) can be evaluated by contour integration for $t > 0$ by integrating around a contour closed in the upper-half plane. The double pole at $\omega = 0$ is conveniently evaluated by writing $\omega^2 = (\omega + iw)(\omega - iw)$ and setting $w = 0$ after evaluation. Writing

$$|\Delta|^2 = (\omega + i\Gamma + \Omega)(\omega + i\Gamma - \Omega)(\omega - i\Gamma + \Omega)(\omega - i\Gamma - \Omega) ,$$

where $\Gamma = \frac{1}{2}\Gamma_1'$ and $\Omega = \sqrt{\Omega_R^2 - \Gamma^2}$, we see that there are also poles at $\omega = \omega_\pm \equiv i\Gamma \pm \Omega$, which are out at the relaxation oscillation sidebands. The pole at $\omega = 0$ gives

$$\langle[\Delta\phi(t)]^2\rangle\bigg|_{\omega=0} = \lim_{w\to 0}\left[\frac{2\langle D_{\phi\phi}\rangle}{w} + \frac{\alpha^2}{w}\frac{\frac{\Omega_R^4}{E^2}2\langle D_{EE}\rangle - \frac{w^2 G_N^2}{2}\langle D_{NN}\rangle}{\Omega_R^4}\right](1 - e^{-wt})$$

$$= \frac{R_{sp}}{2I}(1 + \alpha^2)t . \tag{16}$$

This contribution results from slow fluctuations that don't depend on the carrier dynamics and affect the spectrum around $\omega \simeq 0$. Notice the presence of the linewidth enhancement factor α^2.

To evaluate the contributions for the two remaining poles at $\omega = \omega_\pm \equiv i\Gamma \pm \Omega$, we define δ such that

$$e^{i\delta} = \frac{\Omega + i\Gamma}{\Omega^2 + \Gamma^2} = \frac{\Omega + i\Gamma}{\Omega_R^2}, \tag{17}$$

in terms of which $\omega_{\pm} = i\Gamma \pm \Omega = \pm(\mp i\Gamma + \Omega) = \pm\Omega_R \exp(\mp i\delta)$. Since $\Omega_R^2 \propto I$, $\langle D_{\phi\phi}\rangle/\omega_{\pm}^2$ is smaller than the other terms by a factor $1/I$ and can be neglected above threshold. On the other hand,

$$\left.\frac{\omega - \omega_{\pm}}{|\Delta|^2}\right|_{\omega=\omega_{\pm}} = \frac{\pm 1}{2\omega_{\pm}\cdot 2i\Gamma\cdot 2\Omega} = \frac{\pm 1}{8i\Gamma\Omega\omega_{\pm}} = \frac{e^{\pm i\delta}}{8i\Gamma\Omega\Omega_R}$$

so that

$$\left.\frac{\alpha^2\Omega_R^4 2\langle D_{EE}\rangle(\omega - \omega_{\pm})}{E^2\omega_{\pm}^2|\Delta|^2}\right|_{\omega=\omega_{\pm}} = \alpha^2 R_{sp}\frac{\Omega_R^4 e^{\pm 3i\delta}}{16i\Gamma\Omega\Omega_R^3} = \alpha^2 R_{sp}\frac{e^{\pm 3i\delta}}{16i\Gamma\cos\delta},$$

$$\left.\frac{\alpha^2 G_N^2\langle D_{NN}\rangle(\omega - \omega_{\pm})}{2|\Delta|^2}\right|_{\omega=\omega_{\pm}} = \frac{\alpha^2 G_N^2\langle D_{NN}\rangle e^{\pm i\delta}}{16i\Gamma\Omega_R^2\cos\delta}.$$

Using these values and the residue theorem in Eq. (9.126), we have

$$\left.\langle[\Delta\phi(t)]^2\rangle\right|_{\omega=\omega_{\pm}} = \mathrm{Re}\left\{\frac{\alpha^2[1 - e^{(-\Gamma\pm i\Omega)t}]}{8\Gamma I\cos\delta}\left[R_{sp}e^{\pm 3i\delta} + \frac{I G_N^2\langle D_{NN}\rangle e^{\pm i\delta}}{\Omega_R^2}\right]\right\}$$

$$= \frac{\alpha^2 R_{sp}[\cos 3\delta - e^{-\Gamma t}\cos(\Omega t - 3\delta)]}{8\Gamma I\cos\delta}$$

$$+ \frac{\alpha^2 G_N^2\langle D_{NN}\rangle[\cos\delta - e^{-\Gamma t}\cos(\Omega t - \delta)]}{8\Gamma\Omega_R^2\cos\delta},$$

which is the same for the poles at $\omega = \omega_{\pm}$. Accordingly multiplying by 2 and adding Eq. (16), we have

$$\langle [\Delta\phi(t)]^2 \rangle = \frac{R_{sp}}{2I}(1 + \alpha^2)t + \frac{\alpha^2 R_{sp}[\cos 3\delta - e^{-\Gamma t}\cos(\Omega t - 3\delta)]}{4\Gamma I \cos\delta}$$

$$+ \frac{\alpha^2 G_N^2 \langle D_{NN} \rangle [\cos\delta - e^{-\Gamma t}\cos(\Omega t - \delta)]}{4\Gamma \Omega_R^2 \cos\delta}$$

$$\simeq \frac{R_{sp}}{2I}\left[(1 + \alpha^2)t + \frac{\alpha^2 [\cos 3\delta - e^{-\Gamma t}\cos(\Omega t - 3\delta)]}{2\Gamma \cos\delta}\right], \qquad (18)$$

where δ is defined by Eq. (17) and the last approximation is valid so long as $G_N \langle D_{NN} \rangle / G \ll R_{sp}$, which is typically the case. For $t < 0$, we close the contour in the lower–half plane, which gives Eq. (18) with t replaced by $-t$.

INDEX